Pollination of cultivated plants in the tropics

FAO
AGRICULTURAL
SERVICES
BULLETIN

118

Edited by
David W. Roubik
Smithsonian Tropical Research Institute
Balboa, Panama

Food
and
Agriculture
Organization
of
the
United
Nations

FIAT PANIS

Rome, 1995

M-11
ISBN 92-5-103659-4

Pollination of Cultivated Plants in the Tropics

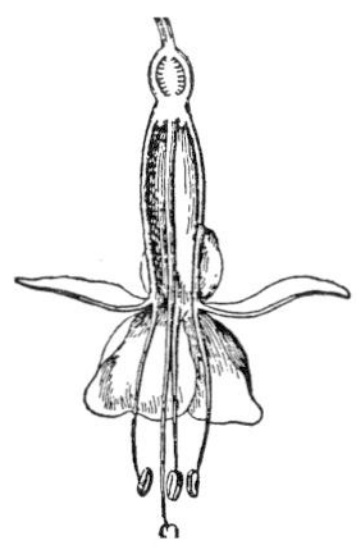

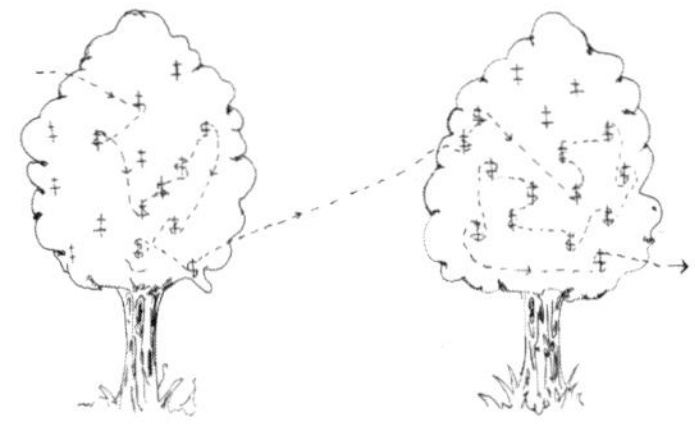

David W. Roubik (editor)

Food and Agriculture Organization of the United Nations

Contributors

Marco Accorti
Istituto Sperimentale per la Zoologia Agraria
Cascine del Riccio
50125 Firenze, Italy

Stephen L. Buchmann
United States Department of Agriculture
Carl Hayden Bee Research Center
2000 E. Allen Road
Tucson, Arizona 85719, USA

Art Davis
Department of Biology
University of Saskatchewan
Saskatoon, Saskatchewan S7N 0W0, Canada

Peter G. Kevan
Department of Environmental Biology
University of Guelph
Guelph, Ontario N1G 2W1, Canada

Rainer Krell
Icon Development
Viale Regina Margherita 239
00198 Roma, Italy

Rod P. Macfarlane
Buzzuniversal
33 Woodside Common
Christchurch, New Zealand

Makhdzir Mardan
Universiti Pertanian Malaysia
Malaysian Bee Research Development Team
43400 Serdang, Selangor, Malaysia

John K. S. Mbaya
National Beekeeping Station
P. O. Box 34188
Nairobi, Kenya

David W. Roubik
Smithsonian Tropical Research Institute
Apartado 2072
Balboa, Republic of Panamá
or Unit 0948, APO AA 34002-0948, USA

R. C. Sihag
Department of Zoology
CCS Haryana Agricultural University
Hisar 125004, India

P. F. Torchio
Bee Biology and Systematics Laboratory
United States Department of Agriculture and
Utah State University
Logan, Utah 84322-5310, USA

Mohd Basri Wahid
Palm Oil Research Institute of Malaysia
B. B. Bangi, Selangor, Malaysia

Figures & Acknowledgments

General Acknowledgments

For providing excellent working facilities and assistance the editor thanks the USDA-Utah State University Bee Biology and Systematics Laboratory and its personnel, particularly P. Torchio, T. Griswold, Twilla Hansen, J. Vandenberg and V. Tepedino, and also the Utah State University Graphics-4 Laboratory (P. Wood) and Publication Design and Production Department. Special thanks are also given for bibliographic assistance by L. J. G. van der Maesen (Agricultural University Wageningen, the Netherlands), R. Villanueva (University of Cardiff, U. K.) and Mr. Chaichanazsuwat (ASEAN-Canada Forest Tree Seed Centre Project, Muak Lek, Saraburi, Thailand).

Figures, Tables, & Acknowledgments

1. p. 2 a hummingbird, Phaethorninae.
2. p. 2 a worker honeybee, *Apis mellifera*. adapted from G. Nixon, 1955. The world of Bees. New York Philosophical Library, New York, illustration by Arthur Smith.
3. p. 3 a fly, Calliphoridae.
4. p. 4 a wasp, *Cerceris*.
5. p. 4 a weevil, Cucurlionidae.
6. p. 5 a hawkmoth, Sphingidae.
7. p. 5 a 'flying fox', *Pteropus*.
8. p. 5 a leaf nosed bat, Phyllostomatidae.
9. p. 11 a flower, showing the reproductive structures.
10. p. 12 a diagram of pollination showing 'selfing' and 'outcrossing' (original drawing by C. Kevan).
11. p. 15 Table. Benefits of different modes of pollination to some plants grown in the tropics.
12 p. 16 cartoon portraying benefits of augmented pollination by bees (original drawing by I. Smith).
13. p. 18 Animal Pollination and the Consumer.
14. p. 22 Rosaceae: pear and peach flowers (*Prunus* and *Pyrus*) reprinted by permission from: Simpson, B. B., Conner Ogorzaly, M. 1986. Economic botany — plants in our world. New York: McGraw-Hill.
15. p. 25 guava flower (*Psidium guajava*) reprinted by permission from: Simpson, B. B., Conner Ogorzaly, M. 1986. Economic botany — plants in our world. New York: McGraw-Hill.
16. p. 27 Papilinaceous legume (*Vigna*) and caesalpiniaceaous legume (*Cassia*), flower and fruit reprinted by permission from: Simpson, B. B., Conner Ogorzaly, M. 1986. Economic botany — plants in our world. New York: McGraw-Hill.
17. p. 27 mimosaceous legume flower (*Mimosa*) reprinted by permission from: Simpson, B. B., Conner Ogorzaly, M. 1986. Economic botany — plants in our world. New York: McGraw-Hill.
18. p. 30 honeybee, *Apis mellifera*, visiting but not pollinating a flower of red clover.
19. p. 31 queen *Bombus* pollinating the flowers of red clover.
20. p. 32 histogram showing improved clover seed yield with increased bumblebee colony number.
21. p. 39 Rosaceae: berry flowers and multiple fruit (*Rubus*) reprinted by permission from: Simpson, B. B., Conner Ogorzaly, M. 1986. Economic botany — plants in our world. New York: McGraw-Hill.
22. p. 40 Table of available bee, bird and bat pollinators on Atlantic and Indian Oceanic islands.
23. p. 42 Table of available bee, bird and bat pollinators in tropical Pacific oceanic islands.
24. p. 44 passion fruit flower and fruit (*Passiflora edulis*) reprinted by permission from: Simpson, B. B., Conner Ogorzaly, M. 1986. Economic botany — plants in our world. New York: McGraw-Hill.
25. p. 50 tamarind fruit and flower (*Tamarindus indica*) reprinted by permission from: Simpson, B. B., Conner Ogorzaly, M. 1986. Economic botany — plants in our world. New York: McGraw-Hill.
26. p. 51 litchi fruit (*Nephelium chinensis*). reprinted with permission from: Bailey, L. H. 1935. The standard cyclopedia of horticulture. New York: Macmillan.
27. p. 51 rose apple flower (*Syzygium jambos*) reprinted with permission from: Bailey, L. H. 1935. The standard cyclopedia of horticulture. New York: Macmillan.
28. p. 52 mango fruit and flower (*Mangifera indica*) reprinted by permission from: Simpson, B. B., Conner Ogorzaly, M. 1986. Economic botany — plants in our world. New York: McGraw-Hill.
29. p. 53 sesame flower and fruit (*Sesamum indicum*) reprinted by permission from: Simpson, B. B., Conner Ogorzaly, M. 1986. Economic botany — plants in our world. New York: McGraw-Hill.
30. p. 54 pepper flower and fruit (*Piper nigrum*) reprinted by permission from: Simpson, B. B., Conner Ogorzaly, M. 1986. Economic botany — plants in our world. New York: McGraw-Hill.
31. p. 56 coconut flower and fruit (*Cocos nucifera*) after Beumée, J. G. B. 1928. De bloemen van *Cocos nucifera*. L. Trop. Natuur. 17:7-12.
32. p. 57 African honeybee (*Apis mellifera scutellata*) on

Contents

1.1 INTRODUCTION

About 600 years ago plant crops entered a phase of increasingly being grown in countries far from their origin, with subsequent massive changes in world production. Few pollinators or pollination needs were considered, despite knowledge that plants often require pollen transfer by animals (rather than by wind, gravity or water). At present, it is evident that crops cultivated in the tropics are often productive with pollinators that did not travel with them to their new homes. Some, like *Vanilla*, are pollinated by hand, lacking the native neotropical pollinating *euglossine bee, Eulaema*. Others, the vast majority, are pollinated by animals similar to those in their place of origin, and these include *insects*, *bats*, and *birds*. A few have benefited dramatically from purposeful introductions of selected native pollinators, such as the *weevil, Elaeidobius* for the *African oil palm* in the plantations of Asia and the Americas.

Certainly many plants need fairly specialized pollinators, such as the *megachiropteran bats* that pollinate the *durian* of southeast Asia. However, seemingly most crops are pollinated by the unspecialized flower-visitors found almost everywhere in the tropics, such as native *stingless bees, honeybees,* and in wide areas of the tropics, introduced *western hive bees, Apis mellifera,* either in wild, unmanaged populations, or in apiaries.

Selective breeding of plants also has made it possible to obtain some fruits and seeds without pollination — for example *Mission* and *Kadota figs* do not need pollinators, but their wild *Ficus* progenitors and almost all other figs need a species of tiny *wasp*. Even so, despite the convenience of extremely domesticated crops that have no pollinator requirement, the need for cross pollination in seed and fruit production predominates and should receive greater attention as more crops are raised commercially. We can already list over 750 plant genera cultivated on a large or small scale in tropical areas, including the cooler highlands (see **Appendix I**). Only fragmentary information is available on pollination needs and pollinators for many of these important plants. Many processes are involved in obtaining marketable crops from the varied plant species used in the subtropics and tropics. Some have been cultivated for thousands of years, and some only now are being converted from wild forest crops to managed, sustained populations under cultivation. How, then, can pollination science best serve the needs of tropical agriculture now and in the future?

Much more work is needed both in research and applied technology. Fortunately, many pollinators and plants are flexible enough to fit into new settings. Within mixed plantings and in cultivated areas that are not extensive monocultures, the natural diversity of tropical organisms (see following section) often supports the reproduction of exotic plants. In terms of pollination, however, only a fraction of economically important plants have been studied anywhere in the world. This is easily appreciated in the confusion that exists about pollination requirements among different cultivars for widespread crops. We now know that some *mango, citrus, mangosteen* and *Shorea* can make mature seed and fruit through 'asexual' processes. For plants that produce without pollinators, the resulting seeds and offspring plants are *clones*, genetically like the female parent. Although vegetative propagation is the rule for some crops, many are established from seed. Seed production is, with few exceptions, greatly improved by *outcrossing*, or pollination whereby pollen is exchanged between different plants. Plants may further require specialized pollinators or costly hand pollination for normally seedless fruits — like *pineapple* and *banana*.

Without reliable information on the *breeding system* and *mechanics* of pollination, explanations for success or failure of a crop will be incomplete. For instance, *cardamom*, a seed crop ginger that is native to Asia and pollinated by *Apis dorsata* and *A. cerana*, was imported long ago to Jamaica, where *Apis* is for the most part absent. Native bees likely take the place of honeybees, but this knowledge is insufficient for pollinator conservation or pollination management.

What does the future hold for tropical agriculture and its pollination needs? Consider the following: About 60 years ago in the United States of America, where many of the fruit, seed and nut crops grown on commercial scale anywhere in the world were under exploitation (particularly in warmer areas of California, Florida and Hawaii), it became very clear that production could increase by careful management of pollination. Size and quality of the crop improved with better pollinator service. Two other observations were also very important. First, agricultural areas were coming under such intensive use that the pollinating insects and other animals living in the more natural habitats were being driven out or destroyed. There were fewer potential pollinators available. Second, large-scale crop production required large populations of appropriate pollinators. The rational answer seemed then to be, and still is, 'bees in boxes'— the familiar European honeybee, brought in by truck by the thousands whenever and wherever needed. Now there are other 'bees in boxes' that do not form large colonies but can be reared in large number and managed — ready to bring to a crop as flowering begins. The age of the 'designer pollinator' is beginning, and it will take many years of dedicated effort to decide how best to employ this tool.

Most developing countries have yet to go through the

pains of super-development — where virtually every square meter of land requires management for its intended use. However, this state of affairs already applies to a part of the tropics. Unless conservation of pollinators and knowledge of their biology are sighted as goals, there is no guarantee whatsoever that pollination needs will be met. Lessons from development of pollination science in all parts of the world are vital, as are those already applied in the tropics.

Study of pollination ecology and pollinator management are forms of insurance. Although no monetary value can be placed on tropical pollinators — it is too enormous to estimate reliably — most plants must have an animal visit their flowers to produce seeds, fruit and more of their kind. It would be a completely different world without pollinators for native plants, and it would be a much poorer world without pollinator service, both native and imported, for the many fruit, seed, fiber, hardwood and other products that are mainstays in tropical economies.

This book is a guide for development of pollination resources in the tropics. A major share of the effort involves beekeeping with *Apis* species, or honeybees. Because only eight species among many thousands of pollinators are honeybees, potentially hundreds more might be managed or 'enhanced'. In coming years this will be essential to insure that agriculture, silviculture, horticulture and conservation can progress together. Without pollinators and needed advances in their study, these vast interests may not progress at all.

1.1.1 Tropical Flower Visitors & Pollinators

Although approximately half of all animals that pollinate tropical plants are bees, the remainder are an extremely varied group. They include far-ranging bats and hawkmoths, as well as tiny agaonid wasps and thrips which are barely capable of sustained flight. Even bees are surprisingly varied — over 95% of all species have *nothing to do with honey or colonies*, and most of them build nests in the ground. The animals that visit flowers for food should include 10,000s of species. This section presents a tabular summary of their groups in the subtropics and tropics, giving estimates of species number in different regions. Many flies, beetles and bats visit flowers only occasionally. They should not be taken as equivalent to full-time pollinators such as bees or some birds.

1) Birds

Pollinator Group	***Approximate No. Species in Given Tropical Region***			
	Neotropics	**Africa**	**S. & SE Asia**	**Australasia**
False Sunbirds (Philepittidae)	-	2	-	-
Flowerpeckers (Dicaeidae)	-	-	30	25
Honeycreepers (Coerebidae)	15	-	-	-
Honeyeaters (Meliphagidae)	-	-	-	160
Hummingbirds (Trochilidae)	319	-	-	-
Leafbirds & Others (Chloropseidae)	-	10	-	-
Lories (Loriinae)	-	-	2	50
Sunbirds (Nectariniidae)	-	60	35	13
White-eyes (Zosteropidae)	-	15	60	10
Others	>100	>75	>50	>50

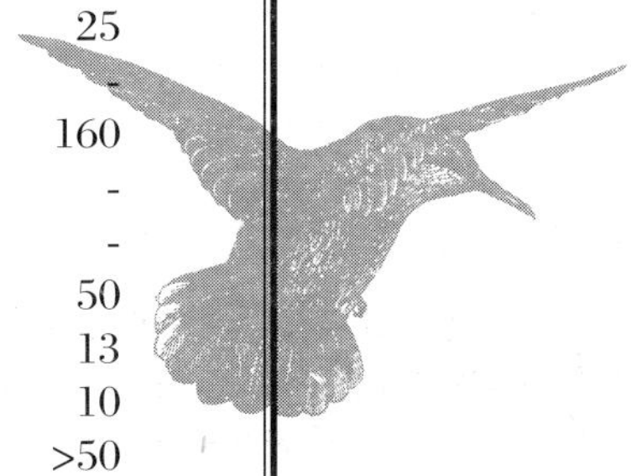

2) Bees (order Hymenoptera, superfamily Apoidea)

Pollinator Group	***Approximate No. Species in Given Tropical Region***			
	Neotropics	**Africa**	**S. & SE Asia**	**Australasia**
Long-Tongued Bees (Apidae*)	2000	1000	1000	20
Long-Tongued Bees (Megachilidae)	1000	1000	1500	50
Short-Tongued Bees (Andrenidae)	200	100	50	-
Short-Tongued Bees (Colletidae)	400	300	100	2000
Short-Tongued Bees (Halictidae)	2000	700	500	500
Short-Tongued Bees (Melittidae)	20	100	50	-

* including bees formerly placed in a separate family, Anthophoridae

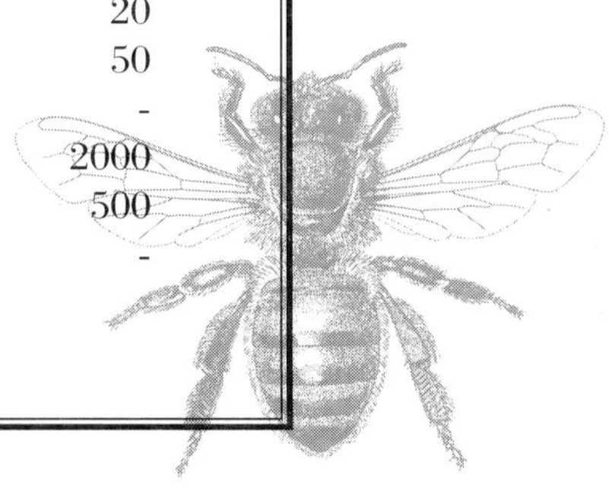

3) Flies (order Diptera)

Pollinator Group	*Approximate No. Species in Given Tropical Region*			
	Neotropics	**Africa**	**S. & SE Asia**	**Australasia**
suborder **Nematocera**				
Anisopodids (Anisopodidae)	>10	>100	>100	4
Black Flies (Simuliidae)	>100	>100	>100	30
Crane Flies (Tipulidae)	>100	>100	>100	70
Fungus Gnats (Mycetophilidae)	>100	>100	>100	220
Gall Midges (Cecidomyiidae)	>100	>100	>100	110
March Flies (Bibionidae)	>100	>100	>100	25
Midges (Chironomidae)	>100	>100	>100	130
Minute Scavengers (Scatopsidae)	>10	>10	>10	20
Mosquitoes (Culicidae)	>100	>100	>100	230
Moth Flies (Psychodidae)	>10	>10	>10	70
Root Gnats (Sciaridae)	>100	>100	>100	60
Sand Flies (Ceratopogonidae)	>100	>100	>100	200
suborder **Brachycera**				
Apiocerids (Apioceridae)	>100	>100	>100	75
Bee Flies (Bombyliidae)	>100	>100	>100	410
Bigheaded Flies (Pipunculidae)	>100	>100	>100	50
Chyromyids (Chyromyidae)	>10	>10	>10	3
Dance Flies (Empididae)	>100	>100	>100	90
Flesh Flies (Sarcophagidae)	>100	>100	>100	70
Frit Flies (Chloropidae)	>100	>100	>100	135
Fruit Flies (Tephritidae)	>100	>100	>100	130
Horse Flies (Tabanidae)	>100	>100	>100	240
House Flies (Muscidae)	>100	>100	>100	200
Hover Flies (Syrphidae)	>100	>100	>100	170
Humpbacked Flies (Phoridae)	>100	>100	>100	90
Lauxaniids (Lauxaniidae)	>100	>100	>100	230
Long-Legged Flies (Dolichopodidae)	>100	>100	>100	130
Micro-headed Flies (Acroceridae)	>10	>10	>10	30
Mydaids (Mydaidae)	>10	>10	>10	35
Nemestrinids (Nemestrinidae)	>10	>10	>10	60
Pelecorhynchids (Pelecorhynchidae)	>10	>10	>10	30
Pyrgotids (Pyrgotidae)	>10	>10	>10	70
Robber Flies (Asilidae)	>100	>100	>100	380
Shore Flies (Ephydridae)	>100	>100	>100	55
Smoke Flies (Platypezidae)	>10	>10	>10	10
Snipe Flies (Rhagionidae)	>10	>10	>10	65
Soldier Flies (Stratiomyidae)	>100	>100	>100	100
Spear-Winged Flies (Lonchopteridae)	>10	>10	>10	1
Stiletto Flies (Threvidae)	>100	>100	>100	160
Tachinids (Tachinidae)	>100	>100	>100	520
Thickheaded Flies (Conopidae)	>10	>10	>10	75
Upside-down Flies (Neurochaetidae)	>10	>10	>10	3
Vinegar Flies (Drosophilidae)	>100	>100	>100	70

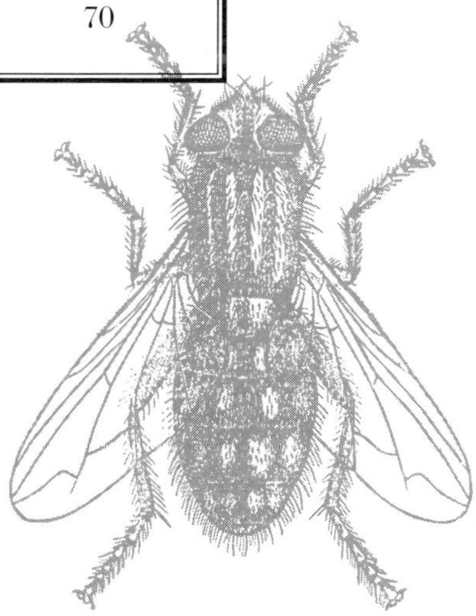

4) Wasps (order Hymenoptera)

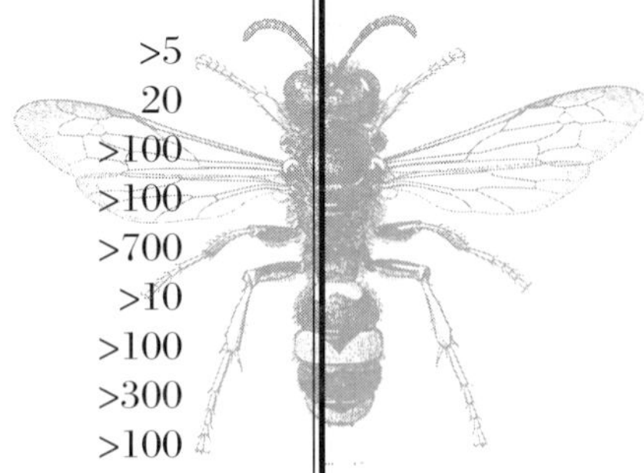

Pollinator Group	*Approximate No. Species in Given Tropical Region*			
	Neotropics	**Africa**	**S. & SE Asia**	**Australasia**
Ichneumonoidea: Ichneumonidae				
Ichneumons (subfam. Pimplinae)	?	>5	>5	>5
Chalcidoidea: Fig Wasps (Agaonidae)	>100	>100	>100	20
Vespoidea: Spider Wasps (Pompilidae)	>100	>100	>100	>100
Tiphiids(Tiphiidae)	>100	>100	>100	>100
Scoliids (Scoliidae)	>100	>100	>100	>700
Vespids (Vespidae)	>100	>100	>100	>10
Eumenids (Eumenidae)	>100	>100	>100	>100
Ants (Formicidae) or Formicoidea	>100	>100	>100	>300
Sphecoidea: Sphecids (Sphecidae)	>100	>100	>100	>100

5) Beetles (order Coleoptera)

Pollinator Group	*Approximate No. Species in Given Tropical Region*			
	Neotropics	**Africa**	**S. & SE Asia**	**Australasia**
Ant-like Flower Beetles (Anthicidae)	>100	>100	>100	170
Blister Beetles (Meloidae)	>100	>100	>100	60
Carpet Beetles (Dermestidae)	>100	>100	>100	90
Checker Beetles (Cleridae)	>100	>100	>100	300
Click Beetles (Elateridae)	>1000	>1000	>1000	610
Darkling Beetles (Tenebrionidae)	>1000	>1000	>1000	200
Ladybirds (Coccinellidae)	>100	>100	>100	260
Leaf Beetles (Chrysomelidae)	>1000	>1000	>1000	2100
Lizard Beetles (Languriidae)	>100	>1000	>1000	10
Longhorns (Cerambycidae)	>1000	>1000	>1000	1040
Marsh Beetles (Helodidae)	>100	>100	>100	50
Metallic Woodborders (Buprestidae)	>1000	>1000	>1000	800
Net-winged Beetles (Lycidae)	>100	>100	>100	200
Oedomerids (Oedomeridae)	>100	>100	>100	85
Pythids (Pythidae)	>100	>100	>100	5
Rhiphiphorids (Rhipiphoridae)	>100	>100	>100	60
Rove Beetles (Staphylinidae)	>1000	>100	>100	650
Sap Beetles (Nitidulidae)	>100	>100	>100	120
Scarab Beetles (Scarabaeidae)	>1000	>1000	>1000	2100
Scraptiids (Scraptiidae)	>100	>100	>100	15
Shining Flower Beetles (Phalacridae)	>100	>100	>100	70
Silken Fungus Beetles (Cryptophagidae)	>100	>100	>100	20
Soft-winged Flower Beetles (Melyridae)	>100	>100	>100	250
Soldier Beetles (Cantharidae)	>100	>100	>100	100
Stag Beetles (Lucanidae)	>100	>100	>100	75
Throscids (Throscidae)	>100	>100	>100	5
Tumbling Flower Beetles (Mordellidae)	>100	>100	>100	120
Weevils (Curculionidae)	>1000	>1000	>1000	4000

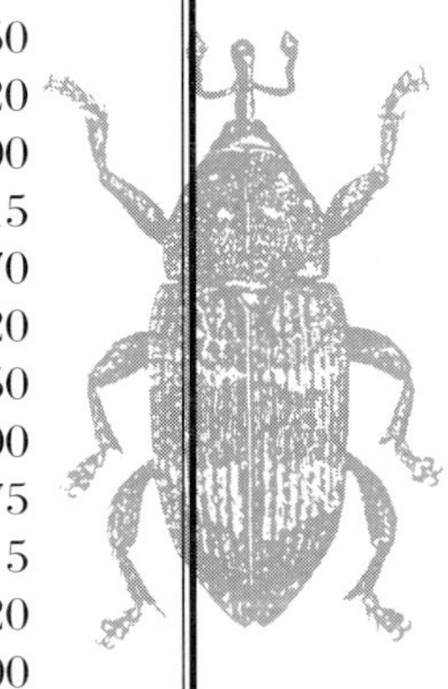

6) Butterflies and Moths (order Lepidoptera)

Pollinator Group	*Approximate No. Species in Given Tropical Region*			
	Neotropics	**Africa**	**S. & SE Asia**	**Australasia**
Arctiids (Arctiidae)	1000	700	600	50
Brush-Footed Butterflies (Nymphalidae)	>1000	>1000	>1000	100
Day-Moths (Uraniidae)	5	10	10	5
Hairstreaks (Lycaenidae)	>1000	>1000	>1000	150
Moths (Micropterigidae)	5	5	5	5
Owlet Moths (Noctuidae)	>1000	>1000	>1000	>1000
Skippers (Hesperiidae)	1000	1000	1000	100
Sphinx or Hawkmoths (Sphingidae)	400	200	100	80
Sulphurs (Pieridae)	400	400	300	50
Swallowtails (Papilionidae)	>500	>200	>400	20
Tinaegeriids (Tinaegeriidae)	20	50	50	-
Yucca Moths (Prodoxidae)	>10	>10	>10	>10
Zygaenids (Zygaenidae)	200	100	50	-

7) Bats, Marsupials & Rodents (class Mammalia)

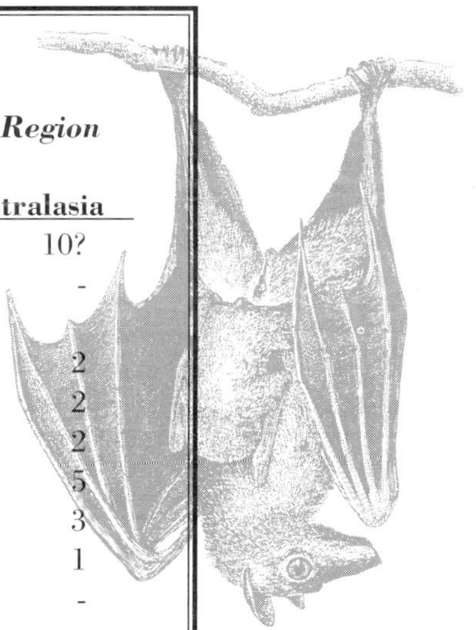

Pollinator Group	*Approximate No. Species in Given Tropical Region*			
	Neotropics	**Africa**	**S. & SE Asia**	**Australasia**
Fruit Bats, Flying Foxes (Pteropodidae)	-	40	174	10?
Leaf-nosed Bats (Phyllostomatidae)	148	-	-	-
New Zealand short-tailed bat (Mystacinidae)	-	-	1	
Flying Lemurs (Dermoptera)	-	-	-	2
Pygmy Phalanger (Burramyidae)	-	-	-	2
Marsupial Mice (Dasyuridae)	-	-	-	2
Gliders (Petauridae)	-	-	-	5
Possums (Phalangeridae)	-	-	-	3
Honey Possum (Tarsipedidae)	-	-	-	1
Opossums (Didelphidae)	9	-	-	-
Shrews (Microbiotheriidae)	7	-	-	-
Ringtails and Gliders (Pseudocheiridae)	-	-	-	8
Bush Rats (Muridae)	-	-	-	2

8) Thrips (Thysanoptera) - families Thripidae, *Aeolothripidae, Phalaeothripidae, and others, worldwide, No. species > 500*

1.1.2 How to Use This Book

This manual defines some major ideas and advances in pollinator selection and the potential for management. Commercial crops and the development of comprehensive pollination programmes are discussed in detail, along with methods and theory. In the first part of the book (Sections 1.1-1.7) we discuss general issues, review applied pollination in the tropics and subtropics, and make recommendations on the present and future study of pollination as applied to sustainable agriculture and conservation. In the second part of the book, Sections 2.1-2.5, we present more detailed information for pollination research and pollinator management. Three appendices follow. Appendix I is a list of cultivated and semi-cultivated plants (those harvested in the wild and to some extent encouraged to grow there). It is not exhaustive, because there are many additional ornamental plants, plants used for medicinal purposes and those used for timber. What is known of the breeding system and pollinator is given for each species in the Appendix, along with botanical data such as family, common and scientific names, and place of origin. The list also contains some major timber species of tropical regions. Appendix II gives pollination contracts and honeybee colony standards that can be used in order to let producers and pollination specialists help each other to benefit from the services rendered through applied pollination technology. Formal or written agreements seem essential for this purpose. In Appendix III all available pesticides are listed with specific regard to their effects on bees and beneficial organisms in general.

We have made efforts to review the whole of pollination in the tropics, and then place it in the perspective suggested to us by our practical experiences in field and applied science. A large and somewhat complicated literature is put in a form accessible to technicians, students and others who may have no extensive training in botany, zoology, agronomy, ecology or animal behaviour — these are the fields that permeate the entire subject of pollination technology and make it one of the most fascinating of applied sciences. It is the combination of many disciplines, bound together in the interest of understanding and applying ecology to improve production, progress toward sustained and self-sufficient agriculture, and conserve what are among the most vital of all resources: the animals that allow plants to reproduce their kind.

This is not a beekeeping manual — many excellent general or regional treatments, for all levels of sophistication, already are available. Nor is this a textbook on plant genetics and breeding. This book is a condensed guide to evaluating pollination requirements of crops and other economically important plants, performing the critical laboratory and field tests, and making rational and realistic choices of crops and pollinators where management decisions are involved. We do provide guidelines for apiculture and introduce several new or less well-explored forms of pollinator 'cultivation'. Examples of applied pollination throughout the tropics are given for the pollinators that we discuss. Agricultural pesticide use is discussed in several sections. Evaluating pollination services also is treated, and specifics for management and pollination are given for major crops selected from over a thousand that are grown on a surprisingly large scale among tropical countries. The issues of post-pollination management and harvest are given due consideration. With a little more thought and a lot more effort, these examples can be expanded by the next generation of pollination scientists to make some needed improvements in the ways that plants and their requirements are understood. The local and world economies will continue to demand their exploitation and improvement. We now are beginning to realize to what extent a contribution can be made based upon knowing all the details of pollination and putting them to good use.

1.1.3 General References

(& additional references for Introduction)

Abrol, D. P. 1988. Ecology and behaviour of three species pollinating Loquat (*Eriobotrya japonica* Lindley). Proc. Ind. Nat. Sci. Acad. B54:161-163.

Aitken, Y. 1974. Flowering time, climate and genotype: The adaptation of agricultural species to climate through flowering responses. Melbourne, Australia: Melbourne Univ. Press., 193 pp. [a treatment of native and introduced crops for the plant breeder, emphasizing ecology and development]

Amaral, E. 1970. Estudos de polinização entomófila em plantas de interêsse econômico para o Brasil. 1° Cong. bras. Apic.:65-68. [focuses on legumes, coffee, sunflower, cucurbits]

Anonymous. 1969. A national program of research for bees and other pollinating insects affecting man. Washington, D. C: United States Department of Agriculture, 33 pp.

Anthakrishnan, T. N. 1993. Bionomics of thrips. Annu. Rev. Entomol. 38:71-92. [bionomics and a review of population dynamics, with mention of pollinators]

Appanah, S. in press. Mass flowering of dipterocarp forests in the seasonal tropics. In: Diversity and flexibility of biological communities in fluctuating environments. D. W. Roubik, T. Inoue, P. S. Ashton, (eds.). J. Biosci. (supplement). [an environmental cue for mass-flowering in Asia is a 2°C dip in night-time temperature for four to five nights; mass-flowering trees are pollinated by flower thrips that multiply rapidly within buds, and also by pollinators coming from outside the forest; plants sharing pollinators flower in sequence]

Arizmendi, Ma. del Coro, Ornelas, J. F. 1990. Hummingbirds and their floral resources in a tropical dry forest in Mexico. Biotropica 22:172-180. [10 of 23 spp. pollinated exclusively by birds, the most abundant resident, *Amazilia rutila* aggressively dominates flower patches]

Armstrong, J. A. 1979. Biotic pollination mechanisms in the Australian flora — a review. New Zealand J. Botany 17:467-508. [ranks the proportion and importance of flower-visiting insects as bee>fly>beetle; for vertebrates, birds, bats, marsupials, placentals and rodents; emphasizes need for cooperative and comparative studies]

Ägren, J., Schemske, D. W. 1991. Pollination by deceit in a neotropical monoecious herb, *Begonia involucrata*. Biotropica 23:235-241. [stingless bee *Partamona grandipennis* pollinates rewardless female flowers because their stigmas look like anthers]

Bailey, L. H. 1935. The standard cyclopedia of horticulture. (2nd Ed.). New

York: Macmillan, 1200 pp. [for the United States, covering many tropical plants; little information on pollination]

Baker, H. G. 1976. "Mistake" pollination as a reproductive system with special reference to the Caricaceae. In: Tropical trees: variation, breeding and conservation. J. Burley, B. T. Styles (eds.). London: Academic Press for the Linnean Society. [review of subject; hawkmoths provide outcrossing for papaya]

Baker, H. G., Bawa, K. S., Frankie, G. W., Opler, P. A. 1982. Reproductive biology of plants in tropical forests. In: F. B. Golley (ed.). Tropical rain forest ecosystems: structure and function. Amsterdam: Elsevier, pp. 183-215. [botanical, behavioural and ecological evaluation of reproduction, primarily Costa Rican plants]

Banaszak, J., Manole. T. 1987. Diversity and density of pollinating insects (Apoidea) in the agricultural landscape of Rumania. Polsk. Pism. Ent. 57:747-766.

Batra, S. W. T. 1967. Crop pollination and the flower relationships of the wild bees of Ludhiana, India. J. Kan. Entomol. Soc. 40:164-177. [97 Punjab bee species, including *Apis*, collected at flowers]

Batra, S. W. T. 1976. Comparative efficiency of alfalfa pollination by *Nomia melanderi, Megachile rotundata, Anthidium florentinum* and *Pithitis smaragdula*. [subtropical species and greenhouse studies]

Batra, S. W. T. 1980. Ecology, behaviour, pheromones, parasites and management of the sympatric vernal bees *Colletes inaequalis, C. thoracicus* and *C validus*. J. Kansas Entomol. Soc. 53:509-538.

Bawa, K, Hadley, M. (eds.) 1990. Reproductive ecology of tropical forest plants. France: Paris, UNESCO, 647 pp. [botanical and pollination inquiries, reviews by tropical forest ecologists and silviculturists]

Bawa, K. S. 1990. Plant-pollinatior interactions in tropical rain forests. Ann. Rev. Ecol. Syst. 21:399-422. [discusses varied breeding systems and pollinators of forest plants]

Bawa, K. S. 1992. Mating systems, genetic differentiation and speciation in tropical rain forest plants. Biotropica 24:250-255. [emphasizes outcrossing and self-incompatibly as the rule in tropical woody plants]

Bawa, K. S., Ashton, P. S., Salleh Mohd, N. 1990. Reproductive ecology of tropical forest plants: management issues. In: K. S. Bawa, M. Hadley (eds.). Reproductive ecology of tropical forest plants. France: Paris, UNESCO, pp. 3-13. [emphasis on reproductive modes of timber, fruit trees in forest settings; very high frequency of dioecy (separate sexes), and apomixis in *Shorea, Garcinia, Mangifera and Citrus*]

Bawa. K. S., Perry, D. R., Bullock, S. H., Coville, R. E., Grayum, M. H. 1985. Reproductive biology of tropical lowland rain forest trees. II. Pollination mechanisms. Amer. J. Bot. 72:346-356. [field study outcrossing, documenting arrival of large bees at flowers]

Bohart, G. E. 1972. Management of wild bees for the pollination of crops. Ann. Rev. Entomol. 17:287-312.

Boonithee, A., Juntawong, N., Pechhacker, H., Hüttinger, E. 1991. Floral visits to select crops by four *Apis* species and *Trigona* sp. in Thailand. In: VI Intl. Symp. Poll. C. van Heemert, A. de Ruijter (eds.). pp. 74-80. [N. Thailand orchards of *Pyrus, prunus* and *Euphoria longan* showed more visits by native *Apis* and *Trigona* than imported *A. mellifera*).

Brücher, H. 1992. Useful plants of neotropical origin and their wild relatives. Berlin: Springer-Verlag, 296 pp. [comprehensive botanical and economic discussion but pollination seldom mentioned]

Campbell, D. G., Hammond, H. D. 1989. Floristic inventory of tropical countries. New York: The New York Botanical Garden. 555 pp. [worldwide summary of species richness; appendices on collection techniques for germ plasm, pollen, wood, bark, etc.]

Compton, S. G., Holton, K. C., Rashbrook, V. K., van Noort, S., Vincent, S. L., Ware, A. B. 1991. Studies of *Ceratosolen galili*, a non-pollinating agaonid fig wasp. Biotropica 23:188-194. [African fig *Ficus sycomorus* visited by two fig wasps but only pollinated by one, *Ceratosolen arabicus*]

Cooper, R. C., Cambie, R. C. 1991. New Zealand's economic native plants. Auckland: Oxford Univ. Press, 234 pp. [historical and economic botany, pollination not emphasized]

Corbet, S. A., Williams, I. H., Osbourne, J. L. 1991. Bees and the pollination of crops and wild flowers: changes in the European community. Scientific and Technical Options Assessment, European Parliament, pp. 1-50. [meagre but convincing data show population declines]

Crane, E., Walker, P. 1983. The impact of pest management on bees and pollination. Intl. Bee Res. Assn. Trop. Dev. Res. Inst., London. 199 pp. [a source book for lists of crops and research, as well as a statement on the negative effects of careless pesticide application]

Crane, E., Walker, P. 1984. Pollination directory for world crops. Intl. Bee Res. Assn., London. 184 pp. [pollination requirements of over 400 species, tabular references]

Cumaraswamy, A., Bawa, K. S. 1989. Gender allocation and mating systems in pigeonpea (*Cajanus cajan* L.). Pl. Syst. Evol. 168:59-69. [pollination science of a common tropical cultivated plant]

Dafni, A. 1992. Pollination ecology — a practical approach. Oxford: Oxford Univ. Press, 250 pp. [a general treatment of pollination in a manner similar in diagnostic form to medical entomology — methods and terminology emphasized]

Day, M. C. 1991. Towards the conservation of aculeate Hymenoptera in Europe: an outline of the case for recognition of the high value of Hymenoptera Aculeata as indicators of biotope integrity and diversity, with relevant examples and proposals for conservation actions. Nature and environment series, No. 45. Council of Europe, Strasbourg. [stinging bees and wasps can and have been intensively surveyed and many also are essential to plants for pollination]

Dhaliwal, H. S., Adlakha, R. L. 1977. Insect pollination of seed cauliflower: foraging behaviour of major pollinators. Proc. XXVI Intl. Beekeeping Congr.: 395-396. [*Eristalis* (Diptera), *Ceratina, Halic-tus, Lasioglossum and Apis cerana* are pollinators]

Dressler, R. L. 1981. The orchids: natural history and classification. Cambridge, Mass.: Harvard Univ. Press, 332 pp. [comprehensive treatment of orchids and, particularly for neotropical spp., their pollination]

Faegri, and van der Pijl. 1979. Principles of pollination ecology (3rd ed.). Oxford: Pergamon Press, 312 pp. [comprehensive and well-illustrated reference work and general textbook]

Falcão, M. de A., Lleras, E., Kerr, W. E., Medeiros Carreira, L. M. 1981. Aspectos geológicos, ecológicos e de produtividade do biribá (*Rollina mucosa* (Jacq.) Baill.). [this sp. of Annonaceae is visited by chrysomelid beetles near Manaus, Brazil]

Ferwerda, F. P., Wit, F. (eds.). 1969. Outlines of perennial crop breeding in the tropics. H. Veenman & Zonen N. V.: Wageningen, the Netherlands. 511 pp. [Comprehensive review of 21 tropical crops, with details of breeding programs, genetics, pollination and general biology and management]

Ford, H. A., Paton, D. C., Forde, N. 1979. Birds as pollinators of Australian plants. New Zealand J. Bot. 17:509-519. [Over 100 bird species visit 250 plants, some, like *Eucalyptus*, are visited both by birds and bees; birds are less specific than in the Neotropics]

Frankie, , G. W., Haber, W. A., Opler, P. A., Bawa, K. S. 1983. Characteristics and organization of the large bee pollination system in the Costa Rican dry forest. In: C. E. Jones, R. J. Little (eds.). Handbook of experimental pollination biology. New York: van Nostrand & Reinhold Co, pp. 411-447. [detailed account of flower visitation by the largest local bees]

Free, J. B. 1975. Observations on the pollination of papaya (*Carica papaya* L.) in Jamaica. Trop.

Agric., Trinidad 52:275-279. [skipper butterflies visit flowers at dusk]

Free. J. B. 1993. Insect pollination of crops (2nd ed.). London: Academic Press. 768 pp. [practical treatment, by plant family, of over 400 crops and harvested wild plants; extensive literature references]

Freeman, C. E., Worthington, R. D., Jackson, M. S. 1991. Floral nectar sugar compositions of some south and southeast Asian species. Biotropica 23:568-574. [62 native species (23 families) display high or low sucrose nectars; the latter usually bee-pollinated; bat-pollinated flowers are richer in sucrose than New World bat flowers]

Fujita, M. S. 1991. Flying fox (Chiroptera: Pteropodidae) pollination, seed dispersal, and economic importance. A tabular summary of current knowledge. Resource Publication No. 2, Texas: Austin, Bat Conservation International. [food, fibre, spice, medicinal, timber products pollinated or dispersed by bats are listed for Asia]

Gibbons, R. W., Tattersfield, J. R. 1969. Outcrossing trials with groundnuts (*Arachis hypogaea* L.). Rhodesian J. Agric. Res, 7:71-75. [peanut pollinators in Malawi and Zimbabwe are *Megachile* and the halictid *Lipotriches*]

Gimenes, M. 1991. Some morphological adaptations in bees for collecting pollen from *Ludwigia elegans* (Onagraceae). Revta. bras. Ent. 35:413-422. [bee hairs helpful in manipulating viscin threads of *Ludwigia* pollen masses; *Tetra-glossula* and *Pseudagapostemon* are common pollinators]

Godley, E. J. 1979. Flower biology in New Zealand. New Zealand J. Bot. 17:441-466. [widespread selfing and variety of visitors characterize many species; some 7 birds, 1 bat, 40 bees and 16 butterflies are among pollinators]

Gómez-Pampa, A., Kaus, A. 1992. Taming the wilderness myth. Bioscience 42:271-279. [tropical forests are not pristine environments; human residents have long made an impact]

Goulet, H, Huber, J. T. (eds). 1993. Hymenoptera of the world: an identification guide to families. Ottawa, Canada: Canada Communications Group. 680 pp. [fully illustrated key, biological data, references; bees, 26 subfamilies and families of superfamily Apoidea, included]

Gryj, E., del Rio, C. M., Baker, I. 1990. Avian pollination and nectar use in *Combretum fruticosum* (Loefl.). Biotro-pica 22:266-271. [passerine (and perching) birds rather than hummingbirds are the main pollinators among 31 bird spp; nectar is dilute and hexose-rich]

Haber, W. A., Frankie, G. W. 1989. A tropical hawkmoth community: Costa Rican dry forest Sphingidae. Biotropica 21:155-172. [hawkmoths pollinate many night-flowering trees, including papaya, and also use flowers visited by bats, birds and bees; they are unusually wide-ranging and not very constant to a floral species; their over-abundant resources provide food for diurnal pollinators]

Heard, T. A. 1988. The requirement for insect pollination by macadamia and the pollinator efficiency of *Trigona* bees. In: Proc. IV Australas. Confr. on Tree and Nut Crops. D Batten (ed) Australia: N.S.W., Lismore. pp. 219-233. [*Macadamia* and *Trigona* come from separate habitats in Australia; when brought together *Trigona* pollinate flowers]

Hedström, I. 1988. Pollen carriers and fruit development of *Pisidium guajava* L. (Myrtaceae) in the Neotropical region. Rev. Biol. Trop. 36:551-553.

Heemert, C. van, Ruijter, A. de. (eds.). 1991. Sixth Pollination Symposium, Tilburg. Act. Hort. 288, 594 pp. [a wide selection of applied pollination research, with some new data from the tropics]

Henderson, A. 1986. A review of pollination studies in the Palmae. *Bot. Rev.* 52:221–259. [reviews the fly, beetle and bee-pollination syndromes of palms]

Hernández, X., E. 1993. Aspects of plant domestication in Mexico: A personal view. In: T. P. Ramamoorthy, R. Bye, A. Lot, J. Fa, eds., pp. 733-753. Biological Diversity of Mexico. New York: Oxford Univ. Press. [lists plants in tropical and temperate Mexico]

Indian Bee Journal. 1981. Work on bee pollination in India. Indian Bee J. 43:140-145. [studies on sunflower, sesame, mustard, safflower, lemon, grape, litchi, horse gram and cucumber; *Apis mellifera* and *A. cerana* frequent visitors]

International Bee Research Association. 1989. Proceedings of the fourth international conference on apiculture in tropical climates (Cairo, 1988). London: IBRA. 529 pp. [research reports including pollination and bee management, with Africa well represented]

Itino, T., M. Kato, M., Hotta, M. 1991. Pollination ecology of the two wild bananas, *Musa acuminata* subsp. *halabanensis* and *M. salaccensis*: chiropterophily and ornithophily. Biotropica 23:151-158. [Sumatran bananas pollinated by pteropoid bats (*Macroglossus*) or by nectariid birds, *Arachnothera longirostris* and *Aethoipyga siparaja*]

Jackson, D. L. D., Jacobs, S. W. L. 1985. Australian agricultural botany. Sydney: Sydney Univ. Press. 377 pp. [account of species, discussion of breeding systems and details of pollination]

Jackson, G. C., Woodbury, R.O. 1976. Host plants of the carpenter bee *Xylocopa brasilianorum* L. in Puerto Rico. J. Agric. Univ. Puerto Rico 60: 639-660. [25 fruit and seed crops included]

Jadhav, L. D. 1981. Role of insects in the pollination of onion (*Allium cepa* L.) in the Maharashtra State, India. Indian Bee J. 43:61-63 [*Apis florea* is very important; other *Apis* elsewhere]

Janzen, D. H. 1983 (ed). Costa Rican natural history. Chicago: Univ. Chicago Press. 893 pp. [section 6 on Agriculture discusses crops and pollination; Table 6.4 provides a list of vegetables grown in varied climates and soils]

Jayarathnam, K. 1965. Preliminary observations on insect pollinators of *Hevea* in India. Bull. Rubber Bd. 8:138-144.

Johansen, C. A. 1977. Pesticides and pollination. Ann. Rev. Entomol. 22:177-192.

Kapil, R. P. (ed.). 1986. Pollination biology — an analysis. India: Inter-India Publications, New Delhi.

Kapil, R. P., Jain, K. L. 1980. Biology and utilization of insect pollinators for crop production. Hissar, India: Haryana Agricultural Univ. 82 pp. [Megachilid bees — 6 *Megachile*, *Nomia*, *Apis florea*, *Pithitis*, *Xylocopa*, *Braunsapis*, *Nomioides*, *Melissodes*, *Halictus* and Diptera discussed in relation to parasites and host plants]

Kearns, C. A., Inouye, D. W. 1993. Techniques for pollination biologists. Boulder: University of Colorado Press, 583 pp. [complete and technical reference for pollination work, emphasizing field studies and tactics for research]

Keller, S., Armbruster, S. 1989. Pollination of *Hyptis capitata* by eumenid wasps in Panama. Biotropica 21:190-192.

Keng, H. 1990. The concise flora of Singapore: gymnosperms and dicotyledons. Singapore: Singapore Univ. Press, 222 pp. [botanical reference, no information on pollination or breeding systems]

Kevan, P. G. (1984). Insect pollination of economically important plants of tropical and sub-tropical Asia. Proc. Exp. Consul. Beekeep. with *Apis mellifera* in Tropical and subtropical Asia. Italy: Rome, FAO. pp. 77-85.

Kevan, P. G. 1986. Pollination and flower visiting insects and the management of beneficial and harmful insects and plants In: M. Y. Hussein, A. G. Ibrahim (eds.). Biological control in the tropics. Malaysia: Universiti Pertanian Malaysia, Serdang, Selangor. pp. 439-452.

Kevan, P. G., Morse, R. A., Akratanakul, P. 1984. Apiculture in tropical and sub-tropical Asia with special reference to European honeybees and development programmes. Rome: FAO, pp. 10-33.

Kevan, P. G., Hussein, M. Y., Hussey, N., Wahid, M. B. 1986. Modelling the use of *Elaeidobius kamerunicus* for pollination of oil palm. Planter [Kuala Lumpur] 62:89-99.

Lal, R., Sanchez, P. A. 1992. Myths and science of soils of the tropics. Madison, Wisconsin, USA: Social Science Society of America, Inc., Spec.

Pub. No. 29, 185 pp. [emphasizes the variety of soils found in the tropics, from extremely rich to poor, and most with potential for growers]

Lee, D. 1980. The sinking ark. Malaysia: Kuala Lumpur, Heinemann Educational Books. [the durian is pollinated by cave-roosting *Eonycteris spelea* bats, which also visit flowers of mango, banana, *Oroxylum*, petai, *Artocarpus*, jambu merah, *Duabanga* and *Sonneratia* — a mangrove]

Liengsiri, C., Piewluang, C., Boyle, T. J. B. 1990. Starch gel electrophoresis of tropical trees: a manual. Muak-Lek, Saraburi 18180, Thailand: ASEAN-Canada Forest Tree Seed Centre Project. 52 pp. [simplified guide to practical population genetics and forest research, some standard recipes given]

Mabberley, D. J. 1988. The plant-book. Cambridge, England: Cambridge Univ. Press, 706 pp. [essential guide to plant genera, common names and origins]

Maesen, L. J. G. van der. 1985. *Cajanus* DC and *Atylosia*, W. & A. Agricultural University of Wageningen, Papers 85-4, the Netherlands. 225 pp. [review of the fourth most important tropical vegetable and 30 related species of value for breeding programs; cross-pollination varies tremendously and 'pre-anthesis cleistogamy' produces selfing; the stigma is receptive longer than pollen is shed; megachilids and xylocopines are pollinators]

McGregor, S. E. 1975. Insect pollination of tropical crops. Proc. III Intl. Symp. Poll.: pp. 47-55.

McGregor, S. E. 1976. Insect pollination of cultivated crop plants. Agric. Handbk. 496. Washington, D. C.: U.S. Dept. Agric. 411 pp. [well-illustrated, helpful for deciding whether crops need bees, biased toward *Apis mellifera*]

Michelbacher, A. E., Hurd, P. D., Linsley, E. G., Jr. 1971. Experimental introduction of squash bees (*Peponapis*) to improve yields of squashes, gourds and pumpkins. Bee Wld. 52:156-166. [*Peponapis* and *Xenoglossa* are the natural pollinators]

National Research Council. (Board on Science and Technology for International Development). 1975. Underexploited tropical plants with promising economic value. Washington, D. C.: Natl. Acad. Press. 186 pp. [plants and research recommendations, but no mention of pollination and almost nothing on breeding systems]

National Research Council. (Board on Science and Technology for International Development). 1977. Tropical legumes: resources for the future. Washington, D. C.: Natl. Acad. Press, 331 pp. [thorough economic treatment for the tropics, but excludes breeding systems and pollination]

National Research Council. (Board on Science and Technology for International Development). 1989. Lost crops of the Incas: little-known polants of the Andes with promise for worldwide cultivation. Washington, D. C.: Natl. Acad. Press.

National Research Council. (Board on Science and Technology for International Development). 1991. Managing global genetic resources: forest trrees. Washington, D. C.: Natl. Acad. Press. 228 pp.

Nelson, B.W., Absy, M.L., Barbosa, E.M., Prance, G.T. 1985. Observations on flower visitors to Bertholletia excelsa H.B.K. and Couratari tenuicarpa A.C. Sm. (Lecythidaceae). Acta Amazonica 15, suppl., 224-234 [large bees are pollinators: Bombus, Epicharis, Centris, Eulaema, Eufriesea, Euglossa]

Newton, L. E., Lomo, A. 1979. The pollination of *Ficus vogellii* in Ghana. Bot. J. Linn. Soc. 78:21-30. [agaonid wasp *Allotriozoom heterandromomorphum* is the obligate pollinator]

Nur, N. 1976. Studies on pollination in Musaceae. Ann. Bot. 40:167-177. [wild bananas pollinated by macroglossine bats, honeyeaters and sunbirds, and even tree shrews]

O'Malley, D. M., Buckley, D., Apsit, V., Prance, G. T., Bawa, K. S. 1988. Genetics of Brazil-nut (*Bertholletia excelsa* Humb. & Bonpl. Lecythidaceae). II. Mating system. Theor. Appl. Gen. 76:929-932. [pollination science of a intensively-harvested wild crop]

O'Toole, C. 1993. Diversity of native bees and agroecosystems. In: J. LaSalle, I. D. Gauld (eds.). Hymenoptera and the biodiversity crisis. Wallingford, U.K.: C. A. B. International, pp. 169-196. [review of domesticated and wild crop-visiting apoids]

Onim, J. F. M., Eijnatten, C. L. M., Pathak, R. S. 1979. Some factors which encourage outcrossing in pigeon pea. Proc. IV Int. Symp. Poll.: 219-225. [*Apis mellifera scutellata* is abundant but relatively inefficient compared to *Chalicodoma*, *Megachile* and possibly *Xylocopa*]

Owens, J. N., Sornsathapornkul, P., Tangmitcharoen, S. 1991. Studying flowering and seed ontogeny in tropical forest trees. Muak-Lek, Saraburi 18180, Thailand: ASEAN-Canada Forest Tree Seed Centre Project. 134 pp. [illustrated guide and current references to seedling growth and flowering, including pollination and laboratory procedures, and research in general]

Pandey, R. S., Yadava, R. P. S. 1970. Pollination of litchi (*Litchi chinensis*) by insects with special reference to honeybees. J. Apic. Res. 9:103-105. [*Apis* and *Trigona* are the important pollinators]

Parker, F. D., Torchio, P. F. 1980 Management of wild bees. In: Beekeeping in the United States. Washington, D. C.: U. S. Dept. Agric. Handbook, 335 pp. 144-160. [describes rearing, parasites and population requirements for some crops, i.e. >4000 *Megachile rotundata* per hectare needed for alfalfa]

Purseglove, J. W. 1968. Tropical crops: dicotyledons (vols. 1 & 2). New York: John Wiley, 719 pp. [concise but extensive; pollination considered]

Purseglove, J. W. 1972. Tropical crops: monocotyledons (vols. 1 & 2). New York: John Wiley, 607 pp. [concise but extensive; pollination considered]

Raawal, K. M., Bryant, P., Rachie, K. O., Porter, W. M. 1978. Cross pollination studies of male-sterile genotypes in cowpeas. Crop Sci. 18:283-285. [Nigerian studies showed *Apis mellifera adansonii* the chief pollinator]

Ramirez, N. 1989. Biología de polinización en una comunidad arbustiva tropical de la alta Guyana venezolana. Biotropi-ca 21:319-330. [62 visitor species seen on 55 plants; 80% said to be pollinators. Plants pollinated 56%by bees and wasps, 11% by butterflies & moths, 10% by flies, 3% by beetles,12% by birds; 8% by wind]

Ramírez, W. B. 1974. Coevolution of *Ficus* and Agaonidae. Ann. Missouri Bot. Gard. 61:770-780.

Rashad, S. E. 1978. Utilization of non-*Apis* bees as crop pollinators (EG-ARS-66); first annual report. Giza, Egypt: Cairo Univ., 48 pp. [*Prosopis*, *Nomia*, *Chalicodoma* and *Halictus* as visitors of medicinal plants —*Carum*, *Mentha*, *Pimpinella*]

Raw, A. 1979. *Centris dirrhoda* (Anthophoridae), the bee visiting West Indian cherry flowers (*Malpighia punicifo-lia*) Rev. Biol. Trop. 27:203-205. [oil-bees, not *Apis*, visit oil flowers]

Rhem, S., Espig, G. 1991. The cultivated plants of the tropics and subtropics: Cultivation, economic value, utilization. Weikersheim, Germany: Verlag Josef Margraf, 552 pp. [most complete reference available on cultivated plants in the tropics, but pollinators not treated]

Richards, A. J. 1986. Plant breeding systems. London: George Allen and Unwin, 529 pp. [a clear and full treatment of the breeding systems of plants, genetics and related evolutionary ecology]

Richardson, R. W., Jr., Alvarez, E. 1957. Pollination relationships among vegetable crops in Mexico. I. Natural cross-pollination in cultivated tomatoes. Proc. Am. Soc. Hort. Sci. 69:366-371. [halictid bees are primary pollinators]

Roecklein, J. C, Leung, P.-S. 1987. A profile of economic plants. New Brunswick: Transaction Books, 623 pp. [catalog of useful plants cultivated in the Hawaiian islands, by scientific name, common name, end-use, summary of economic and botanical information for each; no family names, no pollination or breeding systems discussed]

Roubik, D. W. 1989. Ecology and natural history

of tropical bees. New York: Cambridge Univ. Press. 514 pp. [ecology, behaviour, taxonomy, and literature; discusses pollination]

Roubik, D. W. 1991. Loose niches in tropical communities: why are there so many trees and so few bees? In: M. D. Hunter, T. Ohgushi, P. W. Price (eds.). Effects of resource distribution on plant-animal interactions. San Diego: Academic Press, pp. 327-350. [discusses why pollinator specialization is rare in the tropics and documents variation in space and time in pollination systems]

Roubik, D. W. 1993. Tropical pollinators in the canopy and understory: field data and theory for stratum "preferences". J. Insect Behaviour. 6:659-673 [no canopy specialist bees are known for Neotropical forests, most forage at varying levels, and some avoid the upper canopy]

Roubik, D. W., Moreno, J. E. 1991. Pollen and Spores of Barro Colorado Island. St. Louis: Missouri Bot. Gard. Monog. Syst. Bot. 36, 269 pp. [illustrated guide and key to 1200 Neotropical plants]

Roubik, D. W., Schmalzel, R., Moreno, J. E. 1984. Estudio apibotanico de Panamá: cosecha y fuentes de polen y nectar usados por *Apismellifera* y sus patrones estacionales y anuales. El Salvador: Teh. Report. O.I.R.S.A. [El Salvador], 74 pp. [analysis of important pollen sources used by Africanized and European honeybees; honey also analyzed from pollen types]

Ruggiero, C., Banzatto, D. A., Lam-Sánchez, A. 1976. Studies on natural and controlled pollination in yellow passion fruit (*Passiflora edulis* var. *flavicarpa* Deg.). Acta Hort. 57:121-124. [*Xylocopa* are more effective than *Apis*]

Ruttner, F. 1988. Biogeography and taxonomy of honeybees. Berlin: Springer-Verlag. 284 pp. [most complete reference on *Apis*, which led immediately to the recognition of 7 species in place of 4]

Rzedowski, J. 1978. Vegetación de México. Editorial Lemusa: México, D. F., 432 pp. [reference to vegetation, cultivated and semi-cultivated plants; no discussion of pollination]

Sastrapradja, S., Aminah, S. H., Lubis, I., Sastrapradja, D. 1975. Studies in the Javanese species of *Canavalia*. I. Floral biology and cytolgy. Annls. Bogorienses 6:43-55. [*Xylocopa confusa* pollinates jack bean]

Sazima, I., Sazima, M. 1989. Mamangavas e irapuás (Hymenoptera, Apoidea): visitas, interações e consequências para polinização do maracujá (Passifloraceae). Revta. bras. Ent. 33:109-118. [on *P. edulis*, *Trigona* was an aggressive nectar consumer while *Xylocopa suspecta* and *frontalis were* pollinators]

Scariot, A. O, Lleras, E. 1991. Reproductive biology of the palm *Acrocomia aculeata* in Central Brazil. Biotropica 23:12-22. [potential source of oil is self-pollinated, wind-pollinated, outcrossed by beetles *Mystrops* (nitidulid), *Cyclocephala* (scarab) *Andranthobius* (weevil) and is thus a good colonizer]

Schatz, G. E. 1990. Some aspects of pollination biology in Central American forests. In: K. S. Bawa, M. Hadley (eds.). Reproductive ecology of tropical forest plants. France: Paris, UNESCO. pp. 68-83. [emphasizes beetle pollination in Annonaceae, Araceae, Palmae, Cyclanthaceae, Nymphaeaceae and Mag-noliaceae and floral fragrances in the neotropics; discusses other pollinators also]

Seavey, S. R., Bawa, K. S. 1986. Late-acting self-incompatibility in angiosperms. Bot. Rev. 52:195-219.

Sidhu, A. S., Singh, S. 1961. Studies on agents of cross pollination in cotton. Indian Cott. Grow. Rev. 15:341-353. [three *Apis* compared]

Sihag, R. C. 1988. Characterization of the pollinators of cultivated cruciferous and leguminous crops of sub-tropical Hissar, India. Bee Wld. 69:153-158.

Simpson, B. B., Conner Ogorzaly, M. 1986. Economic botany — plants in our world. New York: McGraw-Hill. 640 pp. [thorough and well illustrated; pollination given some emphasis]

Sirikul, W. 1992. Constraints to seed ontogeny and seed production in Thailand: an overview. Muak-Lek, Saraburi 18180, Thailand: ASEAN-Canada Forest Tree Seed Centre Project. Tech. Publ. No. 14. 11 pp. [flowering times differ between clones, variants and individuals of *Tectona*, *Pinus* and *Dipterocarpus*]

Smith, N.J.H., Williams, J.T., Plucknett, D.L., Talbot, J.P. 1992. Tropical forests and their crops. Ithaca, New York: Cornell Univ. Press, 568 pp. [extensive information primarily for plant breeders; includes some very obscure crops, interesting histories of importation and domestication, some information on pollination, not presented critically]

Soria, S. de J. 1977. Dinâmica populacional de *Forcipomyia* spp. (Diptera, Cerata-pogonidae) na Bahia, Brasil. 2 — variáveis bioticas relacionadas com a polinização do cacaueiro. Revta. Theobroma 7:19-33.

Start, A. N., Marshall, A. G. 1976. Nectarivorous bats as pollinators of trees in West Malaysia. In: J. Burley, P. T. Styles (eds.). Tropical trees: variation, breeding, and conservation. London: Academic Press. pp. 141-150.

Stiles, F. G. 1981. Geographical aspects of bird-flower coevolution, with particular reference to Central America. Ann. Missouri Bot. Gard. 68:323-351. [ecology of major groups through the world; birds are often territorial and many species share floral resources, often understory plants, even with mammals when nectar is plentiful]

Stone, G. N., Willmer, P. G. 1989. Pollination of cardamom in Papua New Guinea. J. Apic. Res. 28:228-237.

Stratton, D. A. 1989. Longevity of individual flowers in a Costa Rican cloud forest: ecological correlates and phylogenetic constraints. Biotropica 21:308-318. [flower longevity of 110 spp. at 1500 m averages 2.7 days, greater than lowland species (1 day), and lower than montane species (4-8 days)]

Tepedino, V. J. 1981. The pollination efficiency of the squash bee (*Peponapis pruinosa*) and the honeybee (*Apis mellifera*) on summer squash (*Cucur-bita pepo*). J. Kansas Entomol. Soc. 54:359-377. [little difference was found between the two bees]

Tomlinson, P. B. 1986. The botany of mangroves. Cambridge, UK: Cambridge Univ. Press, 413 pp. [concise, detailed, menitons pollination where data exist]

van Drunen, B. 1977. Pollination of fruit trees in the Western Cape. In: African bees: taxonomy, biology and economic use. D. J. C. Fletcher (ed.). Pretoria, South Africa: Apimondia. pp. 4-6.

Vavilov, N. I. 1951. The origin, variation, immunity and breeding of cultivated plants. Wkatgham, Mass: Chronica Botanica 13. 364 pp. [little discussion of pollination or breeding systems, but still a standard reference]

von Renis, S., Lipp, F. J., Jr. 1982. New plant sources for drugs and food from the New York botanical garden herbarium. Cambridge: Harvard Univ. Press. 363 pp. [common and scientific names, families, uses, and collection data for three million herbarium specimens at NYBG]

Vovides, A. P. 1991. Insect symbionts of some Mexican cycads in their natural habitat. Biotropica 23:102-104. [*Zamia* and *Dioon* are pollinated by specialist beetles: *Rhopalotria* (weevil) and *Pharaxonotha* (languriid)]

Wiens, D., Calvin, C. L., Wilson, C. A., Davern, C. I., Frank, D., Seavey, S. R. 1987. Reproductive success, spontaneous embryo abortion, and genetic load in flowering plants. Oecologia 71:501-509. [restrictions in seed production are a fundamental consequence of genetic load]

Williams, I. H. 1977. Behaviour of insects foraging on pigeon pea (*Cajanus cajan* (L.) Millsop.) in India. Trop. Agric. Trinidad 54:353-363. [*Apis dorsata* is most common but *Megachile* is the main cross-pollinator]

Winder, J. A., Silva, P. 1972. Cacao pollination: Microdiptera of cacao plantations and some of their breeding places. Bull. Ent. Res. 61:651-655.

1.2 POLLINATION, POLLINATORS & POLLINATION MODES: ECOLOGICAL & ECONOMIC IMPORTANCE

1.2.1 Pollination in General

Angiosperms are credited with the most beautiful gift of nature — these have flowers. The latter, though scientifically of merely reproductive relevance, have much greater significance in the ecosystem and for human society. As Richards remarks (1987, p. 66):

> An outstanding feature of the Angiosperms is the amazing diversity in forms and colour that has been adopted by the inflorescence, sufficient to inspire great art, fuel a major industry and serve as a solace for suffering mankind.
>
> Yet the flower is merely a sex organ, and never has any function except to promote reproduction by seed, usually sexually. The beautiful, weird, sinister, astounding forms that flowers have acquired are strictly pragmatic, and have encouraged the ecological diversification, and dominance, of the flowering plants...

The message clearly signifies the ecological importance of flowers and if the authentic paleontological and present ethological records can prove (in fact these do) that the evolution and perpetuation of this floral diversity is due to the pollinators, this surely demonstrates the importance of pollinators. Likewise, if it is also proved that pollinators help increasing the seed yield of many crops, this should signify their economic importance.

In angiosperms, pollination is an important event that is a prerequisite of sexual reproduction. This is because pollen is to do the same thing in plants that sperm does in animals. Success of pollen transfer is very important. However, pollen is a non-motile spore. It must be transferred from anthers (the seat of their production) to the stigma (the seat of their germination) by some vector.

Different plant species exercise different pollination modes, and the benefits accrued should depend upon the kind of pollen transferred. Where self-pollen normally tends to increase the homozygosity, pollen from other flowers, plants or genotypes should increase heterozygosity. Self-pollination generally sacrifices the plant quality (surely in 'outcrossers') *while outcrossing helps increase hybrid vigor, resulting in more healthy and stronger plants* (although not always). The kind of pollination also determines chances of gene recombination and exchange between individuals. In changing environments, gene recombination (and therefore cross-pollination) should provide an opportunity to produce strains better suited to new conditions, and should verily be an ecological necessity. In the changing pest scenario and with rapidly evolving resistance to pesticides, especially in the tropics, gene recombination and heterozygosity — through cross-pollination — are reliable means for increasing crop yield. Literally, it provides some crops with the chance of keeping up with, or escaping, their enemies.

In this section pollination modes and pollinators are reviewed, along with their ecological and economic importance. I have ostensibly avoided the use of complex terminology of pollination ecology, some of which is introduced in other sections of the present book (see also Glossary). Instead, simple terminology has been used whenever possible for convenience, and for accessibility of the layman, ecologist and seed grower. The discussion has been confined to describing importance of pollinators and pollination modes in crops grown to produce seeds and fruit.

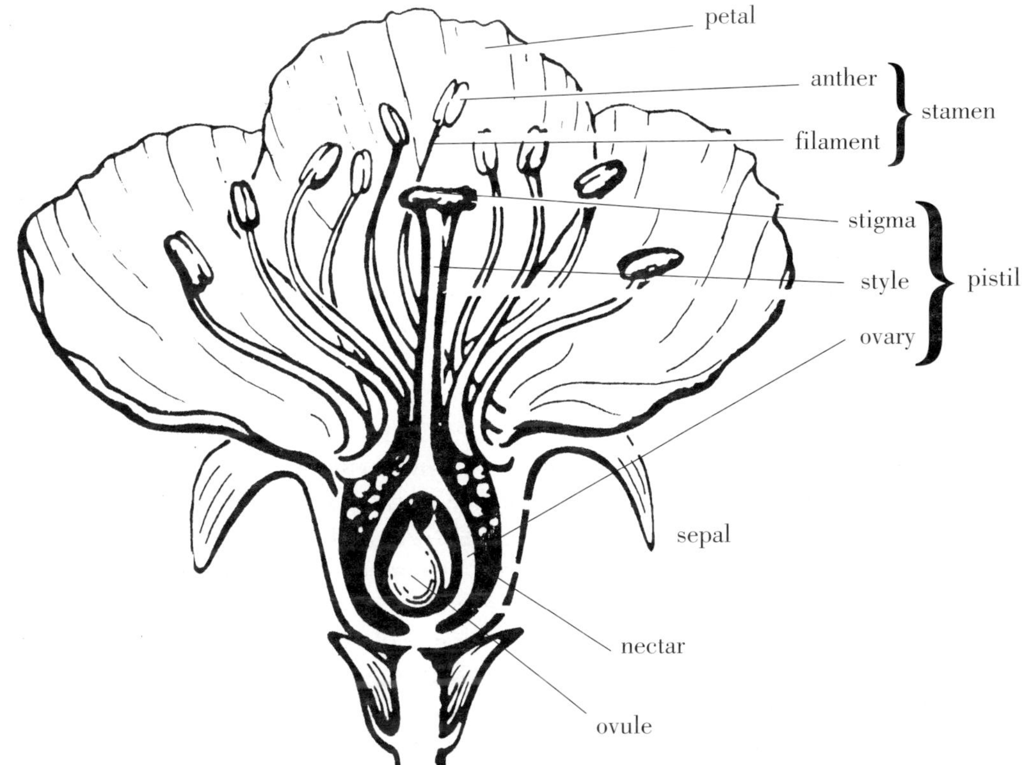

1) The Flower.

Before actually knowing about the benefits of pollination, it is necessary to know about the flower and how it is related to pollination and to pollination modes. These subjects are taken up more extensively in Sections 2.2-2.4.

• ***Flower:*** A typical hermaphrodite (bisexual) flower has four parts:

• ***Calyx:*** The calyx (consisting of sepals) is normally green and provides protection to other floral parts in the bud stage.

• ***Corolla:*** The corolla (consisting of flower petals) is the colored part of the flower that provides primary attraction stimulus to pollinators. This is the advertising agency of the flower.

• ***Androecium:*** This is the male part of the flower, which bears structures that hold pollen — the anthers — the seat of male spores.

• ***Gynoecium:*** This is the female part of the flower and carries the female gamete — the ovule — in the ovary. Pollen is received at the distal tip, called stigma, where the former germinates for fertilization.

2) Pollination Modes and Pollinators.

Pollination is the process involving transfer of pollen from anthers to the stigma. The agent causing this transfer is called the pollinator. Normally, angiosperms exhibit two kinds of pollination modes:

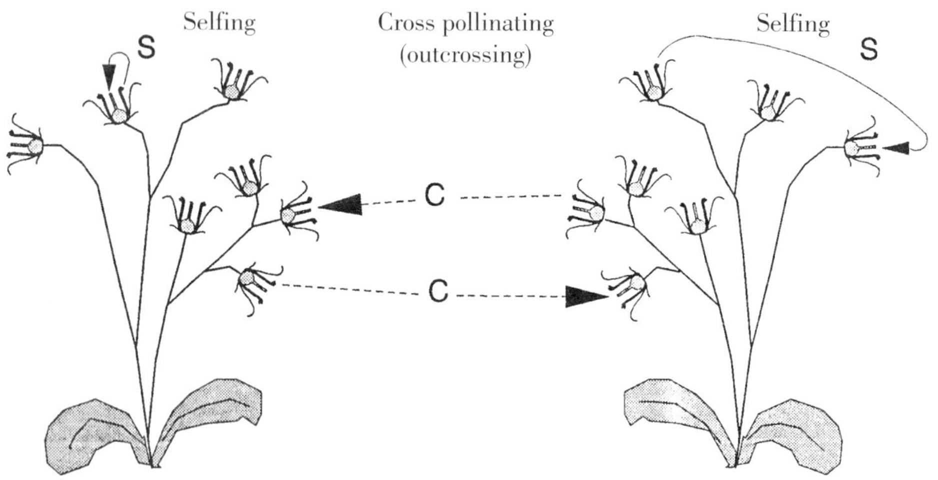

• ***Self-pollination or selfing.***
When pollination takes place within a flower, this is called self-pollination (see Figure). Self-pollination will take place if: ***a)*** Flowers are bisexual and have stigmas and anthers at the same heights (stamens and style are of the same length); ***b)*** both the sexes mature simultaneously (*protandry* or *protogyny* are absent); ***c)*** contact of newly dehisced anthers (releasing fresh pollen) with the receptive stigma is imminent. Proximity of anthers to stigma under the above conditions should result into self-pollination. If selfing results in *fertilization*, this should indicate at least some occurrence of strict inbreeding.

• ***Cross-pollination or outcrossing.***
When pollen from one flower is carried to the stigma of another, this is called cross-pollination. Here an external agent is required for accomplishing the pollen transfer.

Outcrossing is of two kinds: ***a)*** when crossing is between flowers of same plant (see Figure) — this crossing is genetically equal to selfing, however, a foreign pollen vector is required, and ***b)*** when crossing is between flowers of two different plants (see Figure) — this is the true outcrossing. Outcrossing is important in plants where: ***a)*** either flowers or plants are unisexual; ***b)*** anthers and stigma (stamens and styles) of same flower are at different heights (i.e. stamens and style are of different lengths); ***c)*** sexes mature at different times (presence of protandry or protogyny); ***d)*** there is no contact of dehisced anthers with the stigma of same flower during their functional phase; ***e)*** above all, plants are self-incompatible i.e. (pollen from a plant cannot be utilized by its own flowers.

Outcrossing is brought out by two kinds of agents, ***abiotic*** and ***biotic.*** First consider the *abiotic pollinating agents* — these are inanimate physical forces. Thus, abiotic pollination is generally 'random', or at least is not directed specifically between flowers. Different kinds of abiotic pollination have been recognized:

• Gravity-pollination (*geophily*) is found in self-pollinated plants. Here some pollen is expected to fall down on the receptive stigmas of other flowers due to gravity and may pollinate the flowers. However, geophily is highly unreliable and is a rare and insignificant pollinating agent.:

• Water-pollination (*hydrophily*) is found only in some water plants where inflorescences float or are submerged. However, many freshwater plants produce aerial inflorescences.

• Wind-pollination (*anemophily*) is found in many plant families including crop plants — especially grasses. Characteristics of plants using the wind-pollination mode include: ***a)*** a reduced leaf surface area; ***b)*** exposed flowers; ***c***) reduced perianths; ***d)*** long stamens and sometimes explosive anther dehiscence; ***e)*** smooth, dry pollen grains that may bear air sacs; ***f)*** lack of nectaries and nectar in flowers, and ***g)*** flowers having no colour or scent.

Relative disadvantages of the wind-pollination mode are that ***a***) there is little accuracy; ***b)*** pollen concentration declines with the distance from the emitting source; ***c)*** pollen is intercepted by all surfaces in the line of dispersal; ***d)*** pollen availability may diminish with height for flowers of wind-pollinated trees.

In contrast to the above, we now consider *biotic pollinating agents*. Here the pollination is carried out by animals (*zoophily*). Biotic pollination is highly accurate, but it has a higher cost, namely *nectar* in addition to *pollen*must be offered to pollinators by the plant, and the flowers are often large and showy. Pollen vectors like bees show high *floral constancy*.

Characteristics of plants using the biotic-pollination mode include: ***a)*** relatively small amounts of pollen are produced; ***b)*** there is some kind of relationship between the pollen vector and pollination unit (see Sections 2.2-2.4); ***c)*** pollen varies greatly in size and external sculpture but is usually sticky; ***d)*** flowers have attractive colours, odors and also produce nectar. Biotic pollination naturally falls into several distinctive classes:

- pollination of by insects (*entomophily*), e.g. by beetles (*cantharophily*), flies (*myophily*), bees (*melittophily*), butterflies (*psychophily*), moths (*phalaenophily*)
- pollination by invertebrates like snails and slugs (*malacophily*)
- pollination by vertebrates, such as birds (*ornithophily*), and bats (*chiropterophily*).

In the evolution of angiosperms, entomophily has played a great role, the other pollination modes are considered to be secondary derivatives of entomophily. Among these modes, the effectiveness of bee-pollination is highest due primarily to two reasons. First, bees visit the flowers to gather food rewards, thus reliably seek the flowers during all times. Second, flower constancy of bees (their persistence in looking for the flowers of one species) is very high.

In nature, all kinds of pollination modes are present, however, these are scattered in space and time. For example, ornithophily can best be witnessed in Neotropical and Australian forests. Hummingbirds and large nectar-producing flowers are the best example. Different pollinating animals are in fact associated with different sizes and shapes of flowers (see Section 2.2) and are usually the effective pollinators of these flowers, whatever the species happen to be, and wherever they come from originally (see **Appendix I**). These relationships vary from the most specialized to the least specialized types, as illustrated in much of the pollination literature (see 1.1.3) and elsewhere in this book.

1.2.2 Ecological Importance of Pollinators & Pollination Modes

Pollination modes and pollinators strongly influence ecological relationships, genetic variation in the plant community, floral diversity, speciation, plant evolution and ecosystem conservation (see especially Section 1.2.4) Pollination modes (e.g. abiotic or biotic) have a very broad range of effects, some of which are discussed below.

• *The role of selfing.*

Because selfing provides no chance of gene recombination, successful inbreeding over the generations leads to genetical impoverishment (i.e. loss of variability) and limited possibility for adaptation in new situations. Inbreeding may also become a starting point for the formation of a successful inbreeding species. The chances of the former remain larger than the latter. Obligate selfing is a rare event and is found in a small minority of plants. However, individual species may show high levels of selfing, which may exceed 99% of all fertilizations. Examples are *wheat, barley, oats* and *beans*. Selfing is normally found in opportunistic annual plants.

Repeated selfing renders the majority of species relatively less vigorous, when measured as height, weight or reproductive and survival capabilities. Inbreeding depression occurs in organisms that normally are 'outcrossers' and much less in those that have evolved to be 'selfers'. Selfing is a secondary derivative of outcrossing. Environmental changes that made a failure of self-incompatibility among outcrossers led to the evolution of selfing in plants.

• *The role of wind-pollination (anemophily)*

Like selfing, anemophily is also considered a later derivative of a widespread, pre-existing condition — biotic pollination in this case. Retention of floral colour and scent, a well formed corolla, effective wind- and insect-pollination at the same time, and similar features in a wind-pollinated plant indicates its recent development and a connecting link between biotic pollination and anemophily. Sudden environmental change resulting in failure of pollination is considered to be the fundamental cause of anemophily. However, unlike selfing, anemophily provides some chances of genetic recombination through outcrossing. Anemophily might be considered a highly wasteful pollination mode because the pollen falls where it may. A recent analysis show this is not the case, and it is precisely its greater economy that promotes the evolution of anemophily wherever possible. Nonetheless, large amounts of energy and material are used in production of massive amounts of pollen and feathery styles on flowers. In compensation the perianths are highly reduced and rudimentary. Therefore, floral features are greatly restricted. As a consequence, anemophiles exhibit low floral diversity.

Effective anemophily requires dry weather, and either low plant species richness or a large number of individuals in a relatively small area. The frequency of anemophily increases with both latitude and elevation. Wind pollination is generally uncommon in lowland tropical environments, especially in rain forests, and is dominant in temperate deciduous and boreal forests. These latter forests show low plant and floral diversity.

• *The role of animal pollinators (zoophily)*

Most of the extreme floral adaptations are directed towards animal visitors. Animals accurately transport a high proportion of the relatively little pollen produced, taking it over large distances to a tiny stigmatic target. Therefore, zoophily provides the best chances of gene recombination. Pollination by animals goes hand in hand with floral diversity and its perpetuation. In species-rich communities, with a low level of ecological dominance by individual plant or animal species, biotic pollen dispersal predominates. That is why, for example, that alpine grassland, Mediterranean maquis and tropical forests are full of pretty flowers, and show high floral diversity.

In more productive and more stable communities, the proportion of specialist flowers is slightly higher, indicating availability of more reproductive niches. Such communities will tend to have more species in each pollination syndrome. The diversity of reproductive niches available in a habitat is necessarily a major component in floristic richness.

One of the most important roles of pollinating animals is in *speciation* (new species formation). Selfing and wind pollination are considered to have no roles in this process and the generalist animal pollinators have minor roles. In the harmonic relationships of pollination syndromes and the pollinators, the specialized associations, even if only temporary, are vital.

Mutual adaptation of flowers and pollinators and their interdependence are considered to be results of their long and intimate coevolutionary relationship. Many flower forms evolved due to the selective pressure of pollinators. Various paleontological records now clearly reveal that these relationships have evolved through geological time. The non-specialized flower-visiting animals were followed by the highly-specialized visitors, ultimately culminating in the present day specialized blossom and pollinator classes (Section 2.2).

- ***Pollinators and ecosystem conservation.***

As described above, there are several specialized pollinator and blossom classes. These, and the existence of several pollination syndromes reveal the interdependence of pollinators and plants. In such systems, pollinators help perpetuation of plants through making their sexual reproduction a success. This is because successful reproduction is the major currency in the life of an organism, and failure of reproduction is the impairment of individual fitness. Therefore, conservation of pollinators should mean conservation of plant species, and vice-versa. When plants and pollinators are conserved, this must uphold species diversity in the ecosystem. The species-rich ecosystem with high species diversity is considered to be the most stable. This is the normal state of the tropics. Conservation of pollinators and their host plants should therefore mean the conservation of ecosystems.

1.2.3 Economic Importance of Pollinators

The economic importance of pollinators has now been fully recognized and realized in agriculture. There is a vast list of crop plants that either rely completely on pollinators or benefit from their pollinating visits. By increasing their seed and fruit yield through cross-pollination and the fecundity and survival benefits that these brin, .pollinators are also receiving benefits. The relationship is self-sustaining. Since man has a direct dependence on agriculture for food, fibre and other articles, and an increasing human population has exerted a great pressure for these commodities, the importance of pollinators in modern times has increased several-fold. Honeybees and some solitary bees can now be managed successfully and utilized for the pollination of crops. Their necessity has been felt whenever it is established that they increase yield, especially in crops that are self-incompatible or otherwise in need of visitors to their flowers.

The accompanying Table presents a list of crops grown in the tropics — fruits, vegetables, oil seeds, forage, fibers, and spices. *The benefits accrued to these crop plants, and hence the importance of pollinators in agriculture, is indicated in this table as the percentage increase in yield.*

Conclusion.

Pollination involves transfer of pollen from anthers to the stigma. Self-pollination is of little ecological and economic significance to many plant species. When followed by self-fertilization it can cause inbreeding depression. This is a result of homozygosity, which provides no chance for gene recombination. Therefore, variability in the plant species is impoverished. The homozygous individuals have stunted growth and low yield in many of the wild and cultivated plant species. Cross-pollination, on the other hand, leads to heterozygosity and provides chances of gene recombination. This may increase variability in a plant population and provides chances for the evolution of new varieties, strains and even species. Heterozygosity in cultivated crops is expected to increase hybrid vigor, resulting in more healthy plants with higher seed yield. Cross-pollination by wind, water or gravity is of little importance due primarily to their 'random' nature. However, pollination by insects can have great significance in the evolution of flowering plants and many floral, vegetative and genetic traits. All other biotic and abiotic pollination modes are secondary derivatives of zoophily — animal transport of pollen grains. The presence of a wide variety of pollinators and pollination syndromes have contributed to the present-day floral diversity in the tropics and subtropics. Among the animal pollinators, bee pollination (melittophily) is of great significance in agriculture. Bee pollination increases

1) Fruit crops

Acerola	*Malpighia glabra*	1-3% (S), 6.7-55% (H), 6.7-74% (C)
Almond	*Prunus amygdalus*	no bees, no fruit formation
Apricot	*Prunus armenica*	benefited from BP
Blackberry	*Rubus*	benefited from BP
Cashew	*Anacardium occidentale*	55.5% (S), need BP
Cherimoya	*Annona cherimola*	6% (OP), 44-60% (H)
Cherry	*Prunus*	20-35% (s), 49% (H)
Chestnut	*Castanea*	1.3% (S), 68%(OP), 34.9% C(H)
Kiwifruit	*Actinidia deliciosa*	CE
Citrus	*Citrus*	40-60% (H), 80-100% (OP)
Coconut	*Cocos nucifera*	CE
Date	*Phoenix dactylifera*	CE
Grape	*Vitis vinifera*	1.7 seeds/cage, 1.8 (BP), 1.8 (OP) (BE)
Guava	*Psidium guajava*	CE
Jamun	*Syzygium vulgare*	CE
Jujube	*Ziziphus jujuba*	CE
Litchi	*Nephelium chinensis*	0.01-0.03%(BE),0.7-11.2% (BP)
Mango	*Mangifera indica*	Cincreases fruit set
Muskmelon	*Cucumis melo*	1.6 crates/A (BE), 242 crates/A (BP), CE
Pawpaw	*Asimina triloba*	CE
Papaya	*Carica papaya*	CE
Passion fruit	*Passiflora*	CE
Peach	*Prunus persica*	BP increases yield
Watermelon	*Citrullus lanatus*	CE

2) Vegetable crops

Balsam pear	*Momordica charantis*	CE
Beet	*Beta vulgaris*	BP increases seed yield 14%
Cabbage	*Brassica oleracea*	CE
Carrot	*Daucus carota*	128 lb/A (IE), 435 lb/A (TI) 711 lb/A (OP), 840 lb/A (BP)
Chayote	*Sechium edule*	CE
Cucumber	*Cucumis sativus*	CE
Egg plant	*Solanum melongena*	C increases production
Lettuce	*Lactuca sativa*	C increases seed yield
Onion	*Allium cepa*	9.8% (BE), 93.4% (BP)
Pumpkin	*Cucurbita*	6.8% (BE), 61.2% (BP), CE
Radish	*Raphanus sativus*	CE
Tomato	*Lycopsersicon esculentum*	buzz pollination essential
Turnip	*Brassica rapa*	CE
Loofah	*Luffa cylindrica*	CE
White Gourd	*Benincasa hispida*	CE
Bottlegourd	*Lagenaria siceraria*	CE

3) Oilseed crops

Flax	*Linum usitatissimum*	BP increases seed yield 22.5-38.5%
Niger	*Guizotia abbyssinica*	BP increases yield
Rapeseed & Mustard	*Brassica*	64.7 seed set (BE), 95.3% (BP)
Oil palm	*Elaeis guineensis*	CE
Olive	*Olea europaea*	C increases fruit set
Peanut	*Arachis hypogaea*	BP increases seed yield 6-11%
Safflower	*Carthamus tinctorius*	32-47% (BE), 100% (BP)
Sesame	*Sesamum indicum*	BP increases seed yield
Sunflower	*Helianthus annuus*	311 lb/A (BE), 931 lb/A (OP)

4) 'Pulse' crops

Broad bean	*Vicia faba*	BP increases seed yield
Cicer milkvetch	*Astragalus cicer*	2.3% (S), 12.4% (H), 23.1% (C)
Pigeon pea	*Cajanus cajan*	BP increases seed yield 10%

5) Spices, condiments & beverages

Black pepper	*Piper nigrum*	BP essential
Cacao	*Theobroma cacao*	CE
Carambola	*Averrhoa carambola*	C obligatory
Cardamom	*Elettaria cardamomum*	11% (BE), 67% (BP)
Chicory	*Cichorium intybus*	0% (S), 61% (OP)
Clove	*Syzygium aromaticum*	CE
Coffee	*Coffea*	61.7% (BE), 75.3% (BP)
Coriander	*Coriandrum sativum*	C obligatory
Fennel	*Foeniculum vulgare*	BP increases seed yield 7 times
Kolanut	*Cola acuminata*	CE
Methi	*Trigonella corniculata*	0.09 kg/plot (BE) 6.2 kg/plot (BP)
Pimento	*Pimenta dioica*	19 berries (BE), >1000 berries (BP)
Tea	*Camellia sinensis*	CE
Vanilla	*Vanilla*	CE

6) Forage crops

Alfalfa	*Medicago sativa*	0.3 kg/A (BE), 20.3 kg/A (BP)
Berseem	*Trifolium alexandrinum*	0.27-0.64 seed/head (BP) 19.58-70.54 seed/head (BE)
Lespedeza	*Lespedeza*	C level 61.480.9%
Vetch	*Vicia*	BP increases seed production

7) Fiber crops

Cotton	*Gossypium*	2.3-3.4% (BE), 40-53% (BP)
Kenaf	*Hibiscus cannabinus*	C helpful in seed yield
Sisal	*Agave*	C necessary
Sunnhemp	*Crotalaria juncea*	2.6% (OP) 65% (BP)

A=Area, BE=Bees excluded, BP=Bee Pollination Present,
C=Cross-Pollination, CE=Cross-Pollination essential,
H=Hand Pollination, IE=All Insects Excluded,
TI=Tiny Insects Permitted, OP=Open Pollination, S=Self-Pollination.

seed production in many entomophilous and anemophilous crops. For the preservation of floral diversity in the ecosystem, conservation of pollinators and pollination services for plants are essential. For increased crop yield of seed and fruit, managed pollination should receive top priority.

Acknowledgment
I wish to thank Prof. R.B. Mathur for general facilities, Nirmala, my wife for encouragement, Dr. Manju Singh for her critical suggestions, and Anju Rathi for her timely help for the preparation of this article. The funds for this article were utilized from the projects 'C(a)Zoo-2-NP-Agri.' and 'C(g) Zoo-6-OA' sponsored by the State Government of Haryana and Ministry of Environment and Forests, Government of India, respectively.

1.2.4 Conserving Pollinators for Agriculture, Forestry & Nature

Pollination is a pivotal, keystone process in almost all biotically productive terrestrial ecosystems. These include the most remote wildernesses of the Arctic to the most highly managed farming operations such as hydroponics in greenhouses. In the accompanying Figure pollination is at the center of a multi-spoked wheel which has human, livestock, and wildlife consumers around its rim. There are a few other such important relationships in maintaining the ecosystemic health of the planet — fungal and microbial relationships with roots affecting plant growth and nutrition, biophysical interactions in the soil, biophysical interactions of life and the atmosphere, plant propagule (seeds etc.) dispersal by animals, the role of forests, and multifarious pollution problems.

Recently, conservation concerns for pollination have started to take on a greater profile. Some of this concern has come about because of the recognized value of pollination to agriculture. Figures calculated for the U.S.A., Canada, and Australia, mostly in regard to honeybees, show that the value of pollination far exceeds that of hive products like honey. The recognition of the problem in Europe prompted several serious and pioneering studies. Nevertheless, the economics of animal pollination in agriculture within any one country are complex and difficult to assess. That notwithstanding, agriculture and other equally vital economic ventures cannot do without their variety of pollinators, including the most generally important, the honeybees. The total value of animal pollination to world agriculture has not been estimated, and the value to world ecosystemic health is beyond measure!

The demise of pollinators has come about through four major human activities — ***a)*** Pesticide use; ***b)*** Habitat destruction; ***c)*** Diseases; and ***d)*** Competition from introduced flower visitors. Most information comes from temperate regions, but the same problems can be assumed to be equally or more severe in the tropics (see Sections 1.5.2, 1.5.3). The aim of this Section is to review briefly the information available on each factor and to place into perspective the potential seriousness of ignoring what has already taken place.

Another issue in pollinator conservation is an increasing recognition by scientists and others that 'non-honeybees' are important as crop pollinators. Nevertheless, the lack of general acceptance of the greater efficiency of other pollinators for certain crops, and the failure to recognize that some crops are poorly, if at all, pollinated by honeybees have hampered appropriate developments towards pollinator conservation for agricultural productivity.

1) Pesticides. The dangers associated with pesticides, especially insecticides, and pollinators are well documented and understood, especially with regard to European honeybees. Current available information is summarized in **Appendix III** (see also the references, listed after Section 2.1). Johansen and Mayer (1990) have recently written a highly informative book on the subject, with the main emphasis on the U.S.A. For most pesticides used worldwide there is published information on the toxicity to European honeybees, and sometimes to other bees. Many pesticide containers, in fact, bear labels with cautions about the dangers to pollinators.

Recent trends in many parts of the world towards reducing the use of pesticides in agriculture and forestry (see Section 1.6) have gone far toward lessening the incidence of pollinator poisonings. Nevertheless, in developing countries, the problems are still severe. It must also be remembered that pesticides are an integral part of integrated pest management practices (IPM) for crop protection in modern agriculture and forestry. The dangers must still be kept in mind and a constant vigilance maintained.

Most pesticide problems stem from accidents, carelessness in application, and *deliberate misuse despite label warnings and recommendations.* As pesticide applications become more and more regulated and users are required to take safety courses before certification, the problem should diminish. However, in many countries regulations are wanting, lax, or ignored. The general problems are exacerbated by the free availability in developing countries of pesticides that are outmoded or illegal elsewhere. In agricultural settings, pesticide use can be easily monitored and controlled by ***a)*** responsible agents of the agrochemical industries manufacturing and selling pesticides, ***b)*** diligent applicators who pay heed to labels, recommended application rates, and warnings about pollinator poisonings and human health, ***c)*** government extension agents, and ***d)*** other interested persons in agriculture, pollination services, or the general public.

Issues in non-agricultural settings and agro-forestry are more complex because of the importance of a wider diversity of pollinators, both wild and managed. The most well understood situation is in eastern Canada where *fenitrothion*, sprayed against spruce budworms that were defoliating forest trees. It had devastating side effects on wild, native pollinators of commercial blueberry fields. Effects were also immediately felt on the pollinators servicing the sexual reproductive needs of native vegetation. A number of different plant species of the forest and forest margins suffered reduced fruit and seed set, which in turn would be expected to impact wildlife by depriving them of natural quantities of food. The effects on pollinators resulting from other extensive applications of pesticides against other major pests, such as other forest defoliators, locusts, and grassland herbivores have been investigated hardly at all.

2) Habitat destruction. There are three ways that habitat destruction affects pollinator populations, as with populations of any organism: ***a)*** destruction of food sources; ***b)*** destruction of nesting or oviposition sites, and ***c)*** destruction of resting or mating sites.

The destruction of food sources is best illustrated by examples of the removal of vegetation, which provides pollinator food when the crops are not in bloom, from agricultural areas. Very often the vegetation which is removed is regarded as unwanted, as weeds or competition for the crop plants, yet is invaluable to pollinators and other beneficial insects. Kevan (1986) has made special reference to these problems with respect to biological control, IPM, and pollination in the tropics. Documentation of the negative effects on pollinator populations in agricultural areas of removing "unwanted" vegetation come from Europe and North America.

The destruction of nesting and oviposition sites has been documented in central Canada for the demise of populations of leaf cutter bees (Megachildae) left without nesting sites in stumps and logs as fields of alfalfa expanded, in Europe for bumblebees as the amount of relatively undisturbed land in hedgerows and greenbelts declined, and in the tropics for the inadequate pollination of *cacao* by midges in plantations from which oviposition substrates, rotting vegetation, had been too fastidiously removed.

Examples of the destruction of special mating or resting sites would pertain to pollinators with rather special requirements and to those associated with rare plants. Although this problem is suspected to be real, documentation is not available and would be very difficult to obtain without special study.

The general issue of habitat destruction for pollinators has evoked concern on a broad scale. Janzen's 1974 article "The deflowering of America" exemplifies the problem. He points to a vicious cycle of reduced vegetation for pollinators' resources, reduced pollination of the vegetation, the demise of plant reproductive success and reductions in seed and fruit set. These result in the failure of revegetation with a level of biodiversity that would have otherwise come about. This cycle applies to all parts of the world where pollination by animals is an integral part of the ecosystem. Nevertheless, recent books on conservation of insects and other animals give short shrift to pollinators and all but ignore the consequences of their demise. In the context of the present book, this attitude is very difficult to understand.

3) Pollinator diseases. Mite diseases of honeybees have evoked major concern as tracheal mites and *Varroa* have spread at alarming rates. The impact of these diseases on colonies of honeybees is well documented, but little information is available on the effects on pollination. It has been suggested that many amateur and small-scale beekeepers will abandon their activities because of the additional complexities of bee management associated with monitoring for mite diseases and controlling them once detected. Further, chemical control of mites may not be acceptable to producers of pure honey.

The expected changes in the face of beekeeping, which is mostly in the hands of small-scale operators widely dispersed over the agricultural landscape, seems to be resulting in fewer beekeepers and lesser dispersion of free pollination from bees in their hives. Already their are complaints from parts of the U.S.A. about inadequate numbers of honeybees for pollination of pome, stone, and

Animal Pollination & the Consumer

consumers: *H (human), L (livestock), W (wildlife)*

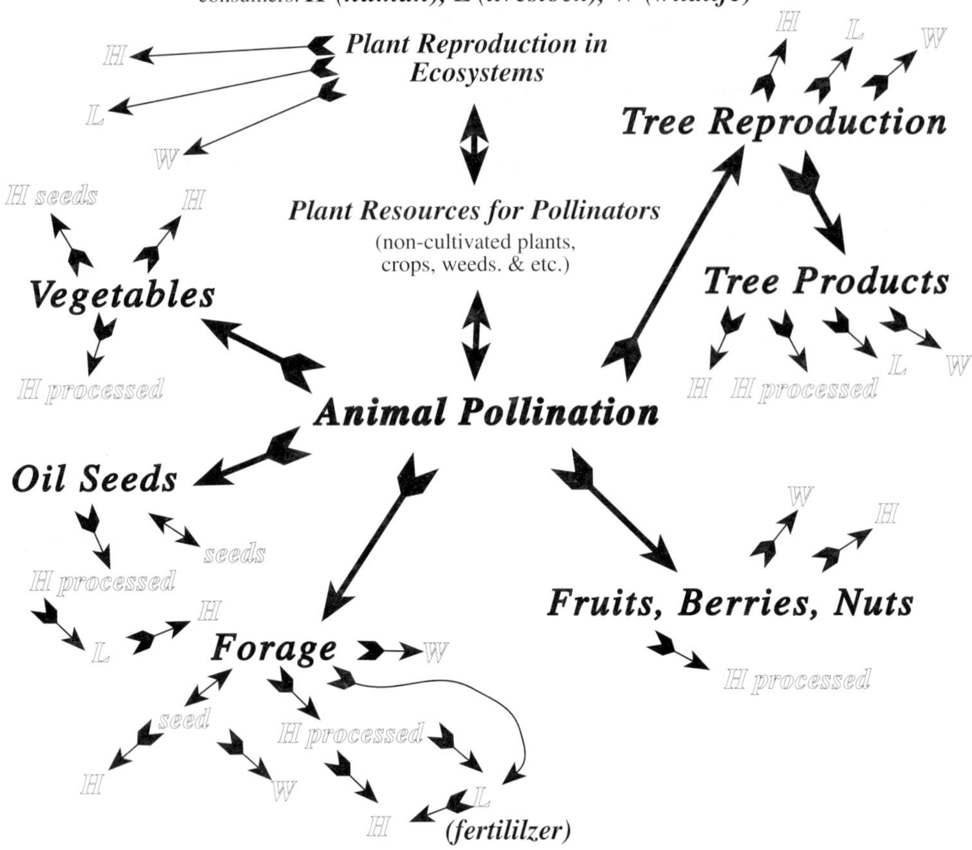

small, soft fruit crops. Pollination services may come to be provided by commercial beekeepers at an additional cost to the grower and consumer (see Section 2.1).

That scenario would apply to beekeeping operations in other parts of the world where non-native diseases have invaded the native stocks of honeybees. In India, the possible transfer of diseases from European honeybees to the Asiatic hive bee (*Apis cerana*) was suggested to have caused the demise of the latter to the detriment of honey production.

Great care is needed for the introduction of honeybees from one part of the world to another. The spread of honeybee diseases, from place to place and between species is mostly attributable to human activity (e.g. *Varroa* in North and South America, Europe and western Asia; tracheal mites in North America; etc.). Quarantine protocols are well established in some countries, but unfortunately lacking in others. Bailey and Ball (1991) provide up-to-date information on honeybee pathology worldwide.

Leaf cutter bees also suffer from diseases. The most important are the many chalk brood fungal diseases, such as that affecting the alfalfa leaf cutter bee, *Megachile rotundata*. This disease has a major impact on the culture of the bees and diagnosis facilities have been established in some places (e.g. western Canada) where these bees are highly important to pollination in agriculture. Research on the diseases other managed pollinators, such as orchard bees (*Osmia*) and bumblebees (*Bombus*) are becoming important as these pollinators take on a role in agricultural crop production.

The importance of disease in the regulation of populations of native pollinators is unknown. The same can be said regarding roles of other natural enemies, such as the many parasitic wasps that attack natural populations of all kinds of bees, but are much more concentrated and capable of creating adverse effects in the commercially established populations of solitary bees (see Section 2.5). However, a wide variety of pathogens, parasites, parasitoids, and predators are known to attack native bees, and other pollinators, in nature.

4) Pollinator competition.

The most studied of the competitive interactions between pollinators as they relate to pollination is that of the effect of the Africanized and African honeybees on native pollinators and European honeybees in South and Central America. Roubik first pointed out the apparent reductions in abundance of native bees in the Neotropics after the invasion of Africanized bees. Subsequently he placed the phenomenon in a broader context, but the issue of the competitive interactions of African bees with native pollinators in South and Central America seems complex. Information in recent articles has been misused and misinterpreted. A detailed field study in the Cerrado of Brazil showed there was a relatively small (about 33%) overlap of African honeybees with native bees on flower species. This article did not deal specifically with competition among the bees, but has sometimes been said to show the African honeybee was not a significant competitor. Besides simply being poor ecology, such interpretations ignore the absolute necessity of recording changes brought about by honeybee invasion — almost all studies focus on phenomena that take place after the fact. This, unfortunately, is the case throughout the world, where *Apis* has been introduced and no studies of any kind were made on possible impact on native or other animal species.

In Australia there has been recent debate on the effects of the introduced European honeybee on the native flora and fauna of pollinators. Some conclude there is justification for the concern that the European honeybees have caused reductions in the pollination of some native plants, especially bird-pollinated ones, by removing the nectar sought by the birds and causing changes in their populations and foraging habits. The issue of the effects on native pollinating insects is not so clear from the botanical side, but the same trends are evident with respect to the native bees.

5) Diversification of pollinators.

Although it must be conceded that honeybees are the most valuable of pollinators in agriculture, they are not the 'be-all' and 'end-all' of crop pollination. The examples of the *greater efficiencies of orchard bees for pome fruit pollination, alfalfa leaf cutter bees for alfalfa pollination, bumblebees for pollination of tomatoes and other solanaceous crops in greenhouses, blueberry bees for blueberries, and carpenter bees for passion fruit illustrate the point.* The lack of pollination brought about by honeybees on oil palm, various annonaceous fruit crops, red clover and other crops with flowers too deep for honeybees to work, and bat-pollinated durian are even more extreme as sound examples of the need to consider alternative pollinators for many crops. The issues in the tropics are especially important because the natural pollination mechanisms of so many plants (crops and others) are not understood.

Conclusions

Conservation of honeybees, other domesticated bees, wild bees, and other pollinators raise an important issue in the global context of agricultural and natural sustainable productivity. It is extremely important that apiculturists expand their horizons to embrace the culture of alternative species and to embrace the importance of other pollinators in agriculture. The importance of pollination, and its being adversely affected on a wide front through habitat destruction, poisonings, disease, and competitive interactions with alien species, needs to be fully recognized by biologists, ecologists, agriculturalists, and the general citizenry of the world in the new spirit of global, environmental sustainability and the conservation of biodiversity

Additional References

Bailey, L., Ball, B. V. 1991. honeybee pathology (2 ed). London: Academic

Beekeeping in the United States. 1980. United States Department of Agriculture, Handbook No. 335. Washington, D. C. 20402: U. S. Government Printing Office, 193 pp.

Janzen, D. H. 1974. The de-flowering of America. Nat. Hist. 83:48-53.

Jones, C. E., Little, R. J. (eds.). 1983. Handbook of experimental pollination biology. New York: van Nostrand Reinhold, 558 pp.

Kevan, P. G. in press. Wind or insects: pollination of coconut (*Cocos nucifera*) in the Maldive Islands. In: L. Conner, T. Rinderer, I. Sylvester, S. Wongsiri (eds.). Asian honeybees and bee mites. Cheshire, Connecticut, USA: Wicwas Press.

Kevan, P. G., Hussein, M. Y., Hussey, M. N. , Wahid, M. B. 1986. Modelling the use of *Elaeidobium kamerunicus* for pollination of oil palm. Planter, Kuala Lumpur 62:89-99.

Kevan, P. G., Plowright, R. C. 1989. Fenitrothion and insect pollinators. In: Ernst, W. R., Pearce, A., Pollock, T. C. (eds.). Environmental effects of fenitrothion use in forestry. Dartmouth, Nova Soctia, Canada: Environment Canada, pp. 13-42.

Menezes Pedro, S. R., Camargo, J. M. F. 1991. Interactions on floral resources between Africanized honeybee, Apis mellifera L. and the native bee community in a natural "cerrado" ecosystem in southeast Brazil. Apidologie 22:397-416.

Mishra, R. C., Sihag, R. C. 1987. Apicultural Research in India. Hisar: Hisar Agricultural Univ. Press, 120 pp.

Osborne, J. A., Williams, I. H., Corbet, S. A. 1991. Bees, pollination and habitat change in the Eiuropean community. Bee Wld. 72:99-116.

Parker, F.. D., Batra, S. W. T., Tepedino, V. J. 1987. New pollinators for our crops. Agric. Zool. Rev. 2:279-304.

Real, L. (ed.). Pollination biology. London: Academic Press, 338 pp.

Roubik, D. W. 1978. Competitive interactions between Africanized honeybees and native neotropical pollinators. Science 201:1030-1032.

Roubik, D. W. 1988. An ovderview of Africanizerd honeybee populations: reproduction, diet, and competition. In: Proc. Intl. Conf. on Africanized Honeybees and Bee Mites. Delfinado-Baker, M., Needham G., Page, R. and Bowman, C. (eds.) Chichester, U.K., E. Horwood, Ltd., pp. 45-54.

Roubik, D. W. in press. Direct costs of forest reproduction, bee-cycling and the efficiency of pollination modes. In: D. W. Roubik, T. Inoue, P. S. Ashton (eds.). Diversity and Stability of Ecological Communities in Fluctuating Environments. J. Biosciences (Supplement).

Sihag, R. C. 1986. Insect pollination increases seed production in cruciferous and umbelliferous crops. J. Apic. Res. 25:121-126.

Southwick, E. E., Southwick, L., Jr. 1992. Estimating the economic value of honeybees as agricultural pollinators in the United States. J. Econ. Entomol. 85:621-633.

Williams, I. H., Corbet, S. A., Osborne, J. A. 1991. Beekeeping, wild bees and pollination in the European community. Bee Wld. 72:170-171.

1.3 APPLIED POLLINATION IN TEMPERATE AREAS

Accounts of crop pollination in temperate climates are given by McGregor for North America, Batra, Kapil, Jain, Ahmad and Dafni for the Indian subcontinent and the middle east, Free for the world, and Crane and Walker for tropical countries. They include illustrations of flower morphology, valuable bibliographies, and accounts of some regional pollinators, their effectiveness, and the need to control insecticide use on crops. The ***21 crop groups summarized in this section focus on major temperate crops grown in highland regions of the tropics.*** The plant families treated here include many major families of crop, weed and indigenous (native) plants. Shrubs, vines, herbs and trees are covered. These species flower in spring to autumn, and their flowering period is short to fairly long. The 21 groups fairly represent the variation in floral structure and breeding system found in crops pollinated primarily by insects or having a mixture of wind/insect and bird/insect pollination. Crops from the less specialized plant families, often suited to general pollinators are considered first. ***Crops that are not well pollinated by honeybees are reviewed in more detail to illustrate the subtleties that can be involved.***

A. Cruciferae *(3,200 species, 380 genera)*

Eight of the 40 *Brassica* species are cultivated as crops. The dwarf honeybee, *Apis florea* fails to pollinate flowers of all cruciferous crops — toria, mustard, cabbage, rape, cauliflowers, turnip rocket cress and radish. *A. cerana* could well behave like the western honeybee by removing nectar from some crops without contacting the anthers and stigma.

1. Swede rape, *Brassica napus*, and turnip, *B. campestris*.

These crucifers are mainly grown for oil seed but the former is used as a greenfodder crop for livestock. Over 60% of the rape oil crop is grown in Asia and the Indian subcontinent in temperate and subtropical areas. Tropical countries grow about 64% of the world crop. The main rape growing countries are China, India, Bangladesh and to a lesser extent Ethiopia, Colombia and Brazil. Rape flowers provide an important source of honey in Bangladesh, China, Pakistan, Japan, Taiwan and Argentina, but honey yields are a modest 7-14 kg per hectare.

Swede rape is self-fertile, but it produces a greater weight of seed from longer pods with more seeds in them — chiefly the result of pollinator visits. Seed set and yield are poor in glasshouses that rely on self-pollination, and seed viability is better with cross-pollination. Flowers of winter rape (sown in autumn) remain open and can be pollinated for one to three days. Spring rape flowers can be pollinated during one or two days. Winter crops flower for about six weeks in late spring, and spring crops bloom in early summer in temperate zones, but during the cool season (late autumn-early spring) in the Punjab. Rape is a useful nectar source for honeybees. Some pollen is gathered from crops too, but 40.0-95.5% of foragers later *discard* the pollen and deliberate pollen collecting is rare.

Brassica is mainly pollinated by insects, bees in particular, but shaking of the plants by wind contributes to pollination. Most insects pollinate the yellow rape flowers. Two beetle pests only carried about 10% of the pollen of honeybees and bumblebees. In spring rape, some honeybees take nectar without pollinating the flowers. Honeybee flower visitation rate increases and the flowers visted per plant decline as temperatures increase, whereby cross-pollination is probably improved. Pollen collection reaches its peak in the morning, thus effective pollination may barely exceed 10 hours even on a fine day. In 10 hours, each honeybee should visit 5,000-8,000 flowers. Hive density of 2-6/hectare is recommended for pollination. The giant honeybee *Apis dorsata* is an effective pollinator and increased yields have been reported when hives of *A. cerana* were introduced to crops.

In Europe, *Bombus* is a consistent pollinator of rape flowers. In India, the solitary bee *Andrena leaena* is a consistent pollinator. Frequent visitors in India and Pakistan include *Apis florea*, which is not an effective pollinator, and *Andrena ilerda*, which may nest in rape fields. Other common visitors include *Halictus*, two more *Andrena*, and less often *Amegilla, Anthophora, Ceratina, Colletes, Lasioglossum, Nomada, Nomia, Nomioides, Pithitis smaragdula, Sphecodes, Thyreus*, and *Xylocopa*.

B. Rosaceae *(3,500 species, 120 genera)*

The rose family have open flowers with stamens near the stigma on each flower. Each flower can usually be pollinated by generalist insect visitors. Flowering is intense and virtually completed in two to three weeks. Pollination is most frequent on the first day of flowering. Almond *(Prunus amygdalus* var. *dulcis*) flowers first in early spring, then peaches and nectarine *(Prunus persica)*, apricot (*P. armenica*), plum *(Prunus domesticus)*, Japanese plum (*P. salicina*), sweet cherry(*P. avium)*, tart cherry *(P. cerasus)* , then European pear *(Pyrus communis)*, nashi pear (*P. serotina*) and finally apple *(Pyrus malus)*. Daytime temperatures are critical for pollinator activity on almonds and rather less important for apples that flower a month later. Peaches, nectarines and apples will be unevenly shaped with inadequate pollination. Seed size is small and fruits elongate in poorly pollinated pears, and part of the apple core will lack pips if not pollinated. Almonds, apricots and peaches have single flower buds, while apples, cherries, pears and plums contain multiple flowers in each bud — giving them 2-10 times the density of flowers. Information is scarce on flower density even though fruit set is discussed (usually based on the proportion of multiple-flower buds that make fruit), which makes all data on fruit-set difficult to interpret. Most orchards are planted with 100 to 400 trees/hectare, while intensively trellised multitier apples and dwarf pears can have up to 1,150 trees/hectare.

2. Almonds, *Prunus amygdalus* var. *dulcis*.

Almond flowers decline in receptivity during the three days they may remain open. Almond has early-, mid- and late-flowering varieties, but most flowering occurs in the month that winter turns to spring. A profitable crop depends on cross-pollination of most flowers and 20-40% of flowers usually set fruit.

The white flowers of almond are relatively attractive to honeybees and open when honeybee colonies are at about their smallest size in the season — when weather often restricts foraging to 1-3 hours per day. Nectar collectors rarely contact the stigma as they perch on the petal and probe nectaries between bases of the stamens. About 150-200 honeybees/tree or 5 to 10 strong hives/hectare are recommended so that honeybees investigate more flowers and more often perform cross-pollination services. California produces about one-third of the almonds in the world and there are not enough honeybee colonies to completely satisfy the need for pollinators.

In the USA, the mason bee, *Osmia lignaria* has been used experimentally in almond orchards, while the orange orchard bee, *O. cornuta* has been introduced for pollination of almonds. *O. cornuta* is being used commercially for almond pollination in Spain, where 60-80% of the pollen collected by these bees comes from the fruit trees. In Kashmir, the temperature threshold for foraging on almond is 8°C for *Apis cerana* and *Xylocopa fenestrata*. In New Zealand, the queens of *Bombus terrestris* freely use almond flowers. Flowering occurs before most nesting has begun, so that queens are the main visitors and nectar the primary resource they seek. Pollinating *Lasioglossum* need both sunny conditions and 13.5°C for flight.

3. Peaches and nectarines.

These need winter chilling, like apples, which restricts them to temperate and subtropical climates. From 11 to 16% of all peaches and nectarines are produced in Asia and India, about 4% in Africa, over 3% in Central America and nearly 3% in tropical South American countries. The main producers from the above regions are China, Japan, Mexico, Korea, South Africa and Brazil.
Most peach varieties are self-fertile, but 'Hale', 'Marsan', 'June Elberta', 'Hallberta', 'Candoka', 'Chinese cling', 'Almarand', 'Crawford', and 'Giant' require cross-pollination. The pink peach flowers are receptive for three days. Wind can dislodge pollen and set fruit. In the field, bees and other visitors contribute to pollination, but show little evidence of improving yields or fruit quality. Fruit set with honeybees can be twice that of bagged flowers. The use of 0.2-2.5 hives/hectare is recommended for pollination.

In Japan, *Osmia cornifrons* is used for peach pollination. In Korea, *Megachile* are common wild pollinators. And in northern India, *Apis dorsata* was the most common honeybee and began visiting the flowers about an hour earlier than imported *A. mellifera*. Flies remained active on flowers in overcast and rainy weather, unlike the honeybees.

4,5. Plums and cherries.

From 21 to 26% of the cherries are produced in Asia, with production in the Neotropics mainly in Mexico, but also Chile and Argentina. 'Stella' is the only one of 30 common sweet cherry cultivars that does not need another variety to serve as a pollinizer. Tart cherries are largely self-pollinating but can yield twice the crop with cross-pollination. Their flowers have a low sugar concentration.

Yearly fluctuations of 200-400% in yields of plums have been reported in New Zealand and the USA. Low availability of pollinizing trees and degree of pollination and fruit formation are considered to cause such variation. With European plums, greengage, about half the purple to blue prune plums, and some yellow egg and lombard (red to pink plums) do not need another pollinizer variety. In all, 19 of the 30 main commercial varieties require a pollinizer to set fruit. Self-fertile plum varieties usually have 20 to 23 stamens on each white flower, while self-sterile varieties have 24 to 32 stamens on a flower. Fruit set with pollinating honeybees can be six times that of bagged flowers, thus plums are more responsive to pollinating bees than are peaches.

Honeybees are effective pollinators of plums and cherries. In Australia, where there are few additional pollinators, fruit set reaches 36% when 7 to 21 honeybees are seen at one time on the flowers of each tree. From 1.3 to 5.0 and 2.5 colonies/hectare are recommended, respectively, for pollination of cherry and European plum. In northern India, *Apis cerana* foraged longer each day than the imported honeybee *A. mellifera* on both plum and apples. A density of 2.5 honeybee colonies/hectare is recommended for pollination in Europe.

The ornamental trees and shrubs, kowhai *(Sophora microphylla)*, *Grevillea* and *Hakea saligna* are more attractive nectar sources for honeybees and bumblebees, while pollen of *Brassica* and gorse compete with plums for honeybee use. In USA and England, cherry pollen has been favored by honeybees, but kale can be a competing pollen source.

The horn-faced bee, *Osmia cornifrons* is being used commercially for pollination of plums in Japan and has been introduced to the USA. Cherries and plums have been favored pollen sources for the mason bee, *Osmia cornuta*, in France. This bee is used to a limited extent in Yugoslavia and Spain for pollination of fruit orchards. *O. cornuta* forages at lower temperatures than the

Pear & peach flowers & fruit (sections) - Rosaceae ***Prunus & Pyrus***

honeybee, and over smaller distances. *Andrena varians* was well-synchronized with cherry and plum flowering, often collecting over 90% of its pollen from flowers. In northern Thailand, *A. florea*, *Trigona*, and *A. cerana* were the main bee visitors to plum flowers. Bumblebees that emerge early in the season and have relatively short tongues — *B. terrestris*, *B. lucorum*, *B. terricola*, and *B. bimaculatus* — are the main species found on plums. Growers especially value their contribution to pollination in seasons with adverse weather during flowering. Cherries are among the favoured flowers for the first of the longer-tongued bumblebee species, e.g *B. hortorum*, that emerge in spring.

6. Pears.

Worldwide, 40-44% of pears are produced in Asia and India, 4% in Africa and < 2% in Central America and the tropical areas of South America. Each white pear flower sheds pollen for two to seven days and each tree flowers for about seven days. Adequate crops require that 2-6% of the flowers set fruit. A 12-ton yield/hectare is possible. Most pear varieties need pollinizers, but under ideal conditions, 'Bartlett', 'Comice' and 'Hardy' will set heavy crops of seedless pears. The use of two to five hives/hectare is recommended for pollination in Europe. During the 10 days of flowering, 80 h of temperatures higher than 16°C are needed for high levels of pollination with *Apis mellifera*. In the western USA, occasional failures of pear crops are attributed to poor weather conditions for honeybee foraging.

Pear flowers have plenty of pollen, but the nectar has as low as 8-10% sugar concentration, less than half that of cherries or apples. Thus pear flowers are often relatively unattractive to bee visitors and may be visited mostly by flies. Honeybees prefer adjacent apple and sweet cherry trees, white mustard *(Brassica alba)* and chickweed *(Stellaria)*, ground cover, and hawthorn hedges *(Crataegeus)* to pear flowers.

At 1,300 m elevation in Northern Thailand, *Trigona, A. florea* and *A. cerana* were the main bee visitors of pear flowers, and the hover fly, *Eristalis cerealis* was a main visitor in Korea (along with the non-wild honeybees). Hover fly abundance on flowers declined by two-thirds when temperatures fell from 19° to 17°, and no flies were present at 16°C.

7. Apples.

These are a major fruit crop in temperate and to a lesser extent the subtropical regions, with at least 25 % of production in Asia and the Indian subcontinent and around 2% in Africa, Central and South America respectively. The main tropical countries producing apples are China, India, Brazil, Mexico, Mauritius and Pakistan. The flowers are an important source of honey, with reported crops of 20-40 kg/hectare.

Most important commercial apple cultivars need pollen from another cultivar to set fruit. However, 'Johnathan', 'Rome beauty', 'Oldenberg', 'Wealthy', 'Golden delicious', 'Newton', 'Grimes,', 'York', 'Red gold' and 'Yellow transparent' will set fruit in large blocks from their own pollen. Even these varieties, with the possible exception of 'Yellow transparent' produce heavier crops with cross-pollination.

Most of the honeybees that visit apple flowers are nectar collectors in N. America and Europe. Nectar-collecting workers are effective pollinators of apple flowers except "Delicous" varieties that have gaps in the stamen ring. Honeybees work from the side around the bases of these apple flowers, while standing on the petals. Apparently the percentage of "side-working" honeybees increases as the season progresses, which is after most flowers have been cross-pollinated. Individual honeybees tend to continue visiting the same foraging area, which discourages cross-pollination. Pollen-collecting honeybees, which effect fruit set better than nectar or nectar-and-pollen collectors, are in a minority unless nectar is unattractive or unavailable. In orchards with about 130,000 flowers/hectare, around 200 pollinating honeybees/hectare were needed to set 5% of apple flowers. However, it was estimated only one in six of the honeybee visits resulted in maturation of a fruit. The needed *revisitation* could double the numbers of pollinator visits required. This explains why 2.5 hives/hectare, which can provide 2,400 pollinating honeybees in a hectare at any one time, are needed for pollination. In the western USA, 20 honeybee visits/tree/minute are recommended for pollination, and 1-5 hives/hectare are recommended for pollination in Europe.

In the USA, oak *(Quercus)* and dandelion *(Taraxacum officinale)* can be important pollen for honeybees while apple is flowering. Dandelion can be a temporary floral source, because it is used for nectar in cool conditions, when apple flowers are unattractive. Pollen from dandelion is collected rather earlier (mainly in the morning) than apple pollen, but very few honeybees collecting pollen from dandelion switch to collecting apple pollen. Mowing dandelion to concentrate pollen collectors on the apple flowers during the earlier part of flowering, when most fruit is set, would seem to be justified. Gorse (*Ulex europeaus*) at the start of flowering and horse chestnut (*Aesculus*) near the end of flowering can be competing pollen sources.

Three species of mason bee, *Osmia cornifrons* (from Japan) *O. lignaria* (from N. America) and *O. cornuta* (from Europe) are valued apple and fruit tree pollinators. All are used commercially in Japan, Europe and USA. One bee can pollinate and cause 2,450 flowers a day to set fruit, causing 50-60% fruit-set within 50-100 m of their nests. Brood cells in nests of *O. lignaria* can contain 84% apple pollen. In France, apple pollen was less common in bee nests, while plum and cherry accounted for 47-87% of pollen stores in nests. Competing pollen sources for *O. cornuta* have been *Salix* and *Brassica*.

The bumblebees that emerge the earliest in spring predominate on apple flowers in Europe — *B. terrestris, B. lucorum, B. pratorum* — in North America, *B. terricola, B. occidentalis, B. affinis, B. impatiens*, in northern India, two species of *Bombus* and presumably others in China, Korea and Japan. Bumblebees do not 'side-work' the flowers and visit them two or three times faster than nectar-collecting honeybees. Bumblebees also forage under cooler conditions, but the effect this has on daily pollination of apples, compared to honeybees, has apparently not been recorded.

In England, *Andrena varians* used more apple pollen (usually over 80%) than seven other bees of this genus, while in Germany *A. albicans* was the most common flower visitor. Like the honeybees, these bees were less active in cooler conditions and visit flowers at about half the rate of *A. mellifera*, making up for their slowness by consistently pollinating each visited flower and carrying more viable pollen. Flies (mainly hover- and blowflies) also visit apple flowers, shown wherever the pollinating guild has been thoroughly investigated. Studies in England and Canada show flies carry about 10% the pollen carried by bees on their bodies. Flies may act as poor supplementary pollinators in the absence of bees, as the author has observed on Chatham Island.

C. Malvaceae *(2,000 species, 88 genera)*

Cottons are the major malvaceous crop, utilizing 4 of the 32 *Gossypium* species. Kenaf *(Hibiscus cannabinus)* and okra *(H. esculentus)* have, like cotton, large flowers and extrafloral nectaries. Honeybees appear unlikely to be efficient pollinators of kenaf and okra.

8. Upland, Asiatic tree cotton, *Gossypium hirsutum, G. arboreum.*

About 50% of cotton is grown in Asia and the Indian subcontinent, 12% in Africa and 10% in South America. The major cotton growing countries in the tropics are India, China, Pakistan, Brazil, Argentina, Paraguay, Egypt, Nigeria and Sudan. Cotton is a minor source of honey in Senegal and El Salvador, producing 30-90 kg/hectare.

Cotton is self-fertile but benefits from cross-pollination by insects. Flowers last one day and begin to wilt in mid-afternoon. Each flower has one floral and four extrafloral nectaries. Nectar secretion is best at 25-35°C; extrafloral nectaries secrete for several days and are difficult for small bees to use until the flower withers. The stigma projects 2.5 mm beyond the stamens.

Honeybees forage from extrafloral nectaries, which contain concentrated nectar (see Section 2.3.1) until they are drained, leaving as few as 6% of bees visiting the floral nectaries. Boll retention (fruit set) on Asian cotton flowers visited by honeybees averaged 57-62%, 50% for solitary bees of similar size, and 30% for self-pollinated flowers. In the USA the highest yields were obtained when one honeybee was present per each 100 flowers. In addition, there was a 500% increase in yield of hybrid seed cotton with 16 honeybee colonies/hectare (compared to no honeybee hives present). Honeybees usually prefer nectar from the male-sterile (pollen-less) flowers. In

Europe, 0.5-12.5 colonies/hectare are recommended for pollination.

In Pakistan and India, *Apis dorsata, Xylocopa, Anthophora confusa, A. quadrifasciata, Megachile monticola*, and scoliid wasps are recorded as the main pollinators of cotton flowers. The flowers are also visited by *Apis florea* which may be less effective pollinators due to their smaller size, and *Andrena ilerda, Coelioxys, Halictus, Nomada, Nomia, Pithitis smaragdula*, and *Xylocopa*. In Egypt, *Xylocopa pubescens* and *Scolia,* and in North America solitary bees, *Melissodes*, and scoliid wasps, *Elis plumipes* always visit the floral nectaries and their bodies become dusted with pollen.

Bumblebees and probably carpenter bees are considerably more effective pollinators than other bees, because they usually visit the floral nectary and touch both the stamens and stigma during foraging. In the USA, a bumblebee averaged 1.2 flower visits/plant and in 9 hours potentially visited 1,700-2,750 flowers.

D. Linaceae *(290 species, 12 genera)*
9. Linseed or flax, *Linum usitatissimum.*

The plants are grown for oil and linseed fibre. Some 28-34% of the world crop is grown in India and Bangladesh, 13-15% in Argentina, 2-4% in China and 1-3% in Africa, mainly in Ethiopia.

The crops flower in summer for several weeks, but the peak occurs after the first week. A crop can produce dense flowers and plant density is up to 400/m^2. In most cultivars the anthers are level with the stigma, but some have a lower or higher stigma. Linseed is self-fertile and self-pollinated. At least 5-6% cross-pollination occurs, notably in varieties with the longest stigmas, and the larger flowering cultivars.

Cage studies with honeybees in USA, Germany and England did not register increased yields, unlike Egyptian and Russian studies that included open plots and mentioned 22-43% increases in seed weight using pollinating honeybees. These different results could be partly due to different cultivars. Honeybees visit linseed most intensively in the morning for both nectar and pollen. Most workers stand on the petals, which are prone to dropping off, and probe the flower from the top. Some workers cling to the stalk and probe between petals, but may still cause selfing as they shake the flower. Linseed is a major honey source in Mozambique and a major pollen source in Egypt. It is a medium honey and pollen source in Argentina and Mexico, and crops may yield a modest 2-12 kg/hectare of honey.

Short- and medium-tongued generalist *B. terrestris, B. lucorum, B. lapidarius* and *B. ruderarius* in Europe or New Zealand are the more common visitors. In Eygpt, the bees *Andrena pseudoorulata, Andrena spp., Xylocopa aestuans, Chalicodoma secula, Tetralonia, Anthophora*, and the wasps *Cryptocheilus discolor, Philanthus abdelkar, Polistes gallica*, and *Vespa orientalis* visit the flowers, as do butterflies, flies and beetles.

E. Liliaceae *(1,200 species, 90 genera)*
10. Onion, *Allium cepa*, shallot, *A. ascalonicum*, spring onions, *A. fistulosum*, leek, *A. porrum*, garlic, *A. sativum*, chives, *A. schoenoprasum.*

These are used for food and onion is produced throughout the world. Garlic and shallot seldom flower, being propagated vegetatively.

The best seed yields of onion are up to 1,500 kg/hectare but yields of 500-700 kg are more common. Each flower head bears 50-2,000 flowers, and the pollen is shed in the first 24-36 h, before the stigma is receptive. Cross-pollination is only needed for plants used in hybrid seed production. Up to 94% of the flowers produce seeds. In the USA, honeybees are considered essential for pollination of hybrid crops, but the erratic use of onion flowers by the bees has made it necessary to stock them at exceptionally high numbers — 10-30 colonies/hectare.

Onion is pollinated by bees and generalist flower-visiting insects, notably flies, with 267 species being recorded from crops in midwestern USA. Shallot flowers can be quite attractive to honeybees and bumblebees. In India, *Apis dorsata, A. florea* and *Trigona iridipennis* were the most common visitors to onion flowers and *Trigona* worked the flowers at about half the rate of the honeybees.

F. Umbelliferae *(3,000 species, 300 genera)*

The flowers of Umbelliferae and upright Compositae have extensive flat platforms and shallow flowers that allow insect visitors to walk, thereby economizing on energy expended when not much nectar is available. Umbelliferae include nine species of vegetables, herbs and spices as follows: anise *(Pimpinella anisum)*, cerelac or celery *(Apium graveolens)*, chervil *(Anthriscus cerefolium)*, coriander *(Coriandrum sativum)*, dill *(Anethum graveolens)*, fennel *(Foeniculum vulgare)*, parsley *(Petroselium crispidum)*, parsnip *(Pastinaca sativa)*, but carrots *(Daucus carota)* are the most important cosmopolitan species.

11. Carrot, *Daucus carota.*

Carrot is the most important cultivated umbellifer species and must be cross-pollinated. Its flowers are mainly used for pollen by honeybees. The crops are in flower for about four weeks. Each white flower remains open at least a few days and flowering within a flower head lasts for up to seven days. The flowers have the reputation of being a good honey source but quantification of yield is difficult, due to the limited areas cultivated for seed production.

Carrot seed crops are mainly pollinated by honeybees, with no stocking rates calculated thus far. Pollen collecting bees of male-sterile lines carry about 10% of the pollen compared to nectar collectors on male-fertile flowers.

Maximum seed production is achieved with 10 honeybees per square meter. Pollination increases yield and evenness of seed size. In India, yields have been two to five times higher in open-pollinated flowers than those in caged plots. In the USA, 860 kg seed/hectare has been produced using honeybees in caged plots.

In Eygpt, *Anthophora* were abundant as the honeybee on carrot flowers. In the USA, 334 insect species, mainly bees (*Andrena, Halictus, Chloralictus, Colletes*), sphecid, vespid, pompilid and ichneumonid wasps (*Cerceria, Lindenius, Nyssa, Philanthus, Sceliphron*), and syrphid, bombyliid, sarcophagid, stratiomyid and tachinid flies (*Syritta, Tubifera, Eulalia, Stratiomys*), visited flowers. Gnats and midges help pollinate carrot flowers.

G. Myrtaceae *(3,000 species, 100 genera)*

Over 523 species of gum trees, *Eucalyptus*, exist in Australia and Indonesia. While birds alone are not exclusive pollinators of eucalyptus, they contribute to pollination of about half the species, and are the main pollinators of perhaps 2%. Nearly 200 species are grown on 3.4 million hectares elsewhere, from highland temperate to tropical climates in Africa, South and North America and the Indian subcontinent, primarily for timber and firewood with added value as windbreaks, shade, erosion control, livestock fodder, and bee forage, depending on the species. Insect visitors of the open brush-cup flowers include honeybees, stingless bees, solitary bees, and birds. At least 20 species are recognized as important sources of nectar or pollen for honeybees. A number of eucalyptus are recommended for planting as bee forage in the tropics and subtropics in Africa, South America, and India. This suggests that honeybees visit their flowers freely, but we still do not know how well the honeybee pollinates these flowers.
Honeybees, stingless bees and other bees are believed to be effective pollinators of species with smaller flowers, such as *Eucalyptus melliodora*. (see **Appendix I**). Birds and some of the larger bees are likely to be more important in the pollination of gum species *E. ficifolia, E. casophylla, E. cosmophylla, E. macrocarpa, E. sideroxylon* and *E. leucoxylon* which have larger red or yellow flowers, and those with little scent and large nectar volumes.

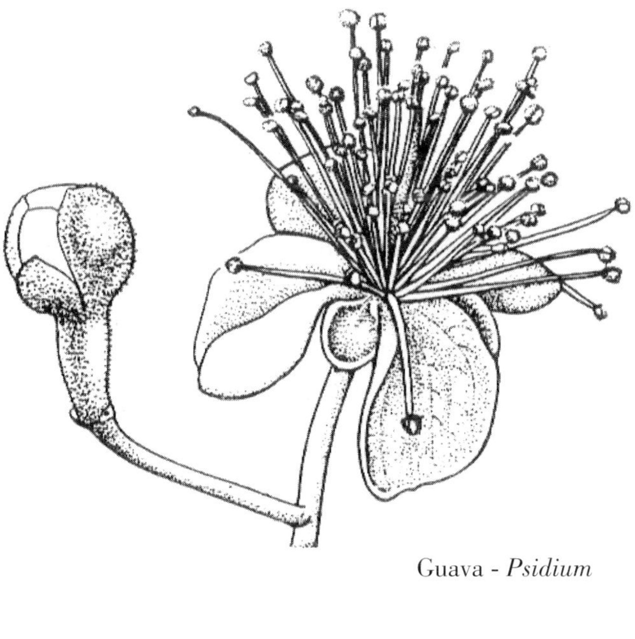

Guava - *Psidium*

Cloves (*Syzygium aromaticum*), guava (*Psidium guajava*), and allspice (*Pimenta dioica*) are shrubs or trees with yellow to white flowers. Bees and other insects visit clove and guava flowers, which are a bit smaller and lack the red of Feijoa, which implies that the bee and other insect visitors are probably more effective pollinators than for feijoas and the large gum flowers. The larger flowers of allspice could limit the effectiveness of some insect visitors. Such predictions need to be verified by more definitive studies on the pollination of these crops, which have so far received little attention.

12. Feijoa, *Feijoa sellowiana*.

This species comes from southern Brazil and Uruguay, where flowers are visited at least by bumblebees, carpenter bees, and birds. Feijoa flowers in the early summer. The stigma of each flower protrudes by about 9 mm from the ring of stamens. With cross-pollination, 88% of the flowers set fruit, which was 2.5 times the level for self-pollinated flowers. Fruit and pulp weight of cross-pollinated flowers were 200-300% higher than selfed flowers and fruit maturity was reduced by 6% (10 days) in the "Apollo" cultivar.

In Italy, New Zealand, Japan, and the USA the larger insect visitors that can at least intermittently transfer pollen include honeybees, *Polistes* wasps, bumblebees and carpenter bees. Smaller halictid bees, ants and longhorn beetles may just collect pollen or chew the petals without accomplishing pollination. The larger blackbirds, *Turdus merula* and mynahs, *Acriodtheres tristis* approach the flower from the top, consume the red petals, and deposit 5-10 times as much pollen on the stigma as honeybees. From 30 to 50% of flowers formed fruit after one visit. However, blackbirds apparently tend to work the margins of crops from shelter belts near their nests, because significantly less fruit was set in the middle of larger blocks (0.5-0.7 hectares) of feijoa. The smaller silver eye, *Zosterops lateralis* approached from below, seldom touched the stigma and deposited only 25% as much pollen on the stigma as honeybees. This was not enough to initiate fruit formation. In the USA, mockingbirds, *Mimus polyglottis* visited the flowers, but their effectiveness in pollination has not been investigated. In two orchards and an enclosure where birds did not visit the flowers, 0 and 3% set fruit, compared to 25-34% for hand pollination. Honeybees approached the flower from the correct angle, but typically they visited older flowers with available pollen, where the stigma was no longer receptive. Honeybees visited an average of 13 flowers per plant (about 10 times more than blackbirds) before moving to the next one, and none of 15 flowers visited formed fruit. Thus in effect, honeybees tend to act as pollen thieves for this crop.

Feijoa flowers were unattractive to honeybees compared to surrounding clovers, *Eucalyptus*, *Ver-bascum*, *Rubus* and *Citrus*. The same was true for bumblebees with sunflower, white clover and lucerne. Carpenter bees and queen bumblebees were likely to be more effective pollinators than honeybees, provided they visited younger flowers, because these bees should more readily contact the stigma. Stocking orchards of feijoa with honeybees has little to recommend it unless more effective pollinators are scarce.

H. Actinidiaceae *(320 species, 4 genera)*

13. Kiwifruit, Actinidia deliciosa.

The kiwifruit is one of at least 60 *Actinidia* species from China, Korea, Japan and Taiwan. Kiwifruit is now widely grown in subtropical and milder temperate regions, but it does require some winter chilling. A kiwifruit orchard flowers for 10-18 days at the end of spring, when each male flower sheds pollen for 2-3 days after opening; female flowers remain attractive to bees for 5 of the 7-9 days they can remain receptive. Kiwifruit represents distinctive pollination features compared to most other crops. Flowers are pollinated by a combination of insects and wind, as are chestnuts *(Castanea)* certain willows *(Salix)* used for windbreaks and erosion control, and grapes *(Vitis)*.

Male and female kiwifruit flowers are on separate vines, there is no nectar, and female flowers have inviable pollen. Female flowers have 30-40 stigmas but are so large that even bumblebee queens may forage for pollen around the stamens with little contact of the stigma. About 200 seeds are needed for the fruit to form. The preferred fruit, 93-110 g in size, contains 900-1,400 seeds.

In New Zealand, five honeybee hives per hectare on average are introduced to kiwifruit orchards. Rain may reduce pollen collection to less than 30% of normal levels. In Italy and New Zealand, kiwifruit accounted for 5-26% of the pollen collected by introduced honeybee colonies, but the percentage of kiwifruit pollen was nearly doubled when colônies were feed sugar syrup. The main competeing pollens were strawberry and white clover, honey suckle, citrus, *Brassica*, poppy, onion, asparagus and, towards the end of flowering, vipers bugloss.

Twenty-six percent of kiwifruit flowers exposed to a single visit by honeybees produced a minimum-sized (72 g) commercial fruit, and 14% a size at least as large as a preferred fruit. This implies that at least four visits to a female flower were needed to produce a high proportion of export quality fruit. Most kiwifruit pollen at commercial stocking rates has been gathered by midday. Therefore most honeybees forage for 3.0-4.5 hours daily on kiwifruit, during which time 740-1,120 flowers could be visited. Kiwifruit orchards have around 500 females flowers per vine. An estimated of two or three honeybees per two female vines are needed to effect adequate pollination, which agrees with the initial studies on fruit formation and bee densities per vine.

Bumblebees, particularly the larger queens, are about 10-15 times more effective per bee than honeybees. Ferguson and Pusch found that bumblebees deposited seven times more viable pollen than honeybees (per visit) and yields rose when bumblebee numbers increased on the flowers. The working rate and ability to forage under cooler conditions means that queens, in particular, can visit about twice as many flowers per day as the honeybees — provided that pollen is not depleted. In addition, the alternative flowers were less important competitors for their visits. Bumblebees have been used to a limited extent commercially for kiwifruit pollination in New Zealand and are being used for pollination of kiwifruit in Europe.

Kiwifruit flowers are an attractive pollen source for diurnal and nocturnal insects, evident by the over 150 species found on flowers among the small floral visitor fauna native to New Zealand. Hoverflies and solitary ground-nesting bees, *Leioproctus* were the other common, larger and active flower visitors. The flies carried only 10% the amount of pollen that solitary bees carried on their bodies. In China, bumblebees, and apparently carpenter and megachilid bees in addition to honeybees forage on the flowers. In wild kiwifruit populations wind may well have acted as the basic background pollinating agent that ensured fruit set on all but the plants most remote from a male vine. Insects, primarily social and solitary bees, carried pollen to the more remote female plants.

I. Vitaceae *(700 species, 12 genera)*

14. Grapes, Vitis spp.

The European grape, *V. vinifera* has generally been bred to be self-fertile (self-pollinating), while American grape, *V. munsoniana* ranges from self-fertile to self-sterile, and varieties of the muscadine grape, *V. rotunidifolia* are self-sterile. Grapes are often thought to be wind-pollinated, but their dependence on wind-pollination has not been investigated as thoroughly as kiwifruit and other species. Grape flowers are generally unattractive to bees and other flower-visiting insects. The flowers produce little nectar and, despite their extensive production, are not a significant nectar source for honeybees anywhere in the world. Recommendations for pollination are 2-5 hives/2 hectares.

Grapes are more important as a source of pollen for honeybees. When other flowers were scarce in Russia, 68-84% of the pollen gathered into hives was from grapes, but this is an exceptional level of pollen collection. With some grape varieties, Russian studies have claimed yield increases of 23-54% with a maximum improvement when honeybees were fed sugar.

J. Fabaceae *(19,000 species, 650 genera)*

The four legume crops chosen for review are herbs that produce both pollen and nectar for insect visitors. Available summaries on pollination, the structure and response of the flowers to visitors and the importance as nectar sources for honeybees, taken together, demonstrate that honeybees

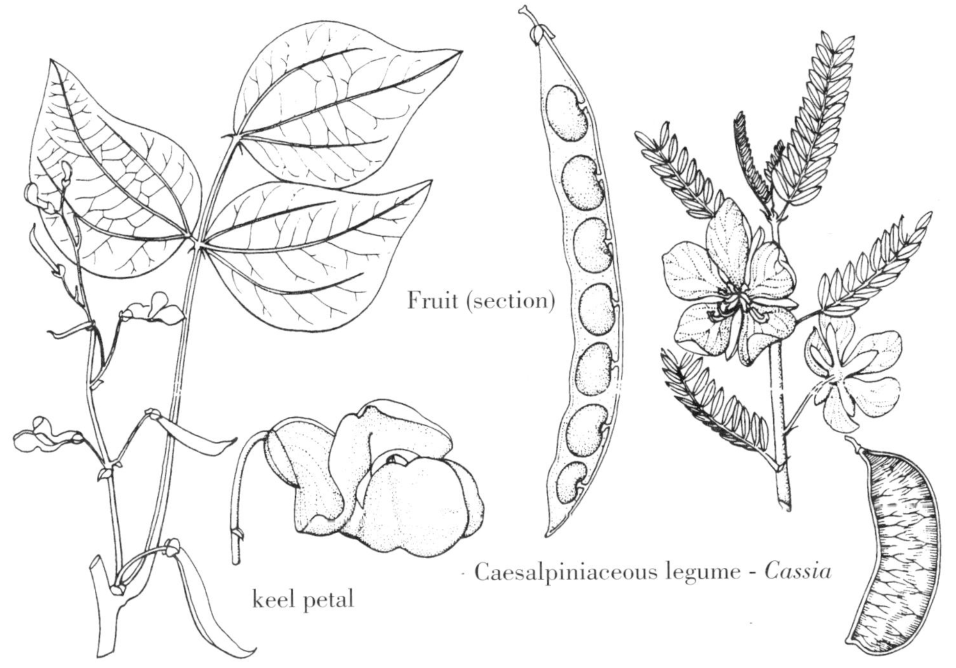

Caesalpiniaceous legume - *Cassia*

are satisfactory pollinators of trefoils (*Lotus*, *Lespedeza*), sweet clovers *(Melilo-tus)*, and sanfoin *(Onobrychis vicifolia)*. European and Canadian pollen analysis of *Bombus*, direct observations of eight legume species in Denmark, and the author's observations in New Zealand and Canada shows t short-tongued *B. teirrestris, B. lucorum*, and *B. terricola* include *Lotus* as a favored pollen and nectar source. Sweet clover is also a favored nectar source. Conversely, some of the vetch species *(Vicia)* and everlasting peas *(Lathyrus)* are favored by longer-tongued bumblebees of Europe and North America. Most temperate zone studies of pollination in legumes have little relevance to tropical legume crops, because the major tropical food crops are self-pollinating (**Appendix I**).

In the tropical and subtropical savannas and deserts of the USA and South America, bee species with generalized food sources form a common group of insect visitors on leguminous trees. This applies at least to the mesquites, algarroba, tamarugo *(Prosopis)*, to nectar-bearing wattle *(Acacia)* in Australia, the Americas, Africa and Asia, and tamarind *(Tamarindus indica)*. Information on honeybee forage in Australia and other parts of the world shows that some of the widespread *Acacia* mainly yield pollen and little or no nectar from the flowers. Other regionally important multiple-purpose, livestock forage, timber and erosion control trees or shrubs are the genera *Albizia*, *Cassia* [now *Senna* in large part] *Ceratonia*, *Cordeauxia*, *Dalbergia*, *Desmodium*, *Entero-lobium*, *Gleditsia*, *Intsia*, *Leucaena*, *Mimosa*, *Pterocarpus*, *Samanea*, *Schizolobium*, *Sesbania*, *Tipuana* or creepers or shrubs, *Phaseolus*. At least attempts at cultivation have been made with most of these genera, but *Instia*, *Schizolobium*, *Enterolobium*, *Tipuana* and *Samanea* have barely gone beyond local trials, despite their rapid growth and valuable wood. Of the cultivated plants, sissoo *Dalbergia sissoo* (India and Pakistan), honey locust, *Gleditsia tricanthos* (Pakistan, South Africa) bracatinga *(Mimosa scabrella* — Brazil), round-leaved bloodwood *(Pterocarpus rotundifolius* — Mozam-bique, South Africa) and tipa *(Tipuana tipa* — Bolivia) are rated as important honey sources. Thus they are likely to be visited often by honeybees. Central American studies of pollination on *Cassia* and *Cochlospermum* showed that some of the tree legumes are pollen-rich flowers that are buzz pollinated by *Centris* and *Xylocopa*. The principles derived from pollination of temperate legumes seem to have the greatest potential application for studies of tree crops in the tropics, which are in need of much research effort to place them on a par with herbaceous species.

Mimosaceous legume - *Mimosa*

15. White clover, Trifolium repens. This herb is mainly used in temperate regions but grows well in subtropical climates. White clover is of particular value in well-stocked pastures, where its stolons enable it to survive repeated heavy grazing. White clover and other legumes have high digestibility and protein for grazers. White clover also has a high capacity to increase soil nitrogen levels. Crops can produce 320-520 million flowers/hectare.

Observations on white clover in the field on Cha-tham island showed that seed is not formed in the absence of honeybees, bumblebees or solitary bee pollinators, which confirms studies in cages. White clover flowers from late spring until autumn, depending on when it is grazed. White clover is an important honey source in China, Iran, Colombia, Ecuador, Argentina and parts of Brazil, while it is a minor source in Algeria, Lebanon and the Azores. Crops usually produce 100-200 kg honey/hectare.

White clover flowers are three to four mm deep and can readily be pollinated by honeybees and short-tongued bees. Seed yields of *T. repens* can be increased 200-300% when crops are stocked with honeybees. Abundance of all pollinating bees is low on white clover flowers at less than 18°C, because nectar secretion ceases at about this temperature. Honeybees are normally effective in white clover pollination. No benefit may accrue from having more than 1.2 honeybees/m^2 — unless the crop flowers profusely, i.e. 430 million flowers/hectare. In this case 5.6-6.0 honeybees/m^2 may be needed. Under field conditions 2.5-3.0 honeybees/1,000 flowers can cause seed-set of 85-95%. One to three colonies/hectare are recommended. Usually there is little need to use other pollinators. However, alternative pollinators can provide backup pollination under adverse weather, on smaller crops, and in pastures understocked with honeybees (less than one hive per 500 ha). European and New Zealand studies show that other pollinators include *B. terrestris, B. lucorum, B. lapidarius, Melitta leporina, Anthidium punctatum*, and presumably *Andrena ovulata, A. labialis, Eucera longicornis*, and *E. clypeata*. *Bombus lucorum* and *B. terrestris* are useful for pollination of breeding lines in screen cages.

16. Berseem, Egyptian clover, *Trifolium alexandrium*.

This forage crop is used more in drier and warmer climates. Berseem clover is important in northern Africa, the Middle East, Pakistan and India. The flower is 7-8 mm deep and may remain receptive for 10-12 days. Crops flower for several weeks in summer. Flowerheads can set over 70 seeds/head and crops can produce up to 600 kg/hectare of seed. Comparisons of caged and open plots demonstrate the need for cross-pollination. Berseem crops are reported to be visited avidly for nectar and pollen by honeybees. Berseem clover is rated as an important source for honey in the Nile, Indus and Ganges plains. Pollen trapping in the Nile valley confirms the importance off Berseem as a pollen source. For maximum seed-set, 2-4 honeybees/m^2 are deemed adequate. In India, *Apis dorsata* and *A. cerana* were the commonest visitors, but the small honeybee *A. florea* only used the flowers when berseem nectar was plentiful. The crop does not have a significantly higher flower density than white clover and presents no problems for honeybee pollination. One to three colonies/hectare should provide satisfactory pollination.

In Eygpt, an alkali bee, *Nomia unidentata* includes beerseem clover in its food sources, while *Cha-licodoma mucorea, Osmia latrielli* and *Andrena ovulata* are considered valuable pollinators. *Anthidium, Andrena, Osmia* and *Xylocopa* visited the flowers and probably have long enough tongues to effect pollination. This should be verified, because the behaviour of these pollinators and *Halictus* on the crop has not been studied.

17. Lucerne, alfalfa, *Medicago sativa*.

This highly productive hay and pastoral crop is more drought resistant than the clovers. The mauve to yellow-streaked flowers have corollas about 4 mm deep and flower during summer. Crops may produce 0.1 to 0.2 million flowers/m^2 over the 4-6 week flowering period. Under favorable conditions lucerne is an attractive nectar source for bees and yields 100-200 kg honey/hectare. Lucerne is a major honey source in Argentina, China, Mozambique and Mexico, but it is a poor pollen source.

Solitary bees regularly trip the flowers and at least 13 genera have efficient pollinators. The lucerne leaf cutter bee *Megachile rotundata* has become the leading pollinator on the American continent, but it is still much less important in Europe, Russia, the Middle East and India. The threshold for flight is around 18°C and it only forages freely over 20°C. Periods of cool weather restrict the effectiveness of this bee. In Canada, the USA and New Zealand seed yields of lucerne have been increased 500-1000%, to 500-1,200 kg/hectare, by using the lucerne leaf cutter bee. Lucerne leaf cutter bees have been introduced to South America, but inadequate steps were taken to screen out natural enemies from stocks imported from North America. The alkali bee of the USA, *Nomia melanderi,* forages under lower temperatures and stronger winds than the lucerne leaf cutter bee. However, the special 'beds' needed for nesting in the ground restrict pollinator placement within crops, compared to the lucerne leaf cutter bee. Bee abundance of 5,000-6,000/hectare tripped around 50% of flowers and yielded 700-1,150 kg/hectare. Introductions have succeeded in producing naturalized populations but limited management options in New Zealand. The alkali bee could be of value in lucerne pollination in parts of southwestern Europe, northern India, Africa, and South America. Diseases and parasites are relatively scarce among the lucerne leaf cutter bee and the alkali bee in New Zealand.

Tripping rates of 50-100% and faster flower visitation mean that many of the solitary bees and bumblebees are 40 to over 100 times more effective (per bee) than honeybees in lucerne pollination, even without taking a longer working day into account. In North America, unmanaged and effective solitary bee pollinators of lucerne include *Anthophora, Osmia, Megachile, Hoplitis, Andrena prunorum, A. wilkella* and *Calliopsis andreniformis*. In Europe, the middle East and Southern Russia, *Rhophites canus, Osmia latreillei, Melitta leporina, Melitturga clavicornis, Megachile* spp. (short adult activity periods), *Eucera longicornis, E. clypeata, E. cinerea* (medium adult activity periods) and

Andrena flavipes, A. ovulata (two adult generations and long activity) are the major pollinating bees. In France, important alternative pollen sources were white clover, red clover, lotus and vetch for solitary bees, *Eucera, Melitturga, Melitta, Andrena labialis*, and *A. ovulata*. St. John's wort, *Hypericum* and Compositae including sunflower were important for *A. flavipes*. *Rhophites canus* are farmed by securing food supplies and protecting the ground nest sites in the Czech republic. Initial management studies have been made on *Osmia latreillei* in Israel.

European *Anthophora parietina* can be partially managed at nesting sites as a general legume seed pollinator, and in Eygpt the mason bee *Chalicodoma mucorea* is a potentially domesticated species for hotter climates. Medium and shorter-tongued bumblebees are generally more effective pollinators of lucerne, because they trip lucerne more consistently and prefer lucerne more than do longer-tongued species. In India, *Megachile nana* and *M. flavipes* have been used for pollination. Similarly, in South Africa, *Xylocopa* and *Megachile gratiosa* have been tested as potential lucerne pollinators. In Argentina, *Megachile pallefacta, Xylocopa ordinaria, X. splendidula, Melissodes nigroaenea* and possibly *Bombus robustus* are efficient pollinators of lucerne, as are colletid bees in Chile. Nectar-collecting honeybees and shorter-tongued bumblebee species are largely ineffective pollinators. These bees rapidly learn to insert their tongue through the side of the flower and only trip the pollinating mechanism accidentally. They trip as little as 0.2-5.0% of flowers, with the higher percentages in hot dry regions. In New Zealand, where pollination relied largely on honeybees, an average of only 75 kg/hectare seed production was achieved without control of mirid bugs. This is about 5% of the potential 1,500-2000 kg/hectare lucerne yield under ideal conditions.

Pollen-foraging bees on lucerne are few, because lucerne does not produce much pollen per flower and the pollen of other crops like mustard, *Brassica nigra* is often more attractive to honeybees. Hence honeybees are only moderately effective pollinators on large crops in Mediterranean climates, where there is very little alternative pollen available. Conversely, on crops that are largely served by leaf cutter and alkali bees, the removal of nectar by honeybees may reduce the propagation of these bees and thus perhaps also reduce seed yields.

18. Field, faba, tick or horse beans, *Vicia faba*.

These beans are the sixth most important world 'pulse' crop. The beans remain a staple food in Arab countries. China produces about 60% and Egypt and Ethiopia about 6% of world supply. Mexico, Morocco, Brazil and Tunisia each produce about 1% of the world crop. In 1981, Chinese yields were 40%, while Ethiopian to Brazilian yields (358 kg/hectare) were 11-25% of the best European producing countries (3,200 kg/hectare). Yields with pollinating bees may reach 4,520 kg/hectare.

Faba bean is partially selfing and partially outbreeding. It represents a legume crop that forms seed readily without a pollinating bee, provided it comes from cross-pollinated seed. Plants from self-pollinated seed form virtually no seed unless the flowers are cross-pollinated. Crops have 20-80 plants/meter, 50-80 flowers/plant and 13.5 million flowers/hectare at peak flowering. Flowers can be pollinated for up to six days and pollen up to five days old will fertilize ovules. Flowers start to open at 12°C and only open freely above 13°C. The extrafloral nectaries are on the underside of the stipules, often with dark spots. In England 8-20% of flowers on the better crops will form seed pods. Cross-pollination allows for setting of pods lower on the stem, accelerates setting, pod maturation and reduces the duration of crop flowering and harvesting losses. Selfing increases plant losses in winter-sown crops. Outcrossing of 30-40% (range 4-54%) apparently represents an equilibrium and selfing increases on later flowers up the stem. The randomness of crossbred seeds within pods suggested that, in England, 67% of the flowers were visited by cross-pollinating bees.

Some shorter-tongued bumblebee species of the subgenus *Bombus*, carpenter bees and apparently some ant species (at least *Cataglyphis bicolor*) make holes at the base of the broad bean to rob nectar. Nectar-collecting honeybees use these holes and subsequently do not pollinate the flowers. Honeybees visit the extrafloral nectaries before any flowers open and continue to use extrafloral nectaries more intensively as flowering progresses. Honeybees collecting pollen make pollinating visits mainly as the new flowers open between 13,00 and 16,00 h. Pollen collection can decline after peak flowering. The daily percentage of bean pollen collected in honeybee hives near the crop varied depending on competing crops and inherent colony preference for bean pollen, with an average of about 20% (range 1-88%). In the Nile valley, broad bean is a primary pollen source for honeybees, suggesting there is are appreciable numbers of effective pollinating workers on the crop. Rape is an important competing pollen source for honeybees.

Low temperature during bean flowering can restrict both foraging by honeybees and the formation of seed in the early flowering crops of temperate zones. In England, crops up to two hectares are considered to have enough wild pollinators, but crossing decreased with larger crops from 12-32 hectares. Larger crops may need 2-5 colonies of honeybees/hectare for pollination.
Cross-pollination at the margins of fields can be almost twice the levels within the field. Stocking with honeybees can therefore improve the distribution of yield within the crop. Australian experience suggests that crops may not need stocking with extra honeybees in a Mediterranean climate, because the flowers open in late winter and early spring when honeybees must search for pollen. Milder

temperatures also allow honeybees to be more effective pollinators. Longer-tongued bumblebees, *B. ruderatus, B. hortorum, B. subterraneaus, B. distinguendus, B. pascuorum, B. sylvarum, B. lapidarius* (in Europe); *B. borealis, B. griseocollis, B. vagans* (in N. America), and solitary bees visit the front of the flower and pollinate, whether or not the bean flower has a hole made at the base. The latter three European species and *B. rufocinctus* may sometimes use the hole at the base of the corolla. *Bombus ruderatus* is the only species extending to Mediterranean climates in northwest Africa and subtropical climates in southern Europe and northern New Zealand.

The longest-tongued bumblebees visit the flowers about 2.5 times faster, with other bumblebees about 2 times faster, compared to pollinating honeybees. With *Bombus hortorum* about 1,800 flowers could be visited in a six-hour day. Employing this figure, a density on the order of one longer-tongued *Bombus* or three to four pollen-collecting honeybees per square meter (given the likely shorter working day of honeybees) should fully pollinate a crop with 13.5 million open flowers per hectare. This prediction needs to be verified under field conditions. Among at least 13 bumblebee species from Central and South America of the subgenus *Fervidobombus*, and subgenera *Thoracobombus* and *Megabombus* from the middle east to Asia, most have tongues that are long enough to permit effective pollination of faba bean. Queens of *B. lucorum* and *B. terrestris* have long tongues, thus at the beginning of flowering the bean flowers are not pierced by them, and even the shorter-tongued species can be effective pollinators. Bumblebees forage at temperatures as low as 10°C. However, temperatures above 15°C and no more than light winds favor foraging and nectar secretion.

In Eygpt, long-tongued *Anthophora senescens* and *A. aegyptiaca* were common pollinators; the crop was also visited by *Andrena ovulata,Osmia submicans*, and *Tetralonia lanuginosa*. Related solitary bees in France pollinated the flowers only slightly faster than honeybees. Male carpenter bees *Xylocopa aestuans* make a hole at the base of the flower to obtain nectar. They are not effective pollinators of faba bean.

19. Red clover, *Trifolium pratense*.

This species is a perennial legume with high yield in temperate to subtropical climates having a reasonably assured annual rainfall above 600 mm. Red clover is mainly used for cattle forage and as a component in dairy pastures. It averages 100-120 flowers/flowerhead (raceme). The flowers have a slender tube 8-10 mm long, depending on whether it is a diploid or tetraploid variety. Average to excellent crops produce 2-4 million flowering racemes/hectare during the peak of flowering — 3,600-7,200 flowers/m^2 need pollinating each day. If crops are not cut or grazed, they flower in early summer for longer (about 8-10 weeks) than crops that were used for fodder (4-8 weeks flowering). Crops that begin flowering about midsummer normally have a higher intensity of peak flowering, more easily determined optimal harvesting, and a better potential for high seed yields. Fully pollinated diploid crops can form seed on 80-90% of the flowers. At best, tetraploids tend to form seed on 70-80% of the flowers due to ovule abortion. Diploid crops yield seed more reliably because the flowers can be pollinated more satisfactorily with honeybees and lucerne leaf cutter bees, which can reach the nectar in the shorter diploid flowers.

Most temperate regions hove shorter-tongued bumblebees of the subgenera *Bombus* or *Mendacibombus*, which bite a hole at the base of the flower to reach the nectar. There are no shorter-tongued bumblebees of these subgenera in highland regions in central and South America, Africa or the subtropical highlands in southwestern India, but carpenter bees *(Xylocopa)* and wasps *(Ropalidia)* make

Honeybee, *Apis mellifera ligustica*, visiting but not pollinating flower of red clover

holes in flowers. Thus, red clover flowers would probably be vulnerable to nectar robbing in highland continental African, Indian, Asian and Australian areas.

The action of nectar-robbing bees can be readily observed: Their head is obscured among the flowers as they use the hole. Nectar-collecting honeybees also use these holes, and when a hole is used no contact is made with stamens or stigma. All longer-tongued bumblebees and pollen collecting shorter-tongued bumblebees forage from the front or top of the flower and contact the stamens and stigma. The whole body can be seen. The most effective bumblebee pollinators of red clover, with the longest tongues, are in 7 of about 30 *Bombus* subgenera as follows:

Megabombus: *hortorum, ruderatus, argillaceus* (Europe-Asia); *Fervidobombus*: *fervidus, pennsylvanicus, dahlbomi* (N. or S. America): *Thoracobombus*: *pascorum, sylvarum, pseudobaibalensis* (Europe-Japan); *Diversobombus*: *diversus, montivagus* (E. Asia); *Subterranobomus*: *subterraneus, borealis, distinguendis, apposistus* (Holarctic);*Bombias*: *nevadensis, auricomus* (N. America), *Seperato-bombus*: *griseocollis, morrisoni* (N. America); *Pyrobombus*: *vagans, ardens* (Holarctic — but only species with the longer tongues are effective in red clover pollination). Species such as *B. dahlbomi* and *B. subterraneus* that start colonies late in the season tend to form small colonies. They become abundant during only part of the red clover flowering.

At the start of flowering, often 80-100% of honeybee foragers collected pollen from red clover. At the time of the main flowering, however, only 5-11% of honeybees and shorter-tongued bumblebees were pollinating the crop. The decline was due to an influx of nectar-collecting honeybees from adjacent white clover crops and weeds which stop flowering as the red clover comes into flower. Honeybees visited an average of 12 flowers/minute — less than half the rate of longer-tongued species. Honeybee foraging on red clover declined below 25°C and ceased altogether at 16-17°C. Persistent low temperatures reduced honeybee foraging to 5.6 h/day, compared to 10 hours for bumblebees. The cumulative effect was such that longer-tongued bumblebees visited about four times more flowers per day than honeybees. In the field, seed formation improved until bee numbers reached levels at which they could pollinate each flower during a day. A predictive model was used to estimate that 2,000-4,000 longer-tongued bumblebees/hectare (or their equivalent) can pollinate the average and best crops at peak flowering. The model estimated that the usual combined honey and bumblebee populations could only pollinate up to one-third the flower crop.

Crops that flower in early summer can be too early to help colony development of the late-season emerging bumblebees. Such crops receive competition for honeybees from lotus or white clover. Box gum trees and thistles *(Cirsium, Carduus)* act as complementary flowers for honeybees pollinating red clover, because honeybees mainly use these flowers for nectar. In the Murray basin, Australia, seed yields of up to 300 kg/hectare were achieved on a tetraploid red clover where box gums were common and no bumblebees or carpenter bees existed. Detailed studies on lucerne and red clover crops in New Zealand also show lucerne was a complementary crop for pollination of red clover by honeybees. Foxgloves *(Digitalis)* may be a competing floral source for bumblebees during the early part of red clover flowering, but even foxglove should probably not be removed near red clover crops because it is often used by longer-tongued bumblebees. This 'competitor' acts as a vital food source for colonies before most commercial crops are in flower.

Honeybee hives have about 500-1,000 times more bees than colonies of longer-tongued bumblebees and usually 4-10 hives of honeybees/hectare are recommended for red clover pollination. Studies on the same farm demonstrated that increases of up to five fold in seed yield could be achieved with up to 12 colonies of *Bombus horturum* and *B. ruderatus* per hectare (see figure). *Bombus hortorum* has colonies that on average are 50% larger than those of *B. ruderatus*. Estimated yields of between 500 to over 600 kg/hectare can be produced, when only 5-10 colonies of *B. hortorum* are used per hectare . In Washington state, yields of 1,850 kg/hectare have been achieved.

Bumblebee colonies, mainly of *B. hortorum*, have been commercially supplied to red clover growers in New Zealand. This began in 1982 with hives occupied by queens in suburban sites with a favorable sequence of food supplies. It is technically possible to rear some of the longer-tongued species too, but a colony of less than half that of *B. terrestris*, which is used in tomato pollination, and poorer revenue returns compared to glasshouse crops have so far prevented the development of a major supply of bumblebee colonies for red clover pollination. The best prospects seem to be to develop a farming system for suitable bumblebee species especially in regions that lack the main colony enemies — *Psithyrus*, *Melittobia acasta*, *Vitula* and *Aphomia sociella* — all of which are difficult to control.

Bumblebee, pollinating the flower of red clover

Most regions have some effective solitary bee pollinators of red clover, which include *Osmia coerulescens* (from Europe but also in the USA),*Xylocopa* (in the Middle East and China), *Eucera socialis* in Japan, *Osmia bruneri*, *Melissodes*, and *Tetralonia* in North America and *Centris chilensis* and *Anthophora* in South America. The lucerne leaf cutter bee can increase seed yields on diploid red clovers from 291 to 410 kg/hectare.

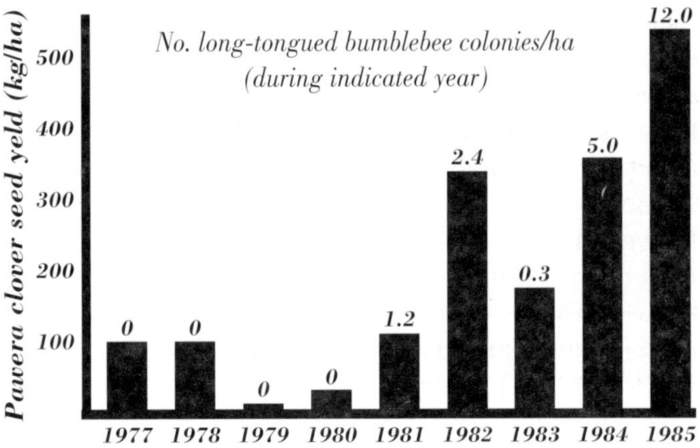

If Southern African nations, Kenya and Ethiopia wish to improve faba bean pollination and to diversify legume production, then the introduction of *Bombus ruderatus* or *B. hortorum* may be appropriate. Although *Bombus hortorum* extends to Spain and Turkey, bees are most abundant in northern Europe and New Zealand. These regions and the Sudan do not have bumblebees. However, the effectiveness of *Xylocopa*, at times the pollinators of faba beans and red clover, remains to be investigated.

K. Compositae (Asteraceae)

(25,000 species, 1,100 genera)

The close association of the tubular florets and the outer ray florets make the inflorescence more conspicuous to bee, wasps, beetles, flies and moths that visit the flowers. Hummingbirds are among visitors to some composite species, often in Neotropical highlands and páramo. Nectar of composites, unlike that of Umbelliferae, is protected from rain by its position in the small, tubular florets. The weedy *Artemesia, Ambrosia* and *Xanthium* are wind-pollinated, but Bathurst burr (*X. spinoosum*) provides a little nectar and pollen for honeybees. The sunflower, *Helianthus annuus* is the most extensively grown composite. Other Compositae crops such as artichoke *(Cynara scolymus)*, chicory *(Chicorum intybus)*, endive *(Chicorum endiva)*, and lettuce *(Lactuca sativa)* have smaller to much smaller flower heads, which are also produced in summer. The small flower heads should allow for more movement with fewer visits per head and, thus, better cross-pollination than with sunflowers.

20. Sunflower, *Helianthus annuus*. Sunflowers are grown in subtropical and highland tropical South America, Asia, India and Africa. The major countries growing sunflower are India, China, Turkey, South Africa and Sudan. Sunflower was cultivated first in the southern USA and Mexico — mainly grown for oil, but also used for livestock feed and human food.

Sunflower has 1,000-2,000 flowers/flower head, 100,000 flower heads and 2-40 million flowers/hectare. Each head flowers for 6-10 days; stigmas may remain receptive for up to 14 days, but receptivity declines each day. A crop flowers for three to five weeks in summer to early autumn. At least 70-90% of the flowers can form seed. In the USA, male-sterile, apomictic sunflower varieties are mainly used, even although male-fertile varieties can be more productive. Pollen and nectar are produced only during the male phase of the central disk florets.

The diversity of flower visitors to sunflower and its importance as a source of honey in parts of Africa, South America and the Indian subcontinent are evidence that sunflower is a relatively attractive flowering crop for bees. Higher levels of nectar lower the efficiency of individual pollinators, because they spend more time on each head and floret. The major producing states in the USA have 8-10% of the recommended 2.5 honeybee colonies/hectare for maximum pollination. A linear model predicts seed oil content to increase 10% with 15 honeybees per 100 heads. Honeybees seldom collect pollen from sunflower and may often clean and discard sunflower pollen from their bodies. A total of five to six honeybee visits/flower are needed to achieve maximum seed formation. At least five honeybees/flower head are needed for average yields (1,500 kg/hectare).

In the highlands of southern India, *Apis cerana, A. dorsata*, and *A. florea* accounted for 85% of flower visitors on sunflowers. Approximately 9% of the 412 bee species that have been recorded visiting sunflower in western USA restricted all pollen and nectar collection to sunflower. Yields of up to 4,960 kg/hectare have been achieved where solitary bees were the main pollinators, compared with the USA average of 1,460 kg/hectare. Less than 1,000 sunflower leaf cutter bees, *Megachile pugnata* per hectare are needed for sunflower pollination. Four solitary specialist pollinators averaged 1.02-0.24 million pollen grains on their bodies, compared to 0.06-0.002 million for non-specialist honeybee, bumblebee and solitary bee pollinators.

Specialist sunflower pollinators, *Melissodes agilis, Andrena helianthi, M. pugnata, Svastra* and *Diadasia*, were more active earlier in the day than the generalist pollinators *Apis, Bombus* and *Halictus*. *Megachile agilis* and *Andrena helianthi* can cause a seed set 5-11 times higher per visit than the honeybee. Bagged flower heads with visits restricted to as few as 7 visits produced 540 seeds, compared to 240 for heads receiving over 100 visits per day. Seed-set declined as the season progressed and the sunflower specialists formed a greater part of the pollinating guild. Specialist solitary species seemed to displace generalist pollinators from sunflower to other flowers — onion, carrot, red clover, and white sweet clover.

In South Africa, the long-tongued *Tetralonia* and short-tongued *Lasioglossum* visited sunflowers, while maize, grasses, *Xanthium*, and *Clematis* were competing pollen sources for honeybees. In Spain, 31 bee species were recorded on sunflowers. In India, 21 genera of insects were found on the flowers, and in Pakistan, *Andrena fulvicrus*, *Ceratina viridissima*, *Xylocopa dissimilis*, *X. rufescens*, and

X. fenestra were the main pollinating bees. Sunflower is a favored food source for *Bombus lucorum* or *B. terrestris*, which extend to northern Israel, northern Iraq and Kashmir, while *B. terrestris* extends to northwest Africa.

L. Solanaceae *(2,000 species, 90 genera)*

Solanaceous crops, tomato *(Lycopersicon escu-lentum)*, sweet pepper and chillies *(Capsicum annum* and *C. frutescens)*, tobacco *(Nicotiana tabacum)*, eggplant *(Solanum melongena)*, potato *(S. tu-berosum)* and tamarillo *(Cyphomandra betacea)* mainly have pendulous flowers that produce little or no nectar for foraging bees. They originated in the Americas. Bees visit the flowers to gather pollen which has the highest protein and nitrogen content among pollens (see Section 2.3.2). Honeybees do not 'buzz' or 'milk' the anthers of any flowering plant, thus are ill-adapted to extract the food offered to bees by the Solanaceae. In the field, honeybees and bumblebees are among the most frequent insect visitors on crops. Other frequently-recorded visitors at crops are the bee genera *Amegilla, Centris, Exomalopsis, Xylocopa, Caupolicana, Hylaeus, Ptiloglossa, Stenotritus, Augochloropsis, Lasioglossum, Nomia*, and *Protoxaea*.

21. Tomatoes, *Lycopersicon esculentum*.

One-third of all tomatoes are grown in Asia and India, 17% in Africa, 6% in Central America and 6% in South America. The highest yields are obtained in Europe and Japan where tomatoes are grown hydroponically in glasshouses. Major producers in the tropics are China, Egypt, India, Mexico, Brazil and Indonesia.

In the glasshouse, European and North American studies on tomatoes and potatoes have found that shorter- or medium-tongued bumblebee, *B. terrestris*, (subgenus *Bombus*), and *B. impatiens, B. ternarius* and *B. vosnesenskii* (subgenus *Pyrobombus*) — which are more generalized in their flower usage — are better pollinators than longer-tongued bumblebees. The last group often depend more on flowers from a few plant families. Honeybees can be somewhat useful in pollination of glasshouse tomatoes in winter, but as soon as flowers become available outside the glasshouse the honeybees soon pursue foraging there. They cannot ordinarily collect the pollen and there is no nectar, thus they are not avid foragers on these or any other solanaceous crops. Mechanical vibration of anthers is expensive and tedious for the operators, who may miss or damage flowers or suffer from allergies to the pollen. Progress with pollination by bumblebees can be monitored, because within an hour a brown ring appears on the stamens where the bees has held the anther in its mandibles while vibrating the flower (see Section 2.3.2). Each bumblebee can set more than 96% of fruit, which averages 105-120% heavier fruit than fruit set with honeybees. About 8-10 colonies of bumblebees/hectare of the crop are used in glasshouses. In the field, bumblebees and other bees pollinate tomato flowers, indicated herein by region: (Peru): *Bombus funebris, Centris surinamensis, Xylocopa brasilianorum, Anthophora arequipensis, A. trin-cincta, Exomalopsis bruesi, Augochlora matucanensis, A. nigromarginata* and *Lonchopria*.; (North America): *Anthophora urbana, Augochloropsis ignita*; (Puerto Rico); *Exomalopsis glubosa*; (Guadeloupe): *Exomalopsis*. With one pollen-collecting bee seen during each minute, 40% of flowers set fruit in Peru, compared to a 10% fruit set observed in California. In Guadeloupe, 67% of pollen gathered by *Exomalopsis* came from the solanaceous crops of tomato, egg plant and sweet pepper.

Additional References

Abrol, D. P. 1987. Interrelationship and path coefficient analysis of environmental factors influencing pollination activity of *Apis dorsata* F. on *Prunus persica* L. Trop. Biol. 28:147-154.

Abrol, D. P. 1987. Activity time budgets and pollination of *Apis cerana indica* workers. Entomol. Ecol. 4:707-7CB. [carrot, sunflower, watermelon, cauliflower]

Abrol, D. P. 1988. Pollination studies in almond (*Prunus amygdalus* L.). Current Sci. 57:397-398.

Ahmad, R. 1987: Honeybee pollination of important entomophilous crops. Pakistan Agric. Res. Council, Islamabad, 103 pp..

Anais, G., Torregrossa, J. P. 1979. Possible use of *Exomalopsis* species in pollination of solanaceous vegetable crops in Guadeloupe (French Antilles). Proc. Int. Symp. Poll. 4:321-329.

Anaewicz, A., Warakomska, Z. 1977. Pollen food of the bumble-bees (*Bombus* Latr., Hymeonptera) and their association with the plant species in the Lublin region. Ekol. Polska 25:309-322.

Arretz, P.V., Macfarlane, R. P. 1988. The introduction of *Bombus ruderatus* to Chile for red clover pollination.Bee Wld. 67:15-22.

Asensio de la Sierra, E. 1979. Report on alfalfa pollination studies in Spain. Proc. Int. Symp. Poll. 4:103-107.

Asensio de la Sierra, E. 1984. *Osmia cornuta* Latr. (Hymenoptera; Megachilidae) as a potential pollinator of fruit trees in Spain (French). Int. Symp. Poll. 5:461-465.

Atallah, M. A., Aly, F. K., Eshbah, H. M. 1989. Pollen gathering activity of worker honeybees, on field crops in Minia region, middle Egypt. Proc. Int. Conf. Apic. Tropical climates 4:109-115.

Baker, M.J., Williams, W. M. (eds.) 1983. White clover. CAB international, 53 pp..

Baltazar, c. R. 1966. A catalogue of Phillipine Hymenoptera (with a bibliography 1758-1963). Pacific Insects Monogr. 8, 488 pp.

Banda H.J., Paxton, R. J. 1991. Pollination of greenhouse tomatoes by bees. Acta Horticulturae 288:194-198.

Barskii, Ya S. 1956. Training bees to pollinate grape vines (Russian). Sad Ogrod 4:64.

Batra, S. W. T. 1977. Bees of India, their behaviour, management and a key to the genera. Orient. Ins. 11:289-324.

Batra, S. W. T. 1979. *Osmia cornifrons* and *Pithitis smaragdula*, two Asian bees introduced into the United States for crop pollination. Proc. Int. Symp. Poll. 4:307-312.

Benedek, P. 1968. The flight period of wild bees (Hymenoptera, Apoidea) pollinating lucerne, and its plant protection aspects. Acta Phytopath. Acad. Sci. Hungary 3:59-71.

Bhalla,O. P., Verma, A. K., Dhaliwal, H. S. 1983. Foraging activity of insect pollinators visiting stone fruits. J. Ent. Res. 2:91-94. [almond, plum]

Blanchet, P., Douault, Ph., Pouvreau, A. 1991. Kiwifruit (*Actinidia deliciosa* Chev.) pollination; honeybee behaviour and its influence on the fruit. Acta Horticulturae 62:376-380.

Board on Science and Technology for International Development. 1979. Tropical legumes: resources for the future. National Academy of Sciences, Washington, D. C., 331 pp. [at most refers to self- or cross-pollinating needs, key plant species reference source]

Bohart, G. E. 1957. Pollination of alfalfa and red clover. Ann. Rev. Ent. 2:2:355-380.

Bohart, G. E., Nye, W. P. 1960. Insect pollinators of carrots in Utah. Utah Agric. Exp. Sta. Bull. 419, 16 pp.

Bohart, G. E., Nye, W. P., Hawthorn, L. R. 1970. Onion pollination as affected by different levels of pollinator activity. Utah Agric. Exp. Sta. Bull. 482, 5 pp.

Boonithee, A., Juntawong, N., Pechhacker, H., Huttinger, E. 1991. Floral visits to select crops by four *Apis* species and *Trigona* sp. in Thailand. Acta Horticulturae 288:74-80.

Boyle-Makowski, R. M. D. 1987. The importance of native pollinators in cultivated orchards: their abundance and activities in relation to weather conditions. Proc. Ent. Soc. Ontario 118:125-141.

Brooks, R. W. 1993. Report on third PCAM expedition. Melissa 6:1-4.

Brown H., Scott, R. R., Macfarlane, R. P. 1992. An assessment of *Vicia faba* and *Trifolium pratense* as forage crops for *Bombus hortorum*. N. Z. Ent. 15:42-47.

Butler, G. D., Todd, F. E., McGregor, S. E., Werner, F. G. 1960. *Melissodes* bees in Arizona cotton fields. Ariz. Agric. Exp. Sta. Tech. Bull. 139, 11 pp.

Chambers, V. H. 1946. An examination of the pollen loads of *Andrena*: the species that visit fruit trees. J. Anim. Ecol. 15:9-21.

Childers, N. F. 1983. Modern fruit science (9 ed.). Gainesville: Horticultural Publications, 583 pp.

Clemson, A. 1985. Honey and pollen flora. Melbourne: Inkata Press, 262 pp.

Clifford, P. T. P. 1979. The effect of closing date on seed production of red clover cultivars. N. Z.J. Exper. Agric. 7:369-374.

Clifford, P. T. P., Anderson, A. C. 1980. Red clover seed production research and practice. In: P. North (ed.).Herbage seed production. Grasslands research and practice, Ser. 1. N. Z. Grassland Assoc., pp. 76-79.

Clifford, P. T. P. , Scott, D. 1989. Inflorescence, bumble bee, and climate interactions in seed crops of a tetraploid red clover (*Trifolium pratense* L.) J. Appl. seed Prod. 7:38-45.

Corbet S.A., Williams, I. H., Osborne, J. L. 1991. Bees and the pollination of crops and wild flowers in the European community. Bee Wld. 72:47-59.

Cortopassi-Laurino, M., Gelli, D. S. 1991. Pollen analysis, physico-chemical properties and bacterial action in Brazilian honeys from Africanized *Apis mellifera* and meliponine bees. Apidologie 22:61-73.

Craig, J.L. , Stewart, A. M. 1988. A review of kiwifruit pollination: where to next? N.Z. J. Exper. Agric. 16:385-399.

Crane, E. 1991: *Apis* species of tropical Asia as pollinators, and some rearing methods for them. Acta Horticulturae 288:29-48.

Crane E., Walker, P., Day, R. 1984. Directory of important world honey sources. London: Int. Bee Res. Assoc., 384 pp.

D'Ambrosio, M., Intoppa, F., Piazza, M. G. 1986. Preliminary researches on the pollinating activity of honeybees on *Actinidia chinensis* Blanch. Apicoltura 2: 155:164.

D'Arcy W. G. (ed.). 1986. Solanaceae: biology and systematics. New York: Columbia Univ. Press, 603 pp.

Delbrassinne, S., Rasmont, P. 1988. Contribution to the study of pollination of rape, *Brassica napus* var. *oleifera* in Belgium (French) Bull. Recerch. Agron. de Gembloux 23:123-152.

DeGrandi-Hoffman, G. 1987. The honeybee pollination component of horticultural crop production systems. Hort. Rev. 9:237-272.

DeGrandi-Hoffman, G., Hoopinger, R., Pulcer, R. 1987. REDAPOL: pollination and fruit set prediction model for 'Delicious' apples. Ennviron. Ent. 16:309-318.

DeGrandi-Hoffman, G., Roth, S. A., Loper, G. M. 1989. AIMOPOL: a cross-pollination and nut set simulation model for almond. J. Amer. Soc. Hort. Sci. 114: 170-176.

Delande, A., Tasei, J. N., Rollier, M. 1979. Pollination of male sterile lines for hybrid seed production. Proc. Int. Symp. Poll. 4:29-40.

Dennis, B., Holm, S. N., 1977. Recent trends in red clover pollination. Pszczel. Zeszyty Nauk. 21:149-157

Dhaliwal, H. S., Sharma, P. L. 1974. Foraging range of the Indian honeybee. J. apic. Res. 13:137-141.

Dhuyo, A. R., Munshi, G. H., Naqvi, S. M. S. H., Rizvi, S. N. H., Rustamani, M. A. 1988. Insect pollinator complex of cotton crop *Gossypium hirsutum* L. at Tandojam. Pakistan cottons 30:45-47 .

Donovan, B. J. 1979. Importation, establishinent and propagation of the alkali bee, *Nomia melanderi* Cockerell (Hymenoptera: Halictidae) in New Zealand. Proc. Int. Symp. Poll. 4:257-288.

Donovan BJ, Read P E C 1985: Efficacy of lucerne leafcutting bees as pollinators of lucerne in New Zealand. Australasian Conf. Grassland Invert. Ecol. Proc. 5: 257-264.

Donovan, B. J., Read, P. E. C. 1991. Efficacy of honeybees as pollinators of kiwifruit. Acta Horticulturae 288: 220-224.

Drescher, W., Engle, G. 1976. Influence of pollination by honeybees on the yield of sour cherries (German). Erwerbsobst. 18): 17,19-20.

Duke, J. A. 1981. Handbook of legumes of world economic importance. Plenum press, New York, 346 pp. [Pollination and bee visitors mentioned irregularly; cultivation, ecology, uses, distribution & agronomy summarized]

Dulanto, B. A. 1958. The importance of the "bumble bee' *Melitoma euglossoides* Lep. & Serv. in the pollination of the tanguis cotton plant (Spanish). Rev. Peruana Ent. Agr. 1:6-11.

Dulta, P. C., Verma, L. R. 1987. Role of insect pollinators on yield and quality of apple fruit. Ind. J. hort. 44:274-279.

Dylewska, M., Jablonski, B., Sows, S., Bilinski, M., Wrona, S. 1970. An attempt to determine the number of bees (Hym. Apoides) needed for adequate pollination of lucerne (Polish). Pol. Pismo Ent. 60_371-398.

Edwards-Anderka, C. J., Pengelly, D. H. 1969. Pollen analysis in the ecology of bees of the genus *Bombus* Latyr. (Hymenoptera: Apidae) in southern Ontario. Proc. Ent. Soc. Ontario 100:170-176.

Eisikowitch, D., Lupa, A. 1989. Wild flowers as competitors for pollinators in almond orchards. (Hebrew) Alon Hanotea 43:1307-1312.

Fairey, D.T., Leftkovitch, L. P., Lieverse, J. A. C. 1989. The leafcutting bee, *Megachile rotundata* (F.).: a potential pollinator for red clover. J. Appl. Ent. 107:52-57.

FAO Yearbook. 1990. Production. Vol 44: 123 pp., Trade Vol 44, 379 pp.

Ferguson, A. M., Pusch, W. M. 1991.

Development of mechanical dry-pollen application to kiwifruit. Acta Horticulturae 297:299-303.

Ferreres, L. P. 1947. Pollination of almonds with insects. (Spanish) Rancho Mex. 18:17-20,75.

Ford, N. A., Paton, D.C., Forde, N. 1979. Birds as pollinators at Australian plants. N. Z. J. Bot. 17:509-519.

Free, J. B. 1959. The effects of moving colonies of honeybees to new sites on subsequent foraging behaviour. J. agric. Sci. 53:1-9. [faba bean, apples, opium poppies, pollen preferences]

Free, J. B. 1962. The effect of distance from pollinizer varieties on the fruit set in plum and apple orchards. J. Hort. Sci. 37:262-271.

Free, J. B. 1968. Dandelion as a competitor to fruit trees for bee visits. J. Appl. Ecol 5:169-178.

Free, J. B., Ferguson, A. W. 1983. Foraging behaviour of honeybees on oilseed rape. Bee Wld. 62:22-24.

Free, J. B., Smith, M. V. 1961. The foraging behaviour of honeybees moved into a pear orchard in full flower. Bee Wld. 42:11-12.

Free, J. B., Williams, I. H. 1974. Influence of the location of honeybee colonies on their choice of pollen sources. J. Appl. Ecol. 11:925-935. [cherry, faba bean, kale, red clover]

Friese, I., Stark, J. 1983. Measuring the importance of honeybees in rape seed production. J. Apic. Res. 22:272-276.

Frusciante, L., Monti, L. M. 1980. Direct and indirect effects of insect pollination on the yield of field beans (*Vicia faba* L.) Zeitschrift Pflanz. 84:323-328.

Fussell, M., Corbet, S. A. 1992. Flower usage by bumble bees: a basis for forage plant management. J. Appl. Ecol. 29:451-465.

Fussell, M., Osborne, J. L., Corbet, S. A. 1991. Seasonal and diurnal patterns of insect visitors to winter sown field bean flowers in Cambridge. Aspects Appl. biol. 27:95-99.

Garafalo, A. C. 1981. On the biology and importance of pollinators (Portuguese) Correio do Apicultura 5:22-24.

Gary, N. E. 1979. Factors that affect the distribution of foraging honeybees. Proc. Int. Symp. Poll. 4:353-358.

Gary, N. E., Witherell, P. C. 1977. Distribution of foraging bees of three honeybee stocks located near onion and sunflower fields. Environ. Ent. 6:785-788. [honeybee foraging range, methodology too]

Gerling, D., Velthuis, H. H. W., Hefetz, A. 1989. Bionomics of the large carpenter bees of the genus *Xylocopa*. Ann. Rev. Ent. 34:163-190.

Gladwin, F. E. 1937. Pollination with particular reference to the grape. Amer. Fruit Grow. Mag. 57:9,24-25,35; 3:16,35.

González, R. H. 1980. Insect pollinators and the ecology of pollination (Spanish). IDIA 391/392:44-49. [solitary bees main pollinators of lucerne in Chile]

Goodwin, R. M. 1986. Increased kiwifruit pollen collection after feeding sugar syrup to honeybees within their hive. N. Z. J. Exper. Agric. 14:57-61.

Goodwin, R. M. 1986. Kiwifruit flowers: anther dehiscence and daily collection of pollen by honeybees. N. Z. J. Exper. Agric. 14:459-452.

Goodwin, R. M., Houten, A. T., Perry, J. H. 1991. Feeding sugar syrup to honeybee colonies to improve pollen collecting: A review. Acta Horticulturae 288:265-268.

Griffin, H. J., Macfarlane, R. P, van den Ende, N. J. 1991: Rearing and domestication of long tongued bumble bees in New Zealand. Acta Horticulturae 288:149-153.

Hassanein, M. H. 1953. Studies on the effect of pollinating insects, especially the honeybee, on the seed yield of clover in Egypt. Bull Soc. Fouad. Ent. 37:337-344. [Berseem clover]

Hawke, J. G., Lester, R. N., Skelding, A. D. 1979. The biology and the taxonomy of the Solanaceae. London: Academic Press, 738 pp.

Hawthorn, L. R., Bohart, G. E., Toole, E. H. 1980. Carrot seed production as affected by insect pollination. Utah Agr. Exper. Sta. Bull. 422.18 pp.

Hebblewaite, P. O. (ed.). 1983. The faba bean (*Vicia faba* L.). London: Butter-worths, 573 pp.

Heinrich, B. 1979. Bumblebee economics. , Cambridge, Mass.: Harvard Univ. Press, 246 pp

Heywood, V. H., Harborne, J. B., Turner, J. B. 1977. The biology and chemistry of the Compositae. Vol. 1. London: Academic Press, 619 pp.

Hobbs, O.A.,Virostek, J. F., Nummi, W. O. 1962. Food gathering behaviour of honey, humble and leaf cutter bees (Hymenoptera: Apoidea) in Alberta. Can. Ent. 92:409-418.

Hoffmann, M., Wittmann, D. 1987. Wild bee community in an agricultural area of Rio Grande do Sul, southern Brazil, and its impact on pollination of beans and sunflowers. In.: J. Eder, H. Rembold (eds.). Chemistry and biology of social insects.

Holm, S. N. 1962. The most important legumes competing for visits by bees (Danish). Spertryk 18:3-8. [*Bombus*, honeybees]

Holm, S. N. 1966. The ulilization and management of bumble bees for red clover and alfalfa seed production. Ann. Rev. Ent. 11:55-182.

Hopping, M. E., Jerram, E. M. 1979. Pollination of Japanese plums. Orchardist N. Z. 54:104-106.

Hurd, P. O., LaBerge, W. E., Linsley, E. G. 1980. Principal sunflower bees of North America with emphasis on the southwestern United States (Hymenoptera: Apoldea) Smithsonian Contrib. Zool. 310, 158 pp. [400 bee species]

Ibrahim, S. H., Selim, H. A. 1962. Studies on pollen collection from *Eucalyptus* spp. and Compositae plants by the honeybee. Agric. Res. Rev. [Cairo] 40:116:123.

Intoppa, F., Piazza, M. G. 1989: Feijoa: a plant in search of pollinators (Italian). Riv. di Fruttic. 7:73-75.

Jacobs,. M. R. 1979. Eucalypts for planting. FAO forestry series, Rome 11, 677 pp.

Jadhav, L. D. 1981. Role of insects in the pollination of onion (*Allium cepa* L.) in the Maharashtra State, India. Indan Bee J. 43:61-63.

Jay, D., Jay, S. C.. 1984. Observations of honeybees on Chinese gooseberries ('Kiwifruit') in New Zealand. Bee Wld. 65:155-166.

Jones, H. A., Emsweller, S. L. 1934. The use of flies as onion pollinators. Amer. Soc. Horti. Sci. Proc. 31(suppl):160-164.

Kambal, A. E., Bond, D. A., Toynbee-Clarke, G., 1976. A study on the pollination mechanism in field beans (*Vicia faba* L.). J.agric. Sci. camb. 87:519-526.

Kapil, R. P. 1986. Pollination biology — an analysis. New Delhi, 300 pp. [cucurbits, sunflower, lucerne, brassicas, lemons]

Kendall, D. A., Solomon, M. E. 1973. Quantities of pollen on the bodies of insects visiting apple blossom. J. Appl. Ecol. 10:627-634.

Kendall, D. A., Smith, B. D. 1976. The pollinating efficiency of honeybee and bumble bee visits to field bean flowers (*Vicia faba* L.). J. Appl. Ecol. 709-7171.

Kenrick, J., Knox, R. B. 1989. Quantitative analysis of self incompatibility in trees of seven species of *Acacia*. J. Hered. 80:240-245.

Kevan, P. G. 1986. Pollinating and flower

visiting insects and the management of beneficial insects and plants. Proc. Symp. Biol. control in the tropics 1:439-452.

Kevan, P. G., Baker, H. G. 1983. Insects as flower visitors and pollinators. Ann. Rev. Ent. 28:407-453.

Klug, M., Bunemann, G. 1983. Pollination: wild bees as an alternative to the honeybee? Acta Horticulturae 139:59-64.

Ko, K.-C., Woo, K.-S., Kim, W.-C. 1977. A study on the activities of insect pollinators frequenting various fruit trees near Seoul area. Seoul Univ. Coll. Agric. Bull. 2:407-421. [pollination of apple]

Krunic, M. D. Brajkovic, M. M., Mihajlovicx, L J. S. 1991. Management and utilization of *Osmia cornuta* Latr. for pollination in Yugoslavia. Acta Horticulturae 288:190-193.

Kumar, J., Mishra, R. C., Gupta, J. K., Dogra, G. S. 1985. Pollination requirements of some peach cultivars. Indian Bee J. 47:3-6.

Langridge, D. F., Goodman, R. D. 1981. Honeybee pollination of sunflower cultivars Hysan 30 and sunfola. Aust. J. Exper Agric. Anim. husb. 21:435-438.

Langridge, D. F., Goodman, R. D. 1973. The role of honeybees in pollination of cherries. Aust. J. Exp. Agric. Anim. husb. 13:193-195.

Latif, A., Qayyum, A., Manzoor, A. M. 1956. Role of *Apis indica* F. in the pollination of Egyptian clover (*Trifolium alexan-drinum* Linn.). Pakist. J. Sci. Res. 8:48-50.

Lecompte, J., Turgari, S. 1985. On some pollinators of forage legumes (French). Ann. Abeille 8:83-92. [red, white clover, lucerne]

Lederhouse, R. C., Caron, D. M., Morse, R. A. 1972. Distribution and behaviour of honeybees on onions. Environ. Ent. 1:127-129.

Liu, H. J., Macfarlane, R. P., Pengelly, D. H. 1975. Relationships between flowering plants and four species of *Bombus* (Hymenoptera: Apidae) in southern Ontario. Can. Ent. 107:577-588.

Lomnbard, P. B., Callan, N. W., Dennis, F. G., Looney, N. E., Martin, G. C., Renquist, A. R., Mielke, E. A. 1988. Towards a standardized nomenclature, procedures, values, and units in determining fruit and nut tree yield performances. Hort. Sci. 23:813-817.

Lupo, A. 1984. *Osmia latreillei iberato-afracana* Pert. (Megachilidae: Hymenoptera) as a potential pollinator. Proc. Int. Symp. Poll. 5:465-476.

Macfarlane, R. P. 1980. Bees, bee forage and pollination on the Chatham Islands. N. Z. Beekeeper 41:5-6.

Macfarlane, R. P, Griffin, R. P. 1985. An assessment of bumble bees and honeybees as pollinators of red clover. Australasian Conf. Grassland invert. Ecol. 4:114-119

Macfarlane, R. P., van den Ende, H. J., Read, E. C. 1983. Bumblebee management options to improve 'Grasslands Pawera' red clover seed yields. Proc. N. Z. Grasslands Assoc. 44:47-53.

Macfarlane, R. P., van den Ende, H. J., Griffin, R. P. 1991. Pollination needs of 'Grasslands Pawere' red clover. Acta Horticulturae 288:399-404.

Macfarlane, R. P., Ferguson, A. M. 1984. Kiwifruit pollination: a survey of the pollinators in New Zealand. Int. Symp. of Pollination Proc. 6:367-373.

Maciel Correia, M. L. 1984. Notes on the nesting biology and distribution of wild bees in the northwest of Portugal (Hymenoptera, Megachilidae) (French). Proc. Int. Symp. Poll. 5:237-242.

Maciel Correia, M. L. 1991. Selection of pollinators from the Portuguese fauna of wild bees. Acta Horticulturae 288:69-73.

Maeta, Y. 1979. Studies on the biology of *Megachile japonica* Alfken (I). Nesting habit and nest structure (Japanese). Bull. Tohoku Natl. Agric. Exp. Stn. 61:59-68.

Maeta, Y., Kitamura, T. 1981: Pollinating effeciency by *Osmia cornifrons* (Radoszkowski) in relation to required number of nesting bees for economic fruit production. Honeybee Sci. 2:65-72.

Mahmood, A. M., Furgala, B. 1983. Effect of pollination by insects on seed oil percentage of oilseed sunflower. Amer. Bee J. 123:213-219.

Mann, G. S., Singh, G. 1980. Period of activity and comparative abundance of flower visiting insects on pear at Ludhiana (Punjab). Entomon. 5:65-66.

Mann, G. S., Singh, G. 1981. Activity and abundance of flower visiting insects of peach (*Prunus persica* L.) at Ludhiana. Prog. Hort. 13:25-27.

Mann, G. S., Singh, G. 1981. Activity and abundance of flower visiting insects of almond (*Prunus amygdalus* Batch) at Ludhiana. J. Bombay nat. hist. Soc. 78:617-618.

Mann, G. S., Singh, G. 1983. Activity and abundance of pollinators of plum at Ludhiana (Punjab). Amer. Bee J. 123:595.

Manzoor-ul-Haq, A. M., Fiaz, M. 1980. Insect pollination of sunflower (*Helianthus annuus*) in Pakistan. J. Apic. Res. 19:83-87.

Maraboldi, A., Ruiz, M. A., Aballay, E., Arretz, P. V., Guerrero, M. A. 1986. Evaluation of insect pollination in red clover (*Trifolium pratense* L.) seed fields in Cautin. Inv. Agrícola [Chile] (Spanish). 9:67-73.

Matheson, A. 1991. Managing honeybee pollination of kiwifruit (*Actinidia deli-ciosa*) in New Zealand. Acta Horti-culturae 288:213-219.

Menke, H. F. 1954. Insect pollination in relation to alfalfa seed production in Washington. Washington State Univ. Bull., 24 pp.

Milliron, N. E. 1973. A monograph of the Western Hemisphere bumblebees (Hymenoptera: Apidae; Bombinae) II. The genus *Megabombus* subgenus *Megabombus*. Mem. Ent Soc. Can. 89:81-237.

Minano, A., Patetta, A., Marletto, F. 1991: Investigations on chestnut pollination. Acta Horticulturae 288:335-339.

Morrison, K. J., Foeppel, W. 1984. Red clover seed production. Washington State Coop. Extx. E. B. 1242, 4 pp.

Moustafa, M. A., El Berry, A. A. 1974. A study on the wild bees, *Osmia latreillei* Spin and *O. submicans* Mor. as pollinators to some economic crops in Egypt. Ann. Agric. Sci. 3:233-240.

Murakami, K. 1964. Studies of pollination of red clover by Hymenopterous insects (Japanese). Hokkaido Nat. Agric. Exper Sta. Report 85, 75 pp.

Narayanan, E. S., Sharma, P. L., Phadke, K. O. 1951. Studies on the requirements of various crops for insect pollination - insect pollinators of berseem - Egyptian clover (*Trifolium alexandrinum*) with particular reference to honeybees and their role in seed setting. Indian Bee J. 23:23-30.

Neff, J. L., Simpson, B. B. 1990. The roles of phenology and reward structure in the pollination biology of wild sunflower (*Helianthus annus* L., Asteraceae). Israel J. Bot. 39:197-216.

Ochoa, L. H. 1980. Production of lucerne seed in the province of Santiago del Estero. (Spanish) IDIA 391-391:87-96. [*Xylocopa*, *Melissodes nigroaenea* most effective pollinators in Argentina]

Olmstead, A. L., Wooten, D. B. 1987. Bee pollination and productivity growth: the case of alfalfa. Amer. J. Agric. Econ. (1987) 56-63.

Olmo, H. P. 1943. Pollination of the Almeria

grape. Proc. Amer. Soc. Hort. Sci. 42:401-406.

Olsen, L. G., Hoopinger, R., Martin, E. C. 1979. Pollen preferences of honeybees sighted on four different cultivated crops. J. apic. Res. 18:196-200 (apple, blueberry, strawberry, cucumbers).

Ortiz-Sánchez, F.J., Tinaut-Rivera, A. 1987. Potential pollinators of sunflower in the Grenada. The role of the honeybee. (Spanish). Vida Apícola 26:20-21.

Ozbek, N. 1976. Pollinator bees on alfalfa in the Erzurum region of Turkey. J. Apic. Res. 15:145-148.

Ozbek, H. 1980. Apoidea pollinating red clover (*Trifolium pratense*) in eastern Anatolia (Turkish). Temel. Bilm 4:61-66.

Palmer-Jones, T., Clinch, P. G. 1974. Observations on the pollination of Chinese gooseberries variety "Hayward". N. Z. J. Exper. Aric. 2:455-458.

Panchabhavi, K.S., Devaiah, M. A., Basavanna, G. P. C. 1975. The effect of keeping *Apis cerana indica* F. colonies on the seed set of sunflower *Helianthus annus* Linn. Mysore J. Agric. Sci. 10:631-636.

Parker, F. D. 1981. How effecient are bees in pollinatlng sunflowers? J. Kans. Ent. Soc. 54:61-67.

Parker, F. D. 1981. Sunflower pollination: abundance, diversity and seasonality of bees and their effect on seed yields. J. Apic. Res. 20:49-61.

Parker, F. D. 1981. Sunflower pollination: abundance, diversity and seasonality of bees on male sterile and male fertile cultivars. Environ. Ent. 1012-1017.

Parker, F. D. & D. R. Frolich 1983: Hybrid sunflower polllnation by a manageable Compositae specilaist: the sunflower leafcutter bee (Hymenoptera; Megachilidae). Environ. Ent. 12:576-581.

Patterson, K. J. 1990. Effects of pollination on fruit set, size and quality in feijoa (*Acca sellowiana* (Berg) Burret). N. .Z. J. Crop. Hort. Sci 18:127-131.

Petri, J. L., Pasqua, M. 1980. The pollination of apples along the upper reaches of the river Peixe, Santa Catarlna. (Portugese). Bol. Tech. EMPASC, 40 pp. [includes review, and discussion of pollination]

Pijl van der, L. *Xylocopa* and flowers in the tropics III. Observations on som Papilionaceae, *Melastoma, Calotropis, Cassia* and some orchids, with general considerations. Proc. K. Ned. Akad. Wet. Ser C., Biol. Med. Sci. 57:552-562.

Plowright, R. C., Hartling, L. K. 1981. Red clover pollination by bumblebees: a study of the dynamics of a plant-pollinator relationship. J. Appl. Ecol. 18:639-647.

Plowright, R. C., Laverty, T. M. 1987. Bumble bees and crop pollination in Ontario. Proc. Ent Soc. Ont. 118:155-160.

Pomeroy, N., Heath, A. C. G., Goodwin, R. M. 1988. Forum comments on a review of kiwifruit pollination: where to next? N. Z. J. Exper. Agric. 16:392-399.

Pomeroy, N., Stoklosinski, S. R. 1990. Measuring the foraging strength of bumblebee colonies. Proc. Congr Int. Stu. Soc. Ins. 11:469.

Pringle, O. L., Murray, B. O. 1991. Reproductive biology of the tamarillo *Cypho-mandra betacae* (Cav.) Sendt. (Solanaceae) and some wild relatives. N. Z. J. Crop. Hort. Sci. 19:263-273.

Ptacek, V. 1979. Effeciency of some bee species (Hymenoptera: Apoidea) in pollination of lucerne (*Medicago sativa* L.) (Czech). Sborn. Vedec. Praci 5/6:91-98.

Ptacek, V., Nedbalkova, B. 1988. The effect of bee pollinators (Apoidea) on the seed yield of lucerne under practical farming conditions. (Czech) Rost. Vyroba 32:821-832.

Ptacek, V. 1991. Lucerne seed production with *Rhophitoides canus* Ev., (Hymenoptera Apoidea). Acta Horticulturae 288:443.

Punchihewa, R. W. K., Keoniger, N., Kevan, P. G., Gadawski, R. M. 1985. Observations on the dance communication and natural foraging ranges of *Apis cerana*, *Apis dorsata*, and *Apis florea* in Sri Lanka. J. Apic. Res. 24:168-175.

Rasmont, P. 1983. Annotated catalogue of the bumblebees of the west Palearctic region (French). Notes Foun. Gembl. 7, 71 pp.

Rahoo, G. M., Lohar, M. K., Munshi, O. H., Nizamani, S. M. 1988. The insect pollinators of sunflowers, *Helianthus annuus* L. in Sind, Pakistan. Pak. J. Agric. 1:79-80.

Randhawa, G. S., Yadav, I. S., Nath, N. 1963. Studies on the flowering, pollination and fruit development in peach grown under subtropical conditions. Indian J. Agr Sci. 33:129-139.

Rao, G. M., Lazar, M. 1983. Studies on bee behaviour and pollination of onion (*Allium cepa* L.) Proc. 2nd Int. Conf. Apic. Trop. Climates, pp. 580-589.

Rashad, S. E., Mohamed, M. I., El-Shakaa, S. M. A. 1980. Behaviour of honeybee workers on major pollen sources in Giza region. Annals Aric. Sci., Moshtohor 12:379-384.

Ravestijn, W. van. 1991. Use of bumble bees for the pollination of glasshouse tomatoes. Acta Horticulturae 288:204-209.

Read, P. E. C., Donovan, B. J., Griffin, R. P. 1989. Use of bumblebees, *Bombus terrestris*, as pollinators of kiwifruit and lucerne in New Zealand. N. Z. Ent.12:19-23.

Ribbands, C. R. 1951. The flight range of the honeybee. J. Anim. Ecol. 20:220-226.

Richards, K.W. 1984. Alfalfa leafcutter bee management in western Canada. Apic. Canada Publ. 1495E, 53 pp.

Rick, C. M. 1950. Pollination relations of *Lycopersicon esculentum* in native and foreign regions. Evolution 4:110-122.

Rizk, G. A., Atallah, M. A. 1979. Pollen gathering activity of honeybee colonies in relation to some weather factors (Hymenoptera: Apidae). Proc. Int. Congr. Apic. 27:293-297.

Rohita, B. H., Klinac, D. J. 1990. Relationships between seed set and fruit weight and shape in nashi (*Pyrus serotina* Rehder var. *culta* Render). N. Z. Crop hort. Sci. 18:133-136.

Ruskowski, A. 1971: Food plants and economical importance of *Bombus terrestris* (I.) and *B. lucorum* (L.) (Polish). Pam. Pul. Zest. 47:215-250.

Sanford, J. C. , Hannemann, R. E. 1981. The use of bees for the purpose of inter-mating in potato. Amer. Potato J. 48:481-485.

Santis, L. de. 1970. Note on two species of *Megachile* that pollinate lucerne (Spanish). Ciencia y Abejas 5:44-48.

Santis, L. de. 1981. Notes of some insect pollinators of alfalfa from Buenos Aires and their parasites (Spanish). Rev. del Museo de La Plata. Ser. Zool. 122:227-237.

Satyanarayana, A. R., Seetharam, A. 1982. Studies on the method of hybrid seed production in oilseed sunflower (*Heli-anthus annuus*). 3: role and activity of insect visitors in pollination and seed set. Seed Sci. Tech. 10:13-17.

Severson, D. W., Parry, J. E. 1981. Preliminary studies of pollen collection by honeybees (*Apis mellifera*) in a Wisconsin apple orchard. Amer. Bee J. 121:255-257.

Sharma, P. L. 1981. The honeybee population among insects visiting temperate-zone fruit and flowers and their role in setting fruit. Bee Wld. 42:6-8. [peach, plum]

Sharples, O. C., Todd, F. E., McGregor, S. E., Milne, R. L. 1965. The importance of insects in the pollination and fertilization of the cardinal grape. Proc. Amer. Soc. Hort. 86:321-325.

Shelar, D. G., Suranararauana, M. C. 1983. Effect of pollination by honeybee (*Apis *cerana indica* Fabricius) on the yield of lucerne. Indian J. Agric. Sci 53:190-191.

Sherry, S. P. 1971. The black wattle (*Acacia mearnsii*) de Wild. Pietermaritz-burg: Univ. Natal Press, 402 pp.

Sidhu, A.S., Singh, S. 1961. Studies on agents of cross pollination in cotton. Indian Cotton Grow. Rev. 15:341-353.

Sihag, R. C. 1986. Insect pollination increases seed production in cruciferous and umbelliferous crops. J. Apic. Res. 25:121-126.

Sihag, R. C. 1987. Effect of pesticides and bee pollination on seed yield of some crops in India. J. Apic. Res. 27.

Sihag, R. C. 1988. Characterisation of the pollinators of cultivated crops of subtropical Hissar, India. Bee Wld. 69:153-158.

Simpson, B. B., Neff, J. L. 1987. Pollination ecology in the arid southwest. Aliso 11:417-440.

Singh, J. P., Dharamwal, S. S. 1970. The role of honeybee in seed setting of onion at Plant nagar, dist. Nainital,Uttah Pradesh, India. Indian bee J. 32:23-26.

Snow, A. A., Roubik, D. W. 1987. Pollen deposition and removal by bees visiting two tree species in Panama. Biotropica 19:57-63. [Buzz pollination by *Xylocopa*, *Centris* on *Cassia reticulata*, *Cochlospermum*]

Stephen, W. P. 1958. Pear pollination studies in Oregon. Oreg. Agr. Exp. Sta. Tech. Bull. 43, 43 pp.

Steshenko, F. N. 1958. The role of honeybees in cross-pollinating grape vines. (Russian) Pchelovodsto 35:37-40.

Stephens, S. G., Finkner, M. D. 1953. Natural crossing in cotton. Econ. Bot. 7:257-269.

Steven, D. 1988. The Chinese bee hunt continues. N. Z. Kiwifruit 19:22.

Steward, K. M., Craig, C. L. 1989. Factors affecting pollinator effectiveness in *Feijoa sellowiana*. N. Z. J. Crop. Hort. Sci. 17:145-154.

Stoddard, F. L. 1991. Pollen vectors and pollination of Faba beans in Southern Australia. Aust. J. Agric. Res. 42:1173-1178.

Stoddard, F. L., Bond, D. A. 1987: The pollination requirements of the faba bean. Bee Wld. 68:144-152.

Tanda, A. S. 1983. Assessing the role of honeybees in a field of Asiatic cotton (*Gossypium arboreum* L.) Amer. Bee J. 123:593-594.

Tanda, A. S. 1984. Bee pollination increases yield of 2 interplanted varieties of Asiatic cotton. (*Gossypium arboreum* L.) Amer. Bee J. 124:539-540.

Tasei, J. N. 1973. The nesting behaviour of Osmia (*Osmia cornuta* Latr. and *Osmia rufa* L., Hymenoptera, Megachilidae). (French). Apidologie 4:195-225.

Tasei, J. N. 1976. Pollen collection and nest provisioning by *Osmia coerulescens* (Hymenoptera: Megachilidae). (French). Apidologie 7:277-300.

Tasei, J. N., Picard, M., Carre, S. 1978. Insects pollinating lucerne (*Medicago sativa* L.) in France (French). Apidologie 9:175-194.

Tepedino, V. J. 1981: The pollination efficiency of the squash bee (*Peponapis pruinosa*) and the honeybee (*Apis mellifera*) on summer squash (*Cucur-bita pepo*). J. Kans. Ent. Soc. 54:359-377.

Testolin, R., Vixxotto, G., Costa, G. 1991. Kiwifruit pollination by wind and insects in Italy. N. Z. J. Crop. Hort. Sci. 19:381-384.

Thies, S. A. 1953. Agents concerned with natural crossing of cotton in Oklahoma. Amer. Soc. Agron. J. 48:481-484.

Thompson, M. M., Liu, L. J. 1972: Pollination and erratic bearing in 'Italian prunes'. Amer Soc. Hort. Sci. Proc. 97:489-491.

Thorp, R.W. 1979. Honeybee foraging behaviour in California almond orchards. Int. Symp. Poll. 4:385-392.

Toit, A. P. 1988. The pollination ecology of commercial sunflower *Helianthus annuus* L. in South Africa with special reference to the honeybee *Apis mel-lifera* M. Sci. thesis, Pretoria Univ.,143 pp.

Toit, A. P., Coetzer, L. A. 1991. Difference in potential attractiveness to bees of seventeen south African sunflower (*Helianthus annuus*) cultivars. Acta Horticulturae 288:288-293.

Torchio, P. F. 1981. Field experiments with *Osmia lignaria propinqua* Cresson as a pollinator in almond orchards I. 1975 studies. J. Kans. Entomol. Soc. 64:815-823.

Torchio, P. F. 1984. Field experiments with the pollinator species*Osmia lignaria propinqua* Cresson in apple orchards: III. 1977 studies. J. Kans. Ent. Soc. 57:517-521.

Torchio, P. F. 1985. Field experiments with the pollinator species *Osmia lignaria propinqua* Cresson in apple orchards: V. (1979-1980), methods of introducing bees, nesting success, seed counts, fruit yields (Hymenoptera: Megachilidae). J. Kans. Entomol. Soc. 58:448-464.

Torchio, P. F. 1987. Use of non-honeybee species as pollinators of crops. Proc. Ent. Soc. Ont. 118:111-124.

Torchio, P. F. 1990. *Osmia ribifloris*, a native bee species developed as a commercially managed pollinator of highbush blueberry (Hymenoptera: Mega-chilidae). J. Kans. Entomol. Soc. 63:427-436.

Vaissiere, B. E. 1991. honeybee stocking rate, pollinator visitation, and pollination effectiveness in upland cotton grown for hybrid seed production. Acta Horticulturae 288:359-363.

Verma, L. R., Dutta, P. C. 1986. Foraging behaviour of *Apis cerana indica* and *Apis mellifera* in pollinating apple flowers. J. Apic. Res. 25:197-201.

Vesely, V. Titiera, D. 1985. The ability of the Carnolian bee (*Apis mellifera carnica* Pollm.) to gather pollen from the field bean (*Vicia faba* L.) under the conditions of Czech Socialist Republic (CSR). Pszel. Zesz. Nauk. 29:187-190.

Vroege, P. W., Stelleman, P. 1990. Insect and wind pollination in *Salix repens* L. and *Salix caprea* L. Israel J. Bot. 39:125-131.

Wafa, A. K., Ibrahim, S. H. 1980. Effect of the honeybee as a pollinating agent on the yield of beans and clover in Egypt. Bull. Fac. Agric. Cairo Univ. 206.

Wafa, A. K. & M. I. Mohameed 1970. Anthophorinae pollinating on broad bean flowers. Bull. Soc. Ent. Egypte 54:213-220.

Warrington, L. J., Weston, G. C. 1990. Kiwifruit: science and management. Wellington: N. Z. Soc. Hort Sci., 576 pp.

Watmough, R. H. 1987. A leaf-cutter bee (Megachilidae) and a carpenter bee (Anthophoridae) as possible pollinators of lucerne (*Medicago sativa* L.) in the Oudtshoorn district. S. Afric. Bee J. 59:114.

Webb, M. C. 1961. The biology of the bumble bees of a limited area in eastern Nebraska. Ph.D. thesis, Univ. Nebraska, 337 pp.

Westwood, M. N. 1988. Temperate-zone pomology. Portland, Oregon: Timber Press, 428 pp.

Williams, I. H. 1985. The pollination of swede rape (*Brassica napus* L.). Bee Wld. 66:16-22.

Williams, I. H. 1988. The pollination of linseed and flax. Bee Wld. 69:145-152.

Williams, I. H. 1991. The pollination of linseed. Acta Horticulturae 234-238.

Woyke, H. W. 1981. Insect pollination of carrot and onion. Int. Bee Res. Assoc. bibliography 28,11 pp.

Woodward, D. R. 1990. Food demand for

colony development, crop preference and food availability for *Bombus terrestris* (L.) (Hymenoptera: Apidae). Ph.D. thesis, Massey Univ., N. Z., 253 pp.

Wratt, E. C. 1968. The pollinating activities of bumblebees and honeybees in relation to temperature, competing forage plants, and competition from other foragers. J. Apic. Res. 2:61-66.

Wyborn, M. H., Matteoni, J. A. 1992. The pollination of greenhouse tomatoes with two western North American bumblebee species. Intl. Workshop on non-*Apis* bees and their role as crop pollinators. Utah State Univ., Abstract, p. 81.

Yakhlef, A. 1986. Effects of insect pollination on production in almond (*Prunus amygadalus* Batsch), pear (*Prunus communis* L.) and apple (*Malus pumila* Mill.) (French) Annales Inst. Nat. Agron. (El-Harrach). 10:115-131.

Yalden, P. E. 1983. The pollen collected by *Bombus lucorum* (L.) (Hym., Apidae) in the peak district, England. Ent. monthly Mag. 105-109.

Yamada, I., Ebara, K. 1952. Studies on the fertilization of red clover, *Trifolium pratense* by bumblebees and honeybees (Japanese). Hokkaido Nat. Agric. Exper. Sta. Bull. 45, 33 pp.

Yokozawa, Y., Yasul, A. 1957. Studies on the insect visitors of peach. 1. Insect visitors to the flowers of peach. (Japanese) Hort. Assoc. Jap. J. 26:185-191.

Yoshida, A., Maeta, Y. 1988. Utilization of *Osmia cornifrons* (Radoszkowski) as a pollinator of apples in southwestern Japan (Japanese) Honeybee Sci. 9:1-6.

blackberry - Rosaceae: *Rubus*

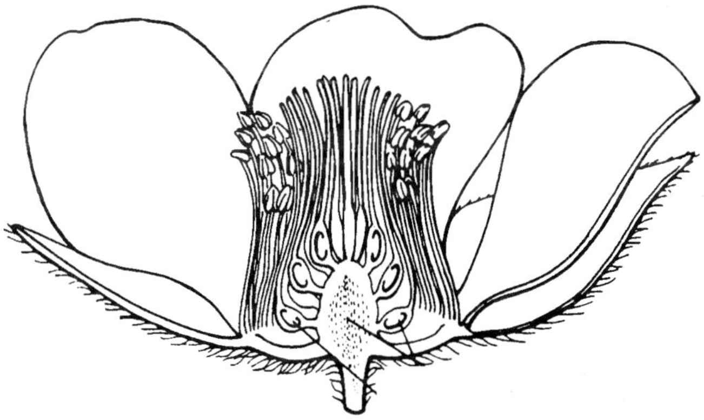

flower (section) & aggregate fruit

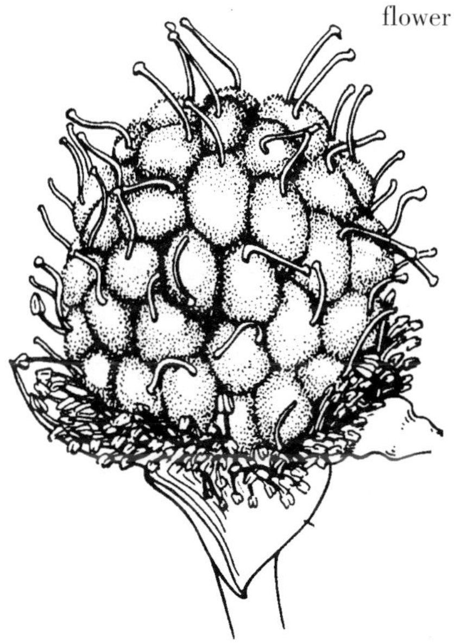

1.4 APPLIED POLLINATION ON TROPICAL & SUBTROPICAL OCEANIC ISLANDS

One can learn much from tropical islands and their pollination systems. Island studies have shown, repeatedly, that plants and pollinators can adjust to sometimes drastically new pollination environments. Self-compatibility and autogamy evolve relatively rapidly, or floral traits change to accommodate new flower visitors and potential pollinators. Animals confronted with a new range of floral resources often show surprising flexibility in their choice of plants, belying past assumptions about their supposed preferences. Islands also provide, in the clearest manner possible, examples of the importance of 'stocking density' that may guide management decisions for mainland areas — and more complicated 'systems' in general. Beginning with an island having no pollinators, one can see the difference that importing a pollinator can make, or the needed pollinator abundance to successfully pollinate a crop of given size. Where pollinators are already present, or after they have been introduced, it is also possible to detect ways that weedy plant species are spread by pollinators — a needed consideration when assuming the responsibility for stocking pollinators in managed habitats.

Island	*Area (km2)*	*Climate*	*Date of Intro.*	*Apis dorsata*	*Apis cerana*	*Xylocopa or Bombus*	*Mega-chilids*	*Other Bees*	*Birds*	*Bats*
St. Helena	419	arid	?0	0	0	0	-	?+	0	0
Asencion	88	arid	0	0	0	0	-	?0	0	0
Azores	2344	wet	+	0	0	1B	-	?+	0	0
Bermuda	54	wet	1600s	0	0	-	-	?+	0	0
Trindade	30	wet	?0	0	0	-	-	?0	?+	0
Madiera	798	humid	+	0	0	2B	?+	?+	0	0
Cape Verde	4033	arid	+	0	0	0	?+	?+	0	0
São Tome	964	wet	+	0	0	? 0	?+	?+	3-8*	0
Canary	7575	dry	+	0	0	1B	3	79	0	0
ATLANTIC TOTAL	16305									
Rodrigues	109	wet	+	0	0	-	-	?+	0	1*
Chagos Archipel.	20	wet	0	0	0	0	0	?0	0	0
Seychelles	453	humid	+	0	0	-	-	+	3	1*
Cocos	14	wet	0	0	0	0	1	0	0	0
Mauritius	2040	humid	1721	0	0	-	?+	?+	0-4*	2*
Reunion	2504	wet	1666	0	0	-	?+	?+	0-3*	2*
Christmas	155	wet	+	0	0	0	3	4	0-1*	1*
Maldives	298	wet	0	0	1988 ?	1X	?+	?0	0	2*
Comoros	2171	wet	?+	0	0	-	?+	?+	1-6*	2*
Nicobar, Andaman	8249	wet	0	+	+	-	?+	?+	3-12*	2*
Lacadives	32	humid	0	0	?+	-	?+	?+	0	0
Socotra	3582	dry	+	0	0	-	?+	?+	1	0
INDIAN TOTAL	19627									

Table 1. Availability of bee, bird and bat pollinators on Atlantic and Indian oceanic islands. Islands farthest from the continents are listed first. Climate and annual precipitation: *Wet:* > 2000 mm, no prolonged dry season; *Humid:* 300-999 mm; *Arid:* 130-260 mm; *Dry:* 100-199 mm. Pollinators — date when introduced given for *Apis;* other information in table: + = present, 0 = absent, ? = insufficient records, number shows number of species. Bats and birds include only flower-visiting species, * = maximum potential number

1.4.1 Available Pollinators & Native Flora

Oceanic islands, those relatively far from the mainland or other islands, are a distinctive tropical habitat. Their combined area, excluding the West Indies, (see Tables 1 and 2) exceeds Iceland, Cuba or the South island of New Zealand. With the exception of the Canary Islands, these 47 tropical and subtropical islands are separated from adjacent continents by 200 to 5,000 km. Even the closer Canary Islands were about 60 km away from Africa when sea levels were lower in the last ice age. The mountainous, older, and remote large islands — Hawaii, Canary, New Caledonia — have up to 89% endemism in the flora, while the younger islands like the Galápagos, Vanuatu and Ascension, and all relatively flat islands have smaller floras. Their endemism usually ranges between 5 and 40%. New Caledonia, with 3,500 indigenous species, has the largest distinctive flora. The oceanic islands retain relatively many endemic plant species. The average on 22 island groups is one endemic species per 17 km^2 (range 1.3 to 88.0 km^2).

Oceanic islands contain only vestiges of the bee, bird and bat pollinators found on the neighbouring continents. For instance, Vanuatu has 12 species of native bees, Hawaii 62 species, New Caledonia 28 and Fiji 7. The more remote atolls like Wake, Midway, and Arno Island in the Marshalls have an extremely limited fauna of flower-visiting insects. Diurnal flower visitors of Wake island include only imported honeybees, a hover fly, two flies of the genus *Lucilia*, some muscid flies, one butterfly and one sphingid moth. Faunal diversity is influenced by rate of arrival by potential colonists. Colonists are affected by the distance of the island from the continent and direction, persistence and strength of prevailing winds and sea currents. There are also significant differences among the long-range dispersal and colonization abilities (including fecundity especially) of different animals. Establishment is greatly influenced by island size and topography, patterns of rainfall, and floral diversity.

The only islands that naturally had generalist social bee pollinators were the Andaman, Nicobar and Laccadive islands, with giant and Asian honeybees which can now be at least partly managed. Stingless *Trigona* bees (subgenus *Heterotrigona*, or the '*Tetragonula*' group) occur on the Caroline Islands and on the Solomons — *T. iridipennis* on Truk and Solomon, and *T. fuscobalteata* on Paulus of the Caroline islands. Colonies of stingless bees can be transferred to wooden hives.

In the Atlantic and Indian oceans, *Apis mellifera* of Malagasy and European origin were likely introduced to the larger islands 200-300 years ago for honey production, although few actual dates have been recorded. Since the 1830s repeated efforts were made to establish honeybees on Ascension and presumably the large St. Helena island. On Ascension island, at least, these efforts failed. Failure was attributed to the climate, but inbreeding is likely an important, unacknowledged factor. Inbreeding prevented the establishment of honeybees up until as late as the 1960s on the Chatham Islands, and it also delayed their establishment on Kangaroo island off the coast of Australia. Most of the crops and fruits grown there, such as bananas, plums, guavas, pears, papayas, mangoes, and granadillas, have at least some varieties that set fruit without honeybees or any pollinators at all.

Hawaii, Fiji, Vanuatu, the Solomons, and presumably New Caledonia and Tahiti apparently have had European honeybees for up to 150 years. Honeybees are on at least some but not all (e.g. Vanuatu, Fiji) of 27 and possibly 5 other oceanic island groups. Hive density in the South Pacific is relatively low (> 35 km^2/hive to 2 hives/km^2) depending on whether there is commercial beekeeping (Fiji, Cooks, Niue, Papua New Guinea, Solomons) or virtually a cottage industry (Tonga, Tuvalu, Vanuatu, W. Samoa) with few hobby beekeepers. Commercial beekeeping was established on Niue in 1967), Tonga in 1971, Samoa in 1977, and Solomons in 1989.

Niue is the only island nation where human disturbance to the habitat is low enough, and beekeepers active enough, to export honey. Sugar cane, which provides no bee forage, is a major crop on cultivated land in Mauritius, Reunion and to a lesser extent Fiji. However, only on Reunion are imports significant.

Honeybees on many of the oceanic islands lack diseases such as European foul brood, chalk brood and the important mite enemies *Varroa jacobsoni* and *Acarapis woodi*. *Varroa* is spreading in Papua New Guinea. Protection against the spread of this mite and diseases ranges from an informal arrangement (Vanuatu) or the need to implement beekeeping and quarantine legislation for Fiji, Samoa and the Cook islands, similar to the 1991 legislation to control honey and bee imports in the Solomon Islands. Pollination on oceanic islands is vulnerable to fluctuations in the number of hived bees. There could well be eventual disruption of beekeeping and managed pollination if the worst mites and pathogens become established.

Maintaining and developing beekeeping on small islands faces a combination of difficulties not experienced in larger areas. Road access on mountainous oceanic islands is often limited. Honeybee populations cannot be controlled on smaller islands, where a commercial beekeeper is unable to make a living. Hobby beekeepers may need to import expensive wooden hives and mated queens or otherwise inbreeding will lead to colony failure. Fluctuations of up to 300% in hive numbers have been recorded in Bermuda, West Samoa and the Tonga Islands. These have a maximum of 60,000 inhabitants and at best one commercial beekeeping enterprise. Thus, islands with less than one or two thousand inhabitants can quite readily lose the skills or interest in keeping bees. In addition, the insecurity of apiary sites, hives and the honey crop can cause local problems. Islands with extended dry seasons and restricted floras e.g. Ascencon surely possess a severely

Island	Area (km2)	Cli-mate	Date of Intro.	Xylo-copa spp.	Megachi-lidae	Other bee spp.	Birds	Bats
Pitcairn	49	humid	0	0	0	0	0	0
French Polynesia	4000	wet to humid	+	1	5	0	2*	0
Hawaii	16765	wet to arid	1857	1	6	63	6-11*	0
Easter	119	humid	?+	0	0	0	0	0
Wake	11	humid	+	0	0	0	0	0
Midway	5	humid	0	1	1	0	0	0
Cook Isl.	236	humid	+	0	1	0	1	0
Marshall Isl.	181	wet to humid	1979	?0	2	0	0	0
Mariana Isl.	702	wet to humid	+	1	0	1	1-8*	0
Guam	541	wet	+	1	4	4	1-8*	1*
Kiribati	712	wet to dry	0	?0	1	0	0	0
Carolina Isl.	2150	wet	+	0	6	6 ++	4*	6*
Tuvalu	26	wet	1983	0	1	0	0	0
Nauru	21	humid	?0	0	-	-	0	0
Wallace & Fatuna	255	humid	?+	0	0	0	0	0
New Caledonia	19079	humid	+	0	3	25	8-12*	3*
Fiji	18333	wet to humid	+	0	2	5	4-7*	3*
Amer. Samoa	189	wet	+	0	4	6	4-5*	2*
Niue	263	humid	+	0	2	0	0	0
West Samoa	2842	wet	+	0	4	6	4-5*	2*
Tokelau	12	wet	0	0	1	0	0	0
Galapagos	1045	arid to dry	0	1	0	0	0-2*	0
Tonga	699	wet to humid	+	0	3	1	2	0
Bonins	72	wet	?+	1	-	-	1-3*	0
Vanuatu	12189	wet	+	0	3	10	4-5*	2*
Solomons	29800	wet	+	0	12	18++	9-11*	2*
Ryukyu	4790	wet	?+	3	14	28	0-3*	2*
PACIFIC TOTAL	115086							

Table 2. Availability of bee, bird and bat pollinators in tropical Pacific oceanic islands. Climate and annual precipitation: *Wet:* > 2000 mm, no prolonged dry season; *Humid:* 300-999 mm; *Arid:* 130-260 mm; *Dry:* 100-199 mm. Pollinators — date when introduced given for *Apis;* other information in table: + = present, 0 = absent, ? = insufficient records, number shows number of species. *Bats and birds include only flower-visiting species; the upper number includes those using nectar only periodically. ++ =*Trigona iridipennis or T. fusco-balteata* present

limited variety of productive floral resources during the dry season. Such factors presumably make Ascension, Trinidade, Cocos, Christmas, Pitcairn, Easter, Tokelau unsuitable or at least vulnerable for long-term viability with beekeeping. Beekeeping on small atolls with few people is often impractical, but with determination it can be achieved, as shown on Kwajalein in the Marshalls, where the need for a general pollinator was the main reason for honeybee introduction.

The tropical American carpenter bee *Xylocopa sonorina* has spread to Hawaii, Guam and Saipan in the Marianas. This speices, with its brown males and black females, was recovered in a tree stump in Auckland in 1987, but has apparently not established a population in New Zealand. *Xylocopa darwinii* is present on all but the four northwest Galápagos islands as an endemic species. In 1983 the author saw a *Xylocopa* speices on Tahiti (which could have been *X. sonorina*) visiting flowers including *Hibiscus*. Thus other large islands with frequent Pacific trade or military traffic from Hawaii may now have carpenter bees. Surveys of insects from New Caledonia, Cook, Tonga, and Niue in the 1980s have confirmed the absence of carpenter bees there (pers. obs. at the Bishop Museum and New Zealand arthropod collections). Israeli, Malaysian and Indian studies (see Section 2.5) show that carpenter bees are entering the early phase of domestication. In the Atlantic other large bees occur: the long-tongued *Bombus ruderatus* (Azores, Madiera) as well as the relatively short-tongued *Bombus madieriensis* (Madiera), and *B. canamensis* (Canary Islands).

Most Pacific islands have a few flower-visiting wasp and bee species. One to six species per island of leaf cutter bees (*Megachile* and *Lithurge*) are recorded from the main high islands as far east as the Marianas and American Samoa. Then there are fewer megachilid bees until Tahiti. Nothing has been recorded on the flowers they use, but legume and composite flowers are among the favored food sources of other megachilid bees. In the Pacific there are very few species of other solitary bees. Those recorded are ground-nesting genera of Halictidae. New Caledonia also has a *Nomia* species (Halictidac) that nests in the ground. Virtually nothing is reported on the flower species these bees visit. In New Caledonia, some of the 19 species of sphecid wasps make up a conspicuous part of the generalized pollinator guild (Donovan, pers. comm.), and one *Polistes* species is widely dispersed in the Pacific, extending to Easter Island.

No information has been presented on the effectiveness of flies, wasps, beetles or moths in pollination on oceanic islands. In the Pacific there are at least 55 species of Syrphidae with 6 species introduced to Hawaii. The 'drone fly' is present on Easter Island, but apparently no syrphids extend yet to the Pitcairn and Galápagos islands. The largest island groups have 11 syrphid species in Hawaii, 2 in French Polynesia, 5 in Samoa, 10 in Fiji, 11 in New Caledonia and 6 in Vanuatu. The oceanic islands in the Pacific have an impoverished bombyliid (bee fly) fauna, partly because of the limited host species of ground-nesting bees and wasps needed as hosts by these parasites, and also the wet climate. In the Pacific, bee flies have only been recorded from Vanuatu (4 species), New Caledonia (4 species), French Polynesia (2 species) and the Bonin islands (3 species), with one species introduced to Hawaii. Presumably this also applies to the Atlantic and Indian oceans.

In Australasia, some stratiomyid and tabanid flies are among the commoner, largest and hairiest insects on umbellifer and composite flowers. Stratiomyids (*Odontomyia*) are found in Vanuatu, New Caledonia and Fiji, while *Philolche* on New Caledonia is in the same tribe (Philolchini) as the flower-visiting tabanids of New Zealand. Even the smallest and most remote oceanic islands of Bonin, Micronesia, Pitcairn and Easter Island have several species of Muscidae, Calliphoridae and Sarcophagidae. Studies in temperate zones show flies can transfer limited amounts of pollen on their bristly bodies as they move quickly from flower to flower. Other abundant, large and hairy diurnal flower-visiting flies, e.g. Bibionidae, at least in the temperate zone, may often walk from flower to flower. Observations by the author on Chatham island showed that walking reduced the average rate of flower visitation three to six times, compared to flies flying between flowers. Bibionids are apparently only found in Vanuatu and New Caledonia. Most of the other nematoceran fly families have some flower-visiting species (see Introduction), and are active more at dusk or at night.

In the Australasian region buprestid, languriid, oedomerid, scraptiid and trixiagid beetles exclusively visit flowers, while cantharid, melyrid, clerid, phalacrid, rhipiphorid and dermestid beetles commonly visit flowers adapted to generalist insect pollinators. However, many of these beetle species are relatively small. Small size would restrict their ability to pollinate flowers where the stamen and stigma are well separated. Generally, Scarabaeidae (Cetoninae) Trichiinae or Melolonthinae) and Elateridae form a conspicuos but less mobile part of the insects visiting flowers of at least the composites, umbellifers and some Myrtaceae. Oceanic islands tend to have limited species or subfamily representation of soil-inhabiting scarabaeids, compared to the adjacent continents. Conversely, flower-visiting insects that inhabit twigs or wood, e.g. longhorn beetles and the small almost hairless bees *Hylaeus* have dispersed farther or more freely in the Pacific, being able to float over large expanses of ocean in their nests. Some sphingid moths are active daytime visitors of flowers and a total of 15 species has been found in New Caledonia, 5 in American Samoa and 6 in Hawaii. Only *Hippottion celerio, H. velox, H. scrofa, Theretra nessus, T. clotho*, and *Gnathothibus eras* extend into the southwest Pacific. The butterfly fauna of Fiji, Micronesia and Hawaii consists of 10 to 19 species. The Lepidoptera are not known to have any effect on crop pollination but they are significant as pollinators of *Lantana,* which is a major weed in the Pacific islands, and they also visit wild ginger.

In many parts of the tropics small nectar- and pollen-feeding bats pollinate banana mango, hemp, kapoc, balsa, durian and are most important in pollination of Bombacaceae, Bignoniaceae, and Sapotaceae. Some species of Cactaceae, Gesneriaceae, Sonneratiaceae. *Mucina* and *Musa* depend on bats for pollination. Some bats are generalist pollinators and fruit-eating species that consume nectar primarily in the dry season. In all, 24 species of *Pteropus, Pteralopex macrodonta* and *Notopteris macdonaldi* are found on many islands in the Indian ocean and as far east as the Caroline islands and Samoa in the Pacific (Tables 1 and 2). Most of these fruit bats or flying foxes are basically fruit eating species that visit flowers opportunistically and may also consume them. The Indo-Asian to western Pacific bats are not well-adapted to hovering flight for pollination, compared to bats from Southern USA to South America. In the Solomon Islands, the flowers of *Heliconia* (Musaceae) are visited at night by *Melanycteris*. Perhaps other *Heliconia* as far east as Samoa may be bat-pollinated.
Thus on western Pacific and Indian ocean islands, bats are likely to contribute to the pollination of only a slight portion of the flora, which also applies to New Zealand and Australia.

The most persistent and specialized flower-feeding birds that occur on tropical oceanic islands are Meliphagidae, some Drepanidae, Nectariniidae, Dicaeidae and brushtongued honey parrots or lorikeets (Trichoglossidae, Loriinae). More opportunistic flower visitors include Zosteropidae, parakeets and perhaps parrots (Psittidae), tropical bulbuls (Pycnonotidae), orioles (Oriolidae) and two Galápagos finches. There are one to five species of honeycreepers that forage on flowers on each island group from Vanuatu, the Marianas, Carolines, and to Samoa.

The role of birds in pollination on tropical islands has been studied on Hawaii. In Hawaiian forests five of the surviving species of Hawaiian honeycreepers include nectar as an important food item, and for five more species nectar is used occasionally. The honeycreepers are important for pollination in lobelioid flowers. On *Meterosideros collina* (Myrtaceae) three honeycreeper species and the white eye carry pollen much more often

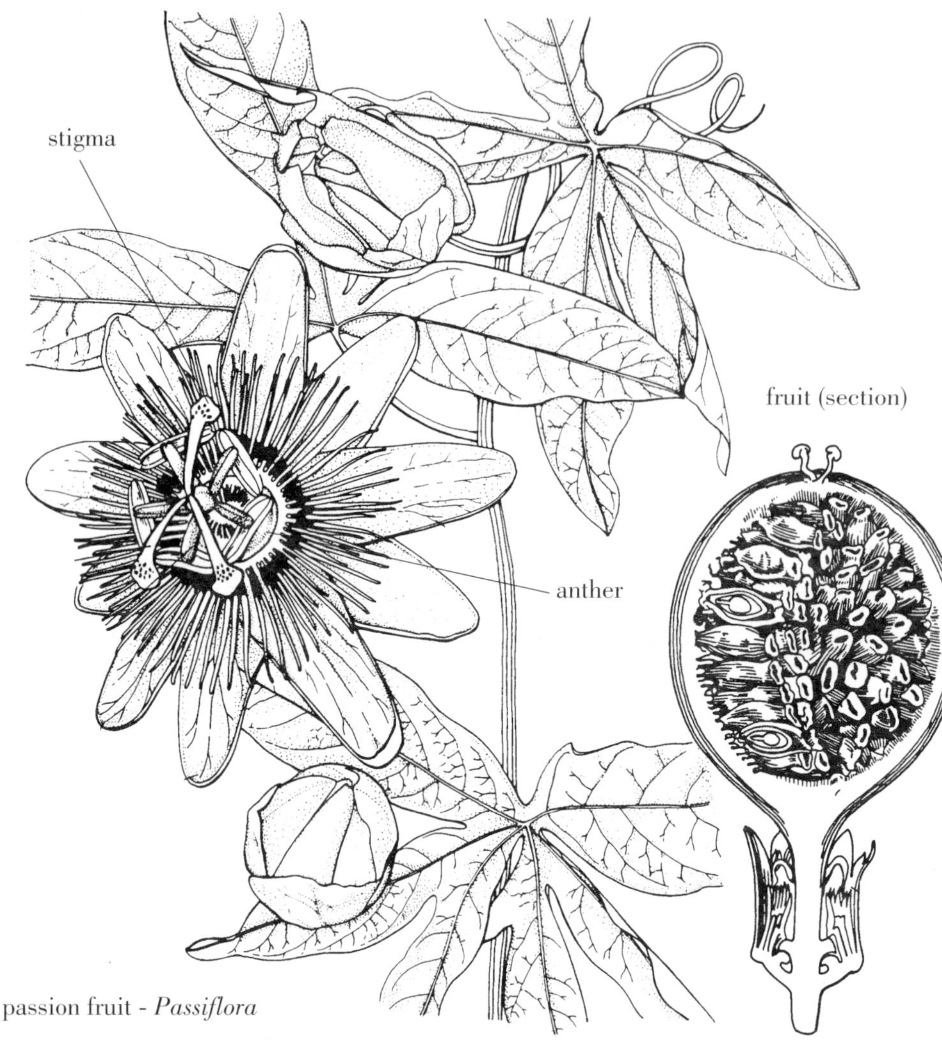

passion fruit - *Passiflora*

producers after Indonesia, India and the Philippines. Lack of soil on coral atolls makes coconut virtually the only land-based export commodity. Coconut palm groves provide a major crop in Fiji, Samoa, Vanuatu, French Polynesia, Wallace and Fatuna, Guam, Andaman and Nicobar islands, Comoros, São Tome and Principe. Coconuts are declining in importance on New Caledonia, Tonga, and the Cook and Seychelles islands.

Coconuts do not require pollinators; wind is equally important (Kevan, Section 1.2). In Java, the white unbagged flowers produced four to seven times more nuts than bagged flowers. And in India, yields at best were doubled with the introduction of *Apis cerana*. It is not uncommon for 50-70% of the coconut buttons to shed, leaving no more than 25-40% mature female flowers. Coconut racemes produce the male flowers first and then female flowers, therefore effective pollinators need to move between racemes frequently. In Hawaii, insects flying between the flowers included honeybees, wasps, flies and earwigs. In west Java, honeybees carried more pollen grains than the other common beetle and fly visitors. Honeybees forage freely on coconut flowers, which are an important source of pollen and nectar in the tropics. Pollen is mainly presented in the morning and most honeybees forage nectar from female flowers in the afternoon. Hence they will not often carry pollen from male flowers (at the end of the spikelets) to female flowers (at the base) — a factor that has not been quantified in the literature. In Costa Rica and Ecuador *Trigona* was the most common pollen collector, which implied they might be effective pollinators. However, *Trigona* inter-ferred with the activity of honeybees by attacking them at the flowers.

on their bodies than other birds visiting the flowers of this dominant forest tree. Two sunbird species are found on the Andaman islands and one flowerpecker species and one sunbird species on the Comoros. There are 17 other species of white eyes with up to three species on each island from São Tome and Principe, Comoros, Seychelles, Mauritius, Reunion, Andaman, Nicobar, the Carolines, Guam, Vanuatu, New Caledonia and Fiji.

Six species of parakeets live on Mauritius, Nicobar or New Caledonia. A different subspecies of the red- fronted parakeet feeds among flowers in the New Zealand region, but general bird texts do not mention whether these other parakeets feed on nectar too. New Zealand and Australian studies of pollinating birds confirm that red flowers are often used, but some yellow and white flowers with large amounts of nectar are also sought by birds. In New Zealand at least 2% of the native flower species from over 18 flower families benefit from bird pollination. Eventually, a similar pattern may be recognized when pollination by birds on the oceanic islands in the tropics has been adequately studied.

1.4.2 Crops, Pollination & Productivity

1) Coconut, *Cocos nucifera*.

The combined islands in the Pacific are the major coconut

2) Cocoa, *Theobroma cacao*.

In São Tome and Principe, cocoa plantations occupy 48,000 hectares, and at the turn of the century these islands were the major cocoa producers in the world. The yields of around 100 kg/hectare are only 10-20% of those achieved by some African nations. Cocoa originated in South America and was distributed to Pacific oceanic islands largely from Africa. Production and yields on the oceanic islands have stagnated or declined. Current yields in Fiji and Samoa are a modest 200-300 kg/hectare. Cocoa is quite important on the Comoros and is a minor crop on the Seychelles and Vanuatu. The flowers mainly produce a sticky pollen. Elsewhere in the tropics the main pollen depositing visitors are *Forcipomyia* midges and gall midges. Seed formation per flower is often low but in Central America it varied as much as 1400% between locations. In Malaysia, less than 1% of the flowers formed pods due to the combined effects of poor fruit set and disease. Elsewhere in the tropics the midges breed in rotting vegetation.

None of these six island countries have a problem of *Trigona* bees robbing pollen from flowers; these bees are rare or absent (Table 2). Fortunately, several species of *Forcipomyia* midge exist on many of islands in the Pacific. But which flies are effective pollinators on these islands, and where they breed, remain to be investigated.

3) Yellow passion fruit, *Passiflora edulis.*

This fruit remains a major export item for Niue, while Fiji and Hawaii are among the six main producing countries. Cultivation of this Neotropical vine extends to the more isolated mountainous islands in Micronesia, Tahiti and Pitcairn.

Yellow passion fruit has a large, primarily white flower so that only larger insects regularly contact the stigma when they seek nectar. In Hawaii, the commonest, largest and hairiest flower visitors were the carpenter bee, the honeybee, the drone fly and a green blowfly. In the Caribbean most honeybees used the flowers only for nectar. Nectar collecting *A. mellifera* are too small to contact the stigma on a regular basis, although they touch stigmas as they search the flower for rewards (L. Escobar, pers. comm. to D. Roubik). A Malaysian study showed that *A. dorsata* and *A. cerana* usually depleted nectar before the effective carpenter bees came to pollinate the flowers, which discouraged visits by the carpenter bees. Fly visitors also are unlikely to be effective in pollination if they visit the flower for nectar. In Fiji, where there were no carpenter bees, yields of passion fruit were increased 300% with hand pollination.

Hand pollination is largely responsible for passion fruit requiring up to three times as much labor as other crops. This reduces the net margin to low levels for growers. Apparently no effort has been made to introduce carpenter bees to Fiji and production there has declined. On Niue, a scientist that investigated passion fruit pollination recommended that efforts be made to find a sport (variety) of passion fruit with a smaller flower. This would allow the honeybee to be an effective pollinator. When subsequent initial investigations on the suitability of introducing the Hawaiian carpenter bee were made by another scientist the impetus to resolve this inadequacy in production had already been lost. This case reveals the ineffectiveness of trying to make the flower fit the pollinator, rather than providing an effective pollinator for the crop. Eventually the recommendation for carpenter bee introduction was formally recorded, in 1990. In New Zealand the author has seen that queens of *Bombus terrestris* and *B. ruderatus* have potential as consistent pollinators for *Passiflora*. They contact the stigma while visiting the flower for nectar. Thus, an alternative is to experimentally introduce these bee species to Niue.

4) Macadamia, *Macadamia integrifolia.*

This crop originally came from Australia. The small trees have pendulous racemes with small flowers. Hawaii is a major producer of *Macadamia* nuts, and *Macadamia* is the third most extensive Hawaiian crop. In the 1960s, Fiji attempted to diversify into *Macadamia* production but failed due to a variety of problems.

Research in Hawaii and New Zealand has shown that some varieties respond to cross-pollination. During two years nut-set was increased 70, 10 and 0% in three varieties, when one honeybee hive per 60 trees was supplied in Hawaii. Queensland studies in the 1980s showed that *Trigona* were more effective pollinators than honeybees, because they consistently collected pollen from the flowers. Pollen-collecting honeybees were, however, effective pollinators. In New South Wales, beetles were the only other flower visitors to carry significant amounts of pollen among *Macadamia* flowers.

5) Watermelon and other cucurbits.

Watermelons are an African plant widely cultivated in Pacific and Indian oceanic islands as the major cucurbit crop. Only on Tonga and French Polynesia is production as high as 1-3,000 tons. The yields achieved in Tonga, Guam, and French Polynesia are half that of Asian countries having intensive production.

Honeybees are shipped between islands in Tonga for watermelon and squash pollination. The introduction of honeybees to Tuvalu was made so that cucumbers, melons and other crops did not require pollination by hand. A 1991 review of pollination in Europe concluded watermelon was moderately dependent on honeybees and solitary bees — other cucurbits being more dependent (see also Section 1.5.3).

Mauritius, with a population of 1.1 million, is the only minor producer of pumpkins, squashes and gourds among the oceanic island nations. Mauritian yields are about 33% of those achieved in Chile and other large producers, where the native specialist squash and gourd bees, *Peponapis* and *Xenoglossa,* are found. Cucumber yields in Guam and Mauritius are average for Africa and Oceania, respectively. Worker honeybees do not transfer pollen very well in the larger cucurbit flowers, because they do not necessarily contact the stigma when they collect nectar from the female flower. In pumpkin patches, the author has seen that individual honeybee workers often go between female flowers and may even discard the sticky pollen from male flowers. In the New Zealand region, larger bumblebees contact the stigma more often. On islands without bees, cucurbits (squashes, gourds, cucumbers) need to receive pollination by hand. Thus it is likely that the frequency of fruit set, or even cost of production and fruit size may be improved on oceanic islands if more effective pollinators are made available.

6) Additional crops.

Commercial vanilla (*Vanilla planifolia*) is an orchid vine that came from southern Mexico. It now accounts for over 60% of the export income in the Comoros, and French Polynesia has resumed limited exports. Vanilla has become naturalized in Niue (*Vanilla tahitensis* is native to the South Pacific). Vanilla usually needs to be pollinated by hand where it is grown away from its native euglossine bee pollinators (*Eulaema polychroma, E. nigrita, E. cingulata* and *E. speciosa*, see also Section 1.5). Cloves (*Syzygium aromaticum*) are the second most important export crop of the Comoros. Few of the bisexual flowers form fruit, thus flowers need cross-pollination. Honeybees visit the flowers, where even nectar-collecting workers should be large enough to frequently contact the stigma on these shrubs.

Samoa and the Cook islands produce 1-2,000 tons of avocados (*Persea americana*), but they are grown elsewhere too, on the Canary Islands, in Micronesia and on Easter island. In Israel the size and shape of the flower does not suit foraging honeybees, which prefer pollen of other flowers. The pollen-collecting honeybees are effective pollinators, unlike nectar collectors. The lack of records concerning pollen use for these islands makes it impossible to evaluate whether citrus, coconut or pasture legumes could have an adverse effect on avocado production by competing for their pollinators.

Reunion has become an important producer of the Malaysian tree Ylang-ylang (*Cananga odorata*) and it is the third most important export crop for the Comoros. Honeybees collect nectar from the bisexual flowers, but otherwise nothing seems to have been recorded on its pollination. Soursop, custard apple and sweetsop, *Annona* species, are minor fruits on Niue and other tropical islands. In Brazil, dynastine scarab beetles were important pollinators of large-flowered *Annona* species, while in Israel, nitidulid beetles were the most important pollinators.

Cotton (*Gossypium*) was once the major crop on Mauritius. Thirty-six endemic cotton species are found throughout the world, including one on the Galápagos and Hawaii. Cotton, sunflowers (*Helianthus*) and litchi (*Syzygium litchi*, often given as *Nephelium litchi*) are among some of the crops attractive to bees and not grown in any quantity on oceanic islands. Freedom of these crops from pests on oceanic islands may be of some advantage. São Tome and Principe produce palm oil, but Malaysian experience (see Section 2.5) has shown that improved production can be achieved with introduction of a pollinating weevil. Other oceanic islands may diversify in agriculture by including this palm, if they also introduce the weevil.

1.4.3 Selecting Potential Pollinators for Introduction

Production of commercial crops and subsistence cropping on the oceanic islands has proceeded with little or no thought of the need for the original pollinators (Section 2.2). Introduction of beekeeping to many of the oceanic islands and the action of local generalist insect pollinators has allowed at least moderate production to be achieved for many crops. Nevertheless, most island communities rely heavily on a few crops for export earnings. The lack of other generalist pollinators such as the stingless bees, carpenter bees and bumblebees has hindered production by yellow passion fruit and may be a contributing factor to the indifferent performance of some other crops.

Cross-pollination and pollinator reliability are influenced by the speed of movement between flowers, movement between plants and preferences for surrounding flowers, and yet these factors have not been assessed for local pollinators on oceanic islands. Neither has the role of competitive bloom on honeybee foraging been assessed, except for macadamia in Hawaii.

The introduction of selected bee species and management of others could allow for diversification or improved yields for islands, especially in the southern and eastern Pacific. Seven critical criteria required for candidate species for importation include:

1) a similar habitat and climate in the country making the introduction
2) the ability to pollinate the target flowers
3) compatibility with humans, honeybees and other domestic animals
4) freedom from enemies and diseases.

Species for introduction should not:

5) disrupt other arthropods
6) disrupt pollination of other crops or native species
7) contribute unduly to pollination and spread of major weeds, such as the composites, Noogoora burr (Xanthium stumarium), Yasawa tobacco weed (Pseudelephantopus spicatus), guavas (Psidium) and Lantana.

These weeds are unlikely to show enhanced seed production if new pollinators are introduced, because they can either be pollinated by generalist pollinators or butterflies. However, the introduction of large generalist pollinating carpenter bees or bumblebees could affect pollination, particularly of weedy Passifloraceae and Solanaceae. The creeper poke or banana passion fruit, *Passiflora mollisima* is a significant weed in native habitats in Hawaii, but the author has observed that its flowers are less attractive to bumblebees and honeybees compared to other *Passiflora* species. Carpenter bees and to some extent honeybees are likely to effect most of the cross-pollination of banana passion fruit. Carpenter or bumblebees can be safely introduced to Niue, because this weedy *Passiflora* does not occur there, unlike Fiji. On Fiji, the possibility of eradicating this weed should be assessed before trying to introduce the carpenter bee for pollination.

In Australia, *Amegilla* and *Nomia* have been the most effective cross-pollinators of *Solanum* species, while *Trigona* is more significant for within-plant crossing. Flowers of *Solanum* offer only pollen, which bees may collect surprisingly quickly, so that few are seen visiting the flowers. Canadian researchers found *Solanum* was a primary source of pollen in the nests of bumblebees. Thus if bumblebees or *Trigona* were considered for crop pollination on a Pacific island, their possible impact on pollination of *Solanum* weeds should be assessed before making such introductions.

The safest new pollinators to consider for introduction to island communities would be the squash bees *Peponapis* and *Xenoglossa* because they concentrate on the flowers of Curcubitaceae and are not aggressive, stinging bees. So far, the single attempt to introduce squash bees to Hawaii has apparently failed, probably because too few could be released. Releasing weevils for oil-palm pollination seems likely to be a safe introduction. Carpenter bees from Hawaii may be introduced to other islands without risk to timber in buildings or posts, and there are no other apparent risks, apart from pollination of some weeds.

Parasite- and relatively pathogen-free bumblebees, with mild temperaments, might be imported from New Zealand or elsewhere as generalized, unmanaged pollinators. They would pollinate pasture legumes, fruit crops and curcubits in highland regions of mountainous tropical islands or on those such as Ascension or Easter Island where upper ground temperatures seldom exceed 30°C and air temperatures remain in the upper 30s. The extent of possible infections of queen *Bombus* from tropical South America by internal parasites has yet to be determined. Costs of quarantine and removal of parasites and diseases could preclude importations from being made at present, even though rearing enough of these 'pocket-making' species is now technically possible. These *Fervidobombus* form small colonies in protected sites and defend them relatively aggressively, compared with other bumblebees.

The introduction of stingless bees for macadamia and avocado may be supported because of their relative efficiency in pollination compared to honeybees, and their ability to occupy different nest sites as unmanaged pollinators. *Trigona* in Australia are frequent visitors to flowers of papaya, passion fruit, mango, grape and grandilla. South American stingless bee species of the *Trigona* s. str. do have the potential for robbing flowers of nectar and pollen, while aggressively defending them from legitimate pollinators, and such interactions should be assessed before introductions are attempted. Increasing the stocks of honeybee hives could have some adverse effects on seed formation in a few legume weeds, especially gorse in the uplands of Hawaii. Honeybees and, to a lesser extent, *Bombus terrestris* are the main pollinators of gorse in England and from the author's observations, in New Zealand. On Chatham island extensive areas of gorse formed very little seed, unlike patches nearest to honeybee hives. On the Chathams there are few roads and a low density of honeybee hives or feral colonies, and no other bees apart from *Hylaeus*.

Additional References

Ahmad, R. 1989. A note on migration of *Apis dorsata* in the Andaman and Nicobar Islands. Bee World 70:62-65.

Anderson, G. J., Symon, D. 1988. Insect foragers on *Solanum* flowers in Australia. Ann. Missouri Bot. Gard. 75:842-852.

Anonymous. 1982. Does coconut palm benefit from bee pollination? Indian Bee J. 44(1):24.

Anonymous, 1990. Report to Australian national parks and wildlife service: CSIRO entomological survey of Christmas island.

Armstrong, J. A. 1979. Biotic pollination mechanisms in the Australian flora: A review. N. Z. J. Bot. 17:467-508.

Barrows, E. M. 1976. Nectar robbing and pollination of *Lantana camara* (Ver-benaceae). Biotropica 8:132-135.

Barrows, E. M. 1980. Robbing of exotic plants by introduced carpenter and honeybees in Hawaii, with comparative notes. Biotropica 12(1):23-29.

Bohart, R. M., Menke, A. S. 1976. Sphecid wasps of the world. a generic revision. Berkeley: Univ. California Press.

Bramwell, D. 1979. Plants and islands. Academic Press. London, 459 pp.

Callan, McC. E. 1990. Sphecidae of New Caledonia. Sphecos 19:22

Calquist, S. J. 1974. Island biology. New York: Columbia Univ. Press, 660 pp.

Cardale, J. C. (in press). A catalogue of Australian Hymenoptera: Apoidea. CSIRO series.

Carlquist, S. J. 1980. Hawaii, a natural history, geology, climate, native flora and fauna above the shoreline. (2nd ed.) Honolulu: Pacific Trop. Bot. Gard., 468 pp.

Carpenter, F. L. 1976. Plant -pollinator interactions in Hawaii. Ecology 57:1125-1144.

Chandler, P. J. 1979. Flies, bees and butterflies on La Palma, Canary Islands in 1976. Ent. Rec. J. Variation 91:103-107, 117-121

Chandra, S. 1976. A review of recent research on the yellow passionfruit. Fiji Agric. J. 38:41-48.

Chandra, S. 1977. Productive effeciency of Fijian and Indian farming systems in semisubsistent agriculture Part 2. Description of the farming systems and partial productivity results. Fiji Agric. J. 39:57-70.

Chandra, S. 1979. Crop production statistics for Sigota valley, Fiji. Fiji Agic. J. 41. 21-30.

Common, I. F. B. 1990. Moths of Australia. Melbourne, Australia: Melbourne Univ. Press, 535 pp.

Comstock, J. A. 1966. Lepidoptera of American Samoa with particular reference to biology and ecology. Pacific Insects Monogr. No. 11, 74 pp.

Corbet, G. B., Hill, J. E. 1980. A world list of mammalian species. Brit. Mus. Nat. Hist. Comstock Publ., 226 pp.

Corbet ,S. A., Osborne, J. L., Williams, I. H. 1991. Bees and pollination in the European community. Bee World 72:47-59, 99-116, 170-180.

Corbet, S. A., Willmer, P. G. 1980. Pollination of the yellow passion fruit: Nectar, pollen, and carpenter bees. J. agric. sci. Camb. 95:655-666.

Crane, E. 1990. Bees and beekeeping. Science, Practice and world resources. London: Heinemann Newnes, 614 pp.

Cross, T. 1980. St. Helena, including Ascencon Island and Tristan de Cunha. David and charles, London 192 pp.

Davenport, T. L. 1986. Avocado flowering. Hort. Rev. 8:257-289.

Donovan, B. J. 1983. Comparative biogeography of native Apoidea of New Zealand and New Caledonia. Geojournal 7:511-516.

Donovan, B. J. 1988. Occurrence of *Xylocopa sonorina* Smith in New Zealand. Melissa 3:21.

Donovan, B. J. 1990. Selection and importation of new pollinators to New Zealand. N. Z. Ent. 13:26-32.

Douglas, N. 1989. Pacific Yearbook. (16 ed.) Sydney: Pacific publications, 557 pp.

Duffy, E. 1964. The terrestrial ecology of Ascension island. J. Appl. Ecol. 1:219-251.

Ekert, J. E., Bess, H. A. 1952. Fundamentals of beekeeping in Hawaii. Ext. bull. Univ. Hawaii. 55, 59 pp.

Erlandsson, S. 1979. *Bombus canariensis* Perez. 1985 n. stat and *B. maderensis* n.sp. from the Macronesian islands. Ent. Scand. 10. 187-192.

Esaki, T. 1930. Uebersicht uber die Insectfauna der Bonin (Ogasawara) = Inseln, unter besonderer berucksichti-gung der zoogeographiscen fauner charactere. Biogeog. Soc. Japan Bull. 1:206-226.

Evenhuis, N. (ed.). 1989. Catalogue of the Diptera of the Australasian and Oceanic regions. Honolulu: Bishop Museum Press, 1155 pp.

FAO Yearbook 1990. Production. Vol 44. 123 pp. , Trade Vol 44, 379 pp.

Forde, H. A., Paton, D. C., Forde, N. 1979. Birds as pollinators of Australian plants. N. Z. J. Bot. 17:509-519.

Galon, I., Gazit, S., Podoler, H. 1982. Improvement of natural fruit set of annona by increasing the population of pollinating insects. (Hebrew) Alon Hanotea 36 (9):611-614.

Gardner, D. E., Smith, C. W., Markin, G. P. (in press). Biological control of alien plants in natural areas of Hawaii. Proc. Int. Symp. Biol. Control of weeds

Gazit, S., Galon, I., Podoler, H. 1982. The role of nitidulid beetles in natural pollination of annona in Israel. J. Amer. Soc. Hort. Sci. 107:849-852.

Gerling, D. 1983. Nesting biology and flower relatinships of *Xylocopa sonorina* Smith in Hawaii (Hymenoptera. Antho-phoridae). Pan-Pacific Ent. 58:336-351.

Godley, E. J. 1979. Flower biology in New Zealand. N. Z. J. Bot. 17. 441-466.

Goebel, R. L. 1988. Australian native bees. Aust. Bee J. 69. 5-7.

Gottsberger, G. 1989. Comments on flower evolution and beetle pollination in the genera *Annona* and *Rollinia* (Annona-ceae). Pl. Syst. Evol. 176:189-194.

Gressit, J. L. 1954. Insects of Micronesia; Introduction. Bishop Mus. Publ. No. 1, 257 pp.

Hampton, R. E., Thompson, P. G. 1974. Passion fruit production in Fiji. Agric. J. 36:23-27.

Hargrove, R. G. 1975. How many bees in Bermuda? Glean. Bee Cult. 103:7-8, 27.

Hart-Davis, D. 1972. Ascenscion. the story of a south Atlantic Island. London: Constable, 244 pp.

Hawke, J. G., Lester, R. N., Skelding, A. D. 1979. The biology and the taxonomy of the Solanaceae. London: Academic Press, 738 pp.

Heard, T. A. 1987. Preliminary studies on the role of *Trigona* bees in the pollination of Macadamia Proc. Aust Macadamia workshop 2:192-197.

Heard, T. A. 1988. The requirement for insect pollination by macadamia and the pollination effeciency of *Trigona* bees. Proc. Austral. Conf. tree and nut crops 4:219-223.

Heard, T. A. 1988. Establishment and propagation of hives of native stingless bees. Aust. Beekeeper 90:224-225.

Hedstrom, I. 1986. Pollen carriers of *Cocos nucifera* L. (Palmae) in Costa Rica and Ecuador. Rev. Biol. Trop. 34:297-301.

Hilje, L. 1987. Insect visitors and reproductive effeciency of *Lantana camara* (Verbenaceae). Brenesia 23:293-300.

Hirashima, Y. 1958. Bees of the Amami Islands. I. (Hymenoptera, Apoidea). Mushi 32:69-76.

Hobson, J. V. 1982. Concrete hives have a role to play in tropical islands. N. Z. Beekeeper. 43:11-12.

Holloway, J. D. 1979. Survey of the Lepidoptera, biogeography and ecology of New Caledonia. The Hague: junk, 588 pp.

Howard, R., Moore, A. 1980. A complete checklist of the birds of the world. Oxford: Oxford Univ. Press, 701 pp.

Hussien, M. Y., Lajis, N. H., Ali, J. H. 1991. Biological and chemical factors associated with the successful introduction of *Elaeidobius kamerunicus* Faust, the oil palm pollinator in Malaysia. Acta Horticulturae 288: 81-87.

Ibrahim, A. G., Hussein, A. M. 1987. Role of insects in the pollination of cocoa flowers. Pertanica 10:103-106.

Ikudome, S., Yamane, S. 1990. The distribution of megachilid bees in the Ryukyu Is, Japan (Hymenoptera, Apoidea). Bull. Inst. Minami-Kyushu Reg. Sci. 6. 73-93.

Iqbal, M., Hampton, R. E. 1977. Macadamia nut production in Fiji. Fiji Agric. J. 39. 97-103

Ish-an, G., Eisikowitch, D. 1990. The behaviour of honeybees during visits to the avocado flower and their contribution to its pollination. Alon Hanoten 40:591-606.

Jay, S. C . 1974. Nectar and pollen collection by honeybees from coconut flowers. Bee Wld. 55:105-111.

Kaczmarak, V. 1988. Reunion report of the development of beekeeping (French). Rev. Fr. Apic. 46:348-350, 47:387-389.

Kevan, P. G. 1984. Apicultural developemnt prospects for São Tome and Principe. unpub. Rep. Agsi, FAO., Rome, 15 pp.

Kevan, P. G. 1986. Pollinating and flower visiting insects and the management of beneficial and harmful insects and plants. Proc. Biocontrol in the tropics, Penerbit University, Serdang 1. 439-452.

Kevan, P. G. 1990. How large bees, *Bombus* and *Xylocopa* (Apoidea, Hymenoptera) forage on trees: optimality and patterns of movement in temperate and tropical climates. Ethol. Ecol. Evol. 2. 233-242. [Malaysia, Indonesia, Maldives]

Kress, W. J. 1985. Bat pollination of an Old World *Heliconia*. Biotropica 17:302-308.

Krombein, K. V. 1950. The aculeate Hymenoptera of Micronesia II. Colletidae, Halictidae, Megachilidae and Apidae. Proc. Hawaii Ent. Soc. 14:101-142.

Krombein, K. V. 1951. Additional notes on the bees of the Solomon Islands (Hymenoptera. Apoidea). Proc. Hawaii Ent. Soc. 14:277-295.

Kunz, T. H. (ed.). 1982. Ecology of bats. Plenum, New York, 425 pp.

Lands Survey Dept., New Zealand. 1978. Atlas of the South Pacific, 47 pp.

Linsley E G, Rick, C. M., Stephens, S. G. 1966. Observations on the floral relationships of the Galápagos carpenter bee. Pan-Pacific Ent. 42:1-18.

Macfarlane, R. P. 1985. A kiwi on tour to Chile, England and France. Weta 8:18-22. [*Xylocopa* on Tahiti]

Macfarlane, R. P., Grindell, J. M., Dugdale, J. S. 1992. Gorse on the Chatham Islands. seed formation, insect associates and control. N.Z. Plant Protection Conf. 45:251-255.

Mackerras, I. M. 1961. The zoogeography of western Pacific Tabanidae (Diptera). Pacific Ins. Monogr. 2, pp. 101-106.

Mardan, M., Yatim, I. M., Kahlid, M. R. 1991. Nesting biology and foraging activity of carpenter bee on passion fruit. Acta Hortculturae 288:127-130.

Marshall, A. P. 1981. Beekeeping in the Marshall Islands. Glean. Bee Cult. 109:430, 433-434.

Michener, C. D. 1965. A classification of the bees of the Australian and south Pacific regions. Bull. Amer. Mus. Nat. hist. 30, 362 pp.

Muthiah, S. 1986. A social and economic atlas of India. New Delhi: Oxford Univ. Press, 254 pp.

Nashida, G. M. 1992. Hawaiian tropical arthropod checklist. Bishop Museum Tech. Report 1, 262 pp.

Nashida, T. 1958. Pollination of passion fruit in Hawaii. J. Econ. Ent. 51. 146-149.

Nashida, T. 1963. Ecology of the pollinators of passion fruit. Hawaii Agric. Exp. Sta. Tech. Bull. 55, 38 pp.

Owen, R. P. 1977. A checklist of the birds of Micronesia. Micronesia 13:65-81.

Pardede, D., Randriani, E., Marfuarti, A. 1986. The pollination of coconut at the Pakuwon sub research institute for coconut (Indonesian). Apic. Abstr. 42:27-28.

Rao, M. R. N. 1987. Prospects of beekeeping in Andaman and Nicobar Islands. Indian Farming 36:8-9, 12.

Reid, G. M. 1992. A review of Apiculture in the south Pacific. MAF, Apicultural services Report. 79 pp. [Cook, Fiji, Kiribati, Niue, Papua New Guinea, Solomons, Tonga, Tuvalu, Vanuatu, W. Samoa]

Schulter, D. 1986. Character displacement between distantly related taxa? Finches and bees in the Galápagos. Amer. Natur. 127:95-102.

Shigeura, G. T., Lee, J., Silva, J. A. 1970. The role of honeybees in Macadamia nut (*Macadamia integrifolia* Maiden and Betche) production in Hawaii. J. Amer. Soc. Hort. Sci. 95:544-546.

Sholdt, L. L. 1966. Insects associated with the flowers of the coconut palm, *Cocos nucifera* L. in Hawaii. Proc. Hawaii Entomol. Soc. 19:293-296.

Sholdt, L. L., Mitchell, W. A. 1967. The pollination of *Cocos nucifera* L. in Hawaii. Trop. Agr. (Trinidad) 44:133-142.

Silberrad, R. E. M. 1970. Beekeeping in Seychelles. Rep. Seychelles Dept. Agric. 21 pp. (Apic. Abstr. 817L/78)

Skotsberg, E. 1953. The natural history of Juan Fernadez and Easter Island. Vol. Geography, geology, origin of island life. Uppsala, 439 pp.

Stoddard, D. R. 1984. Biogeography and ecology of Seychelles Islands. Amsterdam: Junk, 691 pp.

Sykes, W. R .1970. Contributions to the flora of Niue. Bull N. Z. Dep. Sci. Ind. Res. 200, 320 pp.

Thampan, P. K. 1975. The coconut and its products. Green villa, Kerala, 302 pp.

The Africa Review 1992. Economic and Business report for 1991/92 (15 ed.) Essex: Saffron Walden, 232 pp.

Vithanage, H. I. M. V. 1986. Insect pollination of avocado and macadamia. Acta Horticulturae 175:97-101.

Wagner, W. L., Herbst, D. R., Sohmer, S. H. 1990. Manual of the flowering plants of Hawaii. Bishop Mus. Special publ. 83, 1853 pp.

Walton, G. M. 1983. Aiding the establishment and development of beekeeping in the tropical and subtropical Pacific. Pacific Sci. congr. 7 pp.

Wirth, W. W., Howarth, F. G. 1982. The '*Forcipomyia ingrami*' complex in Hawaii (Diptera. Ceratopoginidae). Proc. Hawaii Ent. Soc. 24:127-151. [distribution, habits for native cocoa pollinators]

Yasumatsu, K. 1939. A review of the genus *Megachile* in the Ryukyu- group (Hym., Megachilidae). Aikitu 24-650.

Yasumatsu, K., Hirashima, Y. 1964. Red-haired leaf-cutter bees of the group of *Megachile bicolor* from Japan, the Ryukyus and Formosa (Hymenoptera, Megachilidae). Kontyu, Tokyo 32:175-187.

Yasumatsu, K., Hirashima, Y. 1965. Bees (excluding Halictidae) of the Ryukyu Islands taken in the 1963 and 1964 surveys (Hymenoptera, Apoidea). Kontyu 32:247-259.

Young, A. M. 1986. Cocoa pollination. Cocoa Growers Bull. 37:5-23.

Zimmerman, E. C. 1958. Insects of Hawaii: Macrolepidoptera. Vol. 7., Honolulu: Univ. Hawaii Press, 542 pp.

Acknowledgements

I thank Jo Berry, Landcare at CSIRO, Canberra, who provided an unpublished record from Cocos island from the Australian National insect collection and help with other unpublished records for Christmas island. Mr. Murray Reid MAF, Hamilton, New Zealand, Barry Donovan, DSIR, Lincoln and Eva Crane provided invaluable guidance or access to literature about bees. The author is indebted to the curators of the New Zealand national Arthropod collection in Auckland and the Bishop Museum collection in Honololu for access to bee collections from islands in the Pacific.

1.5 APPLIED POLLINATION IN THE MAINLAND TROPICS

The requirements for pollination by insects of crop plants throughout the world are presented in two encyclopedic volumes which summarize the state of knowledge on a crop-by-crop basis. These books, one by J. B. Free (1970 and second edition, 1993) and another one by S. E. McGregor (1976), are indispensable as reference material both for development of beekeeping and on the roles of bees and other animals in pollination. What also emerges from these books is the lack of knowledge about tropical and subtropical crops (**Appendix I**). Research on crop pollination in the tropics, and on economically valuable plants in general, is urgently needed to fill gaps in our knowledge, to resolve confusing and conflicting studies and to suggest ways of improving productivity by pollinator management, conservation and other means.

Given the contrasts between the tropical islands (reviewed in Section 1.4) and large continental areas, it is not surprising that few plants currently under cultivation originated on oceanic islands, where the number of potential pollinators is limited. The localities of plant origins (**Appendix I**) show that of over 1,330 cultivated species there are nearly equal numbers from Asia and the Neotropics. Africa has provided half as many as either of these regions, the temperate zone one-fourth as much, and Australia barely an eighth of the number from the Neo-tropics or tropical and subtropical Asia. Furthermore, over half of the economically significant tropical plants originated in areas where honeybee species naturally occur. The Neotropics, South Pacific region and Australia lacked *Apis*.

Our working knowledge of pollination among tropical plants is limited, both considering regions where the plants originated and where they are grown today. **Appendix I** reveals that of approximately 700 plant species for which we have some data or can make reasonable guesses, 600 have some association with bees, and 100 do not — at least as far as we know. Our knowledge concerning plant breeding systems and the relative pollination value of each of the flower visitors is even less complete.

In following sections we discuss pollination requirements of as many economically important plants of the mainland tropics and subtropics as possible. Some species of the highlands are also treated in Section 1.3. Native and important introduced plants, excluding ornamentals and many weeds, are treated by groups that seem logical, regarding experience in each of these regions.

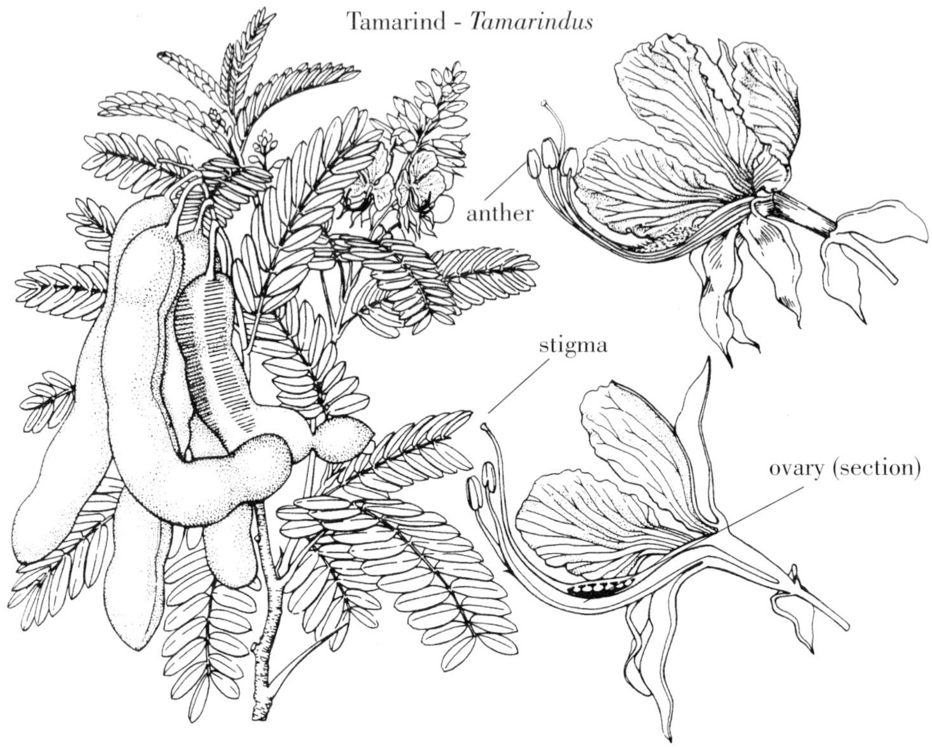

1.5.1 Applied Pollination in Asia

Crops are categorized in the manner followed for **Appendix I**, although not all the categories seemed worthwhile to include in the treatment here. The cultivated plants, their breeding systems, and the pollinators used or depended upon for pollination are discussed as related to: ***a***) *fruits* and *nuts*; ***b***) *vegetables*; ***c***) *cereals*; ***d***) *drug, beverage, condiment* and *spice plants*; ***e***) *oil crops*; ***f***) *forage crops*; ***g***) *timber trees* and *natural vegetation*, and ***h***) *fiber plants* and *isoprenoids* (*rubber*)

1) Fruits and nuts.

Tropical and subtropical fruits are very diverse in Asia. The pollination requirements of many are at least partly understood. In *Citrus* the situation is complex. The *mandarin oranges* are dependent on, and greatly benefit from, cross-pollination. *Pummelo* should be interplanted with other cultivars for cross-pollination and fruit set to take place. Most ranges do not depend on insects for pollination, but may benefit from it, mainly by producing seeds. *Lime* pollination has not been well studied, but cross-pollination may be beneficial. In lemons the data are conflicting, varyng from statements about pollination by insects being completely unnecessary to outcrossing or facilitation by deposition of pollen being required. All of these crops may vary with the particular cultivar. *Grapefruits* do not require cross-pollination; and citrus trees are excellent sources of nectar and pollen.

Apples (see also Section 1.3) generally require inter-plantings of varieties, along with cross-pollination by insects. *Pear* is similar, but is not a good bee plant as the nectar is watery; bees do not collect the pollen. *Plums* and *prunes* are mostly self-incompatible; they require insect pollination and offer good nectar and pollen forage for bees. *Apricots* are variable in their pollinator requirements depending on the variety or cultivar. *Cherries* cannot self-pollinate, but some are self-compatible. Thus, inter-plantings of varieties and the presence of bees for pollination are advised. *Peach* and *nectarine* pollination is not well studied, despite the economic importance of these crops. At least some varieties appear to set fruit using their own pollen, but cross-pollination by bees is needed for good crops. All these Rosaceae are good nectar and pollen plants for bees.

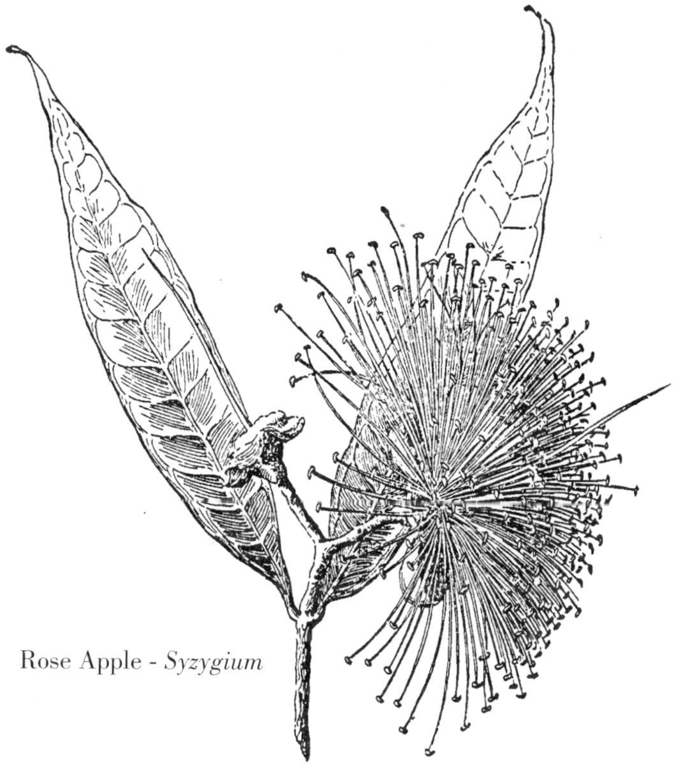

Rose Apple - *Syzygium*

Avocado (*Persea gratissima*) is grown widely throughout Asia and is dependent on cross-pollination by insects. Carambola or star fruit (*Averrhoa carambola*) and ilimbi (*A. bilim-bi*) are similar, as are *Ziziphus* (*Z. jujuba*, jujube and *Z. mauri-tania*, ber). *Nephelium* species: *N. litchi*, *N. lappace-um*, *N. longan*, litchi or lychee, rambutan and longan are excellent honey plants and seem to depend on insect cross-pollination as well. The mango (*Mangifera indica*) generally requires insects for pollination, but cross-pollination between cultivars is not needed. Papaya (*Carica papaya*) has a complex breeding system of male, hermaphrodite (3 types) and female flowers in which the best fruit results from cross-pollinations. Bees are sometimes recommended as pollinators. They collect pollen from male and hermaphrodite flowers. The normal Neotropical pollinators, however, are hawkmoths. Persimmon (*Diospyros kaki*) is dioecious (see Section 2.3.3), and some cultivars produce fruit without pollination, but most cannot. The flowers produce nectar and pollen thst are used by bees. McGregor (1975) lists *Diospyros discolor*, the mabolo, as dependent on insects for pollination. Honeybees are the most important pollinators of phalsa (*Grewia asiatica*) in India. In the loquat (*Eriobotrya japonica*) benefits from visitation by *Apis* and *Bombus*, and is minimally self-pollinating.

Litchi - *Nephelium*

Durian (*Durio zibethinus*) is cross-pollinated by bats. The flowers are also an abundant source of nectar which is used by bees. It has a slimey consistency and an off-smell. The soursops and custard apples (*Annona*) have been studied only slightly in Asia, where, as in South America, they appear to be pollinated by beetles. These flowers have almost no nectar. The cherimoya (*A. cherimola*) of South America requires at least some cross-pollination.

Breadfruit, jackfruit (*Artocarpus*), mombins or hogplums (*Spondias*), jambus, jambolans and related plants (*Syzygium*), bignay (*Antidesma*), otaheite-gooseberry and mirabolan (*Phyllanthus*), lingaro (*Eleagnus philippensis*), wampi (*Clausena lansium*), sapote relatives, star apple, sapodilla (*Manilkara achras*), egg fruit (*Pouteria campechiana*), introduced from South America, all probably benefit from insect pollination.

Mango - *Mangifera*

Flower with one fertile stamen and one stigma

Flower & fruit (section)

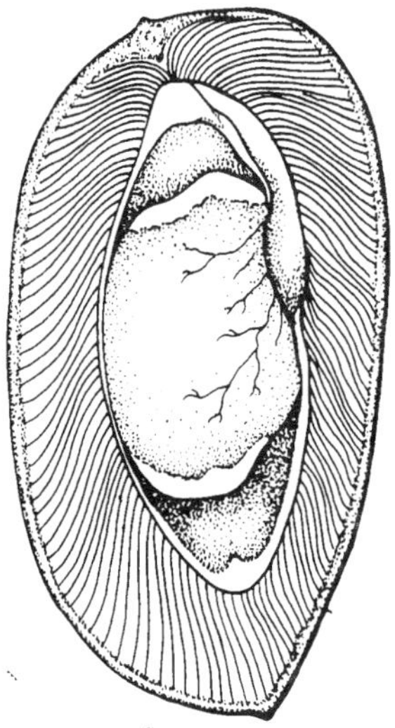

Green sapote (*Calocarpum viride*), Runeala plum (*Flacourtia cataphracta*), kei apple and kitembilla (*Dovyalis*) require insect pollination. Other Indian plum relatives (*Flacourtia*), mangosteens (*Garcinia*), mamey apple (*Mammea americana*), guava (*Psidium*) are known to benefit from cross-pollination by insects but the extent of their dependence is not known. Bananas (*Musa*) in the wild are pollinated by vertebrates, whereas the cultivated plants set fruit pharthenocarpically. Among the viney fruit plants, the Chinese gooseberry or kiwifruit (*Actinidia deliciosa*) is dioecious and requires cross-pollination by pollen-collecting bees. Wild grape (*Vitis*) is similar, but the domestic grape (*V. vinifera*) can set fruit by self- or wind-pollination. Bee pollination seems to increase fruit set in some varieties. Pollination in passion fruits (*Passiflora*) is complicated by the structure of the flower. They require cross-pollination, but honeybees may not always accomplish this. Carpenter bees (*Xylocopa*) seem to be best. The shrubby pomegranate (*Punica granatum*) has not been well studied. The flowers produce only pollen. Pineapple (*Ananas sativus*) requires cross-pollination for seed to be produced, but this is undesirable for table fruit. Honeybees cannot reach the nectar, hidden at the base of the reddish tabular flowers which are pollinated by hummingbirds in South and Central America. Prickly pear (*Opuntia*) flowers provide large amounts of pollen. Their pollination requirements are largely unknown.

From the foregoing, one can see there are weaknesses in our knowledge of fruit pollination in parts of Asia. Incomplete and conflicting reports need resolution. Various fruits, such as *dukus, jambus* and others are mysteries — both considering pollination and their importance as 'bee plants'.

Among the nut plants of Asia, coconut and nutmeg are discussed below. The cashew (*Anacardium occidentale*), imported from the Neotropics (see **Appendix I**), is an important cash crop. India is the world leader in production. Cross-pollination is important, apparently accomplished by small insects. Peanut (*Arachis hypogea*) flowers are visited by insects, including bees that trip their flowers as they collect the pollen. These visits seem to have a beneficial effect in crop production. The almond (*Prunus amygdalus*) is dependent on insects, primarily bees, for fruit set.

2) Vegetables.

In many vegetable plants it is not the fruit that is eaten, and for this reason production does not depend on insect pollination. However, insect cross-pollination is important in obtaining *seeds* of these plants, and their flowers are often excellent forage for bees. Included in the list are carrots (*Daucus carota*), cabbage, broccoli, cauliflower and other cole crops, (*Brassica oleracea* and *B. pekinensis*), chicory (*Cichoriun intybus*) lettuce (*Lactuca sativa*), onions (*Allium*), raddish (*Raphanus sativus*) fennel or saunf (*Foeniculum vulgare*) and others. Sweet potato (*Ipo-moea batatas*) and manioc or cassava (*Manihot esculenta*) are good nectar plants whose flowers are visited by many bees. Bees' roles in pollination are not known. The plants are propagated by cuttings, but seeds are used in breeding programs.

For vegetables in which the fruit is eaten, pollination must be considered — it often determines the size and 'quality' of the fruit, as emphasized throughout this book. The eggplant (*Solanum melongena*) is poorly studied but requires pollination, for which bees can be used. Tomatoes (*Lycopersicon esculentum*) are not automatically self-pollinating but movement of the plant causes pollen to fall from the anthers onto the stigmas. Thus, wind, insects or artificial vibration will bring about fruit set. The hot peppers (*Capsicum*) seem to be able to bear fruit and seed in the same way as tomatoes, but they do better if cross-pollinated. The flowers of solanaceous crop plants (including potatoes, *Solanum tuberosum*, from which seeds are grown to produce seed-potatoes for plants) are not attractive to many bees (see Section 2.3.2) but probably would benefit from their foraging by being more productive in fruit quantity and quality. The cucurbit vegetables, pumpkins and squash (*Cucurbita*) cucumbers and gherkins, (*Cucumis sativus*), balsam pear (*Mormodica charantia*), musk melon (*Cucumis melo*), other watermelons (*Citrullus*) and gourds, including bottle gourd (*Lagenaria siceraria*), ash gourd (*Benincasa hispida*) and sponge gourd or luffa (*Luffa aegyptica*), require pollinators for fruit set. Bees are the principal pollinators. Okra (*Hibiscus esculentus*) is self-pollinating but is well visited by bees. The value of outcrossing has not been assessed.

Of the periodic crops, semi-seasonal crops or *pulses*, cowpeas (*Vigna sinensis*) are highly attractive to bees. Pollination is best effected by large bees, rather than by smaller bees or honeybees. Large bees like *Xylocopa* or *Chalicodoma* are probably the chief pollinators of the horse bean (*Canavalia ensiformis*), at least in Indonesia, although this species and the sword bean (*C. gladiata*) are reported to be self-pollinating. The lentil (*Lens culinaris*), mung bean (*Phaseolus aureas*), and gram (*Cicer arietinum*) are thought to be self-pollinating. Horsegran (*Lablab niger*) requires pollination by insects. However, the pollination requirements of many leguminous crops are unknown, although the wild progenitors almost certainly all needed insects, preponderantly bees. Soybean (*Glycine max*) has been considered to be self-fertile. However, the benefits that bees bring in terms of increased seed-set and oil yield suggest that more research is needed to clearly establish their value.

3) Cereals.

Most cereals are independent of insects for pollination. Wind pollination prevails in the grasses. Maize (*Zea mays*), although a pollen source for bees, is wind-pollinated. Many cereals do not reproduce by pollination — the grain develops apomictically. Pearl millet (*Pennisetum typhoides*) may benefit from insect pollination, as bees are sometimes abundant, collecting pollen from the flowers.

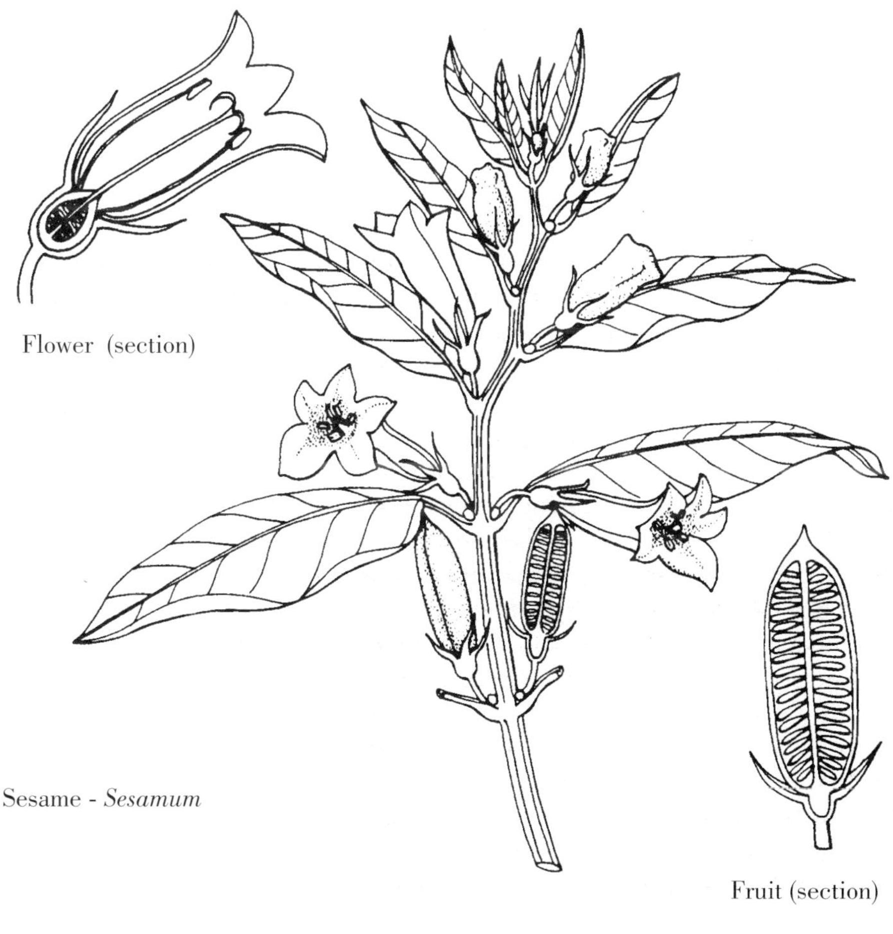

Flower (section)

Sesame - *Sesamum*

Fruit (section)

4) Drug, beverage, condiment and spice plants.

There is no comprehensive account of pollination needs among the wide variety of medicinal plants grown throughout Asia. Many are not economically prominent (yet) and are grown in small patches for specific uses. Poppy (*Papaver somniferum*) is grown for the illicit drug trade. It produces only pollen and is visited extensively by pollen-foraging bees. The number of seeds produced is greater when pollinators are abundant. Hemp (*Canabis sativa*) is wind-pollinated, but the male flowers are visited by bees for pollen. Tobacco (*Nicotiana*) cross-pollination can be brought about by honeybees for hybrid seed production.

Coffee (*Coffea*) production can be considerably enhanced by augmented pollination with honeybees. Coffee flowers are also a good source of nectar, although the honey if bitter and of no commercial value. Tea (*Camellia sinensis*) is not a good bee plant. The small flowers are usually pollinated by flies. Cocoa (*Theobroma cacao*) is also pollinated by small flies (*Forcipomyia*, of the family Ceratopogo-nidae), but some pollen-collecting bees (e.g. *Trigona* and *Lasioglossum*) may also pollinate the small flowers. Condiment and spice plants have been neglected in light of pollination or value as bee plants. The crops of cardamom (*Elettaria cardamomum*) are much improved by honeybee visitation.Pepper (*Piper nigrum*) remains an enigma. Although visited faithfully by both flies and bees, self-pollination, effected by rain, is still suggested as tenable. Coriander (*Coriandrum sativum*) is grown extensively in India and seems to benefit from cross-pollination by bees and other insects in Europe. Sesame (*Sesamum indicum*) is grown for its oil-rich seeds and is reportedly self-pollinating. Bees collect nectar and pollen from the flowers, but their importance in crop production has scarcely been studied. In Egypt the wide variety of flower visitors, mostly honeybees, substantially increased seed yield in open pollinated plants compared with those in cages). *Vanilla* is grown in Asia. Pollination is usually done by hand. The mustards (*Brassica*) are variable in their pollination needs, but they offer good bee forage and benefit from outcrossing.

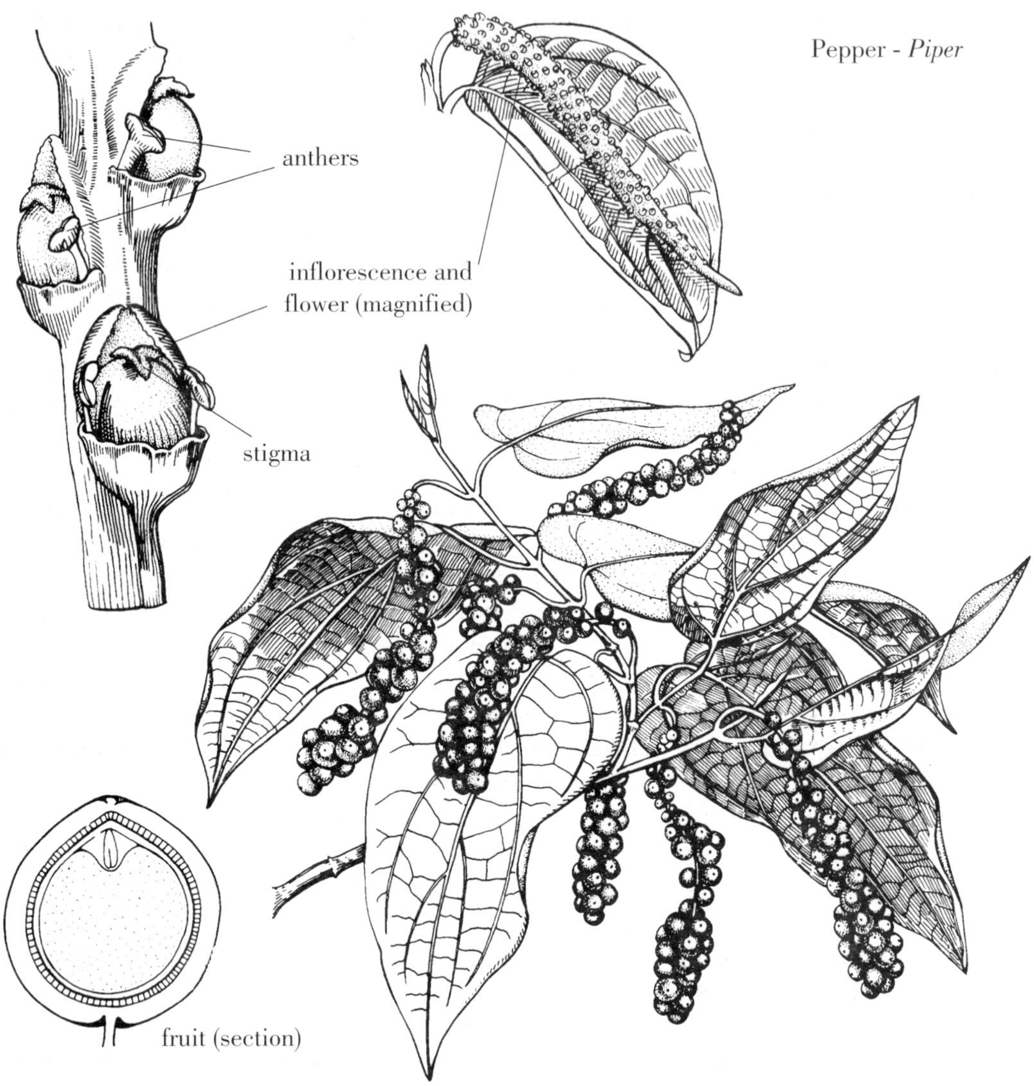

Pepper - *Piper*

Bee pollination of cloves (*Syzygium aromaticum*) requires outcrossing, although the clove of commerce is a flower bud. Bees are probably important, but other insects are also involved and may be more important in some places. The same may be said of saffron (*Crocus sativus*), which is harvested for the floral stamens, but is nonetheless pollinated by bees. Nutmeg and mace are produced from *Myristica fragrans*. This plant is dioecious and so requires cross-pollination. However, how this is achieved is unknown; small insects, bees and wind have all been suggested. Allspice (*Pimenta dioica)* is also dioecious and is effectively pollinated by honeybees in Jamaica (where neither is native). Cinnamon (*Cinamomum zeylanicum*) requires insect pollination.

5) Oil crops.

The rapeseed, mustards and related *Brassica* are grown extensively for oilseed production in Asia. *Brassica campestris* is dependent on insects for pollinaton; it is an excellent source of pollen and nectar for bees. *B. juncea* is also more fruitful if bees are present. Niger (*Guizotia abyssinica*) is a good bee plant and seed yield are greater when beehives are placed in niger fields. Sunflower (*Helianthus annuus*) is being grown more and more in Asia. Even though new self-compatible varieties are available, they benefit from cross-pollination and oil yields are greater. Self-incompatible varieties require insect pollination and bees are the primary agents. The Flowers provide nectar of high quality and also pollen. Flax (*Linum usitatissum*) is grown for oil and fiber and benefits from outcrossing by bees. Safflower (*Carthamus tinctorius*) is a good source of nectar and pollen for bees and yields more oil when cross-pollinated. The oil palm (*Elaeis guineensis*) is a very commercially successful oil-producing plant in tropical Asia (see also Section 2.5). It is pollinated effectively by weevils). While bees sometimes collect the pollen, it is not a good forage plant. The coconut (*Cocos nucifera*) is grown for the fruit which provides food, drink, oil and fuel. Evidence is accumulating slowly that coconut production may be improved by augmented pollination by bees. Wind pollination is still apparently most important, especially in plantations. The coconut offers ample nectar and pollen to bees. Castor bean (*Ricinus communis*) is not understood in terms of its pollination requirements. It is pollinated by bees in Botswana. Tung (*Aleurites fordii*) and tong (*A. montana*) are grown for their nuts, which when pressed yield oil used in varnishes, electrical insulators and protective coatings. *A. fordii* may be dependent on insects for pollination, but this is still not clear.

6) Forage crops.

Alfalfa (*Medicago sativa*) is grown in drier parts of tropical Asia. It requires insects for pollination, honeybees and leaf cutter bees are the most effective. Kudzu (*Pueraria thun-bergiana*) is pollinated by bees. Berseen (*Trifolium alexandrinum*) has a wide variety of pollinators on which it is dependent for seed set. Honeybees are effective pollinators.

7) Timber trees and natural vegetation.

It is very difficult to obtain information on the pollination biology of the more valued timber trees of tropical and subtropical Asia. Chan and Appanah have discussed the role of thrips in pollination of dipterocarps and other trees in Malaysia, and Appanah has made some generalizations on the roles of bees in pollination in Malaysian primary forests. There are a few publications on the pollination of plants of the tropical forests (see General References, e. g. Panayotou and Ashton, Bawa, Appanah, Faegri and van der Pijl, Owens et al.) but, in general, this is a much neglected area of botany, forestry and now, since many formerly forested areas are being replanted in timber species, applied ecology. The value of the forest trees in honey and pollen harvests is well known.

8) Fibre plants and rubber.

The most important fibre giant is cotton (*Gossypium*), however, it is not thought to depend on insects for cross-pollination. In fact, cross-pollination has been considered detrimental, because the resultant seeds and progeny are not necessarily bred exclusively from the desired cultivar. Nevertheless, benefits do accrue from cross-fertilization — larger crops, more oil and hybrid vigor in progeny. Honeybees are used extensively in cotton fields where pure-line seed is not sought. The flowers produce abundant pollen and nectar. Kapok (*Ceiba pentandra*) is a copious producer of nectar and excellent bee forage. In nature, bats are important pollinators but the trees also seem to be self-fertile. Pollination of kenaf (*Hibiscus cannabinus*) is not fully understood but outcrossing by bees may be more important than now thought. Sunn (or sann) hemp (*Crotalaria juncea*) seems to require pollination by bees, at least for maximun fruit set. Carpenter bees (*Xylocopa*) are more valuable than honeybees. *Cannabis sativa* is wind pollinated but is used by bees for its pollen. Information on flax and coconut is given above (see also Section 1.4).

Rubber (*Hevea brasiliensis*) is pollinated by insects, mostly Diptera (midges). However, it is an important honey plant in the tropics as many bees forage for nectar from the extrafloral nectaries on the petioles of the leaves (Sections 2.3.1, 2.1).

With the continued large-scale cutting of tropical and subtropical forests in Asia, attention must be paid to 'reaforestation' programs. In some cases, as with *Calliandra* introduced into Java, beekeeping projects have gone hand-in-hand. Nevertheless, all too little is known of the reproductive biology of most valued timber trees of the forests (**see Appendix I**). To obtain seeds for reaforestation, either by natural processes or aided by human effort, the trees must be pollinated. If the trees are highly specific in their requirements for pollinators, and the pollinators finely attuned to the tree for their own reproduction, then *over-cutting could produce a vicious cycle of decline* in both the plant and pollinator. These issues are urgently in need of consideration. Honeybees or other managed pollinator groups may need to be incorporated in future plans to ameliorate the foreseeable problems in such situations.

Conclusion.

One can readily appreciate that many economically vital plants of tropical and subtropical Asia are 'unknown' insofar as pollination requirements are concerned. Those listed are currently the most important and have received some attention, if only to indicate the lack of information about them. Certainly, of those economic plants for which documentation on breeding systems is available, the number requiring insect pollination is high. If to those are added the plants that certainly benefit from cross-pollination by insects, and those suspected of belonging in this category, only the cereals, bananas and seedless fruits remain outside the influence of pollinators. At the same time, one can see that most of the economic plants for which insect cross-pollination is at least beneficial are also useful to honeybees and other honey-making bees (the stingless bees), which may also provide income to those exploiting these insects. Thus, beekeeping and agricultural productivity are as clearly interwoven in the tropics as they are in the temperate regions of the world.

Additional References

Appanah, S. 1981. Pollination in Malaysian primary forests. Malaysian Forester 44:37-42.

Appanah, S., Chan, H. T. 1981. Thrips: pollinators of some dipterocarps. Malaysian Forester. 44:234-252.

Crane, E. and P. Walker 1983. The impact of pest management on bees and pollination. Tropical development and research institute, London, 129 pp. + Annex A, 79 pp.

Kamal, S., Alchtar, M. 1976. Symposium on pollination, pp. 232-234. Role of insects in pollination of rape (*Brassica juncea* H. F. & T.) flowers. Pakistan J. Agric. Sci. 13:65-72.

Leuck, D. B., Burton, G. W. 1966. Pollination of pearl millet by insects. J. Econ. Entomol. 59:1308-1309.

Nur, N. 1976. Studies on pollination in Musaceae (Bananas). Ann. Bot. 40:167-177.

Rashad, S. E., Ewies, M. A., El-Rabie, H. G. 1979. Insect pollinators of sesame (*Sesa-mum indicum* L.) with special reference to the role of honeybees. Proc. IV Intl. Symp. Poll., pp. 231-234.

Sastrapradja, Aminah, S. H. , Lubis, I. , Satrapradja, D. 1975. Studies in Javanese species of *Canavalia*. I. Floral biology and cytology. Ann. Bogor. 6:43-55.

Syed, R .A. 1979. Studies on oil palm pollination by insects. Bull. Entomol. Res. 69:213-224.

Coconut - *cocos*

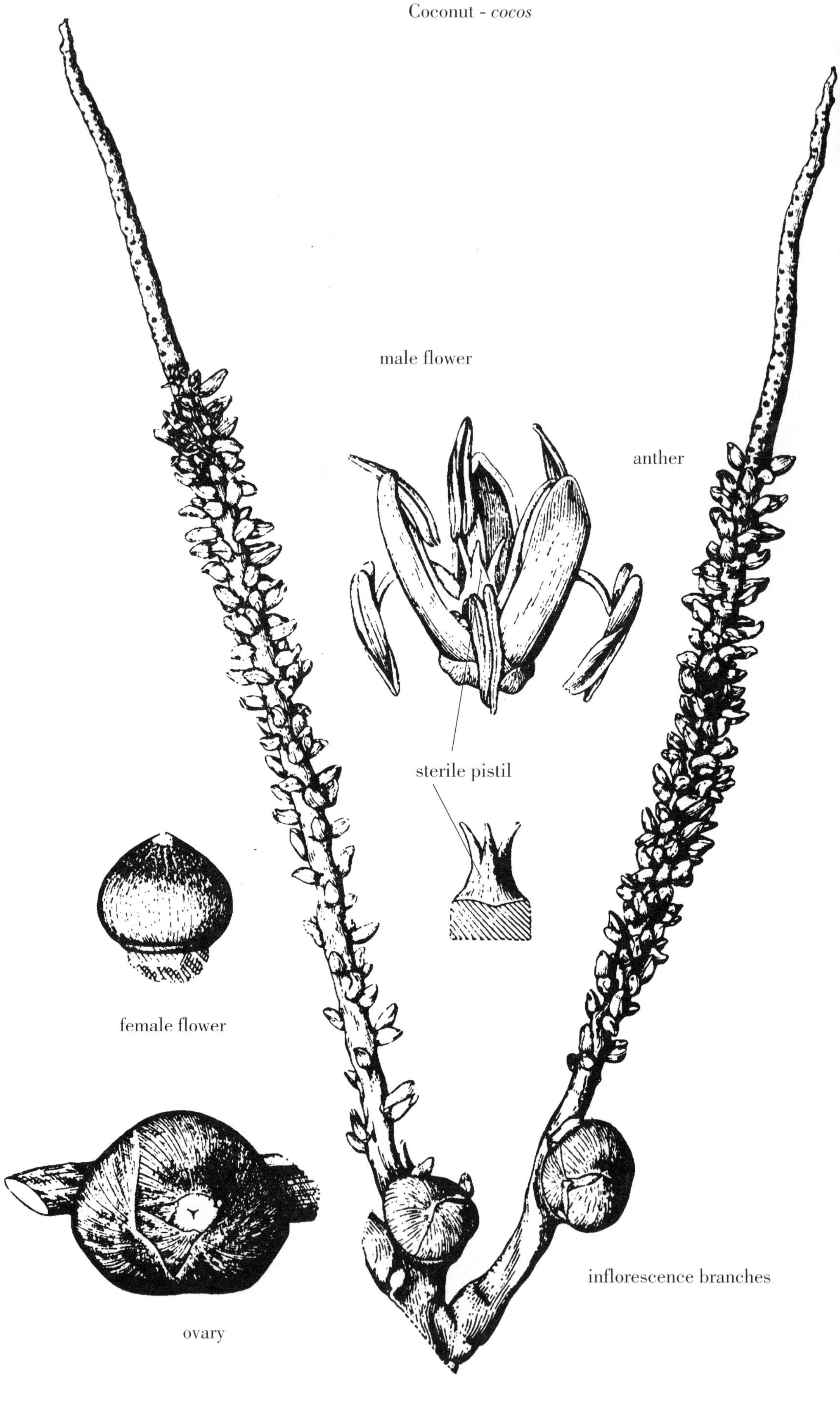

1.5.2 Applied Pollination in Africa

1) Introduction.

The importance of pollination to the yield of most crops grown in temperate climates is known, at least partially, but there is relatively little information about pollination of tropical crops. In Africa and other tropical regions, many studies on crop pollination have been preliminary, mainly because of insufficient facilities.

African honeybee, *Apis mellifera scutellalta*, on a native African inflorescence (coffe)

The great need to increase crop yield to meet food requirements of growing populations all over the world is one of the most important problems in the tropics. In Africa, dependence of pollination of crops is expected to increase as food needs increase. In general, most African studies in pollination were made in response to specific production problems. For instance, in East Africa, research in pollination was, and still is, closely tied to the problems of production arising in export-oriented agriculture of pre-independence governments. Research has been on commercial crops such as *pyrethrum, coffee, cashew nut, sunflower* and *mangoes*. Most post-independence research in agriculture has concentrated on improvement of production of the same commercial crops through breeding, but little has been done on pollination of the improved crops.

Tree crops are common in the tropics and the improvement of any one by breeding is a long-term project. Short-term increases could possibly be secured by raising the density of pollinators around the crops. One way to increase pollinator abundance is by improvement of existing traditional apicultural practices. In some countries however, because of prevailing poverty, the priority in development of apiculture is what can be produced now; research can come later. Beekeeping in Africa is being improved slowly through research, because the latter will not provide honey and beeswax, but it will also increase crop yield through pollination, all of which will raise farmers' income (see also Sections 1.2, 2.1).

2) Pollinator behaviour and management.

Insects recorded visiting flowers of crops are assumed to be pollinators. However, their real value as pollinators, or other types of visitors (see Sections 2.2 to 2.4) may not be known. Many species of insects and small animals visit flowers for nectar or pollen, or both, but few may be good pollinators. The most efficient pollinators carry plenty of pollen, brush against stigmata hence transferring pollen, visit several flowers of the same species in succession and move frequently from flower to flower and plant to plant.

Pollination of few crops depends only on one or a few local insects. Most crops in Africa and elsewhere (**Appendix I**) are visited by honeybees whose pollen carrying capacity and behaviour make them superior pollinators in many instances. Nevertheless, most races of African honeybees are very defensive. Partly because of that, apiaries are located well away from areas of agricultural and other human activity. Colonies also abscond readily after disturbance during handling and transport, migrate during dry periods and swarm often. These features all decrease their value in pollination management and it has been questioned if they are suitable in migratory beekeeping. Moreover, in almost all of Africa except some parts of South Africa, bees are kept in traditional log and other hives, and only more recently in top-bar hives, which cannot be transported without comb breakage (Section 2.1).

Apart from honeybees, very few other insects have been managed for crop pollination in most African countries, partly because of a general lack of knowledge concerning the potential of wild insects as pollinators. Trials in rearing of the Egyptian alkali bee (*Nomia unidentata*), are in progress for pollination of clover (*Trifolium alexandrinum*) in Egypt.

3) Farming systems and destruction of natural vegetation.

Most farming in Africa consists of subsistence agriculture. In peasant agriculture, there is rarely a pollination problem as there are sufficient wild insects living in the surrounding bush. With modernization of agriculture however, large areas of natural vegetation are cleared and often, one particular crop is cultivated in the area. Because of intensive cultivation, there is frequent destruction of plants that are beneficial as forage and nesting sites of pollinators. As a result, there is a strong tendency for populations of pollinators to decrease, until they are too few for even minimum acceptable crop pollination. For instance, in Kenya and Tanzania peasant agriculture left large areas of fallow land, which formerly provided ample sites for wild pollinators. Most of these have now been reduced. It is now apparent that the consequences of natural disasters such as drought can be aggravated by deforestation and unsuitable agricultural and pastoral practices. In addition, the increase in monocultural cropping results in flowering being concentrated for a short period. Larger pollinator populations are needed for those short periods. However, *outside* these flowering periods are times of floral dearth, because no alternative forage is available. In Africa, feeding bee colonies during periods of dearth is not generally practiced. Subsequently, most starve or emigrate from the crop area.

4) Enemies and poisoning of pollinators.

Although many of the known bee diseases have not been reported in Africa, *Varroa* was identified in Tunisia in 1978 and *Nosema* in Libya and Egypt (1977) and South Africa (1972). The importance of bee diseases may not be realized as yet because most colonies of African honeybees are in wild nests or are kept in traditional hives that cannot be inspected readily. Moreover, farmers may not recognize the diseases. The largest threat to honeybee colonies in Africa at present are bee pests and predators. Several types of insects, spiders, birds, mammals (including man) have been recorded as enemies of bees in Africa. There are 17 species of African bee-eaters (*Merops* spp.) whose diet consists of 62-94% Hymenoptera. In the savanna areas of Zimbabwe, Niger and Kenya, honeybees make up as much as 96% the diets of *Merops*. Colonies of bee-eaters can be as large as 25,000 individuals in Niger and an individual bird can eat up to 124,000 bees in a year. Such data indicate how devastating these birds can be. Other pests that can cause great damage to bee colonies include hive beetles (*Oplostomus tuliginens*) in South Africa, calliphorid fly larvae (*Pollenia*) in Egypt, predatory wasps (*Palarus latifrons*) in S. Africa, the hornet *Vespa orientalis* in Egypt, and pirate wasps (*Philanthus triagulum*) in Egypt and Botswana.

In cases where insects are not direct pests to bees or crops, they may compete with honeybees for available nectar. Honey yields from *Eucalyptus gradis*, which form 80% of all *Eucalyptus* grown in S. Africa, vary greatly, probably due in large part to infestation by *Drosophila flavohirta* that may consume nectar and leave none to attract bees. In South Africa, *Eucalyptus* species produced nectar between midnight and sunrise. Ants foraged at night and collected 42% of the nectar before honeybees started to forage. Sometimes, the types of crops adjacent to each may lower pollination efficiency because of crop competition for available pollinators. In one study, from 40 to 46% of introduced bees visited an almond orchard while 49 to 51% visited the adjacent pasture with *Aloe davyana* and *A. marlothii*. The alternative sources of forage attracted bees from the target crop. However, flowering of many crops in the tropics is staggered throughout most of the year. Thus, critical resources for pollinators are available for most of the year.

Continued use of insecticides with the increase in crop plantations accelerates loss of pollinators. Damage to honeybees differs according to many factors (see **Appendix III**), including insecticide toxicity, methods and time of day of application, number of applications, proportion of foragers visiting treated crops, foraging behaviour of bees on the crop and drift of insecticide to the various sources of forage. Most poisonings occur when bees are collecting contaminated nectar, pollen and water. Insecticides imbibed with feed may be transported back to the colony where they poison both immature and adult bees. Several studies have indicated that most insecticides used in agriculture are harmful to honeybees. Spraying of crops, especially cotton, in Egypt, Kenya and Tanzania has made beekeeping impossible in cotton growing areas. The insecticides most toxic to bees include *phosparidon, BHC (lindane), Aminocarb, DDT, trichlorphon, Sevin (Carbaryl), Fenithion, Dimecron (Phosphamidon), Sumithion, Dicofol (Kelthane), Roxion, Toxaphene and Parathion* , among others (**Appendix III**). All insecticides are toxic in nectar on the day of application, thus ordinarily should be applied at night (unless the bees are heavily using bat-pollinated flowers!). Residues on plant parts showed that some insecticides such as hostathion persisted for up to 7 days on leaves. Toxicity of insecticides such as carbamates increase with temperature. It is advisable to close colonies in ventilated hives at the time of application.

5) Pollination studies in African countries.

Studies done in Africa on pollination of crops are here grouped into four regions: a) North Africa, b) West Africa; c) East and Central Africa, and d) South Africa.

North Africa. In this region, the northern part has a Mediterranean climate. Southwards, along the tropic of cancer, the region consists of Saharan desert. Most farming is along rivers and where there are irrigation dams. In the dry parts, the area of cultivated land is being increased through reclamation of desert. Irrigation dams such as Aswan dam in Egypt are thought to be vital for farming. The scarcity of wild pollinators in newly reclaimed lands is mainly because of lack of wild flowering plants for supply of pollen and nectar. Consequently, crops on newly reclaimed areas suffer from lack of pollination. Colonies of *Apis mellifera* have been found suitable for establishment as pollinators of some crops in these areas. However, honeybees cannot pollinate all crops, so introductions of wild bees, such as *Nomia, Andrena* and *Megachile*, are being tried in Egypt. Eventually, established crops and judicious use of pesticides should allow wild insects to become established, thereby improving pollination. Other immigrant insects may become crop and bee pests, suggesting the continued need for insecticides.

Preliminary studies have identified over 30 species of crops that benefit from honeybee visits, although not much

Sunflower - *Helianthus*

work on their pollination requirements has been done. Among the textile crops, most studies have been done on cotton. Of all insects visiting cotton, honeybees were found to be 79.3% their number in Egypt, and 80% in Sudan. In Chad, 80-90% of cotton flowers were visited by honeybees but although only 70% of hybrid capsules were formed, it was thought sufficient for commercial production of hybrid cotton. Pollination of cotton by honeybees increased the weight of seed per boll, weight of cotton per plant and overall yield per plot. A positive correlation between temperature and numbers of honeybees visiting cotton in Egypt has been noted; available nectar in cotton flowers increased with relative humidity and decreased with temperature, and plants contained no nectar at temperatures above 40°C. Most bee visits were during 12,00 to 14,00 h, corresponding to peak nectar secretion and sugar concentration.

In Egypt, flax also benefits from pollination by wasps, field bugs, bees, beetles, flies and butterflies. Honeybees form 90% of all insect visitors and increase both quality and quantity of the crop. Alfalfa (Medicago sativa) and Egyptian clover (*Trifolium alexandrinum*) are among the most important cultivated forage crops in north Africa which benefit from insect pollination. Alfalfa is visited by 37 species of Hymenoptera especially *Andrena ovutula*, *Chalicodoma flavipes*, *C. mucorea*, *Megachile submucida*, *Nomia* and other species of Halictidae, and *Apis mellifera*. The last bee species makes up to 51% of the individual pollinators in Egypt and up to 72% in the Sudan.

Chalicodoma mucorea builds nests in mud houses but trials in transferring to artificial nests failed. The adults did not withstand confinement and transportation. However, cells containing prepupae and pupae were successfully transplanted into artificial cells punched in newly made mud bricks.

In Egypt, alfalfa flowers enclosed with honeybees gave higher yields than did those from which all insects were excluded. The yields were still not as high as those obtained from open pollinated plots. The density of honeybees was higher in fields closer to hives than those 1.5 km away, whereas density of wild bees was higher on distant plots. This suggests that the wild and hive bees compete, and that colonies for pollination should be placed close to the crop. Egyptian clover is self-sterile and honeybees increase its yield. It is visited by Coleoptera and Diptera, halictid bees (especially *Nomia unidentata*) and honeybees. Honeybees made up 59.6% of the total insect visitors in Egypt and accounted for 82% of all pollinators in Sudan. On average, the number of seeds per head were found to be 0.40, 1.21, 22.20 and 43.75, respectively, in plots that were protected from wind, screened against insects, open and caged with honeybees. For the same treatments, the percentage seed set was 0.65, 1.83, 33.80 and 65.87. The number of seedless heads was 82.86, 65.31, 2.80 and 0.40, respectively. As in alfalfa, there was a significant decrease in seed yield as distance from the apiary increased.

Bees and other insects are generally the principal pollinators of a variety of crops. Among seed oil crops in north Africa, sunflower in Sudan is pollinated by *Bombus*, *Nomia melanderi*, *Megachile rotundata*, *Halictus* and *A. mellifera*. The honeybees comprised 75% of all Hymenoptera. Although *Sesamum indicum* (sesame) is usually self-pollinated, insect pollination is important in increasing seed production. The most active insect visitors of sesame in Egypt were Diptera (Syrphidae), Coleoptera and *A. mellifera*. The bees accounted for 30% of all visits to flowers. In Egypt, peanut (*Arachis hypogaea*) is pollinated by Hymenoptera, especially Chrysididae, Halictidae, Sphecidae, Megachilidae, Apidae, and other insects such as Syrphidae.

Date palms in Egypt are hand pollinated. Hand pollinated palms achieved highest total soluble solids in fruit, the largest percentage of sugars, and lower tannin contents than unpollinated ones.

A survey for important pollinators on beans (*Vicia faba*) in Egypt revealed that honeybees made up 77.3% of all its insect visitors. Experiments in which honeybees were excluded reduced yield by 2.8% and enclosing the crop with honeybees increased it by 10.4%. In Sudan, it was estimated that the percentage of buds of beans that dropped before maturity were 86.7% in a local variety and 93.7% in an introduced one. Bud drop was appreciable both before and after fertilization. There were also indications that reduction in pod yield was associated with inadequate insect pollination. Pollination by honeybees and hand tripping of flowers increased pod set. Other studies in

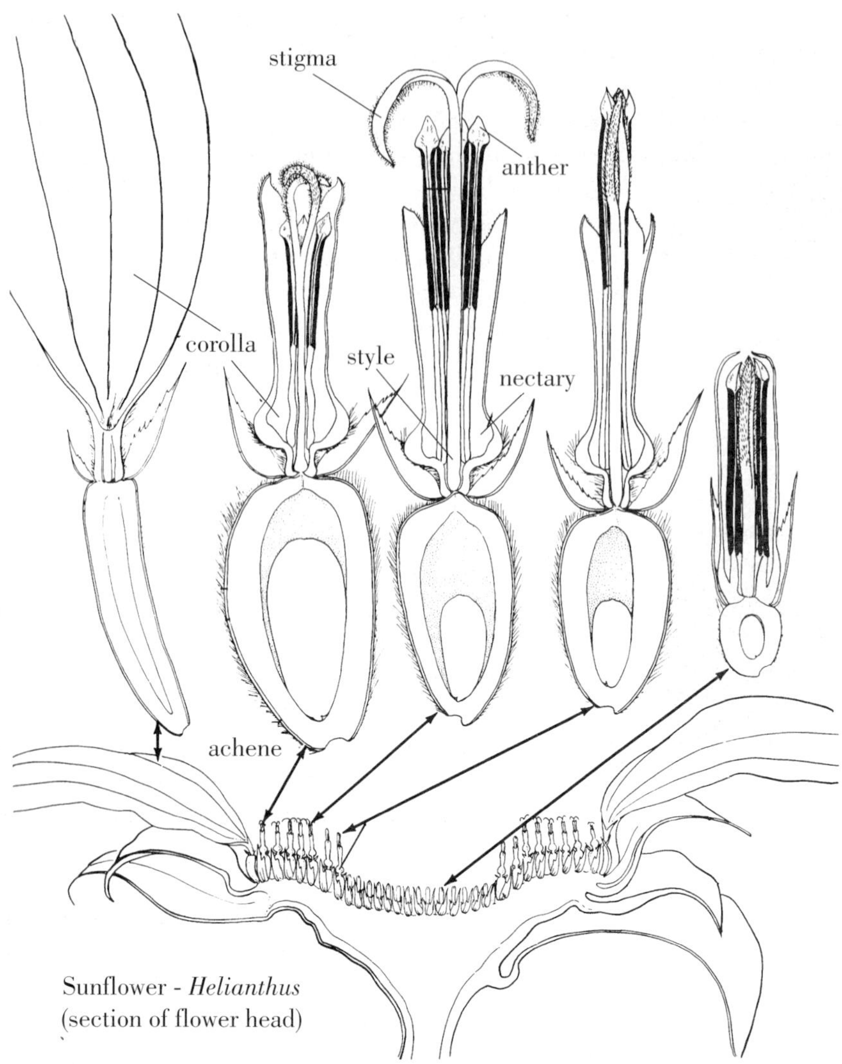

Sunflower - *Helianthus* (section of flower head)

Egypt revealed that *Andrena, Halictus, Nomioides* and *A. mellifera* visited several other crops.

Of all Hymenoptera, honeybees formed 97.9% on *Brassica oleracea*, 68% on *B. sinapis* and 67.7% on *Coriandrum sativum*. On the other hand, wild bees formed 85.7% on *Cynodon dactylon*, 85% on *Raphanus sativus*, 76.9% on *Matricaria chamomilla* and 51.1% on *Daucus carota*. The main pollinators of onion were honeybees whose visitation produced high seed set, weight of seed and yield per plant. Although there has not been much work on pollination on many other crops, honeybees have been found to forage on borage (*Borago officinalis*), fennel (*Foeniculum vulgare*), dill (*Anethum graveoleus*) and squash (*Cucurbita*) in Egypt.

West Africa. This is the tropical to subtropical region on the west coast of Africa, south of the Sahara. Ndiaye surveyed plants of Senegal that are visited by honeybees, more for the investigation of honey production than of pollination. Similarly, a few surveys of pollinators for various plants in Nigeria and Gabon have been initiated, in preparation of a reference collection of honey and pollen plants. Harris (1987) noted that cecidomyiids pollinate cocoa in West Africa. In Ghana, honeybees and *Xylocopa* increased yields and fruit quality of tomatoes (*Lycopersicon esculetum*), peppers (*Capsicum frutescens*), and egg plants (*Solanum melongena*) which are partly self-incompatible and need cross-pollination for fruit set. Honeybees and hover flies constantly visited Rosaceae and pollinated *Prunus spinosa, Crataegus monogyna, Rosa canina*, and *Rubus fruicosus* in Ghana.

Preliminary studies of Locust been tree (*Parkia biglobosa*) and butter tree (*Bytyrospermum paro-doxum*) in Burkina Faso have shown that their fruit production is low but it was not clear whether or not bees act as their pollinators. In Cameroun, oil palm (*Elais guineensis*) was originally thought to be wind pollinated. Syed found that 12 species of insects visited it. *Elaeidobius* and *Atheta* (Curculionidae) and Diptera made up the majority of oil-palm pollinators in Cameroun and Ivory Coast. Pollinator population varied in time and space depending on climatic conditions. From 80 to 85% of fruit bunches were pollinated during the wet season compared to 77 to 82% during the dry season. In West Africa, Barker found that some plants are pollinated at night. He noted that *Yucca* is pollinated by moths (*Tegiticula*) and bats pollinate baobab (*Adansonia digitata*), kapok (*Ceiba pentadra*) and sausage tree (*Kigelia africana*).

East and Central Africa. This comprises equatorial regions with scattered equatorial vegetation. In Tanzania, cashew nut (*Anarcardium occidentale*), coconut (*Cocos nucifera*), *Coffea arabica*, *Pyrethrum* and sunflower (*Helianthus annuus*) benefited from bees. Farmers were encouraged to keep bees in plantations of those crops. Cashew nut flowers are visited by flies, ants and honeybees which transfer the sticky pollen to stigmata. Preliminary studies on lucerne (*Medicago sativa*) in Tanzania revealed that honeybees tripped 53.6% of flowers they visited and increased seed production. Coffee does not require insect pollination but it is benefited by pollinating insects. Coffee plants flower profusely for only a few days and bee colonies may be required for only a week. Flowering varied with altitude and so colonies could be used more than once by moving them from one altitudinal zone to the other. In addition, Tanzania has set research apiaries around coffee plantations for more studies of its pollination. Stingless bees like *Meliponula* (*Axesto-trigona*) *togoensis* have been kept in modern hives in Tanzania but no investigations have been tried on crops that benefit from these bees. Stingless bees frequently visit coffee and other crops elsewhere (Sections 1.3-1.5, **Appendix I**).

Although improved techniques in beekeeping were

introduced on a large scale since the early 1970s in Kenya, few commercial uses of honeybees in pollination have been initiated. Pigeon pea (*Cajanus cajan*) has been found to increase outcrossing of two inbred lines by 25-94.5 % when insects were used. In Kenya, the major insects found visiting the pigeon pea included 10 species of *Chalicodoma*, 5 of *Megachile*, 5 of *Xylocopa*, *Amegilla plumipes*, *Apis mellifera*, *Lampides boeticus* and *Thyreus*. Some trials in Kenya have shown that of all insect visitors of sunflower, honeybees made up 86% and wild bees 33%. It was also found that it was beneficial to place hives in pyrethrum fields to improve seed yield and pyrethrin content. In Madagascar, honeybees visited *Grewia calvata*, *G. fancherei*, *Dombeya elliptica*, lalona (*Weinmamia bojeriana*), *Mimosa* and *Eucalyptus*, in addition to crops such as *coffee*, *cocoa*, *vanilla*, *clove*, *basil*, *ylang-ylang* and *patchouli*.

Oil palm - *Elaeis*

Ficus sycomorus is chiefly found in riverside forests in East Africa and is pollinated by agaonid wasps, especially *Ceratosolen arbicus*, *C. galili* and *Sycophaga sycomori*. Tanzanian work by Smith led to compilation of an important reference to honey plants and their distribution, flowering periods and identification of their pollen. Similar studies have been done by Dale and Greenway (1961) and Mwangi (1990) in Kenya and Lind and Tallantire (1962) in Uganda. Although none of these researchers studied pollination requirements for each plant, such studies are important as reference bases for further research.

South Africa. This region comprises mainly the S. African Republic and adjoining countries. In the north along the Tropic of Capricorn, there is the Kalahari desert and in the south, climate tends again to the Mediterranean. Honeybees are important pollinators of subtropical crops in South Africa (see also Section 1.3). Annual yields from *litchi*, *macadamia* and other fruit or nut crops have been found to be closely related to hive numbers in the orchards. However, in many cases, no recommendation for commercial use of honeybee colonies in pollination exist. Before 1986, less than 5% of Transvaal beekeepers were involved in pollination of subtropical crops although colonies were taken to *litchi*, *macadamia* and *mango* orchards primarily to obtain honey. A recent estimate is that 52,000 bee colonies are needed for pollination of apples, pears and plums in the cape province, but only 12,000 are employed at present. There is great need for cooperation between beekeepers and growers of such crops to increase their production. In Transvaal, some beekeepers move honeybee colonies to crops such as kidney beans for pollination. They pay the growers a small fee or give them some honey.

Nectar production by 17 cultivars of sunflower (*Helianthus annuus*) in South Africa differ significantly. Honeybees and short-tongued Halictidae are unable to reach all the nectar in certain cultivars. Commercial sunflower grown in cages excluding pollinators had an average of 45% seed-set compared to 73.5% seed-set in open fields. Introduction of honeybees to cages increased seed-set to 72%, spotted maize beetles did so by 76%, American bollworm larvae (*Heliothis armigera*) produced an increase of 44% and house flies 38%.

Honeybees studied visiting *Citrus* did not collect its pollen but landed on the flower petals to search for nectar without touching stigmata. However, during takeoff, bees occasionally touched stigmata, resulting into cross pollination. Kiwifruit (*Actinidia deliciosa*) can be pollinated by hand dusting methods using preserved and dry pollen. However, naturally it is pollinated by insects. Honeybees, placed in almond orchards close to a pasture with *Aloe*, collected 40-46% almond pollen and 49-51% aloe pollen but there was no significant reduction in fruit set in almond. These results show that although alternative sources of forage can attract bees away from almonds, successful cross-pollination can still be achieved.

There are 127 species of Aloe in South Africa but their value in beekeeping has not been determined. The floral morphology of *Aloe ferox* suggests it is pollinated by birds, but different bee species, especially *A. mellifera*, *Allodapula variegata* and *Lasioglossum*, formicine ants, chalcid wasps, syrphid flies and five species of birds have been observed to visit its flowers. *Aloe marlothii* is one of the common species. Honeybees collect its pollen and carpenter bees nest in the stems and may also effect pollination. Evidence has been found for hybridization between *Ficus nataliensis* and *F. thonningii* in Natal, where both species are pollinated by *Elizabethiella stuckenbergii*. *Ficus natalensis* in West and Central Africa is pollinated by *Alfonsiella fimbriata* .The adults of *E. baijnathi* were attracted to *F. burtt-davyi* but its attractiveness disappeared after pollination. This indicated that some volatile compound was released before pollination but not afterwards. Agaonid wasps also pollinate other *Ficus* spp. The effect of *Pheidole megacephala* and *Polyrachis schistacea* (Formicidae) visiting fig trees to tend the *Ceratosolen capensis* (Agaonidae), and seed predation by formicids, reduced parasitism of pollinators by *Apocryptas guineensis* and seed predation by *Idarnes*. Insect pollinators have been reared from 70 species of *Ficus* and classification of the pollinators conforms with that of their host figs.

The common plants visited by bees including growth type, flowering periods and usefulness to bees have been studied in Zimbabwe and in South Africa. These studies serve as a reference of identification of African flora for future research.

6) Recommendations for future work.

Although the foregoing reports do not give a wide coverage of most crops in most African countries, they indicate that information on pollination is needed and there is potential to increase crop yields and quality by pollinators. The means to achieve this will primarily be honeybees and wild bees. To harness this potential, the following recommendations are made:

- In some countries, plant and insect collections are lacking because during pre-independence days, colonial governments utilized laboratories and built fine taxonomic collections which were then taken out of Africa. Identification of native species, especially crops and their pollinators, should be encouraged.

- Crop breeding and pollination research should be integrated. In Kenya, studies have found that the highest-yielding cashew trees were the most attractive to honeybees. This suggests that later in breeding programs, it might be necessary to select for high yield and attractiveness to bees (see Section 2.3.1).

- Beekeepers, crop growers and state extension workers should be educated about crop pollination especially, through beekeeping programs. At present in most African countries, the adult literacy level has increased and in addition, younger literate persons are increasingly becoming farmers. It would therefore be easier for them to understand written information in the form of simplified pamphlets, newsletters and booklets. Such booklets have been prepared in South Africa. Farmers can also benefit from radio and television broadcasts.

- Care should be exercised when importing plants and pollinators. For instance *Prosopis* (an arid climate shrub) was introduced as a multipurpose forage crop in the dry part of Kenya. This, however, maybe potentially a hazardous practice since introduced *P. juliflora* may achieve such a rapid growth and dispersal that it attains pest status. Introduced species may also attract native pollinators away from native flora.

- There is a good possibility that there are African honeybees which are naturally less defensive, less excitable and less inclined to abscond and swarm. Selection for such desirable characteristics would allow beekeepers to practice migratory beekeeping for crops requiring pollination. This will also enable them to keep their bees in areas of high agricultural activity.

- To ensure a reliable source of pollinators, both managed and wild, a more comprehensive strategy for management of crop pollination is needed. This can be achieved by educating farmers and beekeepers about pollinators, their value, habitats and conservation. Pollinators should be protected from pests, predators and poisoning by insecticides.

Additional References

Abd-El-Aal, Y. A., El-Sabae, A. H. 1978. Alkaline hydrolysis and insecticidal activity of three carbamates to honeybees. Nippon Noyakn. Gakkaishi, 3:299-300.

Abd-El-Rahim, W. A., Ashry, M. A., Lamie, D, Shawer, M. B. 1975. Preliminary note on reducing poisoning hazards to honeybee *Apis mellifera*. Mededelingen Van de Faculteit Landbouwitenschappen, Rijksuniversiteit Gent 40:345-349.

Abd-El-Roaaf, T. K, Fahmy, M. A. H., Fukuto, T. R, El-Sebae, A. H. 1977. Anticholinerase activity and toxicity of subst.itutedphenyl methylcarbanates to honeybee. J. Econ. Entomol. 70:78-82.

Abd-El-Roaaf, F. K., Yehia, A. I. A., Gahmy, F. M. A. H., El-Sebae, A. H. 1977. The impact of temperature and structure on toxicity of phenyl methyl carbamates to honeybees. In: Pests and Diesease Proc. of the 1977 British Crop Protection Conference, 21st to 24th Nov. 1977, Brighton, U.K, (1977):741-746.

Abd-el-Wahab, A. M., Khalil, F. M., Hussein, M. H. 1973. Toxicity of insecticides to honeybees in laboratory and field (Hymemoplera:Apidae). Bulletin of Entomological Society of Egypt, Economic series (1973) 7:83-89.

Ahmed, H. M. H, Sidding, M. A., El-Sarrag, M. S. A.1989. Honeybee pollination of some cultivated crops in Sudan. I,n: Proc. of 4th Int. Conf. on Apic. in Trop. Climates, Cairo, Egypt. Nov. 1988. London: International Bee Research Association (1989):100-108.

Al-Fattah, M.A.A. 1991. Behaviour of honeybee foragers and their influence as pollinators on summer squash flowers (*Curcubita pepo* L.) Bull. of Fac. Agri. Cairo University 42:1169-1186.

Ambougou-Atisso, V. 1987. Contribution of Palynology to a study of the ecology of social bees in the Makokou region (N.E. Gabon) Travaux Document de Geographie Tropicale 59:16.

Amoako, J. and Yeboah-Gyan, K. 1991. Insect pollination of three solanaceous vegetable crops in Ghana with special reference to the role of African honeybee (*Apis mellifera adansonii*) for fruit set. In: van Heemert, C., Ruijter, A. de, (eds.), The Sixth Int. Symp. on Pollination Tilburg, Netherlands, 27-31 August 1990. Netherlands, Research center for insect pollination and beekeeping, and International Society for Horticultural Science (1991), pp. 255-259.

Anderson, R.H. 1986. Pollination of apples and Pears. Elgin, S. Africa, Elgin Cooperative fruit growers (1986), 44 pp.

Attallah, M. A., Aly, F. K., El-Sbah, H. M. 1989. Pollen gathering activity of worker honeybees on field crops and medicinal plants in Minia region, Middle Egypt. In: Proc. of 4th Int. Conf. on Apic. in Trop. Climates, Cairo, Egypt. Nov. 1988. London: International Bee Research Association, pp.109-115.

Attallah, M.A. 1989. Studies on the major nectar sources in relation to honeybee activity in Minia region, Middle Egypt. In: Proc. of 4th Int. Conf. on Apic. in Trop. Climates, Cairo, Egypt. Nov. 1988. London: International Bee Research Association, pp. 93-99.

Atallah, M.A, Aly, F. K., Eshbah, H. M. 1989a. Pollen gathering activity of worker honeybees on field crops and medicinal plants in minia region, Middle Egypt. In: Proc of 4th Int Conf. on Apic. in Trop. Climates, Cairo, Egypt. Nov 1988. London: International Bee Research Association, pp.109-115.

Baker, H. G. 1961. Adaptation of flowering plants to nocturnal or crepuscular pollinators. Quart. Rev. Biol. 36: 64-73.

Baker, H. G. 1973. Evolutionary relationships between flowering plants and animals in American and African tropical forets. In: Tropical forest ecosystems of Africa and South America: A comparative review. B. J. Meggers, E. S. Ayensu,W. D. Duckworth (eds.). Washington, D. C., Smithsonian Institution Press, pp. 145-159.

Bakhett, B. R. 1989. Pollination and seed set in different genotypes of Egyptian clover (berseem, *Trifolium alexandrinum* L.) Assiut J. of Agri. Sci. 20(1):199-208.

Bartilol, P. 1976. Research and training in Apiculture- "The Kenyan point of view" In: Sec. Int. Con. in Apic. in Trop. Climate, Feb 29-Mar 4 1980:29-31

Baudoin, W. O. 1988. Vegetable production under arid and semiarid conditions in tropical Africa. Food and Agriculture Organization Plant Production and Protection Paper No. 89, Rome: FAO, 434 pp.

Beyleveld, G. P 1968a. Nectar and pollen producing trees and plants of the Transvaal. S. African Bee Journal. 40:10-11.

Beyleveld, G. P. 1968b. Detailed list of nectar and pollen producing plants of Economic Importance in the Transvaal. S. African Bee Journal. 40:13-14.

Botha, J. J. C. 1975. Observations on the banded bee pirate wasp (*Palarus latifrons*). S. African Bee Journal 47:16.

Botha, J. J. C. 1970. About enemies of bees in South Africa Glean. Bee Cult. 98:100-103.

Buys, B. 1987. Competition for nectar between Argentine ants (Tridomyrmex humilis) and honeybees (*A. mellifera*) on black iron bark (*Eucalyptus sidedroxylon*) S. Afri. J. of Zoology. 22:173-174.

Buys, B. 1972. *Nosema* in Brood. S. African Bee Journal 44(6):2-4.

Buys, B. 1977. A *Nosema* disease affecting honeybee brood. In: Biological aspects of Nosema disease. Symposium of bee biology and pathology, Merelbeke, Belgium, July 14-16, 1976. Bucharest, Romania: Apimondia Publishing House, pp. 73-76.

Carricaburu, P. 1972. Effect of Parathion on the frequency-transmission function of the compound eye of the honeybee. Sciences Natururelles 275: 2029-2031.

Chandler, M.F., Mdemu, E. 1975. Pollination research in East Africa: problems and promises. In: Proc. III Intl. Symp. Poll., Prague 15-18 May, 1975, London: International Bee Research Association, pp. 57-61.

Clauss, B. 1985. The status of the banded bee pirate (*Palarus latifrons*), as a honeybee predator in Southern Africa. In as above 1984:157-159.

Compton, S.G, Robertson, H.G. 1988. Complex interactions between mutualisms: ants tending homopterans protect fig seeds and pollinators. Ecology. 699:1302-1305.

Crane, E. 1992. The world's beekeeping past and present. In: The hive and the honeybee, J. M. Graham, (ed.). Hamilton, Illinios: Dadant and Sons, pp. 1-24.

Crisp W.F. 1957. About South African honey flora. S. African Bee Journal 32: 1,3,5, 32: 12-13 and 32:7.

Dale, I. R., Greenway, P. J. 1961. Kenya Trees and Shrubs. Buchanaris Kenya Estates Ltd. and London: Hatchards, 654 pp.

Drescher, W. 1976. Experiences with German apicultural development programmes. In: Apiculture in Tropical Climates. London: International Bee Research Association, pp. 119-124.

Eisikowitch D., Dafni, A. 1988. The use and abuse of introducing honey plants. Bee Wld. 69:12-14.

El-Banby, M.A. 1977. Ecological studies on Nosema disease in the Libyan Arab Republic. In : Biological aspects of Nosema diesease, Symposium of bee Biology and Pathology, Belgium, July 14-16 1976, Bucharest, Romania, Apimondia Publishing House, pp. 53-55.

El-Banby, M. A., Kansouh, A. S. H. 1981. The residue toxicity to honeybees on different parts of cotton plants. In. Proc. of the XXVIIIth Int. Congr. of Apiculture, Acapulco, 1981. Bucharest, Romania, Apimondia Publishing House, pp. 319-329.

Ewies, M. A. and El-Sahhar, K. F. 1977. Observation on the behaviour of honeybees on onion and theri effects on seed yield. J. Apic. Res. 16:194-196.

Fletcher, D.J. 1978. The African honeybee, Apis mellifera adansonii in Africa. Annual Review of Entomology 23:151-171.

Free, J. B. 1980. Research and training in beekeeping development programmes. In: Sec. Int. Cong. on Apiculture in Trop. Climates, Feb 29-Mar 4. London: International Bee Research Association,pp. 10-15.

Froneman, I. J. 1988. Subtropical crops. Pollination by bees. Information Bulletin, Citrus and Subtropical Fruit Research Institute, South-Africa No. 196: 12.

Froneman, I.J. 1988. Pollination (by bees) of subtropical crops. Information Bulletin, Citrus and Subtropical Fruit Research Institute, South-Africa. 194: 8.

Fry, C. H. 1969. The recognition and treatment of venomous and non-venomous insects by small bee eaters. Ibis 111: 23-29.

Galil, J. and D. Eisikowitch. 1967. On the pollination ecology of Ficus sycomorus in East Africa. Ecology 49:259-269.

Guy, R. D. 1976. Commercial beekeeping with *Apis mellifera adansonii* in intermediate and movable-frame hives. In: Apiculture in Tropical Climates. London: International Bee Research Association, pp. 31-37.

Harris, K.M. 1987. Gall midges in tropical agriculture. Insect Science and its Application. 8: 4-6, 441-445.

Hartmann, W.H. 1991. Commercial Pollination in the Western Cape. In: "Bees and Beekeeping in S. Africa.". Proc. of the Int. Beekeepers Symp, Stellenbosch, S. Africa 24-26 Jan, 1990. Eds Anderson, R.H. and Buys, B. Cape Town, S. Africa. S.A. Federation of Beekeepers Associations, pp. 21-25.

Hassanein, M. H.. 1964. The value of pollinating insects to flax seed production in Egypt. Ann. Agric. Sci:773-784.

Hassanein, M. H. 1953. Studies on the effect of pollinating insects, especially the honeybee, on the seed yield of clover in Egypt. Bull. of Soc. Fouad Ento. 37:337-344.

Hermann , J. M. 1984. The *Drosphila* fly in *Eucalyptus* flowers: Who holds the answers to this riddle? S. African Bee Journal. 56(1): 5-8.

Hermann J. M. 1983. Fruit flies and the Eucalyptus nectar flow. S. African Bee Journal 55(2): 25, 27-33.

Hoffman, M. T. 1988. The pollination Ecology of *Aloe ferox*. S. Afri. J. of Bot. 54:345-350.

Hussein, M. H., Abdel-Aal, S. A. 1982. Wild and honeybees as pollinators of 10 plant species in Assiut area, Egypt. Zeitschrift Fur Angewandte Entomologie 93:342-346.

Ibrahim, S. M. 1984. A study on a dipterous parasite of honeybee. Zeitschrift fur Angewandle Entomologie 97:124-126.

Johannesmeir, M.F. 1976. Aloe marlothii in South Africa. S. African Bee Journal 48:10-14.

Kambal, A.E. 1972. Flower drop and fruit set in field beans (Vicia faba L.) J. Agri. Sci. Camb. 72:131-138.

Kroll, M. 1962. The influence of fertilization on the production of pyrethrins in the pyrethrum flower. Pyrethrum Post 6:19-21.

Millogo-Rasolodimby, J. 1989. Burkina Faso: Importance of beekeeping of the butter tree, *Butyrospernium paradoxum* and the Locust bean tree, *Parkia biglobosa*. Revue Francaise d'Apiculture 482:72-74.

Kevan, P.G., Clark, E. A., Thomas, V. G. 1990. Insect pollinators and sustainable agriculture. American Journal of Alternative Agriculture 5:13-21.

Kigatiira, I. K. 1974. Hive designs for beekeeping in Kenya. Proc. Ento. Soc. Ont. 105:118-128.

Kigatiira, I. K. 1976. Keeping bees in fixed comb and movable comb frameless hives. In: Apiculture in Tropical Climates. London: International Bee Research Association, pp. 9-13.

Kigatiira, I. K. 1984. Some aspects of the ecology of African honeybees. University of Cambridge, London. Ph.D. thesis.

Kigatiira, I. K., Kahenya, W. A., Townsend, G. F. 1988. *Prosopis* spp. as multi-use trees in dry zones. Bee Wld. 69:6-11.

Lange, J. H. De, Vincent, A. P. 1972. Evaluation of different cultivaters as cross pollinators for the Washington Navel Sweet Orange. Agroplantae 4:49-56.

Latique, P. 1954. Mother of honey, the bee of Madagascar. Climats, France Et Outre-mer 12954:1 page.

Lind, E. M.,Tallantire, A. L. 1962. Some common flowering plants of Uganda. London: Oxford University Press, 257 pp.

Mariau, D., Houssou, M., Lecoustre, R., Ndigui, B. 1991. Oil palm pollinating insects and fruitset rates in West Africa. Oleagineux-Paris. 46: 43-51.

Mbaya, J. K. S. 1993. PhD thesis, (in preparation).

McGregor, S. E. 1975. Insect pollination of tropical crops. In: Proceeding of the III International Symposium on Pollination IBRA, Prague 15-18 May 1975: 47-55.

Nightingale, J. M. 1976. Traditional beekeeping among Kenya tribes, and methods proposed for improvement and modernization. In: Apiculture in Tropical Climates. London: International Bee Research Association, pp. 15-22.

Ndiaye, M. 1976. Beekeeping in Senegal. In: Apiculture in Tropical Climates. London: International Bee Research Association, pp. 171-179.

Noort, S. van, Ware, A.B. , Compton, S.G., Van-Noort, S. 1989. Pollinator specific volatile attractants released from figs of *Ficus burtt-davyi*. South African Journal of Science. 85(5): 323-324.

Onim, J. F. M, Eijnatten, C. L. M., Pathak, R. S. 1979. Some factors which encourage outcrossing in Pigeon Pea. In: Proc. 4th Intl. Symp. Pollination, D. Caron (ed.). College Park, Maryland: Maryland Agri. Exp. Sta., Misc. Publ. 1, pp. 219-255.

Onim, J. F. M, Pathak, R.S., van Eijnatten, C. L. M, 1978. Influence of insect pollinators on the degree of Outcrossing in Pigeon Pea in Kenya. Proc. IVth Intl. Symp. on Pollination. College Park, Maryland: Maryland Agric. Exp. Sta. Misc. Publ. 1, pp. 211-218.

Onim, J.F.M, Pathak, R.S. and Van Eijnatten, C.L.M, 1978 b. Some factors which encourage outcrossing in Pigeon Pea. Proc. IVth Intl. Symp. on Pollination. College Park, Maryland: Maryland Agric. Exp. Sta. Misc. Publ. 1, pp. 219-225.

Oni, O. 1991. Insect pollinators of the West African Hardwood, (*Terminalia ivorensis*). In: The Sixth Int. Symp. on Pollination, Tilburg, Netherlands, 27-31 August 1990. Eds. Heemert, C and Ruijter, A. Pollination and Beekeeping and International Society for Horticultural Science, pp. 88-92.

Opunte, F.I. 1975. Identification of P-methoxyallybenzene in pollen of oil palm *Elaeis guineensis* Jacq. J. of Expt. Bot. 26:619-623.

Papdopoulo, P. 1970. Rhodesian indigeneous trees and shrubs. Salisbury Dep. Conservation and extension 1970, 15 pp.

Papadopoulo, P. 1969. Bulletin on some *Eucalyptus* spp. (suitable for honey and pollen production) in Rhodesia. Salisbury: Rhodesia Department of Conservation and Extension 1969, 4 pp.

Papadopoulo, P. 1968. Flowering Bulletin No 1. Rhodesian indigenous trees and shrubs flowering during January, Feb and March. Beekeepers observations salisbury, Rhodesia. Dep. of Conservation and Extension 1968, 9 pp.

Papadopoulo, P. 1966. Bulletin on *Eucalyptus* spp. Salisbury: Rhodesian Conservation and Extension, 1966, 4 pp.

Papadopoulo, P. 1966. Flowering Bulletins Nos. 1-4. Rhodesian indigenous trees, shrubs and wild flowers, cultivated crops, trees and flowers. Salisbury: Rhodesian Dep. Conservation and Extension 1966, 32 pp.

Parmer, R. 1969. Bees and bean pollination. S African Bee Journal 39(4):12-13.

Purseglove, J. W. 1991. Tropical Crops (Dicotyledons), Singapore: Longman Singapore Publishers Pte Ltd, 719 pages.

Rashad, S. E. 1976. Experience with pollination of tropical crops. In: Apiculture in tropical climates. E. Crane (ed.). London: International Bee Research Association, pp. 109-112.

Rashad, S.E. and Eweis, M.A. 1985. *Chalicodama mucorea*, a promising Egyptian species for Alfalfa pollination. Proc. 3 Int. Apic. Trop. Climates, Nairobi 1984:134-137

Rashad, S.E., Ewies, M.A. and El-Rabie, H.G. 1979. Insect pollinators of Sesame (*Sesame indicum* L.) with special reference to the role of honeybees. In. Proc of 4th Int Symp. on Pollination, Maryland 1978 (1979): 231-234.

Rashad, S.E., M.A. Eweis and H.G. El-Rabie 1978. Pollination of Peanut (Arachis Hypogaea L.) and the effect of honeybees on its yield. Proc.IVth Int. Symp. on Pollination, Md Agric. Exp. Sta. Spec. Misc. Publ.1:227-230.

Rashad, S.E., M.I. Mohamed and A.M. El Hefny 1978. Some notes on the biology and behaviour of *Nomia unidentata* Oliver. In: Proc. 4th Intl. Symp. Pollination, D. Caron (ed.). College Park, Maryland: Maryland Agri. Exp. Sta., Misc. Publ. 1, pp. 313-319

Remminghoff, V, Ritter, W. 1980. Results of the treatment of Varroa in Tunisia. Allgemeine Deutsche Imkerzeitung 14:150-151.

Reynolds, G.N. 1966. The Aloes of Tropical Africa and Madagascar, Trustees, Aloes Book Fund, P.O. Box 234, Mbabane Swaziland, 537 pp.

Saleh, R. S, El-Dakhakhni, T. N., Shawer, M. B., Hendawy, A. S. 1989. Toxic effects on worker honeybees on seven tropically applied insecticides. In: Proc. of 4th Int. Conf. on Apic. in Trop. Climates, Cairo, Egypt. Nov. 1988. London: International Bee Research Association, pp.231-236.

Salman, A. G. A, Abd-El-Raof, T. K. 1979. Effects of certain pesticides used against cotton pests on three predaceous insects and honeybee workers. Bull. of the Entomol. Soc. of Egypt. 1978/79 11:155-162.

Schneieder, S.S. 1990. Queen behaviour and worker-queen interactions in absconding and swarming colonies of the african honeybee, *Apis mellifera scutellata* (Hymenoptera:Apidae). J. Kansas Entomol. Soc. 63: 179-186.

Shawer, M.B, Salem, M. M., Saleh, R. S. 1989. Survey of insect pollinators of Egyptian clover (*Trifolium alexandrinum* L.) at Kafr El-Sheikh Egypt. In: Proc. of 4th Int. Conf. on Apic. in Trop. Climates, Cairo, Egypt. Nov. 1988. London: International Bee Research Association, pp.134-137.

Shawer M. B, Saleh, R. S., El-Dakhakn, T. N.,Hendawy, A. S. 1989. Attempt to reduce the hazard to honeybee colonies of Pesticide applications. In: Proc. of 4th Int. Conf. on Apic. in Trop. Climates, Cairo, Egypt. Nov. 1988. London: International Bee Research Association, pp. 237-239

Smith, F.G. 1958. Beekeeping observations in Tanganyika 1949-1957. Bee Wld. 39(2): 29-36.

Smith, F.G. 1956. Bee botany in Tanganyika. Scotland, UK.: Univ. Aberdeen: Thesis.17 pp.

Staden, D. F. A. von. 1987. Kiwifruit pollination in South Africa. South African Bee Journal.59(3):59-61.

Steentoft, M. 1988. Flowering plants in West Africa. Cambridge, UK: Cambridge Univ. Press, 344 pp.

Syed, R.A. 1979. Studies on oil palm pollination by insects. Bull. Ent. Res. 69:213-224.

Toit, A.P. du. 1990. Pollination of almonds in the Transvaal. J. of the S. Afri. Society of Hort. Sci. 1:18.

Toit, A. P. du. 1990. The importance of certain insects as pollinators of sunflower (*Helianthus annuus* L) S. Afr. J. of Plant and Soil 7:159-162.

Toit, A.P. du 1990. Pollination research: a missing link in subtropical fruit production. Acta Horticulturae. 275: 239-243.

Toit, A. P. du. 1988. The pollination ecology of commercial sunflower (*Helianthus annuus* L.) in South Africa with special reference to honeybees (*Apis mellifera*). S. Africa: Thesis, University of Pretoria,143 pp.

Toit, A. P. du, Coetzer, L. A. 1991. Difference in Potential attractiveness to bees of seventeen S. African sunflower (*Helianthus annuus*) . In: The Sixth Int. Symp. on Pollination, Tilburg, Netherlands, 27-31 August 1990. C. van Heemert, A. de Ruijter (eds.). Pollination and Beekeeping and International Society for Horticultural Science, pp. 288-293.

Toit, J. T. du, Du Toit, J. T. 1990. Giraffe feeding on *Acacia* flowers: predation or pollination? African Journal of Ecology. 28:63-68.

Townsend, G.F. 1976. The Canadian concept of apicultural development and training programmes. In: Apiculture in tropical climates. E. Crane (ed.). London: International Bee Research Association, pp.115-118.

van Drunen, B. 1977. Pollination of fruit tree in the Western Cape. In: African bees: taxonomy biology and economic use. D. J. C. Fletcher (ed.). S. Africa: Pretoria, Apimondia, pp. 4-6.

Wafa, A. K., Ibrahim, S. H. 1964. The effect of the honeybee as a pollinating agent on the yield of broad bean. Bull. Fac Agric. Cairo University No. 205, 36 pp.

Wafa, A. K., Ibrahim, S. H. 1964. The effect of the honeybee as a pollinating agent on the yield of clover and cotton. Bull. Fac. Agric. Cairo University No. 206, 44 pp.

Wafa, A.K., Ibrahim, S. H. 1960. The honeybee as an important insect for pollination. Bull. Fac. Agric. Cairo University No. 162,13 pp.

Wafa, A. K, Sharkawi, S. G. 1972. Contribution to the biology of Vespa orientalis Fab. (Hymenoptera, Vespidae). Bulletin de la Societe Entomologiques d' Egypte 56:219-226.

Wafa A. K., Ibrahim, S. H., Eweis, M. A. 1975. Insect pollinators of alfalfa (Medicago sativa L.) Agricultural Research Review 53:199-207.

Wafa, A. K, Ibrahim, S. H., Eweis,M. A. 1972. Studies on the effect of honeybee as a pollinator on the yield of alfalfa (*Medicago sativa* L.: Hymemop-tera:Apidae). Bulletin de la Societe Entomologique d' Egypte 56:273-285.

Wafa, A. K, Rashad, S. E., Eweis, M. A. 1974. The Efficiency of honeybees in pollinating alfalfa in Giza region (Hymenoptera, Apidae). Bulletin de la Societe Entomologique d' Egypte. 58:117-121.

Wiebes, J. T. 1989. Agaonidae (Hymenoptera Chalcidoidea) and Ficus (Moraceae): fig wasps and their figs, V (Agaon). Proceedings ofthe Koninklijke Nederlands Akademie van Weten-schappen. Series C. Biological and Medical Sciences 92: 395-407.

Yeboah, G. K. and S. R. J. Woodell. 1987. Analysis of insect pollen loads and pollination efficiency of some common insect visitors of four species of wood Rosaceae. Functional Ecology 1:269-274.

Yehia, Al. A, Fahmy, M. A. H, El-Sebae, A.H, 1977. Molecular modification carbamates in relation to their selctive toxicity. In: Proceedings of the 1977 British Crop Protection, pp.155-159.

1.5.3 Applied Pollination in Tropical America

Constant habitat alteration and the spread of African honeybees have greatly modified pollination ecology in the Neotropics. Prior to the invasion of the African honeybee from southern Brazil, perhaps six million hived colonies of European *Apis mellifera* were kept in the Neotropics. There was no feral honeybee population that was independent of apiaries. European honeybee colony density was very high (1-10 colonies/ km^2) in the Yucatan peninsula, El Salvador, and the subtropics of southern Brazil and northern Argentina. As a rule, these honeybees were completely absent in the Amazon basin, and in all humid or rainy tropical lowlands and mountains.

Now there are, conservatively, an average of several 'wild' African honeybee colonies per square km over an area of roughly 20 million square kilometers — 10 times the number of European apiary colonies that had been maintained up until the 1970s. Virtually no European colonies remain, even though their range was less than 10% the current range of African honeybees. Thus, the number of honeybees has greatly increased, they are a different 'ecotype' and subspecies, and their distribution is general, rather than isolated in apiaries primarily in drier or subtropical agricultural areas. They have caused the demise of some native bees, like species of *Melipona*, social bees similar in size to *Apis*, in northern South America. This happened gradually over a period of 10 years. In the forested areas of Yucatan, where beekeeping with the Mayan honey-making stingless bees, *Melipona* and *Cephalotrigona*, has been important for many centuries, there has been a sudden decline in honey production that coincides perfectly with arrival and proliferation of the honeybees from Africa. Little other documentation exists, however, on impact by African honeybees either on native animals or plants.

The fact that so many cultivated plants used in tropical countries originated in the Neotropics (**Apppendix I**) does not mean that their reproductive biology has been better studied there (see especially Purseglove's treatments of tropical crops). Applied pollination technology in tropical America is mostly devoted to exotic species — *coconut, coffee, citrus, mangoes, teak, melons, apples, peaches* and *cardamom*. European honeybees had been imported for commercial pollination of all these crops. African and 'Africanized' honeybees have replaced them (although often not as managed pollinators). Nonetheless, many Neotropical flower visitors played a part in pollination, but have yet to receive recognition or, in most cases, systematic study. The native crops, fortunately, still seem to have enough native pollinators to service them. Propagation, seed-set and fruit production of these indigenous species do not desperately need to be bolstered by management schemes or other means, at present.

Many forest crops gathered in the tropics come from America. Notable among them are the 'Brazil nut' (*Bertholletia*), rubber (*Hevea*), Pejibaye palm or chonta duro (*Bactris*), kapok (*Ceiba*), vegetable ivory (*Phytelephas*), chicle (*Manilkara*), nanche, nance, or mirici (*Byrsonima*), hog-plum(*Spondias*), prickly pear or nopal (*Opuntia*) and the piquis (*Caryocar*). As elsewhere in the tropics, native palms such as *Bactris, Acrocomia, Attalea, Astrocaryum, Ceroxylon, Chamaedora, Elaeis, Iriartea, Jessenia, Mauritia, Maximiliana, Oenocarpus, Orbygnia, Phytelephas, Scheelea* and *Syagrus* provide many useful products that are not exported. They are among the most valuable of all plants but infrequently produce cash crops or revenue. Their breeding systems often involve dioecy and the pollinators are flies, beetles and bees. Honeybees and stingless bees may visit them in large numbers, but they are often merely thieves — never visiting female inflorescences and therefore not pollinating.

A few native Neotropical crops like *pineapple*, and non-native crops like *banana*, have no pollination requirements — in addition to being propagated vegetatively they are parthenocarpic. However, most crops require pollinators, and native Neotropical crops grown on other continents are numerous. Among the most important are *cacao, sunflower, passion fruit, chili peppers, eggplant, sisal, annatto, cashew, soursops, peanut, star apple, papaya, quinine, chayote, rubber, gourds* and *squash, tomato, guava, cotton, potato, pineapple, vanilla, sweet potato, cassava, tobacco, guayule, avocado*, the *common bean, pimento, mesquite* and *jojoba*.

Agave in full flower

None of these plants was dependent in any way on honeybees, because no *Apis* were present in tropical America until the late 1600s, and almost certainly none were widespread until the last 20 to 30 years due to release of *Apis mellifera scutellata* in Brazil. Natural selection and evolution produced by interactions between plants and a flower visitor such as honeybees could result in very slight changes to plants and their breeding systems, if any, in such a short time.

For native American plants, outcrossing by bees seems the most common mode of reproduction. This is not surprising considering that many more bee species live in the Neotropics than in other equatorial regions. However, modes of crop reproduction vary widely and incorporate all pollinator groups (see Section 1.1.1), as shown below. Pollination of major cultivated plants in the Neotropics is reviewed here, emphasizing the native species of flowering plants.

Agave — sisal, henequen (Agavaceae).
A. sisaliana and several other species are cultivated for fiber in Mexico and Central America, where they originated, and also Africa, Asia, and the Philippines, largely for local use. Flowers are pollinated at night by bats, but also are visited during the day and pollinated by bees. Individual flowers are hermaphrodite but pollen is shed before stigmas are receptive. Selfing within a flower does not occur but, because flowers in an inflorescence open sequentially over a period of days, selfing within inflorescences is possible. Nocturnal pollinators include the bat genus *Leptonycteris*, diurnal pollinators are normally the larger bees — *Xylocopa*, *Mesoxaea*, and others. Because individual inflorescences provide a large amount of resource, introduced social bees such as *Apis* forage profitably and recruit in large number.

Cashew - *Anacardium*

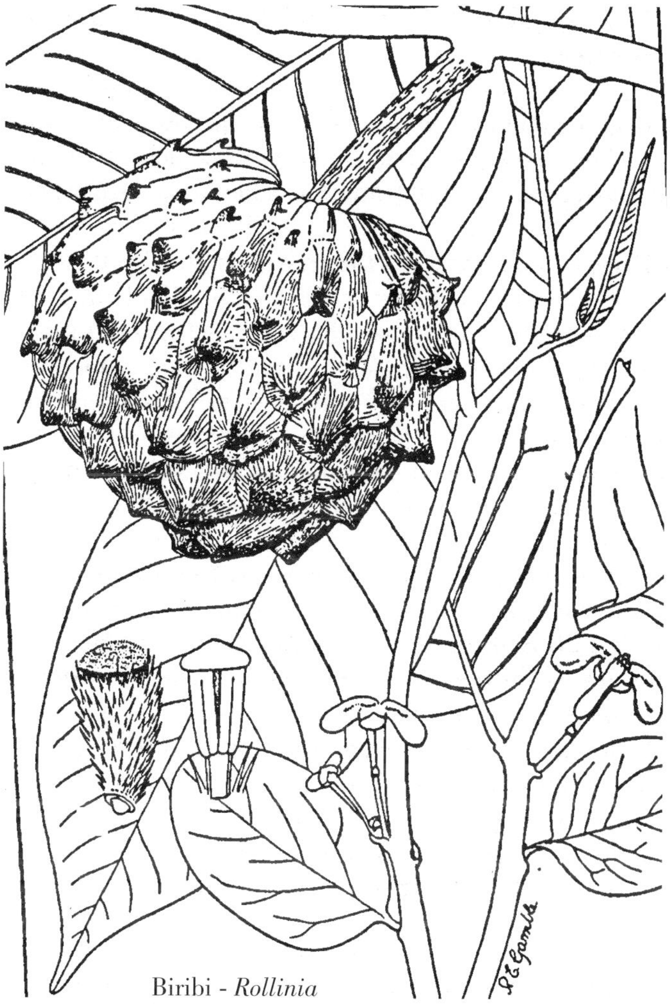

Biribi - *Rollinia*

Anacardium — cashew (Anacardiaceae).
A. occidentale is a small tree producing fruit with a swollen fruit stalk. The former is roasted and exported as a nut crop — extensively in Asia and the Neotropics, while the later is sold locally as an edible fruit. Cashew is insect-pollinated but, because of its tiny flowers and minuscule amount of nectar, it is all but ignored by bees. Flies appear the major pollinators. Flowers are both male and hermaphrodite, with the former shedding pollen earlier in the day. Self-compatibility and selfing have been shown, but cross- pollination greatly improves fruit production, and the breeding system naturally favors outcrossing.

Annona & Rollinia — soursop & biribi (Annonaceae).
Annona and *Rollinia* are small trees that have hermaphrodite, pollen-only flowers in which the stigma is receptive well before anther dehiscence — selfing is rare, flowers are self-compatible, and outcrossing is the rule. The visitors and pollinators have seldom been observed but include beetles, especially the dynastine scarab, *Cyclocephala*.

Arachis — peanut (Legumino-sae).
A. hypogaea originated in Bolivia (30 species exist) but, being tetraploid, is not a wild species and originated from two diploid species. Cultivars of three groups are grown throughout the tropics and subtropics, with Asia surpassing Africa and America in production. Crops are ordinarily self-pollinating and self-fertile, but the original pollinators

are thought to be xylocopine and megachilid bees. In fact, some varieties are mechanically incapable of selfing, and the hermaphrodite flowers must be visited by an insect — thrips serve as pollinators in parts of Africa, but bees of many different kinds are the main cross-pollinators.

***Bixa* — annatto (Bixaceae).**

Bixa orellana is a pollen-only flower, originally from the lowlands of Ecuador and the eastern Andean slopes. The breeding system of this small tree has not been studied in detail, but self-compatibility with outcrossing by bees seems likely. Seeds are coated with a red-orange dye used by the food and cosmetics industries, exported from India, East Africa, Peru, Brazil, Panama, Ecuador and Jamaica. Flowers are visited heavily in early morning by *Melipona eburnea*, *M. fuliginosa*, and a variety of euglossines —*Eulaema* and *Euglossa*, as well as *Epicharis*, *Xylocopa* and small meliponines. Where these bees are absent, halictid bees such as *Augochlora* and *Pseudagochloropsis* come to the flowers. Honeybees do not find this flower attractive.

Annatto - *Bixa*

***Cajanus* — pigeon pea, juandú (Leguiminosae).**

C. cajan is a shrub that grows wild in Africa and is widely cultivated on a small scale in subtropical and tropical America. Flowers are automatically self-pollinating, even before the flower opens, but are later visited by bees and to a small extent outcrossed.

***Capsicum* — chile pepper (Solanaceae).**

C. annuum and the chili peppers (five species and two varieties) are originally Neotropical, and now are exported largely from China, Malaysia and Mexico. Flowers are hermaphrodite but stigma receptivity precedes anther dehiscence, thus outcrossing is favored. The position of the stigma beyond the reach of anthers makes selfing unlikely. Pollinators are solitary bees such as the communally-nesting *Exomalopsis*, and also the social bees, *Bombus*. Pollen is gathered by small stingless bees such as *Trigona (Tetragonisca) angustula*, and also a variety of small halictids but they probably are not effective pollinators.*Capsicum* produces nectar, but the anthers of *Capcisum*, like other Solanaceae, are poricidally dehiscent. Pollen is often the only reward to visitors of flowers with porcidally dehiscent anthers, and it is only effectively collected by some kinds of bees — excluding honeybees. However, the nectar of the chili flowers makes them attractive to honeybees, as well as flies, all of which pollinate.

***Carica* — papaya (Caricaceae).**

C. papaya is a small tree that may produce good crops for a few years. Fruit is consumed locally and seldom exported. Major producers are Brazil, India, Mexico and Indonesia. Both male and hermaphrodite flowers are borne, usually on different trees and also at different times on the same plant. Some cultivars, such as "Solo" are gyno-dioecious, having some individuals that are strictly female — one male tree per every 25 females is recommended for adequate fruit set. Pollinators are hawkmoths, *Hyles* and other species, but some hermaphrodite flowers readily self-pollinate, and apomixis cannot be ruled out for some cultivars. Nectar is provided only by male flowers, which are visited by sunbirds (Old-World tropics), hummingbirds (Neotropics), bees, flies and hesperiid butterflies (*Perichares*). So far, only hawkmoths and butterflies have been observed consistently going between male and female flowers when stigmas are receptive.

***Cinchona* — quinine (Rubiaceae).**

The bark of *Cinchona* and its several cultivated South American species is a source of drugs used against malaria, fever and cardiac arrhythmias. Vegetatively-propagated hybrids are cultivated mainly in Zaire, Indonesia, India, Bolivia, Ecuador and Peru. Flowers are hermaphrodite and dimorphic, having either protruding stamens or stigmas; they require cross-pollination, purportedly by flies, bees or butterflies.

***Citrullus* —watermelon (Cucurbitaceae).**

C. lanatus is native to Africa and now widely cultivated in the Neotropics. Honeybee hives near crops are believed to improve fruit set, but data are inadequate to evaluate the value of *Apis* or other flower visitors in the field. Flowers on a single vine are male or female, although some cultivars have hermaphrodite flowers that are self-fertile, but not self-pollinating (i.e. they need to be visited by a pollinator). Work in India shows that honeybees and *Trigona* (not *Melipona*, as given in the literature, because this genus does not exist in the Old World), frequently collect nectar and pollen at flowers.

Citrus — oranges, etc. (Rutaceae).

These small Asian trees are widely cultivated through tropical and subtropical America. To produce seeds, citrus requires a pollinator, but the same is not always true for fruit production. Agamospermy is common (see Section 1.1.3) Breeding systems vary greatly between species. Endosperm may be initiated by fertilization and normal pollination, but shortly thereafter an apomictic embryo invades the embryo sac, outcompeting the sexually-produced embryo. Flowers may be self-pollinating (see Section 1.5. and **Appendix I**). Citrus farmers in the Neotropics use honeybee colonies, now solely Africanized honeybees, within extensive plantations.

Cucumber - *Cucunis*

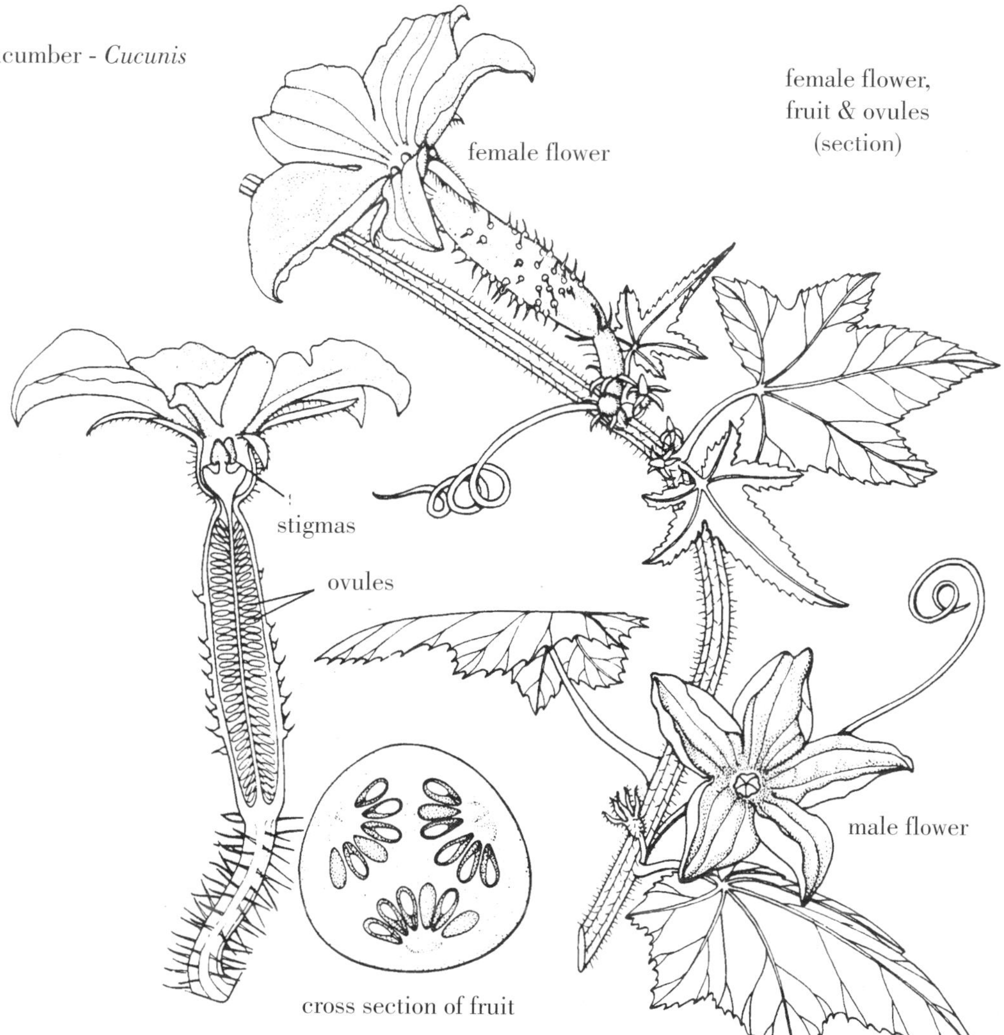

Cocos — coconut (Palmae).

This southeast Asian palm is visited heavily by honeybees and stingless bees. While the bees collect pollen, wasps and other insects also arrive and, along with bees, collect nectar. Separate male and female flowers are borne on the same inflorescence. Hives of honeybees are used for pollination in plantations. Although fruit-set is increased by pollinators, wind is another major pollinating agent and selfing is insignificant.

Coffea — coffee (Rubiaceae).

C. arabica is one of the nine African cultivated species (of 60 species in the genus) and is the only tetraploid, thus likely an artificially-selected hybrid. All diploid species are self-sterile and require outcrossing, done largely by bees but also by wind. Large Neotropical producers of *C. arrabica* are Brazil and Colombia with Costa Rica and Mexico considerably smaller exporters. Hives of honeybees are introduced in the extensive coffee plantation areas and generally are believed to improve yield. One hive each 100 m has been recommended. However, flowers are hermaphrodite, self-fertile and autogamous; they also may be amphicarpic, meaning that some flowers require visitation *and* outcrossing, while others do not. Studies on enhancement of production through cross-pollination by bees are fairly numerous; insect visitors appear not to greatly enhance seed-set and fruit maturation, although many species and genera, with *Apis* and *Melipona* predominating, gather pollen or nectar at flowers.

Cucumis — melon and cucumber (Cucurbitaceae).

Fruits of *C. anguria* (Neotropical) *C. melo* (African) *and C. sativus* (South Asian) are either cooked or eaten raw, all but the first are cultivated worldwide in the tropics. Honeybee hives are introduced to improve yield, and bees in general are strongly attracted to nectar and pollen of the flowers. Flowers are male or female, but some plants have hermaphrodite and male flowers. Plants are self-compatible but insect visitors are required, and honeybees are the most common. One foraging honeybee per each 10 hermaphrodite flowers was recommended to obtain full fruit set. Efficient pollination of cantaloupe and sweet melon in the Neotropics should require 2-3 honeybees colonies/hectare.

Cucurbita — squash and gourds (Cucurbi-taceae).

C. pepo and three other cultivated species are from a group of about 25 species, all Neotropical, originally from Mexico. These herbaceous running-vines dispersed both north and south along with their 20 species of pollinators, solitary bees of the genera *Peponapis* and *Xenoglossa*. The flowers are male or female, apparently not self-compatible among flowers of one plant, and therefore outcrossing by bees is required. Both large and small carpenter bees, *Xylocopa* and *Pithitis*, and *Apis*, among many other bee genera, also pollinate the flowers; 2-4 hives of *A. mellifera*/hectare have been used for pollination.

Elaeis — African oil palm (Palmae).

E. guineensis is primarily a vegetable oil crop cultivated extensively in western Africa and southeast Asia, but now is planted in large areas in lowland Costa Rica, Colombia and Ecuador, sometimes near congeneric species native to America with which it produces fertile hybrids. Bees visit the male inflorescences to collect pollen and are not pollinators, while beetles of the families Curculionidae (*Elaeidobius*, several species) and Nitidulidae (*Mystrops*) visit both male and female flowers and are responsible for almost all fruit set. (see Section 2.5.3).

Gossypium — cotton (Malvaceae).

Neotropical *G. hirsutum* and *G. barbadense* are the world's most extensively cultivated cottons, among the 12 cultivated species and varieties, the oldest of which is African; indigenous species are found throughout the tropics and subtropics. Extrafloral and floral nectar both attract pollinating bees, wasps and other insects. Flowers are largely self-pollinating and outcrossing is minimal, but it does improve yield. *Melissodes* and other solitary bees appear to be the most important native pollinators, but *Apis, Bombus, Xylocopa,* and a few other insects are pollinators.

Helianthus — sunflower (Compositae).

H. annuus is exported largely from Argentina but is grown worldwide, mainly in temperate climates. Like composite flowers in general, it is visited by a wide variety of native bees that may specialize on its pollen and nectar. As pollinators, the larger solitary bees and *Bombus* were undoubtedly the most important pollinators, but all the seven or so *Apis* species visit flowers where introduced. The non-hybrid varieties can self-pollinate but benefit greatly from outcrossing by bees.

Hevea — rubber (Euphorbiaceae).

H. brasiliensis is the most important of all natural, commercial elastomers. From its origin in the Brazilian Amazon it is been widely distributed for commercial use in lowland equatorial wet forests. Flowers are either male or female and self-pollination (strictly dependent on pollinators) within an inflorescence is roughly 10-20% as productive as cross-pollination. Flies at least of the genera *Dasyhela, Artrichopogan, Forcipomyia, Stilobezzia* and *Culicoides* are the main pollinators. Some stingless bees of the genus *Trigona* collect pollen, both in the New World and the Old, but provide minimal pollination, and *Apis* often do not collect the pollen at all, or only visit the extrafloral nectaries, and very seldom visit the female flowers. No species of *Apis* appears to be a reliable pollinator of this crop.

Ipomoea — sweet potato (Convolvulaceae).

I. batatas is widely cultivated, with 1000s of cultivars in three cultivated species. Production is highest in Asia, next highest in Africa and lowest in tropical America, where these crops originated. Hermaphrodite flowers open in the evening and can be pollinated until early the following morning; self-incompatibility and cross-incompatibility are known; cross-pollination (perhaps including nocturnal pollinators) is needed for seed set. Some apparently obligate visitors of this genus in the Neotropics are *Melitoma* and *Ancylosceilis* (Anthophorinae), and *Melissodes* visits the flowers where the former, native bees are absent. Many other bees occasionally visit flowers.

Lycopersicon — tomato (Solanaceae).

L. esculentum is one of the six species, all native to western South America and the Galápagos. Tomato is a pollen-only flower visited by bumblebees, the larger halictids, anthophorines and other buzz-collecting bees. Flowers are self-fertile and may self-pollinate with the action of wind or shaking, but cross-pollination is favored by stigmatic receptivity before anthers dehisce (see also Section 1.3). Parthenocarpic fruits are sometimes produced.

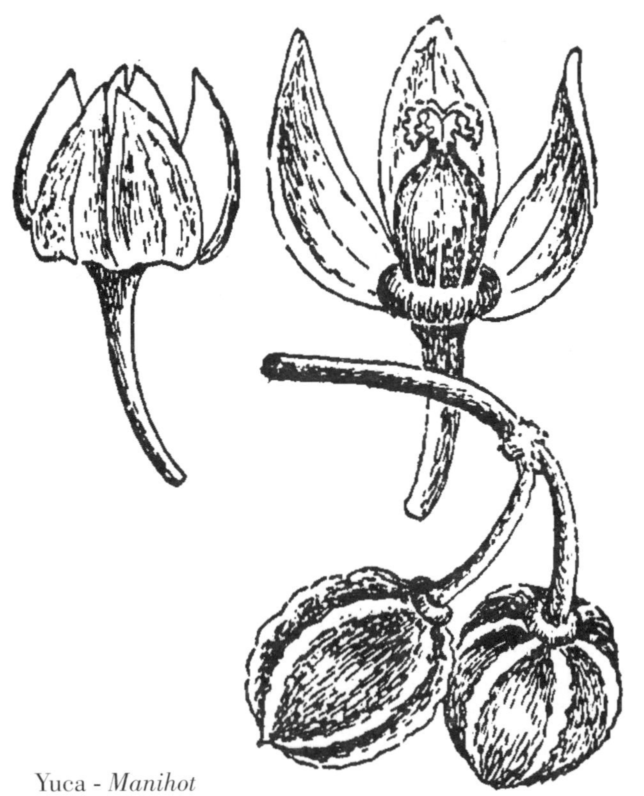

Yuca - *Manihot*

Mangifera — mango (Anacardiaceae).
M. indica is among the more than 60 Asian species; 5 others are cultivated commercially. One or two fertile stamens occur on the hermaphrodite flowers; male flowers have an aborted pistil. Stigmas are immediately receptive and self-pollination occurs, but only when visited by a pollinator or otherwise transported. Like *Citrus*, agamospermy is the rule, but endosperm tissue only results from pollination, whether or not the embryo is a clone of the female parent plant. Some cultivars, at least in Mexico, often are kept with hives of honeybees to ensure adequate fruit set. Flies and bees are the major pollinators, where pollinators are necessary, but visitors are extremely rare on the flowers.

Manihot — cassava, yuca, manioc, tapioca (Euphorbiaceae).
M. esculenta is among the 200 Neotropical species, and is grown throughout the tropics but nowhere known in the wild. Africa is the largest producer, then Asia and next the Neotropics. Flowers are either male or female and outcrossing by bees is enforced by lack of synchrony in opening by male and female flowers on an inflorescence. *Trigona* and other small to medium-sized bees are common at the flowers, collecting nectar and pollen.

Nicotiana — tobacco (Solanaceae).
N. tobacum (and probably the eight other cultivated species) is self-pollinating but hummingbirds, butterflies, bees and other insects seeking its nectar cause outcrossing.

Tobacco - *Nicotiana*

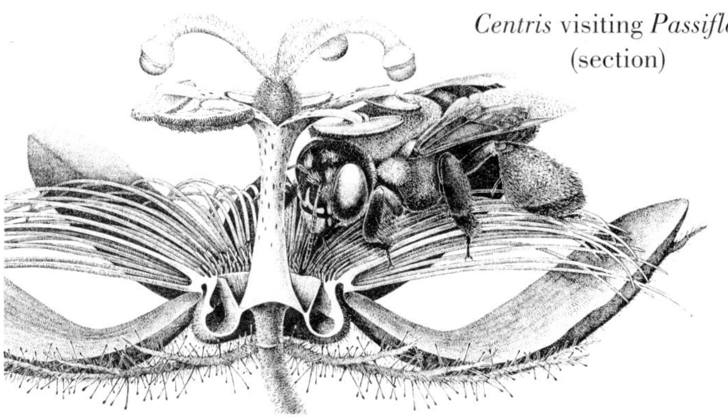

Centris visiting *Passiflora* (section)

Stigma and style lengths vary among cultivars, thus influencing the chances of selfing. Self-compatibility is likely prevalent; cross-incompatibility has also been found. Pollen from the parent plant is at a disadvantage compared to that of other plants, which produce pollen tubes that grow much faster to reach the ovary. Some male-sterile varieties, used for seed production, require outcrossing, provided at least by *Bombus* and *Apis*.

Opuntia — nopal, prickly pear (Cactaceae).
O. ficus-indica is commercially cultivated in drier tropical areas and originated in Mexico. Some species, such as *O. aurantiaca*, are agamo-spermous. Normal pollinators are probably birds and bees, but detailed studies of visitors and breeding system are lacking.

Parthenium — guayule (Compositae).
P. argentatum originated in subtropical northern Mexico and is cultivated in the dry tropics as an elastomer crop. Both rubber content of seeds and seed production increase with cross-pollination. The hermaphrodite flowers are self-incompatible, with *Apis* an adequate pollinator although the most important native pollinating insects are unknown.

Passiflora — passion fruit (Passifloraceae).
P. edulis is Neotropical and the most commercially successful of 500 species, climbing vines native to most of the mainland tropics but nearly all Neotropical. *P. edulis* var. *edulis* (purple granadilla) is grown in tropical highlands and the subtropics, while *P. edulis* var. *flavicarpa* (yellow granadilla) is better suited to lowlands (see also Section 1.4). The large flowers are visited by a variety of bees and other insects, and also some hummingbirds and bananaquits. Larger bees that visit *Passiflora* and pollinate it consistently in the Neotropics are *Xylocopa* and *Ptiloglossa* — *Xylocopa* is also the principal Asian pollinator. Very few flowers are autogamous; self-and cross-incompatibility within clones are common. Cross-pollination greatly improves fruit set and quality. Honeybee hives are sometimes used to enhance pollination and management of *Xylocopa* is not at a technical level, as now in Asia (see Section 1.5.1).

***Persea* — avocado (Lauraceae).**

P. americana is produced mainly in tropical America, where there are three main groups of cultivars — Mexican, with smaller fruits and adapted to poorer growing conditions, Guatemalan, larger and with rough skin, and West Indian, the largest, with smoother skin, but much less resistant to cool temperatures. The hermaphrodite flowers exhibit two flowering schedules: stigmas receptive the morning of day 1, pollen shed on the afternoon of day 2, or stigmas receptive only the afternoon of day 1, pollen shed the morning of day 2. Cultivars of both are needed to ensure pollination, even though the receptive periods and pollen availability overlap slightly for the two flower types. Cultivars vary from self-compatible to self-incompatible, but cross-pollination always improves production. Bees, bats, wasps and flies are pollinators in different parts of the world. Honeybee colonies are stocked at densities of 2-3 colonies/hectare to maximize fruit set.

***Phaseolus* — common bean (Leguminosae).**

P. vulgaris is one of seven Neotropical bean species cultivated widely. Flowers self-pollinate upon opening but remain receptive and can be fertilized with pollen from other plants during eight hours. Extensive studies of native pollinators were made by Búrquez and Saruhkán in Mexico. Many different bees can cross-pollinate the flowers.

***Sechium* — chayote (Cucurbitaceae).**

S. edule is a perennial vine whose main product is the fruit, eaten cooked. Outcrossing is needed because flowers are either male or female, with female flowers more attractive to visiting bees and wasps. Small bees were by far the most frequent visitors in Costa Rica and included primarily meliponines, genera *Trigona* and *Partamona*. Honeybees very seldom visited flowers.

***Simmondsia* — jojoba (Simmondsiaceae).**

Simmondsia chinensis is a native subtropical shrub of southwestern USA that is dioecious and has seeds yielding wax and oil. The largest producers include the southwestern USA, Costa Rica, Australia, Brazil and Paraguay. Flowers are pollen resources for bees, and honeybees contribute to pollen dispersal by enhancing wind pollination when landing on flowers to forage.

***Solanum* — eggplant, potato & naranjilla (Solanaceae).**

Solanum species are pollen-only, hermaphrodite flowers and several from South America are extensively cultivated. *S. quitoense*, lulo and narajilla, are cultivated principally in highlands of Ecuador to Costa Rica. Their pollinators are *Bombus* and perhaps also *Eulaema*, but no detailed studies have been made. Flowers are self-compatible but visitation is required by buzz-collecting bees. *S. melongena* (eggplant) has hermaphrodite flowers displaying heterostyly; some flowers are able to self-pollinate. Certain cultivars are male-sterile, thus hybrids require bee-outcrossing to produce seed. *Exomalopsis* is one bee that effects outcrossing and honeybees, failing to buzz-collect pollen and finding no nectar at flowers, are poor pollinators. *S. tuberosum* (potato) is often self-pollinating, but self-incompatible and male-sterile cultivars are known. For cross-pollination few bees find the amount of pollen offered per flower a sufficient award, but tentative visits by *Bombus* and *Apis* have been seen. Native Andean flower-visiting bees are listed in Section 1.3.

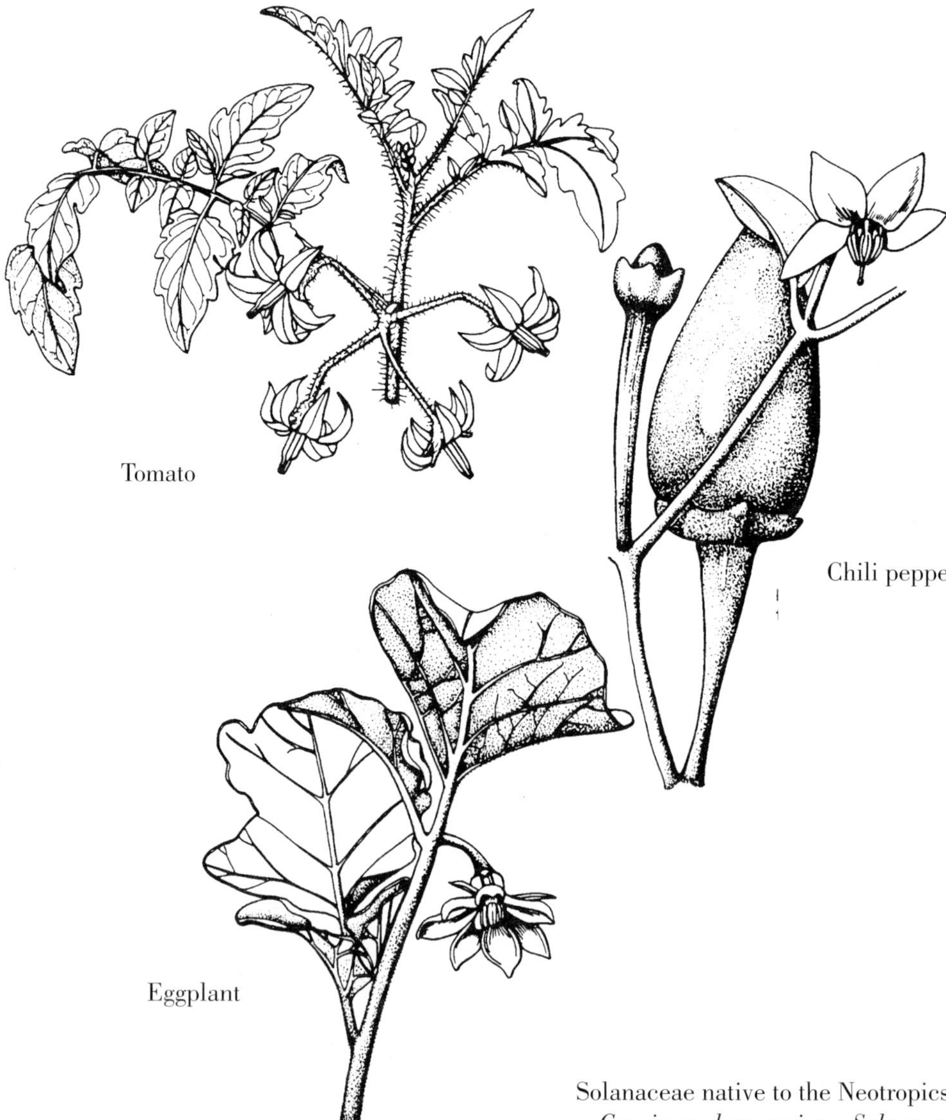

Solanaceae native to the Neotropics - *Capsicum, lycopersicon, Solanum*

***Theobroma* — cacao (Sterculiaceae).**

T. cacao, a small tree among what are interpreted as 4-11 species, is cultivated principally in West Africa, Brazil and Malaysia. Flowers are hermaphrodites but selfing is impossible due to separation of anthers and stigma; self-incompatibility has been shown. Principal pollinators are flies, *Lasioshelea* and *Forcipomyia* (Ceratapogonidae).

***Vanilla* — vanilla (Orchidaceae).**

At least two of the crop species used, *planifolia* and *pompona*, originated in southern Mexico and Central America, where the pollinator is the social euglossine bee, *Eualema* — normally the smaller species — *polychroma, speciosa, cingulata* and *nigrita*. The pollinarium is a triangular wedge carried on the bee's scutellum, and is easily identified in the field. Only outcrossing produces fertile seed. The flowering schedule of this plant, at least in nature, is not well understood and does not occur yearly. Literature reference to *Melipona* as a primary pollinator is based either on incorrect identification or an aberrant occurrence. *Melipona* does occasionally visit the orchids *Maxillaria* and *Xylobium*, but far less than do other stingless bee genera.

Additional References (see also Section 1.1.3)

Búrquez, A., Sarukhán, J. 1980. Biología floral de poblaciones silvestres y cultivadas de *Phaseolus coccineus* L. I. Relaciones planta-polenizador. Bol. Soc. Bot. Méx. 39:5-27.

Búrquez, A., Sarukhán, J. 1984. Biología floral de poblaciones silvestres y cultivadas de *Phaseolus coccineus* L. II. Sistemas reproductivos. Bol. Soc. Bot. Méx. 46:3-12.

Corrêa, P. 1984. Dicionário das plantas uteis do Brasil. Vols. 1-6. IBDF, Rio de Janiero.

Croat, T. B. 1978. The flora of Barro Colorado Island. Stanford, Calif.: Stanford Univ. Press, 943 pp.

Espinz Perez, D., Ordetx Ros, G. S. 1983. Flora apícola tropical. Cartago, Costa Rica: Editorial Tecnologica de Costa Rica, 406 pp.

Frankie, G. W., Haber, W. A., Opler, P. A., Bawa, K. S. 1983. Characteristics and organization of the large bee pollination system in the Costa Rican dry forest. In: C. E. Jones, R. J. Little (eds.). Handbook of experimental pollination biology, NewYork: van Nostrand Reinhold, pp. 411-448.

Hedström, I. 1986. Pollen carriers of *Cocos nucifera* L. (Palmae) in Costa Rica and Ecuador (Neotropical region). Rev. Biol. Trop. 34:297-301.

Heiser, C. B., Jr. 1972. The relationships of the naranjilla, *Solanum quitoense*. Biotropica 4:77-84.

Henderson, A. 1986. A review of pollination studies in the Palmae. Bot. Rev. 52:221-259.

Meeuse, B., Morris, S. 1984. The sex life of flowers. London: Faber and Faber, 152 pp.

Opler, P. A. 1983. Nectar production in a tropical ecosystem. In: The bio-logy of nectaries, B. L Bentley, T. S. Elias (eds.). New York: Columbia Univ. Press, pp. 30-79.

Rodríguez, R. 1979. Principales plantes últiles de la Amazonia colombiana. Bogotá: Centro Interamericano de Foto-interpretación, 263 pp..

Roubik, D. W. 1979. Africanized honeybees, stingless bees, and the structure of tropical plant-pollinator communities. In: D. Caron (ed.). Proc. Fourth Intl. Symp. on Pollination. College Park, Maryland: Maryland Agricultural Sta. Misc. Pub. No. 1., pp. 403-417.

Schatz, G.A. 1990. Some aspects of pollination biology in Central American forests. In: K. S. Bawa, M. Hadley (eds.). Reproductive ecology of tropical forest plants. Man and the Biospher Series, Vol. 7, Paris: UNESCO, pp. 69-84.

Spivak, M, Fletcher, D. J. C, Breed, M. D. (eds.). 1990. The 'African' honeybee. Boulder, Colorado: Westview Press, 537 pp..

Wille, A., Orozco, E. 1983. Polinización del chayote *Sechium edule* (Jacq.) Swartz en Costa Rica. Rev. Biol.Trop. 31:145-154.

Wille, A.1985. Las abejas *Peponapis* y *Xenoglossa* en Costa Rica y su importancia en la polinización de las *Cucurbita* domésticas. Rev. Biol. Trop. 33:17-24.

Cacao - *Theobroma*

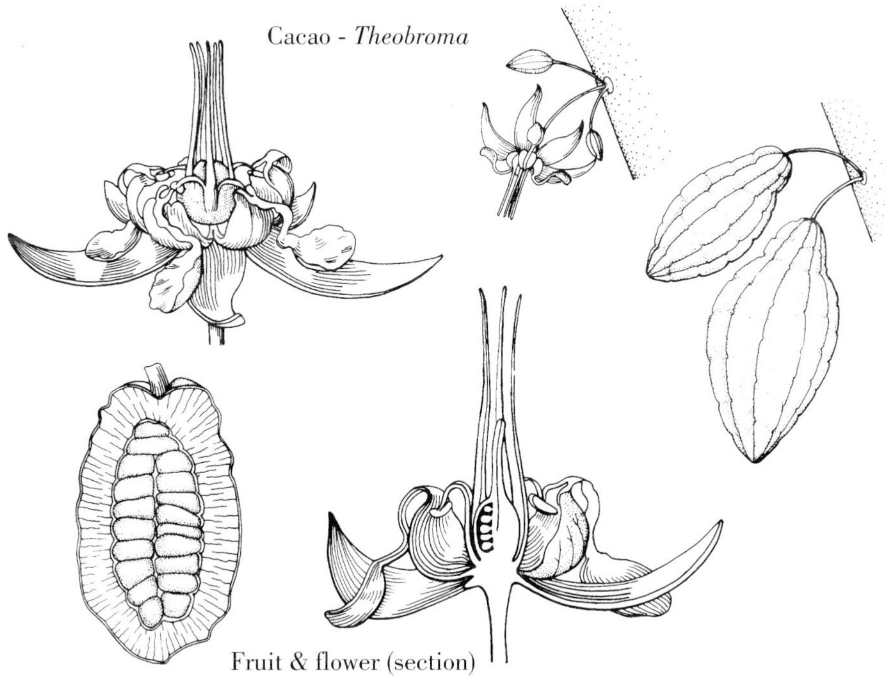

Fruit & flower (section)

1.6 ALTERNATIVES TO ARTIFICIAL POLLINATOR POPULATIONS

Agricultural practices have undergone drastic changes in the last 100 years, and with the push toward mechanization in the past few decades, ever larger areas have been cultivated in single crops to increase profit. Chemical agriculture with its new fertilizers, herbicides and pesticides has been increasingly employed. But at what cost? The application of industrial economics, with higher and higher production goals, did not consider long-term effects on pollinators. Attention was given to pollination requirements of many crops only after the changes in cultivation practices demonstrated new production limits, and thus had to be corrected. In the meantime, natural pollinator populations were already much diminished or lost. If as recently as the turn of this century honeybees were still considered dangerous and damaging to fruit orchards, it is not difficult to accept that the tremendous progress made in understanding beneficial interactions of insects and plants has given rise to many potential applications that have yet to be implemented.

Although agricultural practices similar to those described above have been promoted in both tropical and subtropical developing countries for a long time, many regions still are in the process of being transformed to this type of exploitation. Other areas, of course, have come under different kinds of pressures, such as population growth or desertification, which also lead to drastic habitat changes.

In this chapter, ideas are presented from the perspective of pollinator needs and for improving both degraded habitats and those still to be transformed. That such a discussion cannot neglect social, technical and other environmental concerns should be obvious. At the same time it is beyond the scope of this book to consider all possible aspects. Therefore, emphasis is given to ideas and principles that should be considered by planners, technicians and others involved in making both sustainable agriculture and multiple, sustained, non-destructive use of forests a reality. The concepts necessarily include various subjects — for instance biological pest control and promotion of less capital-intensive farming practices, such as intercropping, rotation, and cover-crop plantings. In the end, these ideas require a new but not necessarily more difficult approach to the improvement of agricultural efficiency. They greatly rely upon a better understanding of the biological, physical and social interactions for all agricultural production.

To suggest practices like the foregoing examples to village communities and countries that have much more pressing problems may seem too idealistic, particularly when such practices are not employed in some of the more stable industrialized countries, which are not facing continuous emergency situations. However, traditional agriculture is often very similar, although in a reduced scale, to modern approaches that minimize dependence on

agricultural chemicals and destructive land use practices. My goal is to point out some alternatives to the mere increase of agricultural production at any cost. Fortunately, there are some ways that can be employed without need of large-scale, long-term scientific studies, huge investments or loss of productivity, by simply using common sense.

It seems unreasonable to place an additional burden on the shoulders of the weakest link of the chain, the primary producer. Therefore a communal or concerted effort might be necessary. Such an effort could be promoted by providing other benefits, such as better prices, more access to markets and privileged access to the omnipresent subsidies. Consequently, it requires also a change in the attitude of local politicians, bankers and merchants, as much as of all those countries and organizations being the primary source of finances, buyers, teaching and the transfer of technology. Thus even the improvement of pollinator availability becomes a 'global' problem in the purest sense. Like all global problems, the solution can only begin with the smallest detail and the attitudes of each and every one of us.

Here are some ideas for those who can test, improve and transfer them to those who need to implement them, and to those who can include them in the mosaic of a more global, more complete plan of development.

1.6.1 Mechanical Pollination

Pollination by hand may be feasible under a few limited circumstances and for small-scale production like home gardens. On a larger production scale it is not profitable. Traditional date palm pollination or sometimes passion fruit and special hybrid seed production, as well as orchid propagation, including *Vanilla*, is done by hand. Increasingly, even for greenhouse production, insect pollinators are used. Mechanical pollination of fruit trees (apples and peaches) with large blowers has been tried, but was never incorporated in commercial enterprises. Thus, hand or mechanical pollination will remain restricted in application and cannot replace pollinators in agriculture on a large scale (see also Section 1.4).

1.6.2 Habitat Management for Wild Pollinators

The natural pollinators of wild plants and also agricultural crops include a wide variety of organisms, not only bees and certainly not only honeybees (as shown in other sections of this manual). But aside from the pollinators whose populations we can manipulate or manage in large numbers, there is a wide array of bee and non-bee pollinators capable of pollinating agricultural crops. Not the least important are a variety of flies. Over a hundred different insect species can be observed on flowers of some fruit trees, though not all contribute significantly to their pollination. Maintaining such a diverse insect fauna increases the chance of sufficient pollination without the need for additional pollinator populations.

In order to have sufficient wild pollinators, their habitats must be preserved and maintained. That means adult and larval stages of the pollinators need to find their food (often highly specific flowers, leaves, other insects, etc.). For many, nesting sites are also required. Some insects require certain soil conditions to survive during one of their life stages. For migratory species like some hummingbirds or Asian and African honeybees, the habitats needed at each extreme of the migratory range should be preserved, in order to assure that sufficient numbers return during the next migratory season. In short, we have to know the life history and requirements of those species we want to conserve or multiply. This is a demanding task even for the much less diverse fauna of the better-studied temperate climates. Fortunately, however, as long as we preserve wildlands with their original plant cover, much of the diversity will maintain itself.

Roadside planted with a diversity of plants - *cocos, hibiscus, glyricidium, capsicum, canna, erika nut.*
The larger plants will grow to provide shade, and must provide forage and refuge for pollinators.

How can one determine the right size of habitat for such purposes? Here we find divided opinions. Because few definitive scientific studies will ever be completed in time, the only safe approach is to conserve the largest possible areas. Minimum requirements for some of the better studied larger animals and some ecosystems are known. Insect populations probably do not need the large habitat required by, for example, mammalian predators. Since many insects do, however, depend on other plant and animal species, they likely need somewhat extensive habitats for their survival. The more we learn about the beneficial insects and other animals, the better we can prepare smaller habitats for them.

In case we are able to maintain only small islands of non-cultivated land, we may have to selectively plant and control species in those habitats in order to maintain the pollinator populations that better suit the needs of our special environments (see also Section 1.4). If the more important natural pollinators for the crops are known, plant species used by these pollinators can be planted or maintained selectively. Thus, availability of flowers would be assured at the correct time. These selected habitats need more advanced planning in land use and they also need more management because they are less stable, being largely artificial. The less that management of an area is possible, the larger the area will have to be in order to maintain the required species diversity and abundance.

In different regions and climates, the composition of reserves or protected habitats must differ, but they should all have a few characteristics in common —

- *as large as possible a diversity of local or locally adapted plants*
- *freedom from exposure to pesticides*
- *connection between habitat 'patches' to allow species exchange, migration, etc.*
- *sufficient numbers and distribution of such habitats in order to provide benefits to many agricultural producers*

The economic benefit of protected habitats cannot be justified only by the provision of pollinators and resulting production increase, particularly if there are few crops planted that benefit from abundant pollinators. Additional values have to be found and a plausible intrinsic value for the local population, since monetary values are

A edge of *Nicotiana*

often of less importance. In order to make the additional effort worthwhile for the farmer, these small pieces of 'unused' or 'unaesthetic' land should preferably have another direct benefit, like the provision of water, firewood, fruits, fodder, windbreaks, soil improvement, or erosion control. If sustainable habitats are to be created or preserved, intrinsic values might include:

- *traditional use of plants and forests for hunting*
- *food reserves for years in which crops fail*
- *medicinal resources*
- *ceremonial or religious uses*

Thus the reserve size or species composition of such habitats might be determined also by intended alternative uses and established values.

Whereas large reserves, such as 'Biosphere Reserves' and 'World Heritage Reserves' can and must conserve whole ecosystems, many small habitats can preserve the natural beneficial pollinator species where they are needed. The smallest such habitats are field boundaries, hedges between fields and forest edges with various stages of successional plant growth. Following in size are fallow fields, planted forest patches for firewood and other communal or private uses, forests along river edges (riparian forests) and other pockets of more or less managed natural forests, preferably all connected to each other.

1) Hedges.

Hedges have played important roles in traditional agricultural systems in extreme climatic or geographical conditions, like steep slopes or windswept plains. Their benefits can be enjoyed in tropical climates as well. Apart from possible aesthetic values, hedges are food and nesting resources for a large variety of animals, including pollinators such as birds, bats and insects. They also include windbreaks and livestock fences, provide erosion control, may stabilize dunes and water runoff and produce firewood, fodder, fruits, and medicinal plants.

Hedge communities can be chosen by observing local habitats and selecting those species most closely matching the desired hedge environment. The woody or shrubby hedge species should be chosen according to major benefits expected from the hedge— among suitable plant species, the ones improving soil, providing rich nectar and pollen sources or having the most diverse use, may be preferred. Orientation of the hedge rows may follow land contours, property boundaries or be positioned to avoid (or enhance) the shading of cultivated plants.

Companion species should be planted or seeded according to the shade the mature hedge will provide. Naturally, shade-tolerant species should be in the center of the hedge and on the side receiving more shade during the hottest part of the day. It may require some maintenance to prevent one species from dominating and eliminating all other species. But weeding by completely destroying any plant cover, so common in tropical countries, cannot be employed. Natural hazards created by providing new sites for poisonous snakes or stinging insects also should be considered. Sensible control of such hazards is usually feasible.

Single or multiple-species hedges are frequently used for erosion control where they directly contribute to increased agricultural production, not only through feeding and protecting beneficial insects, including pollinators, but also through maintaining or improving soil and providing additional crops or food.

Fast-growing species that are easy to establish are preferred if they are nitrogen fixing legumes like ***Gliricidia sepium*, *Calliandra calothyrsus*, *Acacia decurrens*** or ***Desmodium rezonii***. These species give nectar and are actively sought by important pollinators like *Xylocopa* and *Apis*. The trees also can be pruned for mulching, animal fodder and firewood. Hedge pruning often determines whether species come to flower and provide nectar for bees. Selecting woody plants that also are pollinator food sources is very sensible as long as management of the hedges allows for flowering. Width of the hedge may vary with its overall function from a single row of planted sticks to a couple of meters.

A roadside planted with *Glyricidium*, *Musa*, *Cocos*, and 'elephant' grass' for fodder. This could be improved by seeding more legumes among the grass.

agrochemicals. Abundant soil in the water is not only a loss to the farmer, but also a threat to fish and other aquatic fauna. The soil changes the river bottom and the river course, and fills up reservoirs and lakes. Trees on steep slopes or ravines should never be removed, but even on level river banks, borders of 30 to 100 m should be maintained. Again, local conditions of flooding, aquatic life, river changes, land orientation and rainfall patterns need to be considered. In addition, possible alternative use of these areas, as described below for small forest patches, must be taken into consideration in order to plan the size of these borders.

Thus, leaving riparian forests has many ecological benefits, among them providing unusually rich sources of nectariferous plant species and nesting sites for many kinds of pollinators. Where these habitats have already been destroyed it is worthwhile replanting water edges with native tree and shrub species. Selecting the right species constitutes an active area of new research in much of the world.

2) Field boundaries.

Field boundaries, in contrast to hedges, may or may not consist of perennial or woody species. They can be cultivated as boundaries by ploughing, cutting or spraying for maintaining selected beneficial plant species for weed, pest and soil control as well as for providing alternative food sources for pollinator species. Their width and maintenance may change more frequently, with rotation of crops.

Roadsides may cover considerable areas in some countries. These surfaces can be managed by cutting, which is fairly expensive, or by seeding and selective planting in order to maintain growth in certain successional stages. Thus they can serve functions similar to those of field boundaries, hedges or even small forest patches.

3) Home gardens.

Home gardens, due to their size, usually cannot contribute much to feed the larger pollinator populations. However, when entire villages plant flowering hedges around their homes, plant fruit trees and bushes, and cultivate other flowers and some vegetables, these habitats do provide limited support of pollinator populations. Most of all, they supply food while there are few or no wild flowers nearby. This can be particularly helpful for beekeeping with the Asian *Apis*, stingless bees, and many non-*Apis* pollinators.

4) Riparian forests.

Riparian forests — those growing in the immediate vicinity of a creek or river — are different from normal forests and have a special and important ecological function. They prevent soil runoff into the creeks, thus keeping water clear, and also less contaminated by

5) Small forest patches.

Similar to the planting of hedges, forest vegetation can be planted near agricultural fields. Just like natural forests, these patches can have a multitude of uses in addition to that of maintaining pollinators. Selecting only the fastest growing species for firewood or timber production is similar to planting highly-selected monocultures for agricultural production. Application of sustained-yield concepts further considers the benefits of selected species to the soil, alternative uses and the habitats they provide for other crops and healthy populations of plants and animals. Mixed plantings should allow some undergrowth management. Future crop breeding might select for forest undergrowth conditions, thus simulating multilevel natural forests.

The classic eucalypt or pine groves do not present the best solution in most situations, neither short-term nor long-term, since these plants are selected for only some of many important criteria — maximum rate of biomass production. Even though most *Eucalyptus* species provide abundant nectar, their pollen is deficient in nutrients and very few companion plants can grow in the understory of these trees. Thus there are no sources of cover, forage or alternative food for many kinds of animals. Soil quality and the water table are often negatively influenced and no other benefits can be obtained from the barren ground until many years after cutting.

In contrast, there are many fast-growing indigenous tree species that permit various other uses of the land and the tree crop. Carefully selected species can even improve soil conditions through nitrogen fixation and organic matter deposition. More information on species selection, characteristics and requirements is available from a variety of information centers and networks (see *Additional Resources*, at the end of this section). The directory of world honey plants by Crane, Walker and Day (1984) allows cross-referencing of some species that also are known to be good producers of nectar or pollen.

For establishment of small forest patches, a variety of experimental approaches have been employed, mostly with emphasis on multiple use of existing forests, forest conservation, community forestry, agro-forestry, watershed management, and sustained natural forest resource management. Few have considered conservation of beneficial animals like pollinators.

The multiple use of tree plantations should be included in any planting scheme. Selecting highly nectariferous tree species or those that allow nectariferous undergrowth bring additional income sources (beekeeping or native pollinator management) until the tree crop can be harvested. Therefore higher diversity contributes to sustainability of future crops and a higher quality of environmental conditions in general. By planning multiple use wisely, there need not be a loss of income. Instead, it may become an attractive alternative.

6) Successional growth (second-growth habitats).

Not only forests provide a large diversity of resources to nectar and pollen feeding animals. This need is also met by certain savannas and successional regrowth of fields and forests. The latter, in some tropical areas, can sometimes produce more nectar than mature forests. They also form an essential part of the natural and 'mature' ecosystems, harboring many animal species and forming essential habitats for many pollinators, and for other beneficial insects.

Traditional slash-and-burn agriculture continuously creates areas of successional growth. If small enough, and not too dense, these plots might maintain the desired pollinator species. In regions having only such slash and burn agriculture, there should not be pollinator shortages. This is because vast monocultures are lacking. The principle of cutting only small areas and letting them regenerate, or replanting them with forest species, might be practiced even in larger forest plantations. The same may be true in intermediate forest-agriculture zones or some of the park boundary zones where restricted exploitation is permitted. Forest edges provide a narrower, yet similar, habitat that should not be neglected. Through minimal maintenance such as periodic cutting and selective clearing a rich flora and beneficial fauna can be maintained. Fallow fields in crop rotation or land regeneration (dunes, strip mines or eroded soils), like field boundaries, may be left to the natural succession of plant growth. They can also be planted with nectariferous, soil-improving species, or receive minimum management like no-tillage, additional seeding, and periodic cutting in order to maintain successional growth at a preferred stage.

7) Nectar plants cultivated to benefit pollinators.

It is not common practice nor economically feasible in most circumstances to plant crops solely for the purpose of providing nectar to pollinators. The value of honey or the resulting colony population of pollinators is always considered negligible in comparison to the value of the planted crop or the planting cost. For well-planned land use this may still be true in immediately recuperable monetary terms. But in long-term benefits the gap between planting costs and benefits from honey harvests, better pollination, increased natural pest control, lower fertilizer needs and other secondary benefits will become narrower. Eventually, such planting costs may become negligible in comparison to all of the other benefits (when these are properly appreciated).

By proper selection of flower species for their flowering times, pollinator populations can be enhanced. This has been advocated for the maintenance of bumblebees in England, where they are very important pollinators. Thus early-flowering species serve to augment social bee populations or increase solitary bee populations, or next year's population. Late-flowering species may increase the number of reproductive bees for the following season or year. Methods for studying the requirements and the preferred food plants of bumblebees on a countrywide scale were developed for England. Accordingly, school children and volunteers were organized to make many of the basic observations. This worthwhile and affordable effort was educational for the participants, raising their environmental awareness, and also very useful for researchers and farmers.

The abundance of attractive alternative food sources may in some cases reduce the efficiency of artificial as well as natural pollinator populations, if flowering occurs simultaneously with crop flowering. Whether controlling such competing flora will decrease next year's pollinator populations more than it will increase this year's pollination efficiency should be tested, whenever possible. This should be done again alternative choices in pollinator species, crop varieties or timing of planting and pollinator introduction.

8) Cover crops.

The practice of crop rotation allows planting of cover crops during the fallow period. While the soil is recuperating, the cover crop may provide flowers to pollinators needed in neighbouring fields. Self-seeding plants like *Mellilotus* or other nitrogen-fixing legumes enrich the soil while they may even provide a commercial honey crop, very rich fodder to livestock and/or "green manure". A combination of *Mellilotus* varieties can provide flowers for over six months even on poor soils (at <40°C). Some of these varieties have been developed by INTA in Argentina for extreme subtropical climates.

Some problems do arise, similar to those stemming from highly nectariferous successional growth or forests. Attractive nectar producing non-crop flowers can compete with crop flowers for pollinators. In the case of natural pollinators, planting schedules and flowering periods must be synchronized as much as possible. The same problem with artificially-enhanced pollinator populations can also be solved by placing colonies directly in the middle of the crop area, or by providing more pollinators than are usually recommended and/or by introducing the pollinator populations at a time when already 20 to 30% of the crop flowers have opened. In extreme cases competing floral resources may have to be temporarily reduced or eliminated during the crop flowering period.

1.6.3 Crop Selection

It might be possible, as seen for many crops, to select additional varieties that do not require external pollination agents like insects. Those varieties that will continue to require pollinating insects, however, need to be made more attractive to pollinators (see Section 2.3.1). This means that plant breeders have to pay more attention to flowering times and duration, nectar secretion and/or pollen attractiveness.

More emphasis on indigenous crops will reduce the need for exotic pollinators such as *Apis mellifera* in most of the world. Certain pollinators may prove less difficult to manage and propagate than imported honeybees, under local conditions. For example, it is generally well-appreciated that *Apis cerana* is superior to *Apis mellifera* in much of the Asian tropics, due to better resistance to natural enemies and greater tolerance of environmental and resource conditions (see Section 2.5).

Integrated into crop rotations between rice cultivation and cattle grazing, and the planting of *Mellilotus* in northern Argentina shows a promise for profitable honey production (Krell, pers. obs.). A study by Accorti (1992) for Italy also showed substantial savings in fertilizer expenses and petroleum resources for producing honey under improved environmental conditions, rather than using sugar from sugar beets. Further studies on similar subjects most likely will demonstrate that converting to environmentally 'friendlier' cultivation methods can in the end also be more profitable. Maintaining wild pollinators and sustaining imported ones requires careful selection of crop and non-crop (cover crop) species.

Good management practice includes cover crops and perennial crop varieties, including timber species or rubber, are selected among other criteria for their high nectar secretion. Unfortunately, this subject has not been sufficiently considered in the past, nor been given due importance by plant breeders. Particularly, as already mentioned, in forest plantations where harvest and therefore income are realized many years after the initial investment, nectariferous species can provide a "balancing income" — cash flow — and provide for natural as well as managed pollinator species. The selection of nectariferous tree crops is relatively easy because many, if not most tropical tree species are naturally good producers of nectar. Their indiscriminate cutting also drastically reduces the nectar sources available to all pollinator species, not only to honeybees.

The creation or conservation of large wildlands for honey production can have strong secondary effects on pollinator availability in distant agricultural areas. This is demonstrated by an example from Sri Lanka. After the disappearance of most the natural forest suitable for honey production, rubber plantations (*Hevea brasiliensis*) have become the major areas for beekeeping. Recent improvements in bee management techniques are only now starting to permit beekeeping on a larger semi-commercial scale. However, the new varieties of rubber slowly replacing those of old plantations are said to produce little or no nectar. If this proves true, the developing beekeeping industry will have no future. Simultaneously, however — in part due to the same environmental degradation, deforestation and increased pesticide use — the need for moveable pollinator populations is growing. They are needed for increasing seed production requirements and exotic cash crops such as gherkins, i.e. pickling cucumbers. Thus eliminating profitable beekeeping on a commercial scale also eliminates manageable pollinator populations. The latter can only be made available in sufficient numbers through migratory beekeeping (i. e. moving hives into areas where pollinator enhancement is required). In effect, the selection of the new rubber variety might restrict the agricultural cultivation possibilities in parts of the country far removed from the rubber growing areas. This example demonstrates the far-reaching consequences a slight change in cultivar or crop can have on the agricultural productivity of apparently unrelated, distant regions, not only on neighbouring fields.

1.6.4 Pesticides

Apart from habitat destruction, application of pesticides in large quantity, and over large areas, is the primary reason that wild pollinator populations have been reduced

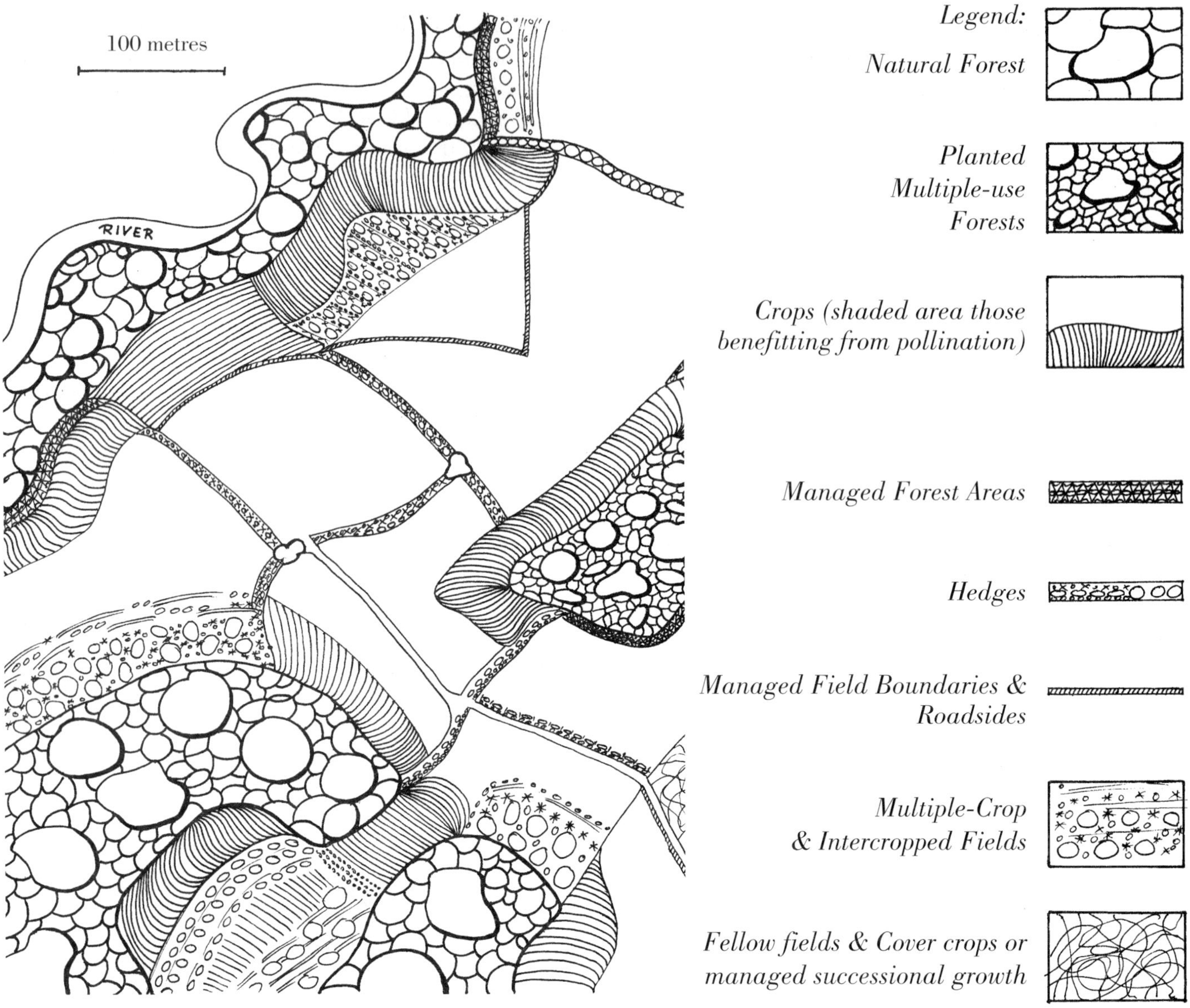

An example of land use for optimizing use and management of wild pollinator population

or completely destroyed. Large aerial applications over 100,000s of hectares of Central American and African tropical forests to control the Mediterranean fruit fly, tsetse fly, and malaria mosquito have undoubtedly made an impact on the pollinator fauna. Documentation of agricultural chemical effects, however, is incomplete (see **Appendix III** for the most recent evaluations regarding bees and beneficial insects). Farm applications are more frequent and widespread, also covering very large areas. Agricultural pesticides are often misapplied and have much more toxic effects on the local animals. A comprehensive list of available pesticides and their known effects is given as an appendix to this Section.

Along the northwest coast of Sri Lanka, pesticides may have led to a production loss involving cucumber cultivation. Initial production, during the first and second year, was fairly high. During the third and fourth year, production strongly declined, and after five years it was only 30% of the first year's production — despite increased fertilizer and pesticide use. During the same time more land was cleared in the dry forest zone and pesticides were applied, not only by the cucumber farmer. The cucumbers are now deformed and of uneven growth — a clear indication of insufficient pollination. Unfortunately there is little that can be done. Together with increased pesticide use, habitat was destroyed which otherwise could have allowed re-establishment of honeybee colonies. Years of replanting will have to precede an increase of the native pollinators.

Over the last decades, pesticides have become more potent, and only recently more specific. The broader a spectrum of pest species a pesticide potentially controls, the more devastating its effect will be on the total fauna, pest and beneficial alike. Its longevity in the environment and application timing and methods may further contribute to its destructiveness.

Although many broad-spectrum pesticides have been banned from the markets of industrialized countries for health and environmental safety reasons, many if not most of them are still being used in tropical and subtropical countries. Less farmer and consumer education and strong political and economical interests permit the continued use of these often cheaper but more dangerous toxins. The newer, sometimes less toxic or more specific pesticides are usually much more expensive and therefore less accessible to the rural poor. At the end of this section is a recently updated list of pesticides that describes their toxicity to honeybees. Toxicity values for other pollinating insects cannot necessarily be obtained from this list, but a general impression is given.

Integrated pest management methods which will reduce pesticide use require very disciplined and well-educated farmers with more technical assistance than is available in most rural areas. Organic farming without the use of artificial or toxic chemicals requires traditional methods and even more education with new crops or at least a different kind of education than the one commonly taught.

1.6.5 Cultivation Practices

Studies of pollinator distribution in crop fields showed very limited foraging ranges of honeybees in situations with many more flowers than foragers. Similarly unsaturated conditions would occur with low natural pollinator populations or exceedingly large surfaces planted with one crop. The overabundance of food (nectar) cannot be exploited, pollinators will concentrate on the areas closest to their natural habitat or nest. Uneven or incomplete pollination is often the result. Smaller field sizes, and shapes following contours of forest edges, are therefore very important for pollination with 'unenhanced' or natural pollinator populations.

Intercropping, or the planting of different crops in alternating rows or otherwise mixed, breaks up the uniform surfaces, reduces the overabundance of one food source and thus increases fruit set across the field. Although there are less plants to produce a crop, the production per plant is increased and the mixture of crops maintains or improves farmer income. Intercropping may also reduce relative production costs due to lessened pesticide and fertilizer requirements.

The most pressing changes to be made in order to preserve natural pollinator populations are the adoption of less toxic and more balanced cultivation practices. Many of the alternatives have already been mentioned, such as reduced and more focused pesticide application (within integrated pest management programs where pesticide-free cultivation is impossible), selection of more resistant locally-adapted or indigenous crops, a larger variety of crops, multicropping systems, crop rotation, less tillage, and more manuring. Last but not least, the soil must be regarded and taken care of as a highly complex living organism — a concept firmly established in many traditional cultures, but so utterly disregarded by most of this century's agricultural development.

Initially some of the suggested changes may mean lower yields than the ones praised by the so-called "Green Revolution", but in the short term they save foreign exchange (pesticides and fertilizers) and farmer's lives (poisoning). In the long term they preserve and likely increase yields for the future and reduce health costs, because of healthier food and water. The modern meaning of the 'green revolution' is no longer equated with 'highest output of biomass by any available means', but instead with the healthiest, least destructive, sufficient output of food.

Conclusions

To generally solve pollination-related problems, the easiest solution would be to switch to crop varieties that do not need pollinators, or to pollinator species that are easily manipulated and multiplied, like some honeybees. This quick fix, often demanding large investment, may be the remedy for some circumstances, but is unlikely to provide a long-term or sustainable solution. Fundamentally, it does not address the need for hybrid seed production, or for outcrossing in the many plant species that must be cross-pollinated to produce seed or fruit. Unless sufficient natural, non-cultivated flora are available, even the ubiquitous western honeybee cannot provide the solution to pollination needs. Only a few highly specialized pollinator species with relatively short lifespans, like the alfalfa leaf cutter bee, may be maintained with one or a few crop species alone.

The next most efficient change would be to increase natural pollinator populations through reduced pesticide use. Alternative cultivation methods, conservation and selective planting will further increase natural pollinator populations and improve environmental conditions, as well as reduce farming costs.

Knowing the requirements, deficiencies and the costs, certain pollinator-limited crops may simply be poor choices for the economics of a given area. This is particularly true for some exotic or export crops which have to meet very specific standards of fruit shape or quality. Taken into account early enough, these conditions can prevent disappointing results, failed projects and farmers' losses.

For any sustainable, and at the end, affordable solution, less destructive cultivation methods are necessary. Conservation efforts and sound agricultural practices are central to this goal. Creation or preservation of diverse environments, not only in National Parks, is also required. This is true to the same extent for natural and managed pollinator populations.

Additional References

Accorti, A. 1992. L'apporto delle api all 'equilibrio energetics in agricoltura — zucchero e mieli: produzioni alternative compatibili. Congr. Int. di Apicoltura, Lazise, Italy.

Agnew, C.,Warren, A. 1990. Sandtrap. The Sciences. N. Y. Acad. Sci. 30:14-19.

Ahmadjian, V., Paracer, S. 1986. Symbiosis: an introduction to biological associations. Hanover, New Hampshire: Univ., Press of New England.

Altieri, M.A. 1987. Agroecology: The scientific basis of alternative agriculture. Boulder, Colorado: Westview Press.

Andrewartha, H,G., Birch, L. C. 1954. The distribution and abundance of animals. Chicago: Univ. Chicago Press.

Baudry, J. 1984. Effects of landscape structure on biological communities. The case of hedgerow network landscapes. In: Proceedings of the lst international seminar on methodology in landscape ecological research and planning, Vol. I. Hoskilde, Denmark.

Boucher, D. H. 1985. Mutualism in agriculture. In: The biology of mutualism: ecology and evolution, D. H. Boucher (ed.). London: Croom Helm.

Brinson, M,M., Lugo, A. E., Brown, S. 1981. Primary productivity, decomposition and consumer activity in freshwater wetlands. Ann. Rev. Ecol. Syst. 12:123-161.

Brown, A.W. A. 1978. Ecology of pesticides. New York: Wiley.

Buchanan-Smith, J. G., Mowat, D. N. 1987. Supplementation of silage with fish meal for growing cattle. In: Proc. 23rd Annual Conference for Feed Manufacturers. Ottawa: Canadian Feed Industry Association.

Bunyard, P. 1985. World climate and tropical forest destruction. The Ecologist 16: 125-136.

Council for Agricultural Science and Technology. 1990. Ecological Impacts of Federal Conservation. and Cropland Reduction Programs. Ames, Iowa: CAST.

Catt, J. A. 1988. Loess — its formation, transport and economic significance. In: physical and chemical weathering in geochemical cycles, A. Lerman, M. Meybeck (eds.). Dordrecht, The Netherlands: Kluwer Academic Publishers.

Coleman, D. C., Hendrix, P. F. 1988. Agro-ecosystem processes. In: Ecological studies 67, L. R. Pomeroy, J.J. Alberts (eds). New York: Springer-Verlag.

Conway, G. R. 1985. Agro-ecosystem analysis. Agriculture Administration 20:31-36.

Conway, G. R. 1987. The properties of agro-ecosystems. Agri. Syst. 24:96-117.

Coupland, R.T. 1977. Grassland ecosystems of the world. Analysis of grasslands and their uses. Cambridge, U.K.: Cambridge Univ. Press.

Dale, T., Carter, V. G. 1965. Topsoil and civilization. Norman, Oklahoma: Univ. Oklahoma Press.

Daly, H.E., Cobb, J. B. 1989. For the common good: redirecting the economy toward community, the environment and a sustainable future. Boston: Beacon Press.

de Jong, E., Kachanoski, R. G. 1988. The importance of erosion in the carbon balance of prairie soils. Can. J. Soil Sci. 68:1ll-119.

Dover, M., Talbot, L. M. 1987. To feed the earth: Agro-ecology for sustainable development. Washington, DC: World Resources Institute.

Edwards, C.A. 1987. The concept of integrated systems in lower input/sustainable agriculture. Amer. J. Alternative Agric. 2: 148-162.

Elliott, E.T., Cole, C. V. 1989. A perspective on Agro-ecosystem science. Ecology 70: 1697-1602.

Evans, J. 1982. Plantation forestry in the tropics. Oxford: Oxford Univ. Press.

Fitter, A. H., Atkinson, D., Read, D. J., M.B. Usher, M. B. (eds.). 1985. Ecological interactions in soil. Oxford: Blackwell Scientific Publications.

Frankel, O. H., Soulé, M, E. 1981. Conservation and Evolution. Cambridge: Cambridge Univ. Press.

Frankel, H., Meiri, A. (eds.). 1986. Soil salinity: two decades of research in irrigated agriculture. New York: Van Nostrand and Reinhold.

Fussell, M., Corbet, S. A. 1992. Flower usage by bumblebees; a basis for forage plant management. J. Appl. Ecol. 29:451-465.

Game Conservancy, 1991. Review of 1990. Fordingbridge, Hampshire, UK. The Game Conservancy.

Georgescu-Roegen, N. 1971. The entropy law and the economic process. Cambridge, Mass.: Harvard Univ. Press. trend. In:

Gimingham, C. H. 1972. Ecology of Heathlands. London: Chapman and Hall.

Gliessman, S. R. 1983. Allelopathic interactions in crop-weed mixtures: Applica-tions for weed management. J. Chem. Ecol. 9:991-999.

Gliessman, S.R. 1989. Agroecology: researching the ecological basis for sustainable agriculture. New York: Springer-Verlag.

Goodman, D. 1976. The theory of diversity-stability relationships in ecology. Quart. Rev. Biol. 50: 237-266.

Gregorich, E. G., Anderson, D. W. 1986. Effects of cultivation and erosion on the soils of four toposequences in the Canadian Prairie. Geoderma 36: 343-364.

Harper, J. L. 1977. Population biology of Plants. London: Academic Press, Inc.

Harwood, H.R. 1990. A history of sustainable agriculture. In: Sustainable agricultural systems, C.A. Edwards, R. Lal, P. Madden, R. H. Miller,G. House (eds.). Ankeny, Iowa: Soil and water conservation society.

Hawksworth, D. L. 1991. The fungal dimensions of biodiversity: magnitude, significance, and conservation, Mycological Research 95:641-655.

Holland, E.A.,Coleman, D. C. 1987. Litter placement effects on microbial and organic matter dynamics in an agro-ecosystem. Ecology 68:425-433.

Holling, C. S. 1986. The resilience of terrestrial ecosystems: local surprise and global change. In: Sustainable development of the biosphere, W.C. Clarke, R.E. Munn (eds.). New York: Cambridge Univ. Press.

Hyams, E.S. 1976. Soil and civilization. New York: Harper and Row.

Jackson, W., Piper, J. 1989. The necessary marriage between ecology and agriculture. Ecology 70: 1691-1593.

Jarman, P.J., Sinclair, A. R. E. 1979. Feeding strategy and the pattern of resource-partitioning in ungulates. In: Serengeti: dynamics of an ecosystem, A. R. E. Sinclair, M. Norton-Griffiths (eds.). Chicago: Univ. Chicago Press.

Kedziora, A., Olejnik, J., Kapuchinski, J. 1989. Impact of landscape structure on heat and water balance. Inter. Assoc. Ecol. 17: l-17.

Kevan, D. K., McE. 1962. Soil animals. New York: Philosophical Library.

Kevan, P.G., Clark, E. A.,Thomas, V. G. 1990. Insect pollinators and sustainable agriculture. Amer. J. Alter. Agric. 5: 13-22.

Krebs, C.J. 1985. Ecology: the experimental analysis of distribution and abundance. New York: Harper & Row.

Likens, G. E., Bormann, F. H., Pierce, R. S., Eaton, J. S., Johnson, N. M. 1977. Biogeochemistry of a forested ecosystem. New York: Springer-Verlag.

Logan, T. J. 1990. Sustainable agriculture and water quality. In: Sustainable agricultural systems, C. A. Edwards, R. Lei, P. Madden, R.H. Miller,G. House (eds.). Ankeny, Iowa: Soil and water conservation society.

May, R.M. 1975. Patterns of species abundance and diversity. In: Ecology and evolution of communities, M. L. Cody, J. M. Diamond (eds.). Cambridge: Belknap Press of Harvard Univ.

Maddock, L. 1979. The "migration" and grazing succession. In: Serengeti: dynamics of an ecosystem, A.R.E.

Sinclair, M. Norton-Griffiths (eds.). Chicago: Univ. Chicago Press.

Margalef, H. 1968. Perspectives in ecological theory. Chicago: Univ. Chicago Press.

McNaughton, S. J. 1979. Grassland-herbivore dynamics, In: Serengeti: dynamics of an ecosystem, A.R.E. Sinclair, M. Norton-Grifflths (eds.). Chicago: Univ. Chicago Press.

Milne, R. 1987, Putting the land out to grass. New Scientist 116:10-11.

Middleton, J. D., Merriam, G. 1988. Distribution of woodland species in farmland. J. Appl. Ecol. 20: 626-644.

Murdoch, W. W. 1976. Diversity, complexity, stability and pest control, J. Appl. Ecol. 12: 796-807.

National Research Council. 1989. Alternative agriculture. Washington, D. C.: Nat. Acad. Press.

O'Connor, R. J., Shrubb, M. 1986. Farming and birds. Cambridge: Cambridge Univ. Press.

Odum, H. T. 1983. Systems ecology. New York: Wiley.

Odum, H. T. 1971. Environment, power and society. New York: Wiley.

Okigbo, B. N. 1990. Sustainable agricultural systems in tropical Africa. In: Sus-tainable agricultural systems, C.A. Edwards, R. Lal, P. Madden, H. H. Miller, G. House (eds.). Ankeny, Iowa: Soil and water conservation society.

Paarlberg, D. 1980. Farm and Food Policy. Lincoln, Nebraska: Univ. Nebraska Press.

Panayotou, T., Ashton, P. S. 1992. Not by timber alone.Washington, D. C.: Island Press. 282 pp.

Paul, E. A., Clark, F. E. 1959. Soil microbiology and biochemistry. San Diego: Academic Press, Inc.

Paul, E A., Robertson, G. P. 1989. Ecology and the agricultural sciences: a false dichotomy? Ecology 70:1594-1697.

Phillips, H. E., Blevins, R. L.,Thomas, G. W., Frye, W. W., Phillips, S. H.. 1980. No-tillage agriculture. Science 208:1108-1113,

Pimentel, D., Hall, C. W. (eds.). 1984. Food and Energy Resources. Orlando: Academic Press, Inc.

Pimentel, D., Warneke, A. 1989. Ecological effects of manure, sewage sludge and other organic wastes on arthropod populations. Agric. Zool. Rev. 3:l-30.

Pimentel, D., Stachow, U., Takacs, D. A., Brubaker, H. W., D̃umas, Meaney, J. J., O'Neil, J. A. S., Onsi, D. E., Corzilius, D. B. 1992. Conserving biological diversity in agricultural/forestry systems. Bioscience 42:354-362.

Postel, S. 1989. Halting land degradation. In: State of the world 1989. New York: W.W. Norton & Company.

Power, J. F. 1987. Legumes: Their potential role in agricultural production. Amer. J. Alter. Agric. 2: 69-73.

Purvis, G., Curry, J. P. 1984. The influence of weeds and farmyard manure on the activity of Carabidae and other ground-dwelling arthropods in a sugar beet crop. J. Appl. Ecol. 21: 271-283.

Risser, P. G. 1958. Diversity in and among grasslands. In: Biodiversity, E. O. Wilson (ed.). Washington, D. C.: National Academy Press.

Ruttan, V.W. 1986. Increasing productivity and efficiency in agriculture. Science 231: 781'

Senate of Canada. 1984. Soil at risk; Canada's Eroding Future. Ottawa: Standing committee on agriculture, fisheries, and forestry to the Senate of Canada.

Smith, R. L. 1974. Ecology and field biology. New York: Harper & Row.

Switzer-Howse, K. D., Coote, D. R. 1984. Agricultural practices and environmental conservation. Publ. 1772/E. Ottawa: Agriculture Canada.

Thomas, V. G., Kevan, P. G. 1993. Basic principles of agroecology and sustainable agriculture. J. Agric. Env. Ethics. 1-19.

Trenbath, B. R. 1974. Biomass productivity of mixtures. Advances in Agronomy 26: 177-210.

Vallentyne, J. F. 1990. Grazing management. San Diego: Academic Press, Inc.

Vandermeer, J. H. 1989. The Ecology ofIntercropping. Cambridge: Cambridge Univ. Press.

Villachica, H., Silva, J. E., Peres, J. R., daRocha, C. M. C.. 1990. Sustainable agricultural systems in the humid tropics of South America. In: Sustainable agricultural systems, C.A. Edwards, R. Lal, P. Madden,R. H. Miller, G. House (eds.). Ankeny, Iowa: soil and water conservation society.

Wall, G.J., Pringle, E. A., Sheard, R. W. 1991. Intercropping red clover with sileage corn for soil erosion control. Canadian Journal of Soil Science 71: 137-146.

Wegner, J., and G. Merriam. 1979. Movements by birds and small mammals between a wood and adjoining farmland habitats. Journal of Applied Ecology 16: 349-357.

World Commission for Environment and Development. 1987. Our Common Future. Oxford, U.K.: Oxford University Press.

Appendix to Section 1.6

A short list of organizations with useful information about muiltiple-use tree crops and alternative farming practices is appended here. There are many more governmental as well as private organizations that may have applicable information. Most likely, none of the organizations will have specific information on alternative agriculture in consideration of pollinator needs as describved above. However, when piecing together information on multiple-use trees, alternative agricultural methods and crops, traditional knowledge of planting and resource management, and about general conservation and the least destructive practices, at least an experimental baseline can be established and certain elements incorporated into on-going activities.

1) Arid lands information network:
274 Banbury Road
Oxford OX2 7DZ
United Kingdom

2)Réseau d'Informations des Terres Arides
C. P. 3 Dakar Fann, Senegal

3) Arbeitsgemeinschaft für Tropische und Subtropische Agrarforschung e. V. (ATSAF)
Hans Böckler Str. 5
5300 Bonn 3, Germany
Tel: 0228-4001313/320
FAX: 0228-4001311 Telex: 886710 DSEBND

4) GATE/GTZ
Dag Hammarskjöld Weg 1-2
Postfach 5180
D-6236 Eschborn
Tel: 06196-790/ 7911
FAX:06196-794-820 Telex:407501-0 GTZ D

5) Food And Agriculture Organizaiton of the United Nations) (FAO)/ Forestry Division
Via Terme di Caracalla
00100 Rome, Italy
FAX: 06-5797-3152

6) Multiple Use Tree Research Network (MUTRN)
Winrock International
1611 N. Kentucky Street
Arlington, Virginia 22209, USA
Tel: 703-525-9430
FAX: 703-522-8758

7) International Council for Research in Agroforestry (ICRAF)
P. O. Box 30677
Nairobi, Kenya
Tel: 02-521-450
FAX: 02-521-001

8) International Institute for Tropical Agriculture (IITA)
Oyo Road
PMB 5320
Ibadan, Nigeria
Tel: 234-2
2-400300/316
— or —
c/o Lambourn & Co. Ltd.
Carolyn House, 26
Dingwall Road
Croydon CR9 3EE, United Kingdom
Tel: 081-686-9031
FAX: 081-681-8583 Telex:946979 LWL G

9) Regional Environmental and Natural Resources Information Centre (RENRIC)
No. 84 Lorensz Road
Colombo 4, Sri Lanka
Tel: 01-582-553
FAX 01-589-369

10) International Union for the Conservation of Nature and Natural Resources (IUCN)
Avenue de Mont Blanc
CH 1196 Gland, Switzerland

11) Tree Aid
Mayfield, The Street Farmborough, Bath
BA3 1AL, United Kingdom

12) International Federation of Organic Agriculture Movement (IFOAM)
General Secretariat
6695 Tholey/Imsbach FRG
Tel: 49-6853-5190
Fax: 6853-30110

13) Organic Crop Improvement Association
Route 3, Box 326
Alachua, Florida 32615, USA
Tel: 904-462-1302

14) New Dawn
Apartado 371-8000
San Isidro, Puntarenas, Costa Rica
Attn.: Dr. E. Bernhardt

15) Rodale Press, Inc.
33 Minor Street
Emmaus, Pennsylvania 18049, USA
[for extensive information on organic agriculture]

16) Forschungersing f. Biologisch Dyna-mische Wirtschaftsweise e. V.
Baumschulen Weg 11
D-6100 Darmstadt, Germany
FAX: 06155-5774

17) Alternative Agriculture
University of Wageningen
Haarweg 333
NL - 6709 RZ Wageningen, the Netherlands, Attn.: Dr. I. C. van Veluw
Tel: 031-8370/83522 or 84676
FAX: 031-84575

1.7 PROSPECTS FOR THE FUTURE

Diversification of Pollination Strategies for U.S. Crops*

P. F. TORCHIO
Bee Biology and Systematics Laboratory, USDA-ARS, Utah State University, Logan, Utah 84322-5310
Environ. Entomol. 19:1649-1656 (1990)

ABSTRACT. The close affiliation between the honeybee industry and commercial pollination of U.S. crops is described, followed by a literature review that demonstrates why insufficient numbers of honeybees may be available to support adequate crop pollination services in the near future. Recent recommendations offering different methods to increase numbers of crop pollinators (continued honeybee research, establishment of management habitats, and development of additional non-honeybee pollinators) are described, and the positive and negative characteristics of each proposal are discussed. The status of crop pollination is then reviewed in some detail, followed by a recommendation to combine pollinator studies into one collaborative pollination programme focused on producing the maximum number of efficient crop pollinators in the shortest possible time frame.
KEY WORDS Insecta, bees, crop pollination, management systems.

The unrelenting drive to maximize efficiency in agronomic practices over the past 80 — 90 years has resulted in a dramatic intensification of agriculture in the United States. These practices have been primarily responsible for the diversification of crop species in commercial production and the rapid expansion of some established crops. Many of the newly cultivated species and cultivars require, or are benefited by, insect pollination provided mostly by bees. In addition, acreages of some well-established, bee-pollinated crops (e.g., apple, avocado, blueberry, almond) have increased rapidly within the past 40 yr in direct response to consumer demand, improved technology in agriculture, and the successful expansion of domestic as well as international markets (Eck & Childers 1966, McGregor 1976, Eck 1988, Robinson et al. 1989). The USDA recognized the need to increase numbers of pollinators nearly 50 years ago, and the Agricultural Research Service (ARS) provided support to the beekeeping industry by establishing additional bee research laboratories and other funded programs on a "need" basis (Levin 1983). Direct and indirect support were founded on the concept that adequate numbers of honeybees, *Apis mellifera* L., would be maintained for pollination services within a thriving industry that derived most of its income from the sale of honey and beeswax (Levin 1983).

The USDA-ARS also established a research unit (Bee Biology and Systematics Laboratory, Logan, Utah) in the late 1940s to determine the efficacy of non-honeybee species as pollinators of targeted crops. As a result of those studies, several bee species have been developed into commercially managed pollinators of various crops (reviewed by Torchio 1987).

As acreages of bee-pollinated crops have increased, inquiries have periodically been made concerning the honeybee industry's capacity to supply additional numbers of pollinators for these crops (Free 1970, McGregor 1976, Prescott & Allen 1986, Parker et al. 1987, Robinson et al. 1989, Levin & Waller 1989, Kevan et al. in press, E. E. Southwick & Southwick in press). The legitimacy of these questions has increased during the past 15 years, because the total number of U.S. honeybee colonies has been decreasing (Robinson et al. 1989). As a result, alternative methods designed to increase numbers of U.S. pollinators have been proposed, and these proposals have been partially reviewed by Parker et al. (1987), Torchio (1987, in press), Kevan (1987), and Kevan et al. (in press).

This paper reviews three programs that are developing different methods to increase numbers of crop pollinators. In addition, the strengths and weaknesses of each programme are evaluated, and the potential of these programs to fulfill future pollination needs for U.S. crops is also discussed.

Why Are Increased Numbers of Pollinators Needed?

When Metcalf et al. (1962) estimated the total value of U.S. bee-pollinated crops at $4.5 billion based on 1957 harvest data, they focused attention on the rapidly increasing dollar value of pollinated crops and the need to produce more pollinators to service those crops. Ware (1973) then increased the estimate to $7.6 billion by simply including additional acreages in commercial production, and Levin (1983, 1984) added indirect beneficiaries (10% of the enormous cattle and dairy industries) in his $18.9 billion estimate. O'Grady (1987), however, used less subjective formulations in amending Levin's pyramidical figures downward to $4.6 billion. Parker et al. (1987), while not offering their own estimates, also found Levin's $18.9 billion figure to be misleading, because it included $12 billion allocated to alfalfa (and a number of its products), a crop that is very effectively pollinated by commercially managed populations of solitary bees (Torchio 1966, 1987). Robinson et al. (1989), by expanding the formula developed by O'Grady (1987), estimated honeybee-pollinated crops to be valued at $9.3 billion. They also suggest that approximately one million honeybee colonies are rented annually for pollination services with some hives moved onto two crops per annum (p.

423). The one million figure is double that offered by Oertel (1983) and 40% greater than the highest estimate given by Levin (1986). Most of these calculations have, however, already been questioned by L. Southwick & Southwick (1989; E. E. Southwick & Southwick in press) who incorporate additional principles of economics to formulate yet another set of lower valued figures.

The estimated dollar value of honeybee-pollinated crops will continue to be debated until such time as: (l) each and all of these crops are evaluated based on uniform criteria (e.g., Robinson et al. [1989] rank crops as "Major" or "Others," based only on known numbers of honeybee colonies rented for each crop rather than the total value of the crop, and [with the exception of rapeseed] they exclude a large number of crops valued at <$10 million [see their table 2]); (2) better methods are developed to determine number of acres under cultivation per crop as well as production per acre per crop; (3) and most importantly, a more objective measure is found to determine percentage of pollination by honeybees versus other pollinating species per crop. All of the published estimates made on crop values to date have, nevertheless, demonstrated general agreement on at least two important points: the acreages of bee-pollinated crops are increasing; and increased numbers of pollinators are needed to service these additional acres.

Until recently, the honeybee industry has been successful in supplying additional bees for the increased acreages of cross-pollinated crops that have come into commercial production. Currently, however, several pressing problems (migratory beekeeping, pesticides, diseases and parasites, reduction of pollen-nectar resources in non-agriculture areas, honey imports and sweeteners, increasing maintenance costs, etc.), have led to a decline in the total managed honeybee colonies available in the United States (Robinson et al. 1989). In addition, the honeybee industry has recently been beset with more serious problems (establishment of the tracheal mite, *Acarapis woodi* (Rennie), the large external mite, *Varroa jacobsoni* Oudemans, and the expected arrival of the African honeybee, *Apis mellifera scutellata* Lepeletier, from Mexico) that may cause a precipitous decline in the remaining honeybee hives available for crop pollination. These problems and their potential effect on the bee industry have been thoroughly discussed in recent honeybee literature, and that literature has been reviewed in detail by Parker et al. (1987), DeGrandi-Hoffman (1987), Robinson et al. (1989), E. E. Southwick & Southwick (in press), and Kevan et al. (in press).

The seriousness of these problems has prompted more recent reviewers to include short summaries that recognize the potential use of native bee species as crop pollinators. A few reviewers (Prescott & Allen 1986, Parker et al. 1987, Kevan et al. in press) have attempted to quantify the importance of non-honeybee species as alternative pollinators of agricultural crops. Prescott & Allen (1986), for example, list seven crops (cashew, mango, highbush blueberry, cranberry, squash, cacao, and cardamom) having a combined average annual value of $1.2 billion pollinated by "wild" insects (p. 345) with honeybees sometimes serving as supplemental pollinators (see their table 9.1, 338-344, and table 9.2, 546-547). These reviewers have not, however, been directly involved in the successful development of solitary bee species as commercially managed pollinators of agricultural crops. As a consequence, the actual and real potential of using managed populations of non-honeybees as an alternative method to increase numbers of U.S. crop pollinators has not been reviewed in adequate detail.

Is Pollination Efficacy Measured Equitably Between Pollinator Species?

The apiculturists (beginning with McGregor [1976] and continuing to the present [Robinson et al. 1989]) consistently mention the potential value of native bees as pollinators, but they have also discounted the importance of this large, worldwide, pollinator force (>20,000 species) through the repetitious use of the following assertion; *population densities of native bee species vary from year to year and, as a result, they are not dependable as crop pollinators* (see Robinson et al. 1989, 418, 421-422, 483, as a recent example). The statement is flawed because a comparison is made between an intensely managed species (honeybee) and a large complex of unmanaged native bee species. Thus, these types of comparisons are appropriate only when parity exists; i.e., managed honeybee populations should be compared directly with managed native bee populations (e.g., alkali bees, leaf cutter bees, mason bees). Unfortunately, rigorous field trials of the type needed to obtain these data have, with few exceptions (Torchio 1979, 1985; Tepedino 1981; Waller et al. 1985), not been undertaken. Equity should, therefore, be incorporated into future studies designed to measure accurately pollination efficacy of honeybees and other pollinating species by crop. Once equity is established, subsequent comparisons on pollination efficacy between honeybees and native bees could be based on scientific evidence rather than on assertions or estimates arbitrarily derived by some apiculturists or economists (see Eckert & Shaw 1960; McGregor 1976; Levin 1983, 1984; O'Grady 1987; Robinson et al. 1989).

The concept of equity incorporated into comparative studies of crop pollination by bees can not be overemphasized because apiculturists have more recently begun to treat estimated numbers as factual data. For example, Eckert & Shaw (1960) stated (again, without evidence) that honeybees account for at least 80% of all insect pollination. Barclay & Moffett (1984) and Levin (1986) then referred to the 80% figure as a legitimate number even though supportive data have never been obtained. O'Grady (1987) lowered the percentage number by deductive reasoning, but Robinson et al. (1989) reestablished the 80% honeybee pollination figure, and L. Southwick & Southwick (1989) provided other estimates. Yet, none of these studies was based on quantitative comparisons of pollination efficacy among bee species. As a consequence, the "data" presented in these reports should be considered as unsubstantiated estimates until such time as quantitative and comparative results are obtained.

Development of Different Strategies for Increasing Numbers of Crop Pollinators

Apiculturists consider the European honeybee the most important pollinator of U.S. crops, and other agriculturists accept the statement as fact (Free 1970, McGregor 1976, Parker et al. 1987, Kevan 1987, Kevan et al. in press). Thus, there is almost total agreement in the agricultural community for a need to increase numbers of honeybee colonies used in pollination rental services. Many agriculturists also recognize the importance of non-honeybee species as pollinators of crops, but others believe the establishment of managed habitats in and around agricultural zones may be the best strategy to increase the total number of pollinators in agricultural environments.

Bohart (1951, 1962) was one of the first to recognize the beneficial and detrimental effects on bees caused by human manipulation of the environment. He recommended the development of habitat management programs as one method to increase numbers of native species in natural as well as agricultural ecosystems (Bohart 1971). He also developed methods for using native bee species as commercially managed pollinators of specific crops and concluded that this latter approach offered greater returns (in pollination effectiveness) in less time compared with results obtained in habitat management studies (Bohart 1972, personal communication).

Stephen (1955) also recognized the decimation of native bee populations resulting from extensive land clearing in Canada. He too proposed conservation and habitat management of areas in and adjacent to agricultural zones as a viable strategy to maintain populations of native pollinator species. His later studies were, however, focused on developing candidate species into commercially managed pollinators of alfalfa seed crops of western North America (Stephen 1959, 1960, 1962). Grower acceptance and support of these programs facilitated rapid growth of the alkali bee, *Nomia melanderi* Cockerell, and alfalfa leaf cutting bee, *Megachile rotundata* (F.), industries, which in turn quickly satisfied pollination requirements of the large acreages of alfalfa seed crops planted in western North America (Torchio 1966, Bohart 1972, Parker & Torchio 1980, Bitner 1982). Much of the baseline information obtained from studies on alternative pollinator species before 1970 has been summarized by Free (1970, 1982) and McGregor (1976), whereas Crane & Walker (1983, 1984), Prescott & Allen (1986), Parker et al. (1987), and Torchio (1987) have reviewed more recent literature.

The most recent reviews of crop pollination by insects (Prescott & Allen 1986, Parker et al. 1987, Robinson et al. 1989, Kevan et al. in press, E. E. Southwick & Southwick in press) have devoted generous discussions to a few subjects that have been previously ignored, summarized as low-priority items, or considered as common knowledge. They include the recognition of (1) a large reservoir of managed or unmanaged bees that pollinate specific crops (see listing compiled by Crane & Walker [1983]), (2) reduction of honeybee and native bee populations exposed to pesticides or bee losses attributed to indirect effects of herbicides, (3) dwindling bee populations in habitats disturbed by human activities, (4) fewer numbers of bee species and smaller populations of remaining bees distributed in and adjacent to areas of intensive agriculture, and (5) the formidable problems currently facing the honeybee industry. Some of these same reviewers have also used these subjects in different combinations to establish proposals outlining methods for increasing numbers of pollinators.

Proposed Recommendations by Apiculturists and Nonapiculturists to Increase Numbers of Pollinators

These recommendations fall into two general categories:

1) The apiculturists (e.g., Robinson et al. 1989) suggest that studies of native pollinators show promise and should continue, but the main thrust of support must be reserved for honeybee research to resolve the most threatening problems plaguing that industry. After these problems are resolved, many in the honeybee industry believe that they could once again supply adequate numbers of bees for the majority of U.S. crops requiring pollination. The apiculturists reason that an established industry, an active research support group, and past successes in pollination services justify the continued use of European honeybees as the primary pollinator of U.S. crops.

2) The nonapiculturists (including those studying native bee pollinators) agree that support of honeybee research should be continued, especially in areas focused on maintaining or increasing numbers of colonies available for pollination rental. One group of non-apiculturists place a higher priority on the developing habitat management programs for native bees as the best strategy to ensure against inadequate pollination of many agricultural crops in the future

(reviewed by Kevan 1989). These non-apiculturists believe that increased numbers of native bees maintained in habitat management programs could be used as alternative pollinators if honeybee colonies continue to decline in the United States. They also believe that a pollination crisis would be averted in most agricultural areas of the United States if managed habitats were established.
A few nonapiculturists (e.g., Parker et al. 1987) have also discussed the real and potential dangers that may occur whenever any one species is used as the only pollinator for the majority of crops. As a result, they support most plans designed to increase populations of managed or unmanaged non-honeybee pollinator species irrespective of honeybee availability at any time.

The effect of these combined proposals has stimulated a growing awareness in the agricultural community for a need to increase pollinator numbers as insurance against inadequate pollination of U.S. crops. These proposals should not, however, be accepted in their entirety, because each is flawed in different respects: The recommendation of apiculturists discounts, by omission, the real and potential value of non-honeybees as crop pollinators; and recommendations by those nonapiculturists to establish habitat management programs do not address the problem of how to introduce and maintain successfully entire guilds of pollinator species within reconstituted habitats.

The concept of habitat management programs is probably unworkable in areas of intensive agriculture because large expenditures would be required to establish sufficiently large habitats needed to maintain adequate numbers of pollinators in and around large farms with high land values; herbicides and pesticides are applied frequently; many cross-pollinated crop species are planted in large tracts of unbroken acreage (monoculture); some crop species included in rotation schedules may not flower during flight periods of any established bee species and/or flowers of other crop species may be unattractive to those pollinators; clean cultivation habits have inadvertently eliminated most of the nesting areas previously used by pollinator species (thus, additional time and expense will be required to reintroduce many native pollinator species into these reclaimed and managed habitats): islands of managed habitats would likely produce reservoir populations of pest species; and irrigation practices often have negative, long-term effects on soil-nesting pollinators. In addition, the most intensive agricultural areas are found in the western sector of the nation, where the majority of cross-pollinated crops are grown (Robinson et al. 1989, especially table l). It is likely that attempts to develop habitat management programs on or adjacent to these large western farms will be cost ineffective or unsuitable for the agro-ecosystems under consideration, or both. Ironically, many farms in the western United States that would be most benefited by increased numbers of pollinators have been depauperated of native bees. The disappearance of native pollinator species from most of these areas occurred soon after intensive agricultural practices were adopted (Torchio 1966).

Use of Non-honeybee Species as Alternative Pollinators

The USDA-ARS has, since 1947, supported an alternative programme to increase pollinator numbers for intensively farmed areas. This ongoing project has pioneered studies that have led to the successful development of a number of non-honeybee species used as managed pollinators of specific crops. In his review of the subject, Torchio (1987) described how the following species were developed: the alkali bee and alfalfa leaf cutting bee established as commercial pollinators of alfalfa seed crops; *Osmia lignaria propinqua* Cresson developed as a pollinator of almond, apple, cherry, pear, plum, meadowfoam, etc.; *Osmia cornifrons* (Radoszkowski) established as a commercial pollinator of apple in Japan and successfully tested as a pollinator of orchard crops grown in areas of higher humidities in the United States; and *Osmia cornuta* Latreille that is established as a pollinator of apple and almond crops in Europe (Spain) and successfully tested in the United States. Another bee native to North America, *Osmia ribifloris biedermannii* Michener,has more recently been used successfully as a highly effective and manageable pollinator of highbush blueberry (Torchio in press).

This particular USDA-ARS programme "development of non-honeybee species as managed pollinators of agricultural crops" has focused on studies of candidate pollinators of crops that are inadequately pollinated by honeybees. In addition, management systems have been developed around candidate bee species that express the greatest potential for population increases when these bees are introduced into intensive agricultural environments. Consequently, research results first obtained in field studies have been subsequently (and sometimes concomitantly) tested in commercial crop environments (e.g., Torchio 1976, 1979, 1982, 1985; Torchio & Asensio 1985; Torchio et al. 1987) through an experimental programme that is outlined below:

1) Flower visitors of targeted crops and related species are surveyed to identify candidate pollinator species.
2) The biologies of candidate species are studied in detail.
3) Biological data are used to develop methods that improve nesting success, immature survival (including control of parasites, etc.), and survivorship of candidate pollinators throughout wintering periods.
4) Methods to establish field-trapped populations in agricultural environments are tested and evaluated.
5) Foraging biology of each candidate pollinator is studied in agricultural and natural environments.

6) Pollination efficacy is determined on the basis of quantitative data obtained for each targeted crop.
7) Pollinator populations of known numbers are introduced into agricultural environments to determine migratory habits of nesting bees and the subsequent increase or decrease of populations based on numbers of live progeny obtained.

Results from these combined studies have then been used to develop management programs for each pollinator species introduced into intensive agricultural environments devoid of managed habitats (Torchio 1966, 1976, 1979, 1985, in press a,b).

It should be obvious from the above review that at least three different approaches can be considered in any programme designed to increase numbers of pollinators used for U.S. crops: (l) additional honeybee colonies can be made available; (2) habitat management programs can be established for the first time to increase numbers of pollinator species and the population size of each species; and (3) additional non-honeybee species can be developed as managed, alternative pollinators of specific crops. It is also possible that these three areas of research interests can be combined into a unified national programme responsible for developing the best pollination strategy for specific crops or cultivars planted in various climatic zones across the country.

There should be no doubt, however, that honeybees (European or African, or both) will continue to be used as the major pollinator force of U.S. crops in the immediate future because alternative programs have not been fully developed: i.e., managed habitats have yet to be established, and commercially managed non-honeybee pollinators have not been developed for many U.S. crops. This situation will not change unless a national pollination programme is implemented to provide recommendations for the best pollination strategies of U.S. crops, If a pollination programme is not established, the long-standing and dominant practice of accommodating one pollinator species to fit the pollination needs of most crops will be perpetuated, and future recommendations on crop pollination will continue to be based on incomparable, qualitative results.

Discussion

When the pugnacious and feared African honeybee enters the United States, the number of managed honeybee colonies reserved for crop pollination is expected to decline rapidly. Inadequate crop pollination associated with reduced numbers of honeybee colonies will likely occur, resulting in a crisis that will produce a flood of papers offering different resolutions to the problem. These proposals will no doubt repeat much of what has already been reviewed herein: apiculturists can be expected to recommend additional support for honeybee research, and the time-honored rationale (too few numbers of other pollinators available) will be used as the major justification for those proposals; a large group of nonapiculturists will cite the growing literature on native bee biology and pollination (mostly basic research) as primary reasons for suggesting the establishment of habitat management programs; and those working to develop commercially managed non-honeybee species will recommend more intensive use of these bees as alternative crop pollinators.

If honeybee numbers do, in fact, decline rapidly in the near future, it is likely that a pollination crisis will first appear in areas of intensive agriculture because (1) acreages of cross-pollinated crop species have increased dramatically, and each of these crops is dependent on the importation of large numbers of pollinators during short flowering periods, (2) one species (the honeybee) is used as the primary pollinator for the majority of U.S, crops, and (3) fewer numbers of honeybee colonies would be available to pollinate these crops.

The status of U.S. crop pollination and the possible changes that may occur in this particular field of interest within the next 5 years can be summarized in the form of three pertinent questions: (l) Will a pollination crisis occur after the African honeybee enters the United States? (2) If so, should the United States reevaluate current pollination strategies? (3) What alternatives are available to increase numbers of pollinators in the shortest time frame possible? A response to each question is given below.

If the African honeybee enters the United States before alternative pollination strategies are established, and if the public's reaction proves to be as negative as now expected by most apiculturists (Taylor 1985, Rinderer 1986), there is little doubt that the U.S. honeybee industry will experience temporarily a rapid decline in the total number of managed bee colonies available for honey production and pollination services. Local ordinances may well be adopted to prevent hobbyist beekeepers from maintaining backyard colonies, migratory beekeeping practices would probably be disrupted through enforcement of restrictive laws, and various lawsuits involving different aspects of the bee industry would probably increase. These negative factors would work in concert to reduce further the availability of crop pollinators, and the first symptoms of inadequate pollination would probably be expressed in the western sector of the country where large tracts of monocultured, cross-pollinated crops are planted (e.g., almond, apple, avocado, citrus, melons). Recovery of the honeybee industry would then occur as the public's negative perception of the African honeybee slowly modified as has happened in other countries already invaded by this pugnacious subspecies (Taylor 1985).

The second question (reevaluate pollination strategies) is difficult to discuss because pollination, unlike other areas of scientific inquiry, has historically been studied at the species or cultivar levels by individuals representing diverse fields of interests (e.g., botanists, horticulturists, plant breeders, entomologists, apiculturists, ecologists, physiologists, geneticists). As a result, the pollination systems of some plant species have been studied in great detail while pollination and fertilization mechanisms of other taxa (including some crop species) are poorly known. If crop pollination enters a crisis phase in the future, preparation can now be made to establish an acceptable policy designed to correct the problem. For example, pollination efficacy indices should be developed for each pollinator species by crop as a first priority. These indexes would then be used as the cornerstone on which pollination recommendations would be based. Accurate recommendations might, in turn, increase the effective utilization of all crop pollinators, and the efficient use of these pollinators would reduce the negative effect of a pollination crisis. If pollination efficacy indices are not established, future recommendations focused on adequate pollination of crop species will continue to vary (e.g., blueberry = 0.5 honeybee colonies/acre, Marucci 1966; or up to 10 colonies/acre, McGregor 1976), and quantitative results obtained for any pollinator species/crop (e.g., pollination efficacy of *Osmia ribifloris* on highbush blueberry [Torchio in press b], or *Osmia lignaria propinqua* on apple [Torchio 1985]) would not be directly comparable to the qualitative results already obtained for other pollinator species studied on the same crops (see "Pollinators" and "Pollination Recommendations" listed under each crop species in McGregor 1976).

The third question (how to increase numbers of crop pollinators) can best be resolved if the three proposals described above are each adequately supported and better focused, For example, a pilot programme can easily be designed to establish a number of reclaimed habitats in a variety of agricultural environments throughout the country. In addition, small populations of large numbers of bee species (including non-honeybees developed as commercial crop pollinators) could be introduced into each managed habitat at least to develop pollination efficacy indices for the maximum numbers of bee species per crop per environment in the shortest time frame possible. Those of us developing non-honeybees as commercially managed pollinators would, during the same period, increase the number of candidate species developed as pollinator species and the number of crops pollinated by non-honeybees. Apiculturists would then be in a position to emphasize research designed to develop methods for increasing numbers of honeybee colonies maintained in agricultural environments. If these programs were coordinated through a single funding source, collaborative participation by apiculturists and non-apiculturists would more easily be accomplished, the establishment of pollination efficacy indexes could become a standard procedure, and the production of adequate numbers of pollinators for U,S. crops would be realized in the shortest time possible.

The rapid disappearance of natural habitats, combined with pollution spreading into existing habitats, are real threats facing future efforts designed to maintain or increase bee pollinators in the United States. Consequently, collaboration between apiculturists and nonapiculturists is necessary as a first step to resolve any pollination crisis that may develop in the United States, because collaboration would be time- and cost-effective, and time is a premium commodity (the African honeybee is expected to arrive in Texas sometime in 1990), In addition, funding sources are limited, and natural environments (from which non-honeybee populations are obtained) are rapidly disappearing or becoming unavailable. It should be obvious, therefore, that collaborative studies would permit the maximum number of researchers to participate equally in developing methods to produce adequate numbers of pollinators for U.S. crops in the shortest time possible.

Implementation of this type of programme should be seriously considered irrespective of any impending pollination crisis that may or may not occur, because it sets standards for comparative and quantitative results in pollination studies by requiring establishment of pollinator efficacy indices. These standards would permit researchers to collaborate in comparative pollination studies of crop cultivars grown in different climatic zones of the United States as an important first step in establishing pollination needs by cultivar and developing management programs for efficient pollinator species (Garcia 1981; Tepedino 1981; Cane & Payne 1988; Torchio 1985, in press b). The pollination of greenhouse crops (an increasingly important industry in the United States) could also be included under the umbrella of a national programme.

Acknowledgment

Appreciation is extended to Edward Southwick (State University of New York, Brockport), Michael Burgett (Oregon State University, Corvallis), Peter Kevan (University of Guelph, Ontario), James Cane (University of Alabama, Auburn), and Vince Tepedino (USDA-ARS, Utah State University, Logan) for their review of an earlier draft of the manuscript. This is a contribution from the Utah Agricultural Experiment Station, Utah State University, Logan, Journal Paper 3931; and Bee Biology and Systematics Laboratory, USDA-ARS, Utah State University, Logan,

References Cited

Barclay, J. S. & J. O. Moffett. 1984. The pollination value of honeybees to wildlife. Am. Bee J. 124:497-498.

Bitner, H. M. 1982. Current management practices with the leaf cutter bee in Idaho, U.S.A., pp. 161-164. *In* Proceedings lst International Symposium on Alfalfa Leaf cutting Bee Management. University of Saskatoon, Canada.

Bohart, G. E. 1951. Alfalfa seed growers of Utah should protect their wild bees. Utah Agric. Exp. Sta. Farm Home Sci. 12: 32-33.

1962. Introduction of foreign pollinators, prospects and problems, pp. 181-188. *In* First International Symposium on Pollination, 1960, Swedish Seed Association, Copenhagen Publ. Comm. 7.

1971. Management of habitats for wild bees. Proc. Tall Timbers Conf. B:255-266.

1972. Management of wild bees for the pollination of crops. Ann. Rev. Entomol. 17: 287-512.

Cane, J. H. & J. A. Payne. 1988. Foraging ecology of the bee *Habropoda laboriosa* (Hymenoptera: Anthophoridae), an oligolege of blueberries (Ericaceae: *Vaccinium*) in the Southeastern United States. Ann. Entomol. Soc. Am. 81:419-427.

Crane, E. & P. Walker. 1983. The impact of pest management on bees and pollination. International Bee Research Association Tropical Development and Research Institute, London.

1984. Pollination directory for world crops. International Bee Research Association, London.

DeGrandi-Hoffman, G. 1987. The honeybee pollination component of horticultural crop production systems. Hortic. Rev. 9:237-272.

Eck, P. 1988. Blueberry science. Rutgers University Press, New Brunswick, N.J.

Eck, P. & N. F. Childers. 1966. Blueberry culture. Rutgers University Press, New Brunswick, N. J.

Eckert, J. E. & F. B. Shaw. 1960. Beekeeping. Macmillan, New York.

Free, J. B. 1970. Insect pollination of crops. Academic, London.

1982. Bees and mankind, George Allen & Unwin, London.

Garcia, L. E. 1981. Attractiveness of rabbiteye blueberry cultivars to pollinating insects and the effects of honeybees and gibberellic acid on fruit set and development. M.S. thesis, North Carolina State University, Raleigh.

Kevan, P. G. 1987. Alternative pollinators for Ontario's crops: prefatory remarks to papers presented at a workshop held at the University of Guelph, 12 April, 1986. Proc. Entomol. Soc. Ont. 118:109-110.

1989. Keeping up with pollination concerns and options for the future. Can. Beekeep. 14:79-80.

Kevan, P. G., E. A. Clark & V. C. Thomas. In press. Insect pollinators and sustainable agriculture. Am. J. Altern. Agric.

Levin, M. D. 1983. Value of bee pollination to U.S. Agriculture. Bull. Entomol. Soc. Am. 29:50-51.

1984. Value of bee pollination to United States agriculture. Am. Bee J. 124:184-186.

1986. Using honeybees to pollinate crops. USDA-ARS Extension Service Leaflet. 549.

Levin, M. D. & G. D. Waller. 1989. The role of pollinating insects in future world food production. Apiacta 24:18-21.

Marucci, P. E. 1966. Blueberry pollination. 33rd Ann. Blueberry Open House Proc. N. J. Agr. Exper. Sta. 33:16-19.

McGregor, S. E. 1976. Insect pollination of cultivated crop plants. USDA-ARS Agricultural Handbook 496.

Metcalf, C. L., W. P. Flint & B. L. Metcalf. 1962. Destructive and useful insects, 4th ed. McGraw-Hill, New York.

Oertel, E. 1983. Estimated honeybee pollination business in selected states. Am. Bee J. 125:200-201.

O'Grady, J. H. 1987. Market failure in the provision of honeybee pollination: a heuristic investigation. M.S. thesis, University Vermont, Burlington.

Parker, F. D. & P. F. Torchio. 1980. Management of wild bees, pp. 144-180, *in* USDA Agricultural Handbook 535.

Parker, F. D., S. W. T. Batra & V.J. Tepedino. 1987. New pollinators for our crops. Agric. Zool. Rev. 2:279-304.

Prescott, C. & B. Allen. 1986. The first resource: wild species in the North American economy. Yale University Press, New Haven, Conn.

Rinderer, T. E. 1986. Africanized bees: the Africanization process and potential range in the United States. Bull. Entomol. Soc. Am. 83:222-227.

Robinson, W. S., B. Nowogrodzki & B. A. Morse. 1989. The value of honeybees as pollinators of U.S. crops. Am. Bee J. 129: 411-423, 477-487.

Southwick, E. E. & L. Southwick, Jr. In press. Economic value of honeybees in the United States. Econ. Entomol.

Southwick, L., Jr., & E. E. Southwick. 1989. A comment on 'value of honeybees as pollinators of U.S. crops.' Am. Bee J. 129: 805-807.

Stephen, W. P. 1955. Alfalfa pollination in Manitoba. J. Econ. Entomol. 48: 543-548.

1959. Maintaining alkali bees for seed production. Oregon State College Agricultural Experiment Station Bulletin 568, Corvallis.

1960. "Artificial" bee beds for the propagation of the alkali bee, *Nomia melanderi*. J. Econ. Entomol. 53:1025-1030.

1962. Propagation of the leaf cutter bee, *Megachile rotundata*, for alfalfa seed production. Oregon State University Agricultural Experiment Station Bulletin 586, Corvallis.

Taylor, O. R., Jr. 1985. African bees: potential impact in the United States. Bull. Entomol. Soc. Am. 31:15-24.

Tepedino, V. J. 1981. The pollination efficiency of the squash bee (*Peponapis pruinosa*) and the honeybee (*Apis mellifera*) on summer squash (*Cucurbita pepo*). J. Kans. Entomol. Soc. 54:359-377.

Torchio, P. F. 1966. A survey of alfalfa pollinators and pollination in the San Joaquin Valley of California with emphasis on establishment of the alkali bee. M.S. thesis, Oregon State University, Corvallis.

1976. Use of *Osmia lignaria* Say (Hymenoptera: Apoidea: Megachilidae) as a pollinator in an apple and prune orchard. J. Kans. Entomol. Soc. 49:475-482.

1979. Use of *Osmia lignaria* Say as a pollinator of caged almond in California, pp. 285-293. *In* Dewey M. Caron ed., Proceedings IV International Symposium on Pollination. Maryland Agricultural Experiment Station Special Miscellaneous Publication 1, University of Maryland, College Park, Md.

1982. Field experiments with *Osmia*

lignaria propinqua Cresson as a pollinator in almond orchards: I, 1975 studies (Hymenoptera: Megachilidae). J. Kans.
Entomol. Soc. 55:l0l-116.

1985. Field experiments with the pollinator species, *Osmia lignaria propinqua* Cresson, in apple orchards:V. (l979-1980), methods of introducing bees, nesting success, seed counts, fruit yields. J. Kans. Entomol. Soc. 58:448-484.

1987. Use of non-honeybee species as pollinators of crops. Proc. Entomol. Soc, Ont. 118:111-124.

In press a. A case history on the development of *Osmia lignaria propinqua* as a managed pollinator of orchard crops. *In* S. L. Buchmann ed., Experimental studies in pollination and pollinator foraging efficiency. University of Arizona Press.

In press b. *Osmia ribifloris*, a native bee species developed as a commercially managed pollinator of highbush blueberry. J. Kans. Entomol. Soc.

Torchio, P. F. & E. Asensio. 1985. The introduction of the European bee, *Osmia cornuta* Latr., into the U.S. as a potential pollinator of orchard crops, and a comparison of its manageability with *Osmia lignaria propinqua* Cresson. J. Kans. Entomol. Soc. 58:42-52.

Torchio, P. F., E. Asensio & B. W. Thorp. 1987. Introduction of the European bee, *Osmia cornuta*, into California almond orchards, Environ. Entomol. 16:664-667.

Waller, G. D., B. E. Vaissiere, J. O. Moffett & J. H. Martin. 1985. Comparison of carpenter bees (*Xylocopa varipuncta* Patton) (Hymenoptera: Anthophoridae) and honeybees (*Apis mellifera* L.) (Hymenoptera: Apidae) as pollinators of male-sterile cotton in cages. J. Econ, Entomol. 78:558-561.

Ware, G. W. 1973. Bees in agriculture: their problems and importance, pp. 28-38. *In* The indispensable honeybee. Conf. Report. Beltsville, Md.

Received for publication 5 February 1990; accepted 30 April 1990.*

Anthophorine bee - *Anthophora*

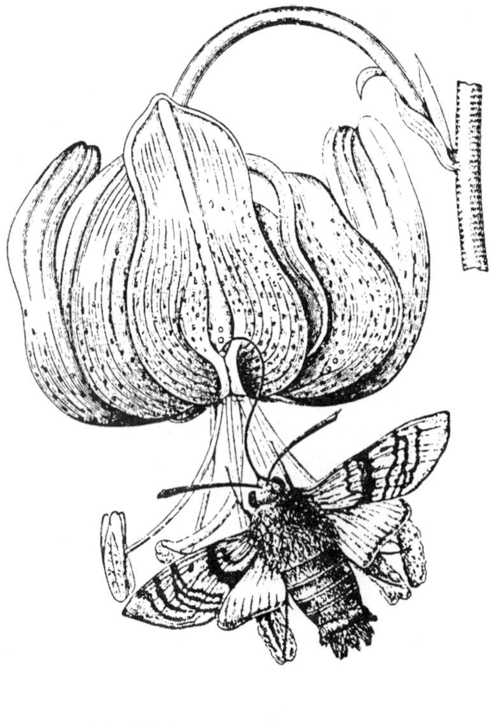

2.1 SUCCESSFUL POLLINATION & ENHANCED POLLINATOR ABUNDANCE

The term '***enhanced pollinator abundance***' is used here in reference to artificially manipulated or managed populations for the purpose of crop pollination (in contrast to wild pollinator populations that have not been manipulated directly, see Sections 1.2, 1.6). In order to use enhanced pollinator populations for agricultural crop pollination, a variety of conditions must be fulfilled. Not only do we need to understand the local requirements of a crop. Additional environmental, economic, infrastructural and social conditions must exist in order to have access to the necessary pollinator populations. This chapter gives a broad discussion of necessary conditions for successful pollination services.

The exceptional advances made with temperate races of *Apis mellifera* allow such observations as those of Free (1993, p. 52):

> When it is know that a crop is difficult to pollinate, usually because it is a species relatively unattractive to honeybees, more than 2.5 colonies (and often 3 to 6) per hectare are sometimes recommended. In contrast, when a crop is attractive, but, its flowers are comparatively sparse, or can be self-pollinated, fewer than 2.5 colonies per hectare are suggested (e.g. *Cucumis melo*, *Trifolium repens*).

Unfortunately, I am not aware that any such generalized figures exist for pollinators other than one or two races of temperate *Apis mellifera* (usually *ligustica*). Some recommendations were recently given for the 'stocking density' of *Apis cerana* in crop fields. These seem similar to the recommended densities for the western hive bee because, although the eastern hive bees maintain much smaller colonies than those of *A. mellifera*, they do not forage as far from the hive, and consequently more will visit the target crop. No studies for the many other tropical pollinators are complete enough to permit such basic estimates of pollinator abundance requirements in croplands or other 'managed systems'.

Before a crop is selected for cultivation in any environment, its pollination needs should ideally be evaluated. Clearly, from the information given in the **Appendix I**, we are ignorant of the pollination requirements of many plants cultivated in the tropics. Frequently, for commercially grown crops it is known whether their fruit or seed production and quality will be increased by abundant pollinators. However, what is not known for most local conditions is whether the pollination need can be taken care of by local, or wild, pollinator populations. This question becomes more difficult when little or no pollination research has been done with a particular crop.

As a guideline, it may be assumed that any crop requiring pollination will produce less without extra pollinators, ***if it is planted in fields wider than 50 to 100 m***, even if the field is near the edge of areas with abundant pollinators. Pollination success of course also depends on many other conditions (see Sections 2.2-2.4). For crops, we must always consider:

- ***the species of natural pollinators and their foraging ranges***
- ***the attractiveness of the crop flowers***
- ***the use of pesticides in the area***
- ***the total size of the field***
- ***the distance to the natural habitat of the wild pollinators***

A Macadamia nut plantation in Malawi, for example, revealed the natural pollinators (mostly *Apis mellifera scutellata*) were effective only in the first few rows of trees, despite abundant honeybee colonies in forested slopes not more than 100 m from the edge of the plantation.

Distribution studies of honeybee foragers in field crops show the importance of saturation of the area with bee colonies and the resulting irregular distribution with lower densities. Depending on the crop's flowering characteristics, lower pollinator populations may be sufficient. A 'rule of thumb' for the density of honeybee colonies that might be needed for successful pollination is that approximately 2.5 colonies per hectare (1 colony per acre) should be distributed evenly through the crop. The number increases with unattractive crops, for hybrid seed production, or in the presence of significant competition from other floral resources.

2.1.1 Selection of Crops & Cultivation Methods

In many countries, food 'self-sufficiency at any cost' often termed the 'green revolution' (Section 1.6) has brought much environmental destruction. Many of the resulting agricultural systems are not self-sustaining or efficient, because they depend to increasing degrees on outside input of imported pesticides, fertilizers, machines and fuel. Large-scale extensive cultivation practices have demonstrated their destructive nature under most tropical climates, soil conditions and management. Yet they are still continued because of the expected high, short-term yields, which are often not sustained. This is a familiar scenario for many enterprises in South and Central America and East Asia. Meanwhile, nondestructive agricultural methods have been demonstrated to be just as profitable as high-technology, high-input agriculture.

Throughout the world, the selection of the right crop or crop variety is very important, as this ultimately determines whether cultivation will be profitable. Local indigenous varieties are usually very resistant to diseases common in their area, but do not have the high productivity of hybrids grown with all the requisite chemical support. In addition, native crops often do not respond to fertilizers as dramatically as hybrids. However, when both are grown *without* all the refined techniques and inputs, indigenous crops in most cases probably out-compete the hybrids. Many local crops could be improved for yield, particularly in conjunction with less chemically-enforced cultivation methods.

Net benefits from managed pollination (in the form of extra fruit or seed set and better quality) will occur only when cultivation practices already are excellent. Given limited investment capital, the efficiency of crop production could be improved through better cultivation methods, storage, processing and shipping, or even advertising. Pollination is just one of the several avenues available to growers seeking to increase profits. Crops grown under marginal conditions cannot be expected to improve much through pollination management unless they depend nearly 100% on pollinators and these are all absent. Growing conditions must be optimal in order to benefit most from pollination enhancement. Normally, there is no need for manipulated pollination services for exotic crops that do not require highly specialized pollinators, when soil, water and weather conditions limit fruit set. Unless the environment has been badly damaged, there should still be a relatively diverse pollinator fauna that can fulfill basic pollination requirements (see Section 1.0). It is in these conditions (an environment that is not severely degraded) that experiments are necessary to demonstrate the benefit of additional pollination and/or improved cultivation methods.

2.1.2 Selection of Pollinators

The selection of pollinators depends not only on the crop's requirements, since there are usually alternative pollinators available (see Section 2.5). Of primary importance are certain technical, practical and economic factors, including basic infrastructures such as reliable transport and good road conditions. Under most circumstances, pollination with honeybees will require migratory beekeeping capabilities, which in turn requires advanced frame-hive equipment and knowledgeable beekeepers.

The type of honeybees present in most parts of Africa and South and Central America are overly sensitive to disturbance and often react with highly defensive behaviour. Such bees can hardly be left on the edge of fields where tractors pass or people regularly work. Vibrations of the tractor or similar diesel engines upset the colonies. Workers in Argentina often have to wear "bee suits" and veils while weeding and irrigating alfalfa fields, surrounded by Africanized honeybee colonies used for pollination.

Maintaining the less defensive European honeybees is often very difficult, resulting in losses due to diseases, queen replacement and low productivity. Importing less defensive queens is relatively expensive and presents a great risk. ***Most bee diseases are spread through such importations.*** And even if the diseases are already present, such new stock may be less resistant to some diseases, which until then had not been a problem. Colonies with imported queens often fail to produce a honey crop for the beekeeper. Imported bees are not adapted to the new area or to the competition with local bees. Therefore, the costs of such importation programs may add up, becoming much higher than justified by economic benefits of better pollination, or the monetary price of the service. Serious diseases not only kill colonies but also drastically reduce the number of active beekeepers and wild colonies.

Where European bees can be bred in isolation and these colonies are maintained at least part of the year in colder areas like mountains or other latitudes, the cost and feasibility of maintaining European colonies in zones of Africanized or African bees (*Apis mellifera scutellata*) may be acceptable. In other areas, selection of less defensive local strains may be possible, but requires time, a well-organized breeding programme, regular re-queening by beekeepers and a sufficient market for such queens.

Colonies of *Apis cerana* in most tropical environments have relatively small forager populations and readily react to disturbances by absconding. However, programs in southwestern India and southern China have demonstrated that large-scale honey production and stable management using modern hives is possible. Well-organized breeding programs combined with improved management may improve the characteristics of *A. cerana* throughout Asia. Already, limited migratory beekeeping with this species is possible. The other species of potential 'hive bee' —the Malaysian honeybee *Apis koschevnikovi* — produces little honey and very readily absconds (D. Roubik, pers. comm.). No breeding or management programs seem useful for this species.

The benefits and limitations of non-*Apis* pollinators are discussed in other parts of this book. Due to their sometimes higher degree of adjustment to specific environmental conditions, they may be better in areas where European honeybees have difficulty surviving. Because many of them are solitary, short-lived species, their life cycles have to be well synchronized with the flowering of the crop (or their diapause and emergence times must be controlled). Meliponine (stingless bee) colonies may present alternative choices because of their usually docile nature and a domestic market for their honey (Section 2.5.3).

Instead of selecting a specific pollinator species and supplying all of the additional training and infrastructure required, the preservation of suitable pollinator habitats and adaptation of cultivation practices may well be a better alternative. Some details are discussed also in Section 1.6.

2.2.2 *Migratory Beekeeping*

Migratory beekeeping, despite the higher cost and need for reliable transport, offers the opportunity for much higher honey yields. Many migratory beekeepers rely upon the income from rental of their colonies for pollination. Most also benefit from installing their colonies at sites that are, even if only temporarily, very productive.

The procedure for hiring commercial pollination services requires agreements in the form of contracts. The most advantageous form and legal structure may be different from country to country, but a few key items should be contained in any verbal or written agreement. Sample contracts that can be used by the grower and beekeeper are provided in **Appendix II**.

Once pollinators have been chosen, it must be possible to transport them to the fields where their services are required. In the case of insects kept in containers this is relatively easy, but where specific nesting sites or domiciles are used, mobility is often more limited. The more frequently pesticides are used, the more important this mobility becomes. Perennial pollinators, those living in colonies or established aggregations, must be protected from poisoning. The future abundance of annual populations is less susceptible to spraying as long as their broods are already provisioned with food and protected from exposure.

Where abundant floral resources are present when the target crops are not in bloom, permanent colonies may be kept, and the pollinators kept from flying when pesticides are applied to crops. Once crop flowering has ceased, the prevention of spray drift, use of less toxic insecticides in lower quantities (less frequently, and application during the evening or at night) will considerably reduce risk to managed pollinators. (see Appendix III).

Where permanently established or immobile colonies are used for pollination, these usually cannot be kept within crop fields and many foragers may visit non-crop flowers. Thus larger numbers of colonies may be required, compared to the usual guidelines mentioned above. If the natural flora does not support such large numbers of colonies, then some additional colonies must be brought in when pollination services are needed. Habitat improvement would diminish the need for migratory beekeeping.

Migratory beekeeping with stingless bees may be possible under conditions similar to honeybees (Section 2.5.3). Some non-bee species such as flies may be reared in large numbers and simply released where needed, when the adult insects emerge (Section 2.5)

Several phases of development apply to migratory beekeeping, outlined and discussed in the following categories:

1) Equipment.

The pollination services of honeybees require colonies domiciled in advanced hives that can be used to provide colonies of the right strength (see 'Conditions of Colonies used for Pollination Service', Appendix II) and at the right time. These can be topbar or frame hive equipment, but for migratory beekeeping, the frames must be wired to support combs during transport, and screening material must be used to allow adequate ventilation during transport (Section 2.5.1).

Topbar hive beekeeping, as widely practiced in most tropical climates, is ill-suited for migratory beekeeping. Comb breakage during transport occurs, with subsequent loss of the colony. Extremely careful transport of topbar hives can nonetheless be attempted, but only when combs contain little or no honey and sealed brood, and when roads are smooth and colonies well ventilated. However, relatively high losses are likely.

Alternatively, smaller colonies, or "nuclei" may be used, which allow easier transport and do not have combs or frames. A larger number of colonies would be needed, however. The use of small, 'disposable' honeybee colonies or "pollination units" was tried experimentally in the USA and may be feasible elsewhere where large numbers of swarms can be easily captured during the time of year when pollinators are needed. This could be particularly useful in Latin America, where during the swarming season of the African honeybee very large numbers of colonies can be obtained. They can be maintained temporarily in almost any kind of container and place. However, absconding by the queen and colony occurs frequently after transport. Two means are available to reduce this tendency — the queen is either confined in a small "queen cage" prior to transport, or "queen excluder" material is placed over the entrance. The latter method has the drawback that, despite its relative ease, a queen that is temporarily starved by the workers may still pass through the excluder, and the colony can escape. Furthermore, African honeybees (*A. mellifera scutellata*)frequently abandon their own queen in such situations, somehow joining other colonies.

Stingless bees (meliponines) and bumblebees require hives or, sometimes for the former, log hives of suitable size and design to accommodate their colonies and for easy access by the beekeeper. They also need to be transportable, in order to concentrate them in large numbers where they could serve as pollinators. Carpenter bees and solitary bees require specially prepared nesting sites. The bees often are kept in holding for several months as immatures, and their use therefore requires advanced planning. Some of these domiciles may not be movable. More research is required to develop appropriate containers for bee species and other pollinators (see Section 2.5).

2) Infrastructure.

For any supply of enhanced or increased pollinator populations, appropriate access to the field is necessary. This usually means road access and not just *any* road or trail, but reasonably smooth road surfaces to avoid destruction of hives, combs, and therefore whole colonies. This is less important where colonies can be left in place permanently. Waterways may be used in some countries for transport to the point that whole apiaries may be installed on barges or boats. Such river transport is less disturbing to the bees than most road transport. In any case, the mode of transport must be safe and dependable. Breakdown may create a hazard to other people or may kill all of the colonies by overheating. Road and vehicle conditions should be highly reliable because most of the transport will take place at night.

Legal regulations should facilitate transport licencing or transit permits, and also regulate sanitary inspection and certification.

Communication (and both contract partners) must be reliable in order to coordinate and time the delivery of pollinators correctly, during the right phase of flowering and so as not to conflict with pesticide applications.

3) Research.

Nectar secretion highly depends on environmental factors and soil conditions. Because flower attractiveness is so important for pollination success, and since the attractiveness depends on the amount of nectar or pollen made available, pollination success strongly depends on soil conditions and weather prior to and during flowering. Thus local conditions may be sufficiently different that the research results from one area may not be applicable to another. Local populations of pollinators, too, may be sufficiently different to change comparative results on pollination of open-pollinated flowers. In other words, where additional pollinators in one area increase production, in other areas the natural pollinator population may be large enough to achieve the same production level without pollinator management. In summary, different crop varieties, or even the same ones, should be tested under local conditions and with different pollinators to determine attractiveness and pollination (see Section 2.3).

With a few exceptions little is really known about non-*Apis* pollinators in tropical climates, and less about wild pollinator populations or their management and effectiveness in agricultural settings. Many possibilities remain for research, and many new practical applications may be the result. Pollination using honeybees, among the races of *Apis mellifera* and to an increasing extent *Apis cerana*, is still the easiest and quickest solution when the crop in question does not demand the use of other pollinators.

The life history and behaviour of many bee species and other possible pollinators has to be studied locally, as is increasingly done in many countries. According to their behaviour some species can be selected for studies on their manageability, possibilities to increase their populations and their suitability for the various crops. Management skills will have to be suitable for the local farming communities or be economically feasible for specialists.

Simultaneously, research and application trials of different cultivation methods need to be conducted. Better choice of crops, including more traditional, indigenous species and varieties is important. In addition, cultivation methods that are less toxic and destructive, and which actually improve or at least maintain soil quality, would result in benefits far exceeding the provision of extra pollinators.

This kind of research is likely to receive growing attention and funding, since it not only improves food production, but also the quality of food and environment. With these anticipated changes, we should not lose sight of pollinator populations and their maintenance. Supporting technology for pollinator management is underdeveloped and research is essential.

4) Training, skills and technical assistance.

Beekeeping with honeybees, contrary to the belief of many — laymen, politicians, agricultural officers and development planners alike — cannot be "learned in a day". For beekeeping to be cost-efficient and applicable to migratory practices, or adequate for the management of population sizes required for crop pollination, years of training and experience are needed. While many countries now have trained and often also experienced beekeeping technicians, beekeeping practices in most village environments are not at the needed levels. This may be due to equipment choices, numbers of colonies managed or the management skills.

To improve the situation, more intensive technical assistance is required. Due to the economical difficulties in most tropical countries it is, however, exactly this technical assistance that is most lacking. Even if sufficient numbers of technicians are available, they are either without transport to get to the villages, do not receive travel allowances for their extra expenses, are overloaded with other extension responsibilities, lack practical training and experience, or any combination of the above. In addition, beekeeping products have to find an attractive and reliable market to entice farmers to invest time and money in more costly management and equipment. The management of non-*Apis* pollinators will require completely new skills that are not traditionally practiced, except in a few instances with the keeping of meliponine bees (Section 2.5.3). This lack of tradition, on the other hand, can make the introduction of new skills *easier*. But most of all it will require specialized service and training.

Therefore, prior to introducing crops that require pollination services, not only an economic feasibility study is required, but also a realistic survey of the possibilities to transfer the required skills; be this for frame-hive beekeeping, management of other pollinator species, or different cultivation methods. Most of all, it will take serious long-term commitment at various levels.

5) Economic feasibility.

Increased production or the higher quality of a crop must justify the additional expense for pollination services. The additional cost first has to be determined. The following have to be included:

- ***cost of improved crop cultivation practices (mechanization, fertilizers, pesticides, etc.)***
- ***fame hive beekeeping or management of other pollinators***
- ***alternative planting schemes***
- ***extra transport requirements***
- ***costs to society or government for skills acquisition, technical assistance, and research into local requirements***

Alternative improvements always merit consideration, such as soil fertility, varietal selection, irrigation, pest damage, and post harvest losses. These alternatives may be less expensive to improve, and may bring larger benefits than improved pollination. An in-depth feasibility study is needed to establish the weight of these factors and priorities.

Adequate cultivation practices are essential with enhanced pollination, otherwise the production increase is sacrificed. (Plants that are starved for nutrients do not produce more seeds and fruit after receiving better pollinator service). This, too, may mean higher investments by the farmer and more chemical imports, unless cultivation methods are changed. The cost to society consequently increases. The higher investment in beekeeping equipment will have to be justified by additional income from pollination fees, and secondarily from higher honey yields. Otherwise, less costly beekeeping methods are usually more adequate for local beekeeping conditions.

Investment capital is an important limitation. It is unlikely that the beekeeper has the required funds. Bank credits are usually very expensive and difficult to obtain. Banks in general are not willing to accept beekeeping equipment as collateral and thus are unwilling to finance beekeepers, unless the beekeeper can offer standard collateral property. Even with sufficient finances available, the establishment of beekeeping operations capable of providing considerable numbers of bee colonies will take time and at best a few seasons with highly experienced beekeepers and managers. Consequently there is a difficult gap to bridge. On the one hand, improvement of beekeeping material and techniques are seldom justified for honey production alone. On the other hand, pollination of crops is not feasible until this more expensive type of beekeeping is widely practised and cultivation practices are adequate. It will be very difficult and uneconomical to introduce crops requiring pollination unless they can, from the beginning, be grown profitably without pollination service (until beekeepers and growers have learned and have put into practice the mutual benefits).

The prospects for managing non-*Apis* pollinators appear to be even less promising because of their more limited use. However, it is exactly their higher degree of specialization or adaptation to more extreme conditions that makes them valuable. More specialized skills will be necessary for their management. The latter again increases the overall cost, but the special applications may render this profitable, not only in an economic sense, but also for the choice of crop.

To summarize, several questions must be asked:

- ***Are there other crops or varieties that produce slightly lower yields, but with less investment and pollination requirements?***
- ***Is the new crop worth the extra costs of additional training, technical assistance and research?***
- ***Is the cost, particularly the 'hidden' or 'secondary cost' of training, research, and so forth, not better spent on improving environmental conditions (Section 1.6)?***

The market must recognize improved quality of the crops with higher prices, or storage losses must be compensated. It also follows that consumer demand must be equal to the increased production.
If these conditions are not met, all the additional costs are not economically sound and cannot be recommended.

6) Social acceptance.

For new methods to be put into practice, the people who are supposed to use them will have to accept them. Will a beekeeper want to move his bees or hire them out? More important, will a farmer be willing to pay for pollination services? Does the one trust the other to comply with a contract or agreement? In several countries it has taken a long time to convince farmers that hiring the additional service for special crops is very much worthwhile. In a Latin American country where I have worked it took several years after the introduction of sunflowers as a crop until farmers actually were willing to pay for honeybee colonies in their fields. Initially, beekeepers were content to have their colonies in or near sunflower fields because of the high honey yields. Larger numbers of colonies in sunflowers however, as necessary for increased seed production, led to lower honey production per hive. To satisfy both sides, the concept of paying for pollination services slowly evolved. It is apparently difficult to convince farmers that the naturally available pollinators or those of the few beekeepers nearby are not enough to increase seed or fruit yields significantly.

2.2.3 Environmental Impact

Manipulating any population of plant, insect or other animal will have some impact on the environment. Increasing honeybee populations will displace native nectar and pollen feeding insects to other flowering species, which may displace other species again, and so on. The same should be true for agricultural systems that have sustained large human disturbances. Lack of natural pollinators in such circumstances is primarily due to the negative impact of large monocultures. Adding pollinators to such a system should lead to an improvement of environmental conditions. In border environments near natural forests the displacement effect, outlined above, may be minimal because the depth of such an effect is likely very limited. In contrast, small plantation or hedge communities might show a large effect.

The general fact that bees or other pollinators are beneficial to the environment may be correct, but the assumption that, therefore, the introduction of additional pollinators is *always* beneficial, is not justified. In areas where manipulated pollinators are native, increasing their populations will probably have local and temporary effects. For exotic species, the impact may range from displacement of local species at resources to their extinction in certain habitats, or favoring the spread of certain plants (including weeds, see Section 1.4), or of diseases.

Improved habitat diversity, which promotes creation of numerous microhabitats such as hedges and planted forest patches, seems to have the least negative impact. This and less destructive cultivation practices generally improve habitat conditions and eventually should contribute to higher production for all crops — whether they are those requiring pollination, or the majority, which do not.

Conclusion

In this chapter I tried to point to some of the essential requirements for improving crop yield and quality through pollination. The basic mechanisms and requirements are known and have been tested under various conditions. What is not known are the individual facts for each country. In industrialized countries the basic infrastructure, including beekeepers, is already present. It has been built over the last century or more. However, beekeepers have become essential for agricultural production only recently, as a consequence of increased environmental degradation and larger monocultures. Creating such an infrastructure in a short period is difficult and expensive. Preventing the necessity through environmental conservation and small-scale, less destructive farming, better planning, logical choice of crops and other practices outlined above is less expensive and more practical — both in the short and long term.

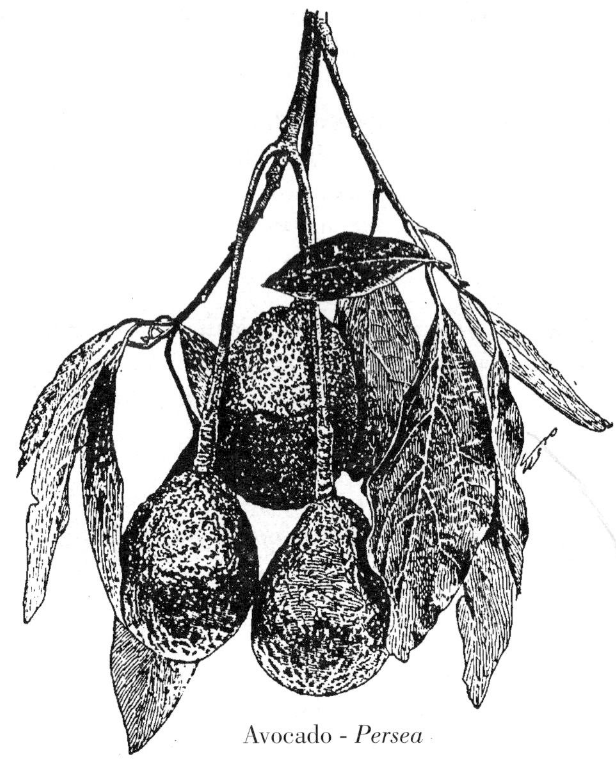

Avocado - *Persea*

Additional References (including pesticide literature)

Accorti, M. in press. Giornata di quali-ficazione dell'assistenza tecnica "Agri-coltura e Apicoltura'. - Ente Toscano Sviluppo Agricolo e Forestale, Ponte a Poppi (AR), 20 guigno 1992.

Adey, M., Walker, P., Walker, P.T. 1986. Pest control safe for bees: a manual and directory for the tropics and subtropics. London: International Bee Research Association, 223 pp.

Akratanakul, P. 1983. On bees and beekeeping. Bangkok [in Thai, translated by IBRA], 182 pp.

Anonymous, 1984. Summary of toxicity to honeybees and bee poisoning hazard of insecticides. Australian Bee J. 65:13-17.

Arzone, A., Patetta, A. 1990. Vent'anni di ricerche sull'azione verso l'ape di antiparassitari agricoli. Apico. mod. 81:111-116.

Atkins, E. L., Anderson, L. D., Nakakthara, E. A., Greenwood, E.A. 1971. In: A. M. SISTO (ed.) 1977 - Repertorio sistematico dei fitofarmaci. S.I.F., (4 ed.), pp. 435-437.

Atkins, E.L., Greenwood, E. A., McDonald, R. L. 1973. Toxicity of pesticides and other agricultural chemicals to honeybees. Laboratory studies. Univ. Califirnia, Agric. Extension M-16, 38 pp.

Atkins, E .L., Kellum, D., Atkins, K.W. 1981. Reducing pesticide hazards to honeybees: mortality prediction techniques and integrated management strategies. Div. Agric. Sci., Univ. California, leaftlet No. 2883, 24 pp.

Charrette, H. 1987. Reglementation. Protection des abeilles et des

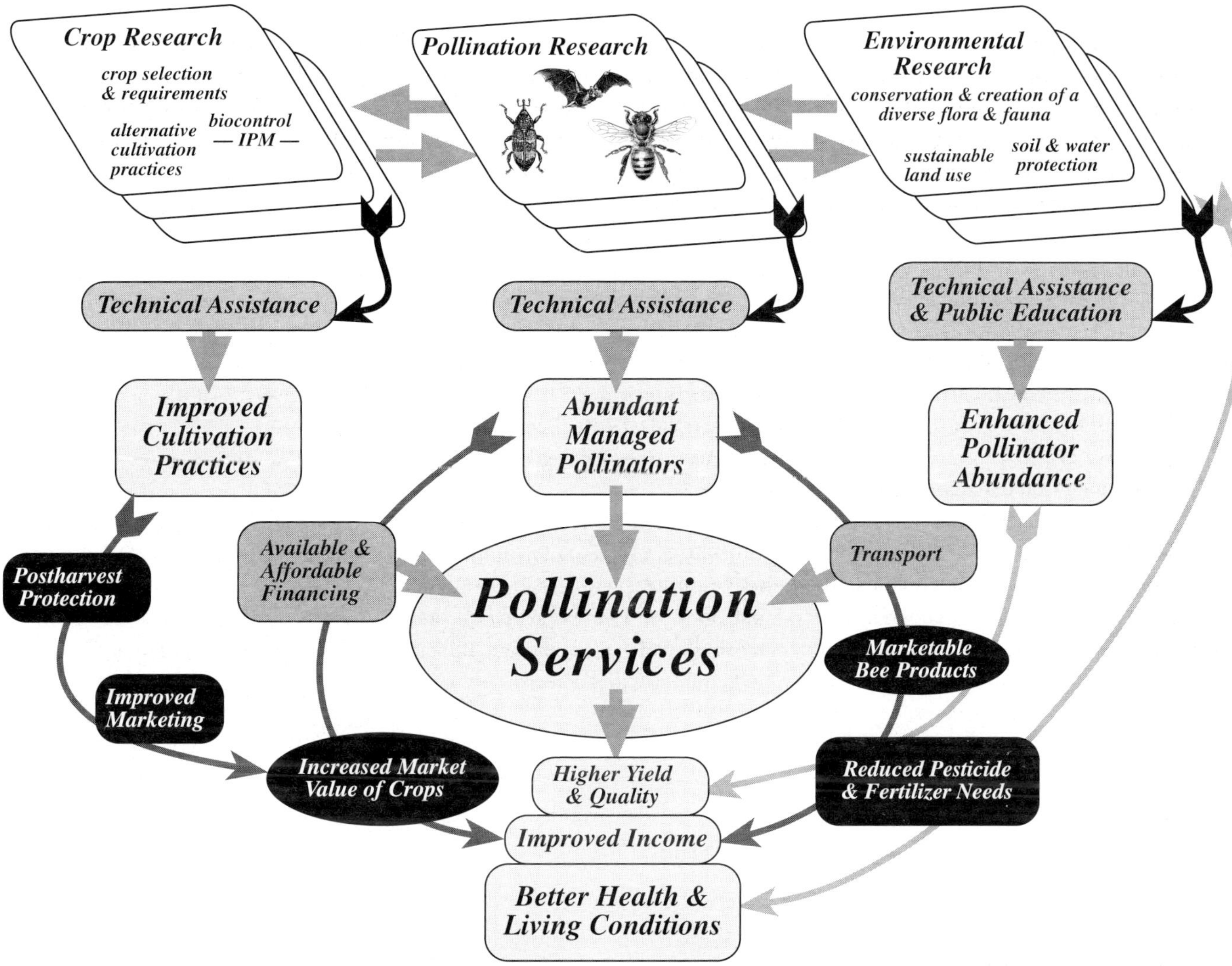

Requirements to improve income from agriculture through enhanced pollination

insectes pollinisateurs contre les produits antiparasitaires. La Defense des Vegetaux 243:15-18

Contessi, A. 1990. In tossicazioni ed avvalenemanti da pesticidi. In: "Le api". Edagricole, Bologna, (2 ed.), pp. 159-168.

Corbet, S. A., Williams, I. H., Osborne, J. L., 1991. Bees and the pollination of crop and wild flowers: changes in the European Community. Draft to Scientific and Technical Options Assessment European Parliament, 50 pp.

Craig, S. M., McCutcheon, D. M., Miller, A. V. 1986. Pesticides and bee poisoning. In: Handbook for pesticide applicators and pesticide dispensers. Ministry of Environment, Pesticide Control Branch, British Columbia, New Zealand, pp.125-130.

Crane, E., Walker, P., 1983. The impact of pest management on bees and pollination. London: Interational Bee Research Association, 129 pp.

Danka, R. G., Rinderer, T. E., Collins, A. M., Hellmich, R. L. 1987. Responses of Africanized honeybees (Hymenoptera: APidae) to pollination-management stress. J. Econ. Entomol. 80:621-624.

Drescher, W. R., Crane, E. 1982. Technical cooperation acvtivities: beekeeping. A directory and guide. Deutsche gesel-lschaft für technische zusammenarbeit, Eschorn, Germany.

Fabbricatorie, A. J.1988. Sostanze attive autorizzate in agricoltura. Rome: Ist. Sper. per la Patologia Vegetale, 517 pp.

Graham, J. M. (ed.) 1992. The hive and the honeybee. Hamilton, Illinois, USA: Dadant and sons, 1324 pp.

Hartmann, W. H. 1991. Commercial pollination in the Western Cape. In: Proc. Intl. Beekepers Symp., R. H. Anderson, B. Buys (eds.). Capetown: South Africa, pp. 21-25.

Huges, A.R. 1984. A selection of information regarding controls on agricultural chemicals. In: A.D.A.B. Course in Apiculture Notes, Queensland Agricultural College, Lawes (Gatton). Queensland, Australia: 4.9.1-4.16.9.

Infor Vie Saine, Nature et progres, les amis de la terra, aves. 1979. La mort des abeilles: un signal d'alarme. Les herbicides, une menace pour l'apiculture et le monde vivant. Infor Vie Saine, Champion (Belgium), 198 pp.

Jaycox, E. R. 1985. Pesticides and honeybees. In: "Beekeeping in the Midwest", Cooperative Extension Service, Univ. Illinois, circular 1125, pp.145-148.

Johansen, C. A. 1966. Digest on bee poisoning, its effects and prevention with an annotated list of 92 insecticides. Bee World, 47:9-25.

Johansen, C. A. 1972. Toxicity of field weathered insecticides residues to four kinds of bees. Environmental Entommology: 1:393-394.

Johansen, C. A. 1979. honeybee poisoning by chemicals: sign, contributing factors, current problems and prevention. Bee World 60:109-127.

Johansen, C. A. 1990. Pollinator protection: A bee & pesticide handbook. Cheshire, Connecticut: Wicwas Press, 212 pp.

Jones, W. A. 1986. Pesticide affecting beekeeping and crop pollination. Dep. Agric. New South Wales, Agdex 481/687,19 pp.

Loper, G. M., Danka, R. G. 1991. Pollination tests with Africanized honeybees in southern Mexico, 1986-1988. American Bee J. March, pp. 191-193.

Louveaux, J., 1984. Les traitements phytosanitaires et les insects pollinisateurs. In: J. Louveaux (ed.), Pollinisation et productions vegetales. I.N.R.A., Paris, pp. 565-575.

Mamood, A. N., Waller, G. D. 1990. Recovery of learning responses by honeybees following a sublethal exposure to permethrin. Pysiological Entomology 15:55-60.

Mazzone, P., Viggiani, G. 1988. Effetti collaterali di fitofarmaci su Artropodi utili. Boll. Lab. Entom. Agraria "F: Silvestri" 45:59-66.

Morse, R., Hooper, T. 1985. The illustrated encyclopedia of beekeeping. Blandford Press, Poole Dorset, 432 pp.

Mountain, P. N., Schehle, J. A., Steinhobel, F., Toit, A P. du. 1989. Pollination and honeybees (Bestuiwing en heuningbye. Brooklyn, South Africa: Pollination Services Association, 21 pp.

Muccinelli, M. 1990. Prontuario dei fitofarmaci. (6 ed.) Bologna: Edagricole, 696 pp.

Sanderson, G., Tazik, P., Killion, E., Steffey, K. 1989. Suggestions for minimizing bee, fish and wildlife losses from pesticides. In: 1989 Illinois pest control handbook. Urbana-Champaign, Illinois: Cooperative Extension Service, Univ. Illinois, pp. 391-405.

Schwartz, P. H., 1982. Protection of honeybees and other insect pollinators. In: Guidelines for the control of insect and mite pests of food, fibers, feeds, ornamentals, livestock, and households. U.S.D.A., Agric. Handbook No. 584:8-9.

Smith, T. A., Stratton, G. W. 1986. Effects of synthetic pyrethrinoid insecticides on non-target organisms. Residue Reviews 97: 93-120

Taber, S. 1986. Pesticides and bees. Gleanings in Bee Culture 114:347,360-361.

Ugolini, A. 1991. Per un corretto impoego degli antiparassitari. Ordinanza Ministeriale 18 luglio 1990. Supple-mento a L'Informatore agrario No. 1.

Worthing, C. R. (ed.). 1991. The pesticide manual — The British Crop Protection Council (9 ed.), 1141 pp.

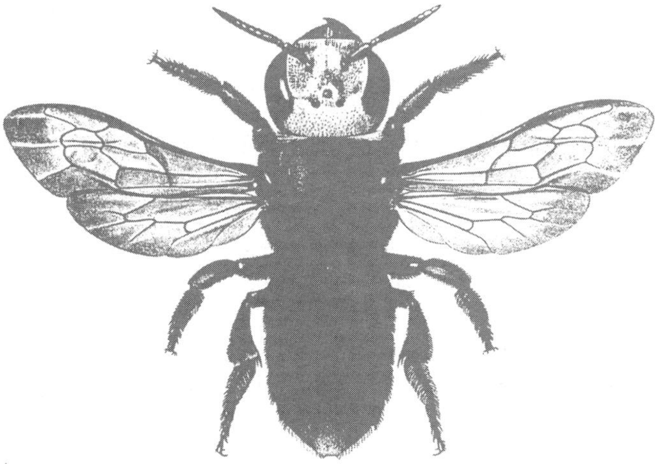

2.2 EVALUATING POLLINATORS

The factors influencing effectiveness of pollinator species must be considered when pollination of any plant is studied. Flower structure provides the framework in which biotic and abiotic pollination agents operate and gives initial clues to the likely effective pollinating agents. By themselves, these considerations are not enough; studies of visitor behaviour and critical field assays are essential, as emphasized elsewhere in this manual and particularly in Sections 2.3 and 2.4.. Studies should also focus on availability of the pollinating agents to determine the most important and manageable species in the particular habitat and circumstances.

2.2.1 Parameters for Measuring Pollination

There are several measurements, both direct and indirect, that can be taken to assess pollination. However, these measurements differ greatly in their dependability as indices of pollination. Particularly in the past, fruit or seed yields (e.g., weights, numbers, size) have been emphasized. Unfortunately, these can be poor estimators of pollination — for several reasons. First of all, fruit production, a measure of flower fertilization, can only be an indirect measure of pollination or the placement of pollen grains on the stigma (Fig. 1b). The problem with counting fruits is that the investigator is at the mercy of *plant-maternal* influences that occur after pollination. That is, pollination can be grossly underestimated in plants that undergo abortion of developing fruits. Such is the case when relatively low nutritive resources are available to the plant (see Section 1.6). And therefore, even though pollination did occur, it is not represented by the fruit. Another hindrance to determining pollination from fruit yields can occur in plant species having low ovule numbers per flower relative to the number of pollen grains placed on the stigma. In such cases, the potential number of ovules that could have been fertilized will never be known from fruit counts, and hence pollination will again be underestimated. Yet another problem is that some plant species can develop fruit parthenocarpically (Section 2.3.10, **Appendix I**), in the complete absence of pollination. In this case, pollination would be overestimated, if it was measured by fruit counts. It is also important to determine whether a plant species is self-compatible or not. Zero fruit counts may not truly indicate lack of pollination, when incompatibility systems are present. And as stated elsewhere, *100% fruit set from flowers fertilized is extremely rare,* even in optimum conditions. A much more direct method to quantify pollination is to actually count the pollen grains placed on the stigma. This can be accomplished by various methods, the suitability of which depends largely on flower size and morphology of the stigma and pollen. One method (see Beattie) involves the mounting of the upper part of the pistil (gynoecium) in acid fuchsin/gelatin on microscope slides in the field, and then counting pollen grains with the light microscope

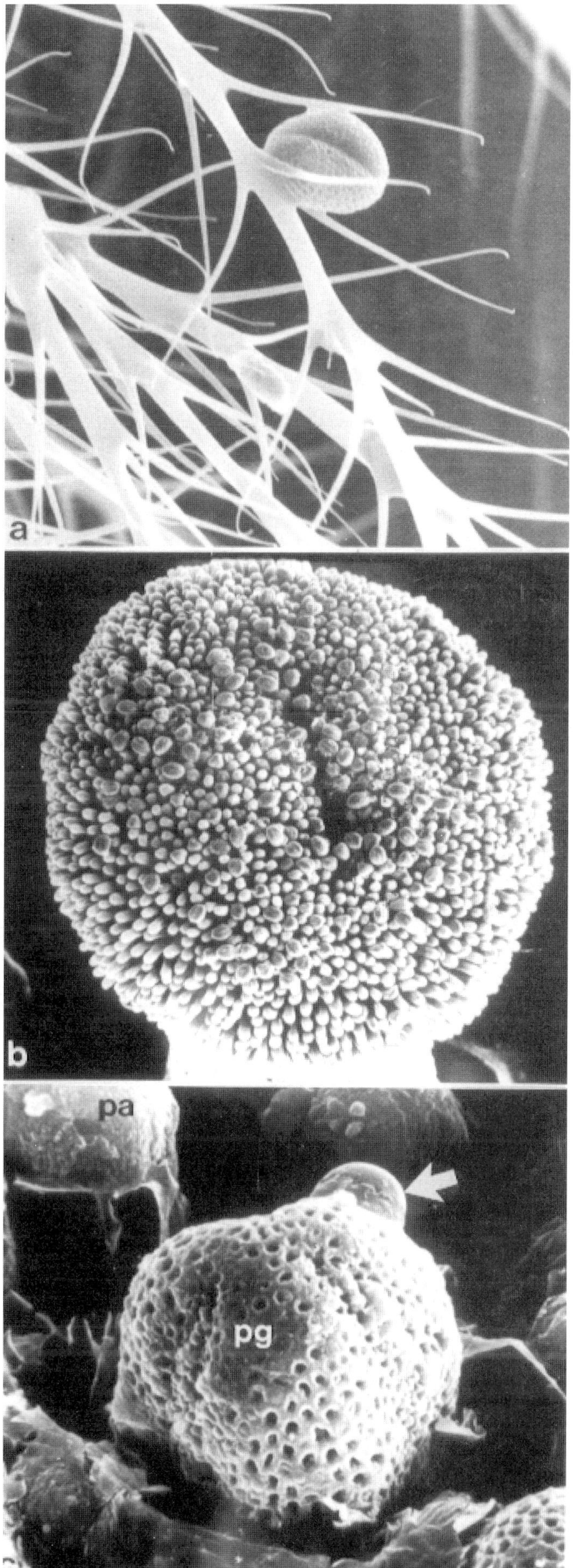

Fig. 1. **a)** SEM of branched body hairs of *Apis mellifera* showing lodged pollen grain of *Hirschfeldia incana*; **b)** SEM of the stigma of *Brassica napus*, showing numerous papillae and several pollen grains; **c)** Higher magnification showing a pollen grain (pg) just germinating a pollen tube (arrow); stigmatic papilla (pa)

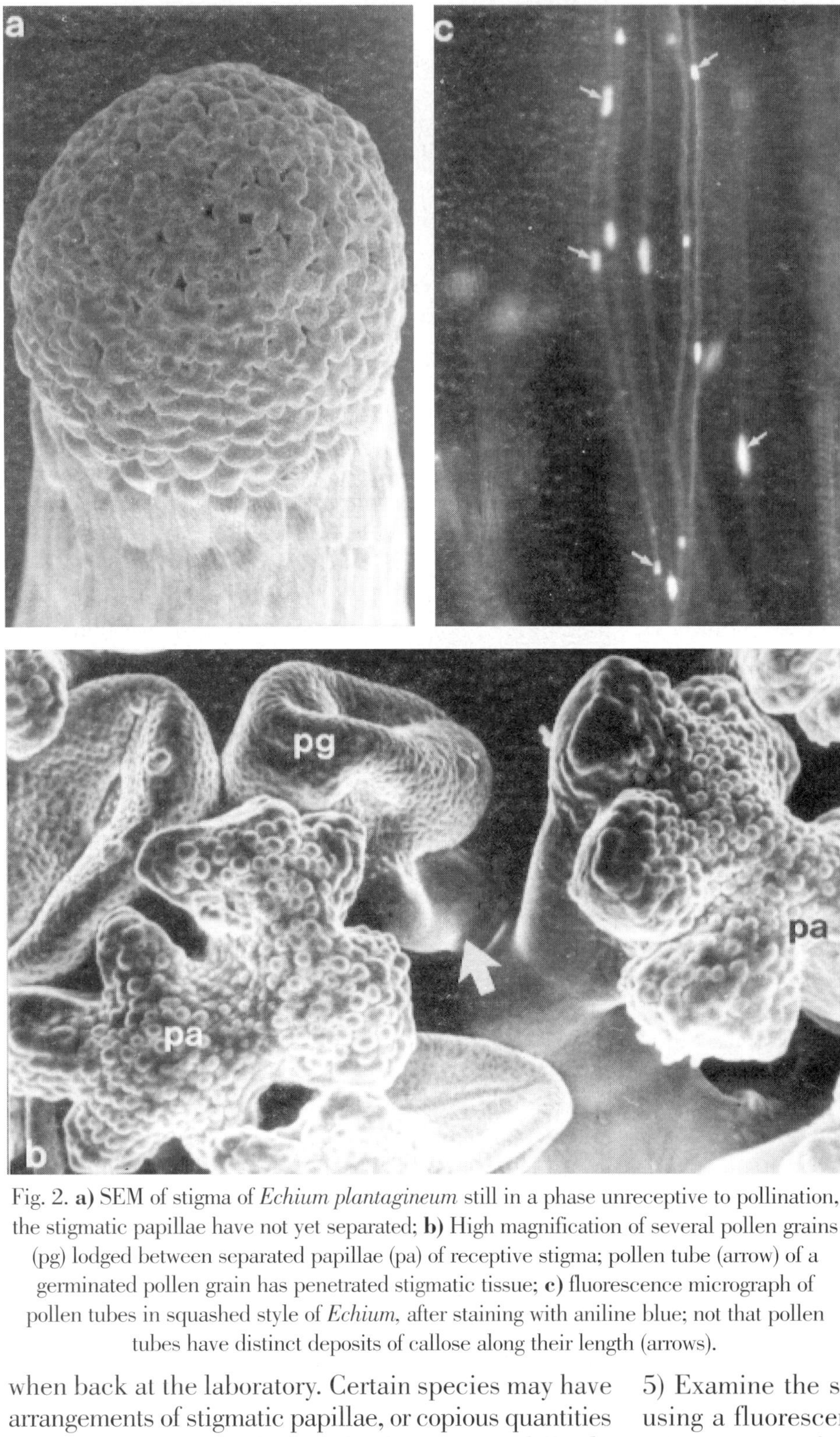

Fig. 2. **a)** SEM of stigma of *Echium plantagineum* still in a phase unreceptive to pollination, the stigmatic papillae have not yet separated; **b)** High magnification of several pollen grains (pg) lodged between separated papillae (pa) of receptive stigma; pollen tube (arrow) of a germinated pollen grain has penetrated stigmatic tissue; **c)** fluorescence micrograph of pollen tubes in squashed style of *Echium*, after staining with aniline blue; not that pollen tubes have distinct deposits of callose along their length (arrows).

when back at the laboratory. Certain species may have arrangements of stigmatic papillae, or copious quantities of stigmatic exudate, which make this practice difficult. There is also a chance of counting pollen grains on the stigma not belonging to the species represented by the pistil. Also, it may not be possible to observe the entire stigma, when mounted on a microscope slide. Scanning electron microscopy may be helpful to visualize the entire stigmatic surface, and may even allow a true assessment of pollen-grain germination and pollen-tube growth (Fig. 1c, 2b), but it involves a more expensive and time-consuming process.

Counting of pollen tubes at the style base (Fig. 2c) in plant species that are self-compatible is another technique useful to measure pollination directly and accurately (see Davis). Each viable pollen grain properly placed on a receptive stigma of the same species sends down one pollen tube, to deliver the sperm-cell nuclei necessary for fertilization to occur. Pollen tubes within the center of the style are relatively protected, and therefore are not nearly as susceptible to loss from the style as pollen grains might be from the stigma, during tissue processing.

The technique can still be useful for assessing pollination of many incompatible species, but prior experimentation is necessary to determine if pollen tubes from self-pollen can reach the style base, or whether instead the growth of these pollen tubes is inhibited in the upper style or on the stigma.

There are various techniques used for pollen-tube detection in the style. The most common one involves the fluorescence of callose, a major component of pollen tubes, after staining of styles with a fluorochrome called aniline blue. One procedure follows:
1) Fix styles in the field in separate, labelled vials containing 1:3 acetic acid: ethanol, for about 4 h, before replacing the fixative with 70% ethanol,in the laboratory.
2) Soften the styles in 10% $NaSO_3$ (sodium sulphite) or other strong base, for several hours, at 60°C in an oven.
3) Immerse the styles in 0.1% aniline blue in 0.1M K_3PO_4, overnight.
4) Remove the styles, mount in a drop of aniline blue on a microscope slide, and gently squash under a coverslip.
5) Examine the style bases for pollen tubes (Fig. 2c) using a fluorescence microscope with exciter filter of 450-490 nm. Callose stains yellow-green. Count the number of pollen tubes. Although an excellent technique for pollen-tube detection, the necessity of a fluorescence microscope may limit the usefulness of this technique. There are other methods, requiring only the ordinary light microscope, for staining pollen tubes at the style base, which may be helpful (see Section 1.1.3).

2.2.2 Assessment of Flower Visitors as Pollinators

Very useful information about pollination can be obtained from experiments that exclude visitors to flowers of the plant species under investigation. Typically, prevention of floral visitation has been achieved in the field by bagging or caging. This practice serves as a valuable control, of course, to determine first of all whether insects or other vectors are even required for pollination. Mesh bags can be easily sewn by hand or machine, and often are made from curtain lining, cheesecloth, bridal veiling, or other porous material. What can be of utmost importance, of course, is the mesh size of the material used, and this can give some very useful information. For instance, in a study of *Echium plantagineum* in Australia, although large insects were excluded, it was found that adult *thrips* could traverse the bagging material used, and the low levels of pollination in emasculated, bagged flowers of this species, has been attributed to their activity (see Davis).

Emasculation, if it can be performed, is a very nice technique when used in conjunction with bagging experiments, because any pollination of emasculated flowers must occur by introduction of pollen from elsewhere. The availability of this technique depends much on the floral morphology. In addition, anther or stamen removal should be conducted with extreme care, in order to prevent selfing during the process. However, if available, it allows the investigator to get a quantitative handle on how much pollen foreign to that flower the insect species 'X' can introduce from its body during each visit. This is very pertinent information from the standpoint of assessing predominant pollinators, and also with the long-term hope of managing and cultivating these particularly useful species. Emasculated, virgin flowers that are bagged and belong to plant species pollinated by airborne grains are also easily assessed for outcrossing by abiotic means.

Fig. 3

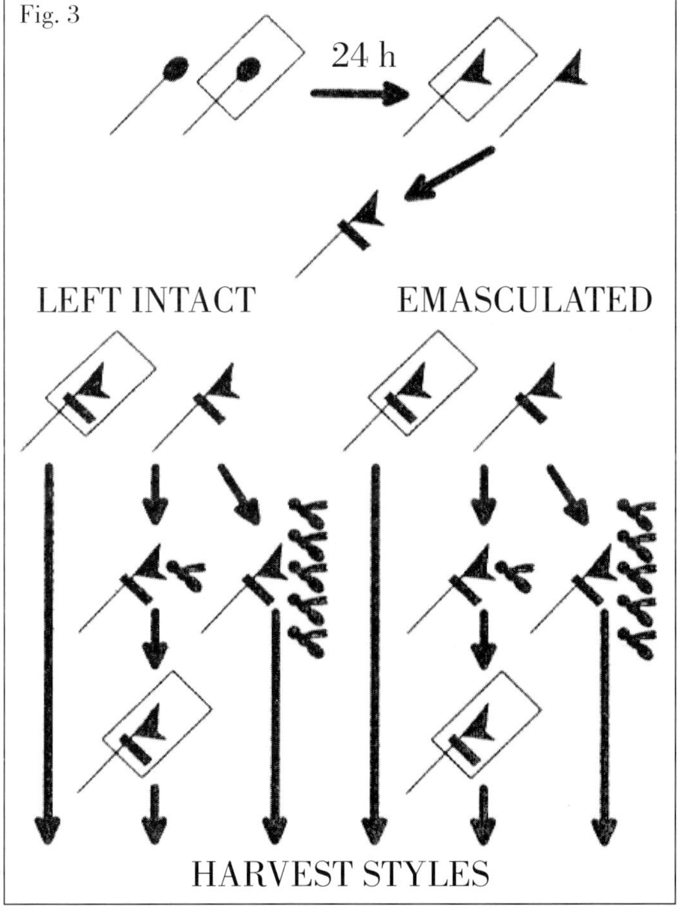

Simple bagging experiments to permanently exclude flower visitors are useful to determine the need for biotic agents of pollination (Section 2.3.10). However, there are even more intricate things that can be done with bagging experiments, which further disclose valuable information. These uncomplicated, additional steps are essential to determine which insects, etc., are responsible for the lion's share of the pollination of that crop in that region. For instance, the procedure of allowing *single* visits to *virgin* flowers allows the discrimination of actual pollinators from mere flower visitors, and also gives the powerful advantage of ranking various insect species according to their efficacy as pollinators of the crop in question. The procedure is outlined in Fig. 3. Virgin flowers are those blossoms initially bagged as unopened, mature buds, to exclude any visitors. Then, after the buds have opened within the bags, they are carefully unbagged, labelled and then continuously watched for the first insect visitor to them. As soon as that initial visitor departs, the flower is carefully rebagged, as it stays for the rest of its flowering lifetime. For comparison, other flowers are bagged for their entire lifetime, or bagged only at night or only during the day or left permanently open to multiple visits. After the necessary, predetermined time period for pollen tubes to reach the style base has elapsed, styles are harvested and processed for pollen-tube counts (see above).

Insect visitors must be identified, and this may involve capture of the insect upon departure from the virgin flower, and its subsequent identification by a specialist, if need be (see 1.1.3). Also, by recording characteristics such as what the insect was seeking (pollen, nectar, both, or other?), how long the visit lasted, or whether the insect carried pollen loads during these single visits allows the investigator to build an even stronger picture of the pollination story. The method even allows one to discriminate pollination efficacy within an insect species: queen vs. worker, male vs. female. Another advantage is that one can determine how many visits by insect species 'X', on average, are necessary to achieve adequate pollination of the crop species in question.

2.2.3 Hallmarks of Plant—Pollinator Associations

For wind, bat, bird and to some extent large bee-pollinated flowers, the concept that flowers have marked features associated with their major pollinators has some utility. However, less efficient pollinators also visit flowers better suited to other animals. Very different kinds of pollinators and flower visitors do in fact occur at each species of flowering plant. For instance, bumblebees and carpenter bees use so-called 'bird flowers', and long-tongued bumblebees use flowers suited to butterflies. In other cases a mixture of wind and large-bee pollination may occur. Flowers pollinated by a guild of generalist insect pollinators like honeybees, stingless bees and flies, have less predictable features. Even so, Umbelliferae flowers, for example, have a characteristic high concentration of fly and short-tongued bees, compared to legume or passion fruit flowers. Subtle floral features, such as chemical composition of pollen and nectar (Section 2.3), may have significant influence on flower-visiting animals.

A number of floral traits are associated with certain visitor types, or 'pollination syndromes' (see Figure). Floral features that can strongly indicate dominant pollinators or mechanisms are as follows:

- ***Wind*** Flowers are inconspicuous, often green or white, and with little or no aroma and no nectar. The flowers have a high pollen-to-ovule ratio. Pollen is abundant, light and dry, lacking pollenkitt. The multibranched stigma and mass-flowering suit both wind dispersal and collection by some insect — grasses, *Casuarina*, and *Plantago* are examples. Honeybees, stingless bees and bumblebees avidly collect pollen from crops like maize and weeds like *Plantago*. Collection of many grass pollens is common in the tropics, particularly for halictids, meliponines and honeybees.

- ***Bats*** Flowers open at dusk or night and have a strong, musty smell. The flowers are generally large, dangle on long hanging stalks, or are arranged as pincushion or pagoda structures, with large-mouthed flowers or strong brush inflorescences. Pollen grains are often gemmate (R. Palacios, pers. comm. to D. Roubik). Flowers are usually whitish, creamy, drab greenish, or purple and rarely pink, with large quantities of nectar that can be dilute (10-30 % sugar) and pollen. Examples are some bananas *(Musa)* and *Agave* species.

- ***Moths*** Flowers have a heavy, sweet perfume and open at night, often closing by the next day. The flowers are horizontal or pendant, white or faint colours, or drab red. Nectar is deeply hidden in long, narrow tubes or spurs.

- ***Birds*** Flowers open during the day and lack a scent. Flowers are deep, tubular or spurred, with hard walls, stiff unit filaments and a well-protected ovary — some examples are Malvaceae, Acanthaceae, Helioniaceae, Myrtaceae and some Rubiaceae. Brush flowers, e.g. *Eucalyptus*, gullet or legume flowers, e.g. firebush *Erythrina* can be pollinated by birds. Flowers are often scarlet or contrasting bright colours and a lip or margin is absent or curved back. Nectar guides may be conspicuous or absent but nectar is abundant and well protected.

- ***Bumblebees and carpenter bees*** Flowers are open during the day and give a fresh but generally not a strong smell. Some pollen producing flowers may smell more strongly (roses, poppies). Gullet (*Impatiens*), smaller tube (*Fuchsia*, *Digitalis*) legume, bowl (Rosaceae, Malvaceae) and even large pendulous flowers (*Actinidia*, *Passiflora*, Solanaceae) seem preferred. Flowers often are *pollen-only* species, lacking nectar and nectaries. Flowers are arranged on a conspicuous rounded flowerhead or spike, so the flowers are more readily seen at a distance (*Digitalis*, *Echium*). A nectar guide is usually present. Nectar of over 30% sugar is preferred. Nectar is hidden at least 4-10 mm within the flower at times in a narrow tube. Flowers with structural depth and an uneven outline are preferred (e.g Lamiaceae, Scrophulariaceae, legumes), at times the flowers have a 'resting platform' for visitors. Colors are lively blue to yellow, pink or white, but not an inconspicuous green. Nectar guides, which strongly reflect light in the ultraviolet end of the visible spectrum, are much less conspicuous to humans than to the bees.

- ***Honeybees and short-tongued bees*** Similar to the larger bees just described, but with less tendency to seek an uneven flower outline. Flowers with conspicuous spikes, and smaller flowers with shorter tubes or green color, and without much petal are satisfactory. The round flowerheads of *Acacia* and *Mimosa* often are attractive to smaller bees.

- ***Butterflies and hawkmoths*** Flowers are open either during day or night and have a weak, agreeable smell. Flowers present a long, narrow tube or spur, ample nectar, an undissected rim with vivid red or purple colors and a simple nectar guide or groove. *Buddliea davidii* and *Lantana* illustrate the type of flower, but other *Buddleia* species are among the flowers favored by bumblebees. Honeysuckle partially fits the criteria for butterfly flowers, but is among the preferred flowers for long-tongued bumblebees.

- ***Flies, wasps*** Unspecialized flies and wasps can be prominent visitors of flowers with a platform and regular, simple flowers without much depth and light, dull colours — *Cissus* (Vitaceae), umbellifers, ivy (*Hedera*), holly (*Ilex*) and some golden rods (*Solidago*). Other flies with long, thin tongues such as some Bombyliidae and Conopidae tend to act like small bumblebees or butterflies, often visiting Lauraceae, Compositae, and some Rubiaceae.

- ***Beetles, some flies*** Some flies (e.g. Sarcophagidae, Calliphoridae, Drosophilidae) and beetles (e.g. Staphylinidae, Scarabaeinae, Nitidulidae) are attracted to carrion or fermenting fruit; these insects tend to seek strong smelling flowers, often presenting the resource in a spadix and, as among Araceae, the odor is dispersed by active heat production within the inflorescence.

Principal Flower Features & Corresponding Flower Visitors

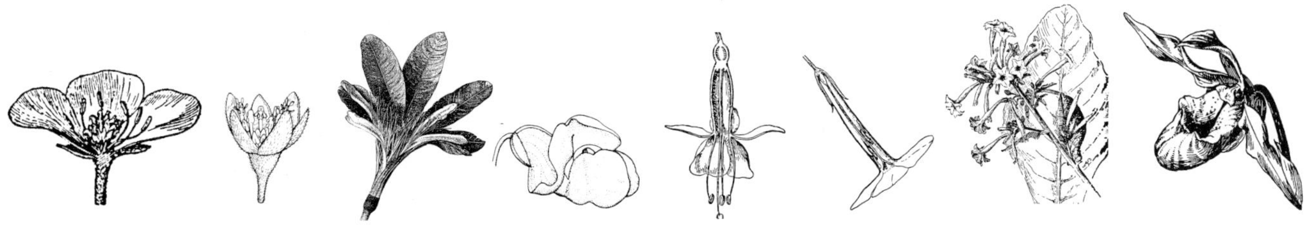

	Bowl	*Bell*	*Brush*	*Flag*	*Gullet*	*Trumpet*	*Tube*	*Trap*
Advertisement *visual & olfactory*	corolla	corolla	stamens	standard petal	lips	limb & tube	corolla	corolla, lip
Pollinator Helpers: *landing surface*	whole flower	flower margin	any part	keel lateral petals (foothold)	lower lip	margin	margin	margin
guiding marks				symmetry & marks on standard	nectar guides	nectar guides		color & pattern
guiding structures				whole flower	none or hairs whole flower	surface structures	narrow tube scales, hairs	hairs
Pollen Cache	exposed	partly hidden centralized	exposed	hidden	hidden	somewhat hidden	hidden	hidden
Nectar Cache	localized		diffuse	centralized (if present)	centralized	centralized	centralized	
Reward Type: *Pollen*	***	*	*-	*	*	*-	*-	
Nectar	*-	*	*	*-	*	**	***	
Main Pollinator	beetle, unspecialized bee, fly, Lepidoptera	short-tongued bee, wasp, fly, settling moth, bird, bat	bee, butterfly, beetle, mammal, bird	long-tongued bee, bird	long-tongued bee, hawkmoth, bird, butterfly	butterfly, hawkmoth, bird	hovering & perching moth, butterfly, bird, long-proboscis fly	carrion fly, beetle, microdiptera, bee

2.2.4 Pollinator Efficacy & the Crop

The effectiveness of a pollination agent depends on:

1) Frequency and kind of stigmatic contact.
This can be a major factor in pollination of large flowers or flowers with nectaries on outer parts. Some flowers need at least 10 stigmas to be contacted for a bee to transfer enough pollen to form a full-sized fruit with one visit. A few bees (*Bombus*, subgenera *Bombus* and *Mendacibombus*) and *Xylocopa* or wasps (*Ropalidia*) make a hole or a slit at the base of flowers of red clover, field beans and blueberries, then *Apis mellifera* uses this hole. Failure to pollinate also results from probing through the side of the flower, for example in crucifers or almonds, or from failure to 'trip' flowers such as alfalfa. This can occur consistently with nectar-collecting *A. mellifera* and *A. florea*. On small and open flowers such as Umbelliferae, Compositae and many of the less specialized flower families, stigmatic contact is achieved easily by most visitors.

2) Amount and availability of pollen carried per pollinator.
On large and relatively unspecialized flowers, bumblebees, honeybees and solitary bees can average 20-90,000 (max. 0.5 million) pollen grains on their body, compared to only a few on thrips. Other bee species, *Hylaeus* and other Hylaeinae carry pollen internally — they are relatively small and hairless. On kiwifruit the pollen carrying ability of *Hylaeus*, with 100-500 pollen grains each, was similar to that of smaller hover flies (Syrphidae).

Some pollinator species frequently carry pollen on part of the head, claws or body that closely contacts the stigma. Flower structure and the grooming habits of the pollinator usually combine to make some part of the body carry pollen available for pollination (other pollen is lost during grooming and pollen feeding or storage in brood cells). The number of pollen grains that can make contact with a stigma is usually much less than the total carried on the body.

3) *Quality of pollen transferred.*

Bees, honey-making wasps such as *Brachygastra* and *Nectarina*, and masarid wasps collect pollen or nectar from flowers to feed their larvae as well as themselves, while other insects sustain only themselves at flowers. Thus female bees, needing a relatively large quantity of pollen to supply their broods, are likely to visit different plants to collect the food, which enhances their cross-pollinating activity far beyond that of all other flower-visitors including the masarid wasps. Frequent movement between flowers on separate plants is important for plants that need cross-pollination or that have distinct male and female individuals. In fruit orchards honeybees mainly forage along rows, which restricts their effectiveness in cross-pollination unless suitable 'pollinizers' are grown in the same row. When there are low nectar levels, bees on red clover and fruit blossoms move more often from flowerhead (or tree) to flowerhead, and over larger distances between flowerheads. With relatively high levels of nectar more flowers are visited on a flowerhead and movements between flowerheads are less frequent and over shorter distances. High nectar production will support greater densities of foraging bees, but in principle cross-pollination may be somewhat reduced, compared to crops secreting less nectar. Differences in this foraging behaviour can be used in the field to assess whether limited nectar is restricting the populations of bees on a crop or encouraging cross-pollination.

4) *Rate of flower visitation and length of 'workday'. (Number of flowers visited).*

Honeybee foraging usually starts at 10-15°C, with a lower threshold in spring than in summer. Flight activity rapidly increases at up to 20-25°. Wind above 10-20 kph curtails foraging. Thus honeybee flight varies more within the day and with temperature variation than for many other bees and flies that visit flowers. Variation in the average hours worked per day has seldom been determined with much rigor.

Heinrich's 1979 text on bumblebees elaborates on the energy economics of foraging and their consequences on flower visitation at different temperatures. Among different bee species, differences in the rate of floral visitation (e.g. flowers visited per minute) quite commonly are on the order of 200-400%. Other flower visitors may be much slower. Insects that walk on the platform of Compositae and Umbelliferae flowerheads are able to economize greatly on energy expenditure. This allows them to continue collecting small amounts of nectar at flowers (when more active pollinators would shun them), and their small rate of return in cooler weather is still sufficient to offset the costs (if any) or warming up. The pollinating rate of visitors that walk on flowers may be only 33% that which occurs when pollinators readily fly between flowers. However, greater time spent on *individual* flowers, as shown recently in studies of flies, can greatly improve effective pollination.

5) *Preference for crop or competing flowers.*

Concurrent blooms within the field or adjacent to the crop may cause competition for pollinators — especially for generalized pollinators like honeybees. However, under certain conditions these 'competing' flowers can complement pollination. This will often be the case for nectar-collecting workers, when the effective pollinating bees collect only pollen. An individual bee can collect both nectar and pollen on one foraging trip, or the impetus for colony foraging is augmented through stimulus from available food. This certainly applies to pollen-only crops, where flowers near the crop and used only for nectar must enhance crop pollination, not hinder it through competition. In addition, flowers used to supply critical nectar sources can support growth or slow a decline in honeybee colonies, which favors continuation of pollen collection for brood rearing.

For example, a four-year trial with supplementary sugar feeding of bee colonies boosted kiwifruit pollen collection 300% throughout flowering, with daily increases over 40 times the amount observed before feeding. If hives fed with sugar perform comparably to those given additional, external nectar sources, it should be clear that some 'competing' blooms do not detract from pollinator service but actually augment it. Hence studies of competing blooms should focus more critically on the intensity of both pollen and nectar usage by pollinators before concluding that adjacent flowers are significant competitors. This can be done by direct observation of pollen on foraging bees. A comparison of the ratio of honeybees to other bees on the crop and adjacent flowers will allow for the alternative flower sources to be ranked according to their relative attractiveness for the main pollinators. Investigation of the other pollen species present on bees foraging on the crop can provide further evidence of what the *potentially* most important competing flowers are (see also Section 2.3.2).

6) *Availability of pollinators.*

Honeybees are often the commonest flower visitors on crops because hives are introduced to the site for pollination or are kept there for honey production. This effect becomes more pronounced as the crops become larger, beyond 2 to 5 hectares, because feral populations of other bees and insects may not forage throughout the field and populations are simply not large enough locally to compare with honeybees (Section 2.1). In Japan and USA, populations of 600 -1,000 solitary bees/hectare (0.6-1.0/10 m^2) have been shown to be enough to pollinate apple, almond, blueberry and sunflower crops with efficient pollinator species. In New Zealand, an estimated peak population of

2,000-4,000 pollinating bumblebees/hectare will fully pollinate red clover crops with 36-72 million flowers/hectare opening per day. Thus low populations of alternative bee species can serve some selected crops if they are efficient in pollination. An assessment of the previous six factors above should allow for studies to focus on more detailed aspects of the impact of a few pollinator species, or the use of mechanical pollinators on seed formation or fruit production.

2.2.5 Assessment of Required Pollinator Numbers

Four distinct methods can be used to assess the impact of a pollinator on crop yields and the number of pollinators or bee hives needed for optimum production. ***Too few*** pollinators can lead to low crop yields, delays in harvesting and poorer quality fruit. ***Too many*** pollinators will result in excessive pollination expenses and over-investment in bee equipment. High populations will virtually force honeybees to seek other food sources more assiduously beyond the crop and possibly cause diurnal periods of pollen depletion, less than optimal honey production, and increased exposure to losses of bees from insecticides and natural enemies. It is desirable to test the effect of alternative native pollinators as soon as adequate pollinators can be managed. Trials on yield responses aid selection of the better species to develop for management. Done carefully (see Sections 2.2.1-2.2.2), they also allow direct checks to be made on the needed quantity of pollinators.

Approaches for determining the needed numbers of pollinators are all based on field trials. After this phase, the information so gained can be handled in various ways. The basic methods are as follows:

1) Yields in cages with bees enclosed or excluded, compared to open fields.

Cages that exclude insects and compare yields with open fields allow appreciation of effective response to pollination. However, the insight gained from such cages is limited — evident now that this approach has been applied for many crops. The enclosure of bees has mainly involved small honeybee colonies. The apparent merit of caging honeybees with a crop is to determine the approximate upper response in plant production of seed or fruit. At times, however, excessive bees in cages have caused poorer production, perhaps because of rapid pollen depletion, or shading that may affect both foraging behaviour and plant resources. Shading reduces nectar production and hours of bee flight. Results with caged honeybees have quite often produced yields that are difficult to interpret. Claims have even been made that honeybees rob (perforate) flowers in such situations, which is not known in nature and suggests that other organisms in the cages (short-tongued bumblebees?) have done the actual robbing. Shading and fine mesh cut wind velocity within the cage and reduce wind flow and its effect on various crops.

Caging bees, especially honeybees, also has the inherent weakness of eliminating competing blooms, which are quickly located by the colony (in contrast to many bees). Thus consistent or high yields may not be obtainable in field conditions. Megachilid bees, carpenter bees or bumblebees forage more comfortably in restricted cages. On fruit and tree crops this may be equal to only a part of the vine or tree. A cage can conceivably disturb parasite populations by excluding or concentrating them, or increasing the humidity — thereby affecting plant diseases that may influence yields. Initial testing of alternative pollinators is usually restricted by the limited supply of bees in hives, nesting material, flies, butterflies or other pollinators. A range of stocking rates can be achieved for small numbers of alternative pollinators in a cage to provide an initial estimate on the numbers of pollinators needed for application to commercial crops. An alternative is to use small plots in open fields while carefully monitoring the visitors.

2) Comparison of yields on large crop fields.

Judicious use of large and small crop areas can allow contrasting populations of the pollinating guild to be studied, where competition occurs in a commercial setting (unlike cage studies). Feral populations of alternative bee pollinators are usually limited, so that in a large crop of at least 15 hectares, the effect of feral bees often becomes minimal compared to pollination from crops stocked with honeybees. (Exactly the opposite, however, occurs where there are large numbers of wild honeybees, or any other kind of pollinator). The introduced pollinators cannot find unused flower patches and must cease to forage or seek resources elsewhere.

Honeybees fly up to 10 km or more to forage, but in doing so they gain little resource for the colony. Honeybees prefer to forage near their hives and the majority may forage within 400-500 meters, provided the crop is attractive, while a minority may freely fly to alternative flowers within 1.5 to 4.0 km. With adverse weather (12-15°C) and winds, activity can be concentrated within 200-300 meters of the hives as foragers normally on more distant flowers remain inside. Studies of production on strips of apple, longan and crops at varying distances from honeybee hives have concluded there was differential pollination by honeybees, but the possible confounding effect of other pollinators has seldom been clearly demonstrated.

The large, active carpenter bees apparently forage up to 1.5 km from their nests from a Malaysian study on passion fruit pollination. Most bumblebees forage within 400-800 meters of the colony, based on marked bumblebees from feral colonies in Nebraska, USA and a study in New Zealand that moved nests and recorded the distances from which displaced bees returned to their former nest sites. In Czechoslovakia, lucerne yields were lower beyond 40 meters of nests of *Rhophites canus*. Other small solitary bees are unlikely to customarily fly farther from their nests than bumblebees or carpenter bees. In New Zealand seed yields of lucerne almost doubled within 20 m of hives in a field with megachilid bee numbers inadequate to permit maximum seed yields in alfalfa.

3) Plots of small crops — tests of different stocking rates.

Replicated small plots can provide unequivocal evidence on the floral preferences of bees, as shown for legumes, cruciferous crops and borage, particularly for bumblebee species in Denmark, Canada and New Zealand. Plots of 0.1 to about 2 hectares can be valuable for testing the actual impact of alternative pollinators provided the species has a preference for the crop being studied. Several small plots spaced at least 1 km apart (beyond the usual foraging distance for most bees and blowflies) allow stocking rates at 12.5, 25, 50, 75 and 100% at peak flowering. The rates tested will depend on the available supply of pollinators and the expected response in yield. A detailed study on a plot can show what pollinator abundance is optimum. In addition, the validity of yields and other parameters used for predictive modelling may be checked under field conditions. Small plots can also attract modest honeybee numbers, especially where there are other attractive crops or weeds in the vicinity.

4) Estimating required pollinator number and the use of predictive models.

A predictive model that makes few assumptions can yield estimates of required pollinator numbers. The model would ideally ensure that the value of each main pollinator species is adequately assessed (see also Section 1.7). Once pollinator populations have been measured in the field, and their efficacy verified, then a national assessment can be made without unduly complicated mathematical formulae or computer simulation. For instance, in New Zealand the typical pollinator populations on red clover crops were only about one-third of those needed for full pollination. Smaller plantings, under 3 hectares, yielded up to twice that of plantings larger than five hectares. The available pollinators were better able to service smaller patches. This shows why some of the large-scale predictive models might fail. Like all models, they are only as accurate as the figures and variables they employ. The variables that are tedious or costly to measure, such as number of flowers opening per day, might still produce no reliable predictions if the model doesn't fit the setting, or if other measurements are not taken.

Two formulas are needed to assess crop (or grower) demand for pollination and the capacity of the pollinators (beekeeper or supplier) to give the required pollination. The first is straightforward:

Flower pollination requirement (or grower demand) = Number of open flowers (standing flower crop) + new flowers opening (per day)

Some flux in daily needs for pollination must occur. Cooler weather, for example, may prolong stigmatic receptivity, as sometimes will fair weather. The response of a crop to pollination will always be below 100% and may depend on weather and growth during the flowering period; it obviously will also be affected by fertilizer or pesticide application. Further response in yield will not be possible above a certain level, and this level will vary from year to year and crop to crop. Nonetheless, a second equation that may be given to show the relation between pollination supply and demand is the following:

Daily pollination delivery (by supplier) = Average number of pollinators per unit area (hectare, tree, flower group, etc.) • flowers visited per pollinator per hour • hours of foraging per day.

Combining and rearranging the two foregoing equations, the number of needed pollinators can be computed as:

$$\text{N pollinators needed} = \frac{\text{N flowers open} \bullet \text{N visits for full seed set}}{\text{N visits per hour} \bullet \text{N hours daily foraging}}$$

A predictive model can provide at least an order-of-magnitude estimate (e.g. whether 1,000 or 10,000 are appropriate numbers) of the required pollinators, so long as accurate figures are incorporated. This particular model would require knowledge of the numbers of visits needed from each pollinator species to produce full seed set, as well as reliable counts of the other three variables.

For apples and almonds, interactive computer-based models have been developed in the USA to predict fruit and nut set from variables including weather, flowering of the crop and honeybee foraging behaviour. This in turn permits colonies to be moved from the orchard as soon as possible to allow pesticide spraying. Such computer models are valuable but have limited value in places having little access to computers.

Additional References *(see also Sections 1.1, 1.3)*

Beattie, A.J. 1971. A technique for the study of insect-borne pollen. Pan-Pacific Entomologist 47(1):82.

Davis, A.R. 1992. Evaluating honeybees as pollinators of virgin flowers of *Echium plantagineum* L. (Boraginaceae) by pollen- tube fluorescence. Journal of Apicultural Research 31(2):83-95.

Heinrich, B. 1977. Bumblebee economics. Cambridge, Massachusetts, USA: Harvard Univ. Press.

2.3 FLORAL BIOLOGY & RESEARCH TECHNIQUES

2.3.1 Nectar

Although it usually contains a number of minor chemical constituents, ***nectar*** is composed largely of water and sugars produced by the photosynthetic machinery in plant leaves. This sweet exudate is secreted from glands called ***nectaries.***

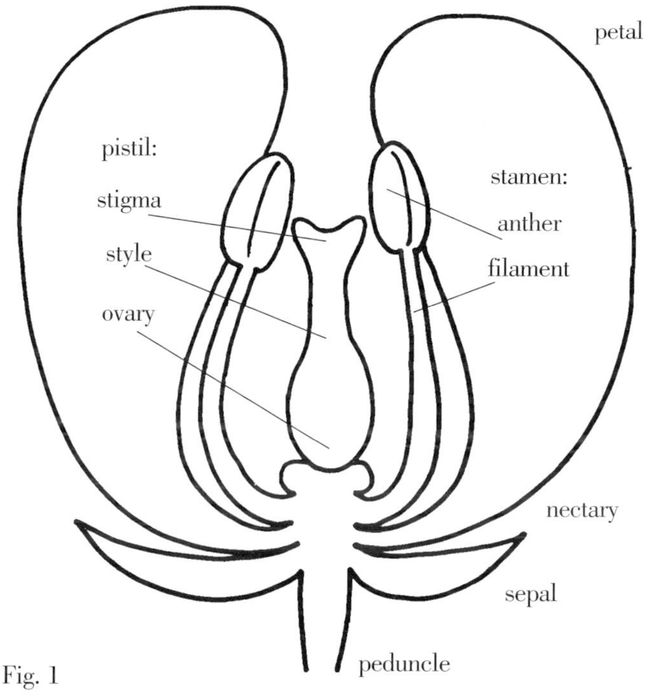

Fig. 1

1) Nectaries and where to find them.

Nectaries can be on many plant parts — cotyledons, stems, leaves, stipules, flowers and fruits — and sometimes in more than one place. For instance, two nectaries of chick-pea (*Vigna unguiculata*) are not on floral parts (rather, they are on stipules and the inflorescence stalk between flower bases), a third type being in the flower, associated with the ovary base. In some exceptional cases, like *Campsis radicans* (Bignoniaceae), five nectary locations exist, four of them on floral parts: sepal and petal lobes, surrounding the ovary's base, and on the surface of developing fruits. A distinction is drawn between such nectaries according to their location. ***Extrareproductive*** nectaries are located on vegetative parts (and as such are *extrafloral* nectaries). ***Reproductive*** nectaries include glands on the inflorescence, flower stem and bracts, as well as those on any floral part. Extrareproductive nectaries frequently provide a mutual benefit to both plant and insect, particularly in the tropics. While foraging nectar, certain tropical ants defend plants by attacking herbivores.

However, it is the reproductive nectaries that are most significant for pollination. Some nectaries are situated on the lower surfaces of outer floral parts, and because they may secrete nectar even before the flower opens, they are not directly involved in attraction of potential pollinators. Glands that encourage pollination by flower visitors are those actively secreting in open flowers and usually located near the base of the flower's reproductive structures — the stamens and/or the ovaries (Fig. 1), or those positioned elsewhere on the flower (e.g., spurs of the corolla, Fig. 2). Their location and the timing of nectar secretion ensure pollinator contact with the anthers and stigma when the gland is probed by a visitor of the appropriate size. Thus, of the four nectaries situated on floral parts in *Campsis*, it is the gland below the ovary that is associated with pollination.

Fig. 2 Flower of hibrid columbine (*Aquilegia*), showing five hollow, tube-like extensions (spurs) at the petal basis (arrow) which produces nectar.

Within taxonomic groups, the position of nectarial tissue is usually consistent. In monocotyledons, nectaries are in septae of the flower's ovary. In the mustard family, Brassicaceae, the glandular tissue is always located at the base of stamens. In *Eucalyptus* (Myrtaceae), nectar exudes around the floral cup, in which it accumulates beneath the ring of stamens.

The location of nectar itself is a way to detect nectaries, but there are species which, under appropriate environmental conditions, produce copious nectar that runs from the flower, thus making detection of glands more difficult. If nectar secretion has not yet begun, or nectar resorption has occurred, nectaries are difficult to find. A difference in colour between the nectarial tissue and that of adjacent floral parts can assist nectary detection. Several species of the mint family (Lamiaceae) display bright yellow glandular tissue below the four green ovules (Figs. 3a, b) or dark purple. Further, pigmentation of some extrafloral nectaries makes them visible from a distance (Fig. 5a).

Morphology of nectarial tissue is another feature used to

Fig. 3. **a)** inflorescences of bugleweed (*Ajuga reptans*), **b)** dissection of flower to reveal yellow nectarial tissue (N) below four green ovules (O); the style (Sy) inserted between ovules, and remainder of calyx (Ca), **c)** SEM of nectary (N) and two ovules (O), **d)** higher magnification of nectary, showing stoma with open pore (p)

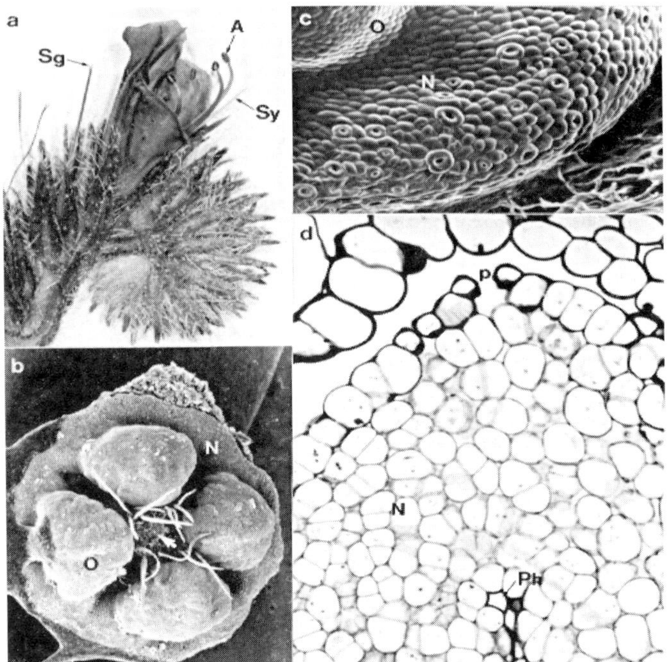

Fig. 4. **a)** inflorescence of Paterson's curse (*Echium planta-gineum*). Note anthers (A), style (Sy) of open flower, and upper left, three persistent styles, each topped by two stigmas (Sg), **b)** SEM showing nectarial disk (N) surrounding four ovules (O), insertion point of dissected style indicated by arrow, **c)** higher magnification of nectary (N) below ovule (O); note numerous stomata with pores on gland surface, **d)** section through nectary (N) showing strand of phloem (Ph) and open pore (p)

detect these glands, although higher magnification is sometimes required. Many nectaries surrounding the base of ovaries take the form of multicellular protuberances or outgrowths. In *Ajuga reptans (*Lamiaceae*)*, nectarial tissue forms a swelling on the lower side of the flower (Figs.3a-c). *Echium plantagineum* (Boraginaceae) has a nectary that is a disk-shaped structure encompassing the bases of the four developing nutlets (Fig. 4a-c). The discoid nectary of *Vicia faba* (Leguminosae) bears a long protuberance below the developing bean (Figs. 5a, b) In each of these examples, scanning electron microscopy reveals that nectary surfaces contain pores of modified stomata (Figs. 3d, 4c,d, 5b) through which nectar can escape. Some nectar glands, on the other hand, are composed of aggregated secretory hairs (*trichomes*), and these are usually only distinguishable by microscopy. In *Hibiscus* (Malvaceae) the inner surface of each sepal bears thousands of multicellular trichomes (Fig. 6a-c). On the stipules of *Vicia faba* the extrareproductive nectaries

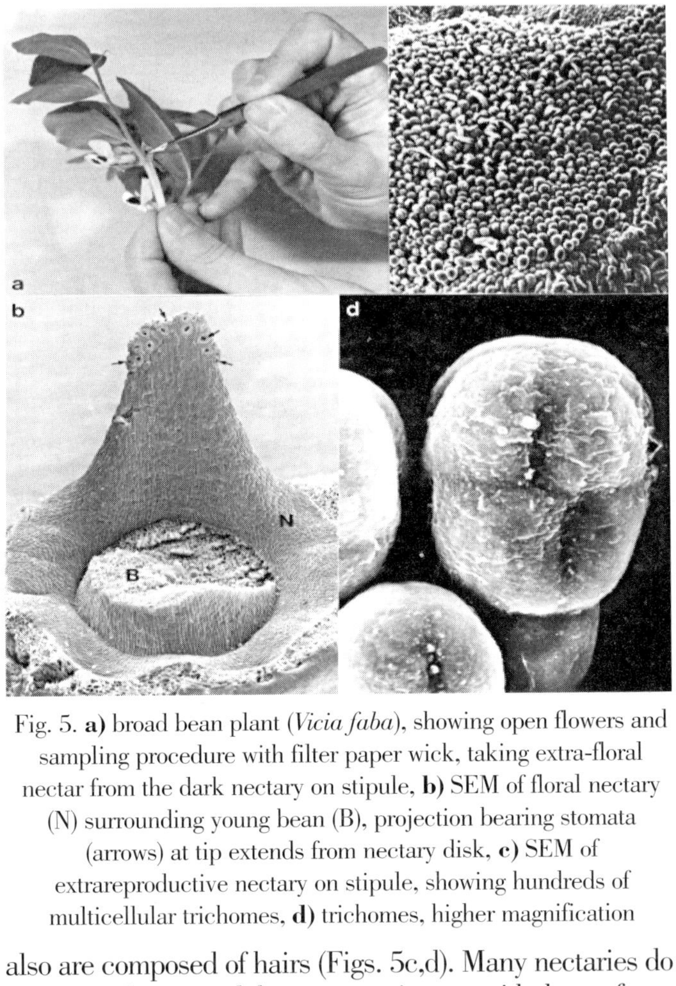

Fig. 5. **a)** broad bean plant (*Vicia faba*), showing open flowers and sampling procedure with filter paper wick, taking extra-floral nectar from the dark nectary on stipule, **b)** SEM of floral nectary (N) surrounding young bean (B), projection bearing stomata (arrows) at tip extends from nectary disk, **c)** SEM of extrareproductive nectary on stipule, showing hundreds of multicellular trichomes, **d)** trichomes, higher magnification

also are composed of hairs (Figs. 5c,d). Many nectaries do not protrude; instead they are continuous with the surfaces of adjacent floral parts and normally consist of relatively few layers of cells (Figs. 7a-c, 8a-c). These types of nectaries often are only recognizable by the presence of nectar itself. In cases like *Medicago sativa* (Leguminosae) and *Eucalyptus leucoxylon*, the surface under the exudate bears stomatal pores (Figs.7c,d,8d). However, other *non-structural* nectaries like that on the outer calyx surfaces of *Paeonia* buds (Ranunculaceae) contain no specialized glandular tissue and are only identifiable by the nectar droplets above them.

2) Nectar volume, concentration, and influencing factors.

Nectar standing in a flower is continually undergoing changes. The types of change depend on the forces acting on it. Not only can a nectary secrete nectar, but nectar

Fig. 6. **a)** flower of *Hibiscus* showing five stigmas (Sg) at tip of style, extending beyond staminal column and anthers (A), **b)** SEM of calyx base (Ca) showing band of nectary secretory trichomes (center right), **c)** close-up of trichomes at gland edge

can in many cases be reclaimed or reabsorbed by the gland. In this way, plants have the ability to conserve nectar constituents and utilize them elsewhere. There are several ways that nectar volume can decrease. Besides resorption or embedment a third means is evaporation. Water evaporation from floral nectar is greater when nectar is relatively exposed to the atmosphere instead of protected at the base of a long corolla tube. When evaporation is intense, nectar volume can decrease despite continuing nectar secretion. Similarly, increases in floral nectar volume can be brought by rain drops that fall or run into the flower and mix with standing nectar. In addition, standing nectar of relatively high sugar concentration, and in humid air, can gain water vapor from the atmosphere. Nectar accumulates whenever net secretion rate exceeds the combined losses of reabsorption, embedment by visitors, and evaporation.

Sugar concentration changes as water evaporates from sugar solution, and volume *decreases* while the concentration of sugar *increases*. This relationship between volume and concentration must be emphasized in nectar analysis because the *quantity of nectar sugar* in the sample is of primary importance. The flowers shown in Fig. 9 have identical nectar volumes, but unless nectar samples are checked for sugar concentration, one might assume, erroneously, that each flower's nectar contains the same sugar. Estimates of a flower's nectar sugar come from multiplying nectar volume by sugar concentration. For example, suppose two flowers have 3.5 µL (microliters, or 1/1,000,000th of a litre). One has nectar of 36% sugar (that is, 36 grams of sugar per 100 milliliters of water), whereas the other has a concentration of 9%. Therefore, although the two have equal volumes, one holds 1.26 mg sugar, and the other only a quarter as much sugar, 0.315 mg. These quantities were calculated by multiplication of nectar volume by sugar concentration (e.g. 3.5 µL x 36 g sugar/100 mL water = 1.26 mg sugar).

Let's take another example. In this instance the volumes of nectar in two flowers differ. In Fig. 10, the flower on the left bears 5.0 µL of 12% sugar. The flower on the right

Fig. 7. **a)** alfalfa (*Medicago sativa*), **b)** SEM of dissected floret base, showing nectary (N), **c)** higher magnification of nectary surface (N), showing open stomata; nectary located between gynoecium (pistil) (G) and base of staminal column (SC), **d)** close-up of nectary surface showing of mature stoma with open pore at left.

Fig. 8. **a)** trees of yellow gum (*Eucalyptus leucoxylon*) in bloom, **b)** mature flowers, composed of rings of stamens around top of floral cup, lack a calyx and corolla, **c)** SEM of floral cup, without stamens, revealing nectary surface (N), **d)** higher magnification of nectary surface, showing pores (arrows)

contains 2.0 µL of 30% sugar. Therefore, both flowers have *exactly the same sugar content* in their nectars (5.0 µL x 12 g sugar/100 mL water = 0.60 mg sugar = 2.0 µL x 30 g sugar/100 mL water), despite the flower on the left having a nectar volume 2.5 times higher. It follows that by measuring two of the three nectar characteristics (volume, sugar concentration, quantity of sugar), an estimate of the unmeasured, third characteristic can be computed from the two other measurements.

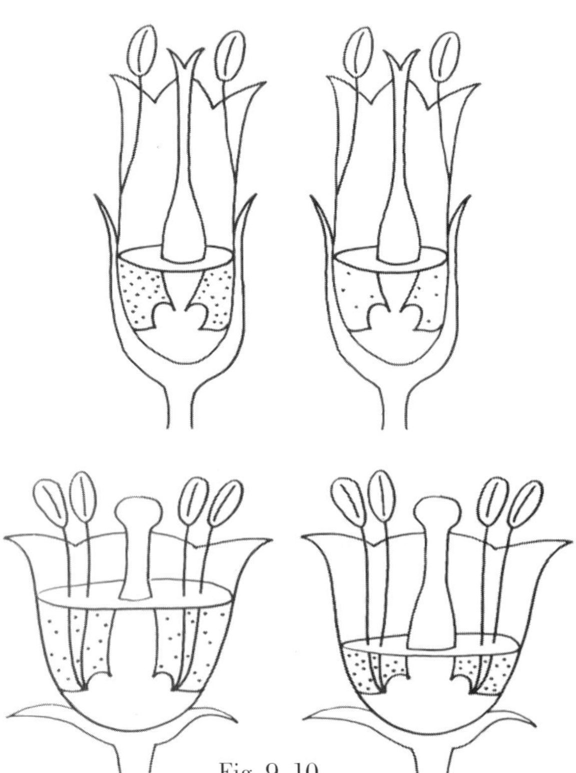

Fig. 9, 10

2.3.2 Collection & Measurement of Nectar Samples

A number of techniques are used to collect floral nectar. Besides depending on the equipment available, methods depend critically on the morphology and size of the flower. In very small flowers, for instance, there may only be minute quantities of nectar. Furthermore, if nectar is released from a nectary located at the flower base and nectar is relatively inaccessible, nectar collection is possible only after careful dissection of the flower.

1) Nectar volume.

Several techniques available for nectar collection can also yield, without too much difficulty, estimates of nectar volume. Accuracy depends of course on the method used. One relatively simple technique does not involve nectar collection — visual observation is coupled with a scale measurement. This technique can be performed most accurately for tubular corollas where the accumulated nectar can be measured within the cylindrical tube. However, in a meaningful analysis of nectar, measurement of the nectar sugar is needed — either nectar sugar concentration (percentage) or nectar sugar content (weight). This involves direct sampling of nectar.

A popular technique to collect nectar and measure its volume involves glass microcapillary tubes. Many types are commercially available, such as those of Fig. 11a. One end of the capillary is inserted into the nectar droplet, and the nectar is drawn into it by capillary action. Using capillaries of a specified inside diameter, marked to show exact volume, (also called *calibrated* microcapillary tubes) allows the length of the sample column to be expressed as a fraction of total microcapillary capacity. Capillary tubes work best when the viscosity of the nectar is not too high; capillary movement of the nector is low when nectar concentration exceeds 60%. In certain cases it is ideal to be able to remove all standing nectar, such as when comparisons of absolute nectar sugar content are being made between cultivars of a crop species. With many flower and nectary morphologies, it is not uncommon that capillaries fall short of removing all accumulated nectar.

A calibrated syringe or pipette also can be used to collect nectar directly, which has prominent application when nectar volumes are very large (e.g. 50-100 µL or more) and collection using several micro-capillaries would be too time-consuming.

Centrifugation is another method used to simultaneously collect nectar and measure its volume. This technique has been used to collectively sample the nectar of a compound inflorescence of several small florets. In the system utilized

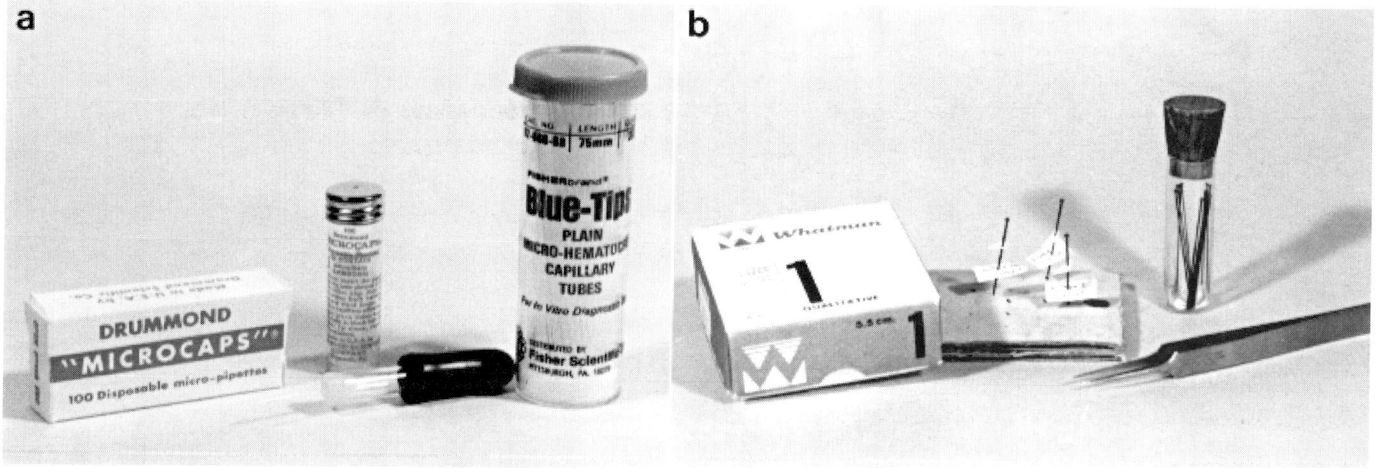

Fig. 11. **a)** assorted micro-capillaries for nectar collection, black bulb used for expulsion from capillaries, **b)** forceps, pins and filter paper wicks for nectar colleciton

in Fig. 12, an inflorescence is inverted into the top of a specially-designed glass centrifuge tube and secured with a stopper. Nectar spinning from the florets during centrifugation flows to the bottom into a calibrated tube, where nectar volume can be assessed. However, a recent study utilizing this system for *Brassica napus* gave a low estimate of nectar sugar, compared to that sampled directly by a capillary. Nectar volume during centrifugation might have been artificially increased when water condensation inside the centrifuge tubes entered the spun nectar. If this system is to be used extensively for measurements of nectar volume, it would first be recommended to determine the time required to extract the nectar from the flowers under investigation and to compare nectar sugar concentration of spun nectar with nectar withdrawn by capillary. Keeping centrifugation durations identical between samples will also help to minimize variation due to technique.

Filter-paper wicks are becoming increasingly popular as tools to collect floral nectar. Small pieces of filter paper can be held at one end with forceps (Fig. 5a) or impaled on pins (Fig. 11b), and their edges dipped into the nectar. The wicks are then air-dried and stored, preferably in a desiccator that will prevent growth of microorganisms on the enriched wicks. When analysis is performed, the nectar sugar is eluted from the wicks by placement in a known volume of water and shaken vigorously and periodically. Although the use of wicks alone does not permit direct measurement of nectar volume, there are instances where volumes of nectar are too low to be determined by the other methods (capillaries, syringes, centrifugation, see section C, below).

Rinsing flowers in a known volume of water is another means to collect nectar. The flower must be shaken vigorously, to ensure wetting of nectaries, and then allowed to soak before it is removed and the rinsate is available for assay. This method is best when nectar is scarce or viscous, and acquisition of at least some data is preferable to none at all. The rinsate is then analyzed for its sugar (see section C, below). One potential drawback of soaking the flower to collect nectar is that other sources of floral sugar (e.g., leakage from flower parts, or leaching of the sugar from the stigmatic exudate, in the case of wet stigmas) may confound the true sugar level of the nectar.

A related technique to rinsing, but not involving submersion of the entire flower in water, is the controlled addition of a measured small volume of water onto the nectar droplet. This technique again has application for small nectar quantities, or when nectar is too viscous to be otherwise collected. Repeated and careful uptake and expulsion of this water using a micropipette will mix it with the residual nectar, thus providing another way to recover nectar sugar. Again, with this technique, nectar volume is not measured directly. Occasionally, a combination of these various techniques may be very useful. For instance, when it is desirable to obtain all available nectar from the base of a flower, such as one possessing a protruding nectary (Fig. 5b), the task can be very difficult to perform with only a microcapillary. It is advisable to draw some nectar first, with the microcapillary, from the edge of the nectar droplet (without probing the nectary, which may be damaged by the glass capillary). Nectar volume and concentration (see section B, below) are determined from this subsample. The nectary is then gently swabbed to gather residual nectar at the flower base, using a filter-paper wick.

Fig. 12. **a)** rack for holding centrifuge tubes of R. W. Shuel, **b)** centrifuge tube (11 cm) with inverted inflorescence of *Medicago* held by cork stopper; arrow shows nectar accumulated after centrifugation

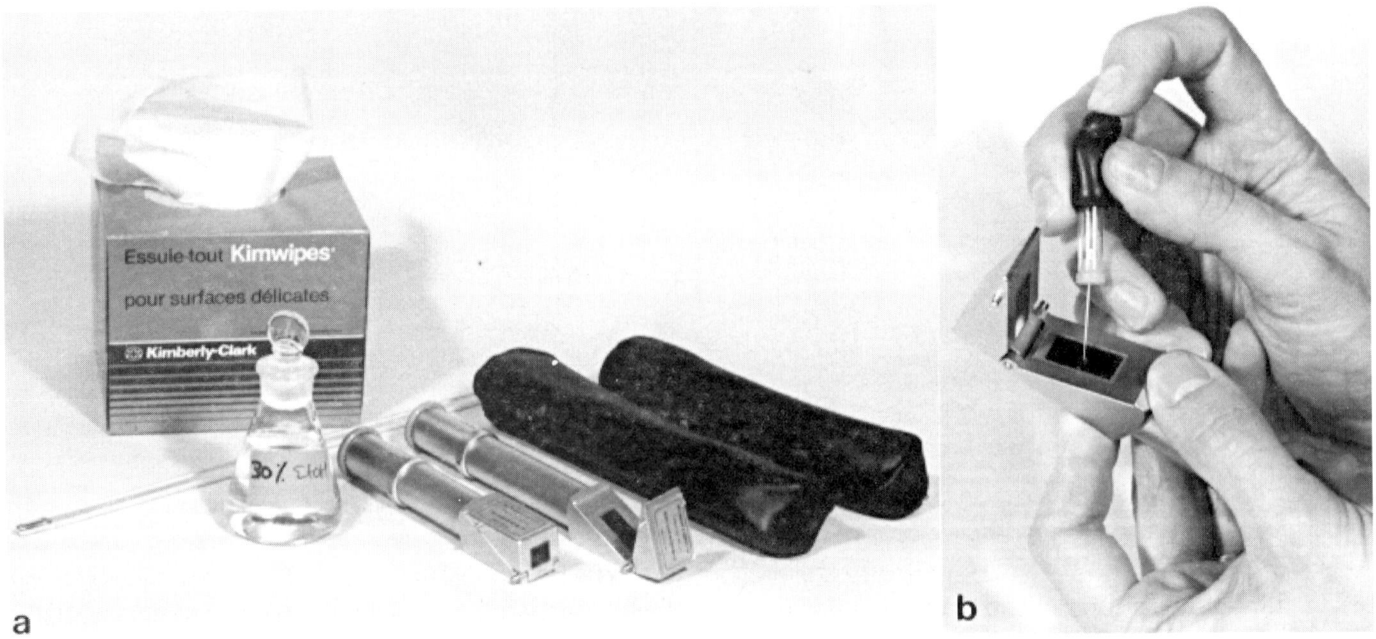

Fig. 13. **a)** instruments for determining nectar sugar concentration — refractometers and protective cases, thermometer, tissues and 30% alcohol for cleaning, **b)** method for applying nectar from microcapillary

Subsequent determination of the sugar content from wick and capillary allows total nectar volume to be calculated, assuming the nectar to be a homogeneous solution. *Such a combined method of nectar collection is recommended where repeated sampling from the same nectary is desired.* Experimental studies of the effect of nectar removal on total nectar production, or of resorption, should utilize a wick and thereby avoid inflicting physical damage to the nectary during a previous collection period by contact of the gland with the probing end of a glass capillary.

2) *Nectar sugar concentration.*

A simple taste of a nectar sample often allows distinction between weak and strongly-concentrated nectars, but this cannot provide sufficient accuracy. Refractometry is the common and most accurate method of measuring solute concentration of a nectar sample. Because sugars make up almost all of the dry weight of a nectar, it is generally a safe assumption that solute concentration represents a close estimate of the sugar concentration. Amino acids, as secondary constituents of nectar, have been shown in certain species to make up 10% of the nectar dry weight. Amino acids are difficult to quantify chemically. When comparisons are being made between nectar samples of the same species, the dependability of refractometry to give a close, relative estimate of sugar concentration is assured.

Particularly in the past, nectar concentration has been determined in the laboratory using Abbé refractometers. Now it is possible to obtain lightweight hand-held refractometers (Fig.13a) that can be carried and utilized easily in the field. For capillary and syringe collection, this flexibility allows nectar concentrations to be determined outside, immediately after volume measurements are taken. Refractometers can be obtained to give concentration readings across various ranges. For instance, the pair shown in Fig. 13a yield concentration measurements for samples in the range of 0-50% (left) or 40-80% (right). If studying only nectar at the base of long-tubed corollas, the smaller refractometer may be all that is necessary, but if the intent is to study nectar from the relatively smaller flowers visited by bees, it often will be necessary to be able to measure nectar concentrations greater than 50%. To obtain a concentration reading, the minimum volume of nectar required depends on the type of refractometer, but can be as low as 1/50th of a µL in those excellent refractometers specially modified by the manufacturer to accommodate small samples. This type of refractometer is preferable, because it allows the measurement of nectar solute concentration on a small nectar sample available per flower, and thus conserves the variation between individual flowers that is lost when pooling nectar from several flowers.

The procedure to measure nectar sugar by refractometry is quite simple. The sample is expelled onto one of the two prisms (Fig. 13b) which oppose each other when the sample area of the refractometer is closed. This closure spreads the sample into a very fine layer, through which light passes. The more concentrated the nectar solution, the greater its refractive index. Refractometers vary with the manufacturer, but types are available that give readings in '%' directly on a sucrose sugar scale. It is important to note that refractive index of the nectar sample is dependent on temperature, and therefore a record of the latter during nectar handling is also important, in order to calculate the true concentration adjusted to 20.0°C. The data tables necessary to allow such temperature conversions usually are supplied by the manufacturer of the refractometer. Other refractometers give concentration readings after automatically compensating for temperature.

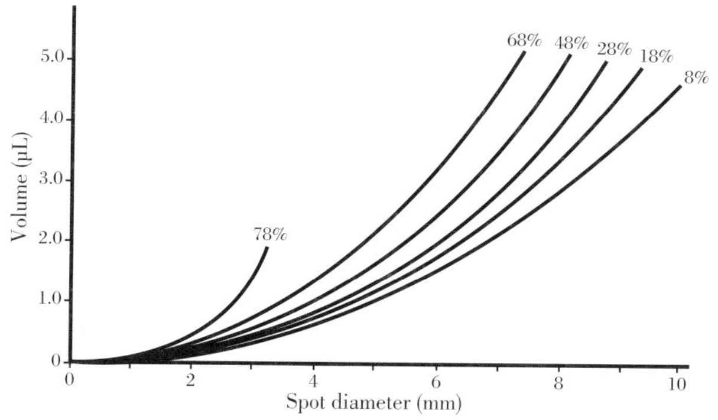

Fig. 14. plot of nectar volume against diameter of nectar spot after application of different concentrations to Whatman No. 1 chromatographic paper (redrawn from Punchihiwa)

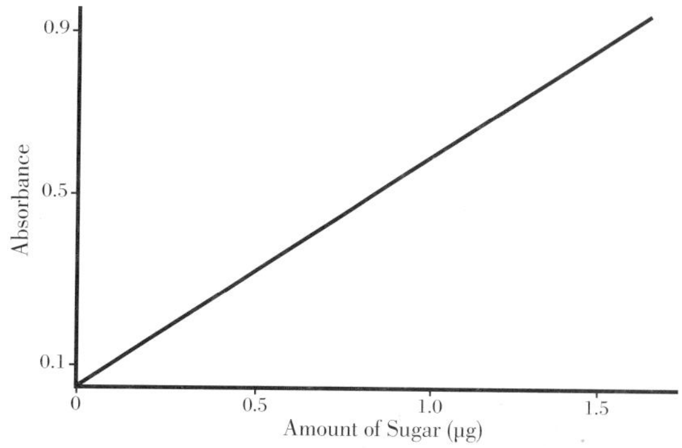

Fig. 15. standard curve of absorbance versus quantity of nectar sugar, determined by the technique of R. B. Roberts

There is another conversion to this refractometer reading of % concentration that needs to be performed before the true solute concentration of the nectar is known. This conversion is necessary because the refractometer reading is expressed as the gram weight of sucrose (or its equivalent as glucose and fructose) in 100 g solution, instead of in the volume measurement of 100 mL of solution. Fortunately, a simple quadratic equation makes the conversion. The equation is:

$$\mathbf{NCV} = \mathbf{NCW}\,(9.224 \times 10^{-3}) + \mathbf{NCW}^2\,(59.6 \times 10^{-6}) + 7.08 \times 10^{-3}$$

where **NCV** is % nectar concentration based on volume, and **NCW** is the refractometer reading of % nectar concentration based on weight, after the correction for temperature.

As an example of these conversions, let's examine a nectar sample that gave a refractometer reading of 56.5%, when measured in the field at 29.0°C. From tables, it is found that, at this concentration, for each degree in temperature above 20.0° one must add 0.73% to the reading. Therefore, the temperature-corrected reading is 56.5 + (9)(0.73) = 56.5 + 6.57 = 63.07%. This is **NCW**. By substituting this value into the equation above, we find that this nectar sample has an **NCV** (sugar concentration based on volume of solution) of 82.6 g per 100 mL solution (i.e., NCV = $(63.07)(9.224 \times 10^{-3}) + (63.07)^2(59.6 \times 10^{-3}) + 7.08 \times 10^{-3} = 0.82592$, or 82.6%). This value of NCV (82.6 g/100 mL) can now be multiplied by nectar volume to determine sugar content of the nectar (see preceding example of Fig. 9).

If refractometers are not available, there is another, but less accurate, indirect method to obtain estimates of nectar concentration. This technique is based on the degree of spreading of a nectar droplet (Fig. 14). Using identical volumes, highly-concentrated droplets of sugar solution spread, when dabbed on filter paper from capillaries, less than those that are very dilute. The maximum diameter of the outline of the nectar spot is measured immediately with calipers, say three times per spot, and the average diameter calculated. In the laboratory, a series of standard curves can be constructed, in which nectar concentrations later can be extrapolated. The curves are created by using freshly-prepared sugar solutions of known concentration, such as in 5% increments. The use of different nectar volumes (µL) (e.g., 0.2, 0.4, 0.6, 0.8, 1.0, 1.2, etc.) of each concentration allows a series of diameters to be plotted against nectar volume, for each concentration. Then, knowing the volume of the nectar sample, it is possible to estimate its sugar concentration from the standard curves.

3) Sugar Content.

A number of sugars have been analyzed from nectars and some sugars are known only from nectar. The three most common sugars in nectar are the disaccharide sucrose and the monosaccharides glucose and fructose. Each sucrose molecule consists of one glucose and one fructose molecule joined together. Generally, flowers with relatively long corollas and a deeply-concealed nectary produce sucrose-dominant nectar, whereas those flowers which are relatively "open" in form produce nectar richer in glucose and fructose. There are many exceptions to this generalization, and not all nectars contain all three sugars. The ratios of the various sugars in standing nectar can change, too, in those plant species that liberate an enzyme called invertase with their nectar. Invertase assists the conversion of sucrose into its separate building blocks, glucose and fructose.

As stated earlier, the sugar content of a nectar sample is of particular interest, *because it remains constant during any change in nectar volume or concentration.* Again, generally an acceptable estimate of nectar sugar content can be obtained indirectly from the product of volume and concentration. However, there are common situations in nectar collection (where wicks, rinsing, etc. are employed) where it is impossible to measure accurately either volume or concentration, or the necessary equipment to do so is lacking. A direct measurement of the quantity of nectar sugar in a sample is then required.

Unlike measurements of volume or concentration, which can normally be performed in the field, determination of total nectar sugar content of a nectar sample can only be conducted in the laboratory. The various chemical methods used to analyze nectar sugar content are based either on chromatography (see section 8 below) or spectrophotometry (colorimetry; see sections 4-7), wherein a change is detected in intensity of color of a sample or subsample of nectar. This color change is a result of a chemical reaction that involves conversion of the nectar's sugar to other compounds.

Other Methods As well as the spectrophotometric and paper-chromatographic techniques detailed below, there are other methods available including *enzymatic analysis*, *gas chromatography* and *high-pressure liquid chromatography*. Because of the specialized instrumentation required, these three methods are much more expensive to apply, and will not be presented here. For further information about these other methods, the reader is referred to Erickson and Severson or Kronestedt-Robards et al, at the end of 2.3.

4) Spectrophotometry.

Spectrophotometric techniques differ in the information that they provide. Some can determine quantitatively the relative amounts of sugar components of a nectar sample. Below we give three spectrophotometric methods for nectar analysis. The first two are the simplest and hence only provide the total quantity of nectar sugar, whereas the third is more complex and can yield additional information about the relative abundance of dominant sugars in nectar. One of the two simplest methods is usually sufficient. An advantage of the third protocol is that knowledge about relative quantities of sugars may be subjected to artificial selection in plant breeding programs.

Nectar sugar content by spectrophotometry can be determined either using freshly collected samples or, fortunately, when a great number of stored samples can be analyzed at once. A large number of small wicks on which nectar has been previously gathered and allowed to dry in air can be stored at room temperature for long periods in a desiccator. This affords greater ease than using the rinsates of diluted nectar recovered from shaking flowers in water. The latter must be stored under refrigeration or, better yet, frozen in vials that withstand freezing temperatures, to ensure the prevention of microbial growth — which normally consumes sugar.

The appropriate volume of water to be used for soaking each flower or eluting the dried sugar from each wick will vary with the flower size and nectar production of the species under investigation, but commonly falls in the range of 1 to 5 mL. Determination of this water volume requires prior experimentation. Generally, use the minimum volume necessary, because it is better to dilute the rinsate by measured addition of water, rather than use sugar concentration too weak to analyze.

- *Using a pipette, a known volume of rinsate is transferred to a cuvette, the special glass container in which the chemical reaction will occur and which will be placed eventually into the light chamber of the spectrophotometer.*

- *Light of a specified wavelength will be passed through the cuvette, and the absorbance of the solution in the cuvette is measured automatically.*

- *Each cuvette must be clean and free from scratches. Each nectar sample should be analyzed at least twice, and preferably, in triplicate, and the mean absorbance then calculated.*

- *Greatest accuracy is achieved if the absorbance reading is kept in the range of 0.10 to 0.90. Prior experimentation involving changes in the total volume in the cuvette, in the amount of rinsate placed into the cuvette, the minimum volume required in the cuvette to achieve absorbance readings, etc., to register routine absorbance in this range, is necessary.*

- *Sometimes after the initial analysis, the rinsate is found to be too concentrated (i.e., absorbance > 0.90). Then it must be diluted to achieve a reading. Remember to include this dilution factor and the initial volume of rinse water used, when working back from the standard curve to find out how much nectar sugar was present in the flower.*

5) Phenol-sulphuric acid test for total sugar.

WARNING: *Both phenol (carbolic acid) and sulphuric acid (H_2SO_4) at high concentrations can be extremely caustic to the skin, and must only be handled with extreme care by authorized laboratory personnel, wearing adequate eye and cloth protection. In case of skin contact, rinse the area repeatedly with water.*

a) Prepare the nectar sample in water:

Filter paper wick — Soak the wick in a known volume of distilled water in a test tube or vial, with periodic and vigorous shaking, over a 5 min period. Because this test is strongly exothermic and non-selective, even sugar in the form of lint or plant tissue, such as pollen grains, or fibres loosened from the wicks during shaking, can be broken down and measured erroneously as nectar sugar. Therefore, it is strongly recommended after shaking to centrifuge the vials or tubes containing the wick rinsate, to cause any such particles of potential contamination to sediment at the bottom, and then to draw the aliquot to be assayed for nectar sugar from the top portion of the centrifuge.

Nectar collection by rinsing — Shake the flower vigorously in a known volume of distilled water in a vial, and let it soak for 45-60 min. Transfer the rinsate to a tube, and centrifuge to remove particulate matter to the bottom. If the rinsate was previously collected and stored frozen, thaw it now.

b) Using a pipette, transfer a known volume of the rinsate to a glass cuvette. For spectrophotometers requiring a minimum volume of 3 mL in the cuvette, 0.5 mL of rinsate works well.

c) Add 0.5 mL of 5% phenol to the rinsate in the cuvette.

d) To the cuvette, now add 2.5 mL of concentrated sulphuric acid, being careful of spattering. In this example, the final assay volume is 3.5 mL.

e) Mix this hot solution thoroughly (e.g., a clean glass rod; care must be taken not to scratch the cuvette) and allow 45 minutes for a yellowish-orange colour to develop at room temperature. Once developed, the color is stable for as long as two hours.

f) Measure the absorbance at 490 nm wavelength compared to a blank containing the same volumes of phenol and sulphuric acid, but having water instead of rinsate, and calculate sugar content from a linear standard curve. The standard curve, like that in Fig. 15, is created by plotting the absorbances obtained against quantity of sugar analyzed after assaying standard sugar solutions (i.e., solutions carefully prepared of known sugar concentrations). For instance, take 0.5 mL of a 10 uM sucrose solution (prepared by diluting 1.0mL of a 10 mM sucrose solution [3.42 g sucrose dissolved in water in a 1000 mL volumetric flask] in 999 mL of water). Add 0.5 mL of 5% phenol (step c), 2.5 mL of H_2SO_4 (step d), and then measure absorbance. For final assay volumes of 3.5 mL, the functional range of this test is 0.1 to 1.5 µg sugar (see Fig. 15).

When preparing the standard sugar solutions, ideally the ratio of sucrose/glucose/fructose used in the standards matches the ratio of these three sugars in the nectar species being analyzed. A semi-quantitative method to determine the approximate ratios of these three sugars is given in **section 8).**

6) Anthrone reagent test for total sugar content.

The anthrone reagent test is less sensitive than that of phenol-sulphuric acid, and hence it is useful for relatively large amounts of nectar sugar, but can fall short of detecting sugar with sufficient accuracy in nectars low in volume or sugar concentration. Like the phenol-sulphuric acid test, the anthrone reagent reacts with all carbohydrates: mono-, di- and polysaccharides, dextrins, dextrans, starches, gums and glucosides. Because the saccharides are by far the predominant carbohydrates in nectar, total carbohydrate content is assumed to represent total sugar content.

a) Prepare the nectar sample in water, as in the phenol-sulphuric acid test. Centrifuge the rinsates to avoid any foreign debris in the analysis.

b) Transfer a known volume of the rinsate to test tubes. Two mL of rinsate works well per test tube. Tubes with screw caps are ideal for this test, because at some stages the tubes will be subjected to a boiling water bath.

c) To each test tube add 4.0 mL of fresh (0.2%) anthrone reagent. This reagent is prepared by dissolving 0.4 g of anthrone in 200 mL of concentrated sulphuric acid. Storage of the reagent in the refrigerator extends its usable life to one or two weeks. However, as the reagent darkens with storage, it is imperative that sugar standards always are analyzed as well.

d) Heat the test tubes for 10 minutes at 100° C (boiling water bath). A blue colour develops.

e) After the test tubes have cooled to room temperature, transfer the contents to cuvettes for spectrophotometry.

f) Measure the absorbance at 620 nm wavelength, after zeroing the spectrophotometer against a blank containing only water and anthrone reagent. The sugar content of the nectar is calculated from a plot of absorbance versus sugar prepared from standard solutions.

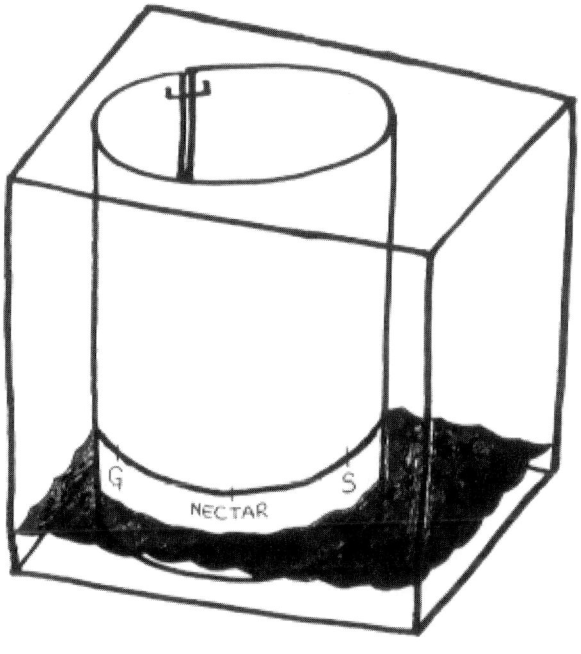

Fig. 16. roll of chromatography paper spotted at baseline with nectar and standard sugars, after placement in sealed glass chamber containing solvent

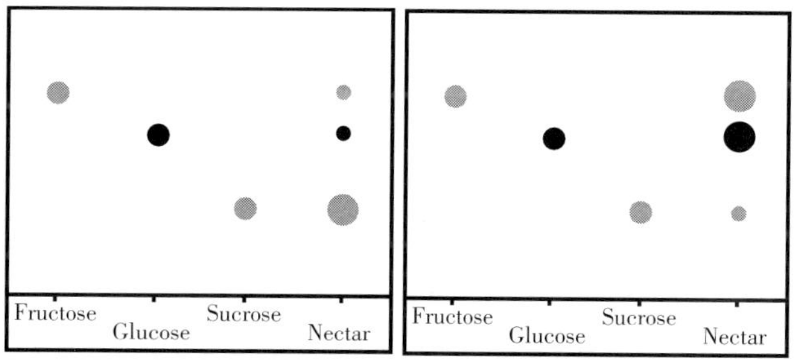

Fig. 17. Examples of paper chromatograms developed according to Block and collaborators

7) Acid hydrolysis test — reducing sugars and sucrose content

With this method the amount of reducing sugars (mostly glucose and fructose, although maltose is found in some nectars) is determined both before and after subsamples are subjected to acid hydrolysis. The apparent sucrose content of the nectar is calculated as the difference in sugar quantity between the two assays.

- *Prepare the nectar sample in water, as in the phenol-sulphuric acid test, using at least 2.0 mL water. Centrifuge the sample.*

• *Prepare the reagent, a solution of 3, 5 - dinitrosalicylic acid, as follows: At room temperature, dissolve 1 g of this acid in 20 mL of 2 N sodium hydroxide (NaOH) and 50 mL of distilled water. Then dissolve 30 g of Rochelle salt (the sodium-potassium salt of d-tartaric acid, $NaKC_4H_4O_6 \cdot 4H_2O$), and make this solution up to a final volume of 100 mL by the addition of distilled water. Cover this solution to protect it from carbon dioxide of the air.*

• *Assay of Reducing Sugars:*
Add 1.5 mL of the reagent to 1.0 mL of the rinsate and mix well in a test tube. Incubate the mixture for 5 minutes at 100° C (boiling water bath); a brown color develops. After cooling to room temperature, transfer the solution to a cuvette, add 2.0 mL of distilled water and measure spectrophotometrically the absorbance at 540 nm (green filter) against a blank consisting of water instead of rinsate. The content of reducing sugars in the rinsate is determined from a standard curve prepared using equimolar solutions of the reducing sugars glucose and fructose.

• *Assay of Total Sugars:*
Add 1.0 mL of 2 N hydrochloric acid (HCl) to 1.0 mL of the rinsate in a test tube and incubate for 5 minutes at 100° C. This procedure converts any sucrose present into its monosaccharide components, glucose and fructose. Then add 1.0 mL of 2 N NaOH and 1.5 mL of the 3,5-dinitrosalicylic acid reagent, to achieve the final volume of 4.5 mL. This solution is then heated for another 5 minutes in the boiling water bath. After cooling, the solution is transferred to a cuvette and the absorbance read at 540 nm wavelength.

• *Calculate the amount of sucrose in the nectar sample by subtracting the content of reducing sugars from the assay of total sugars, above. The method cannot distinguish fructose from glucose. In most nectars analyzed to date, however, the ratio of these two reducing sugars is usually at or near one.*

8) Paper chromatography.

Chromatography is an excellent technique that allows the separation and identification of components of a mixture in a solution. For nectar analysis, paper chromatography has long been a useful tool to detect the various sugars present in a sample. Unlike the acid hydrolysis test, this technique is not able to yield accurate quantitative data (such as the per cent sucrose in the nectar) but can give some indication whether the nectar is dominated by sucrose. The test is also used to determine if the ratio of glucose to fructose is very far from unity.

The method involves "*spotting*", using capillaries of small amounts of nectar. Nectar samples and standard sugar solutions (sucrose, glucose, fructose, and maltose, if available) are 'spotted onto a large sheet (e.g., 30 cm x 45 cm) of chromatography or filter paper, along a labeled

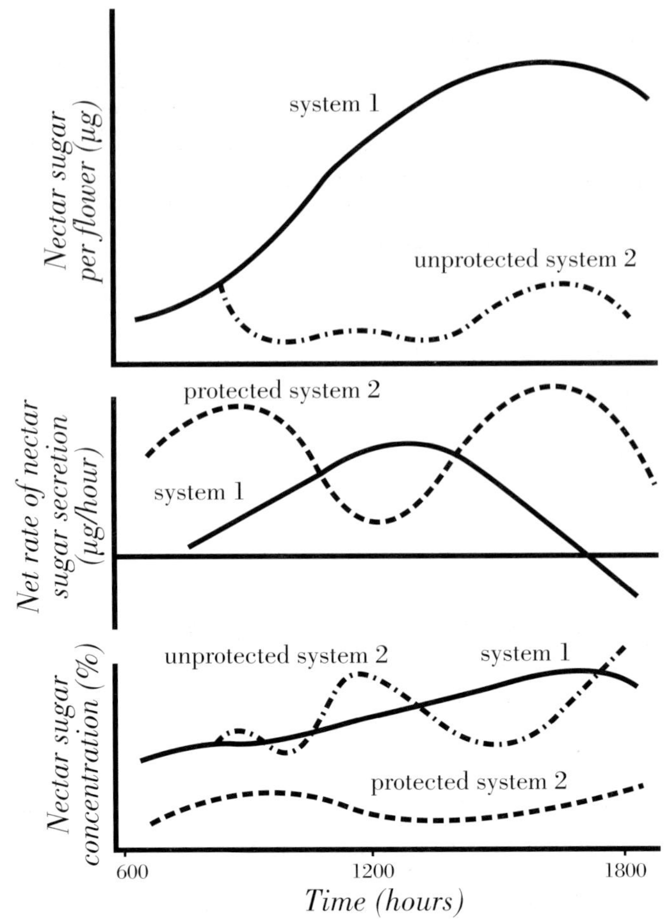

Fig. 18. graphs adapted from Corbet & Delfosse, used to indirectly demonstrate reabsorption (arrow in middle plot) of nectar sugar in flowers of *Echium*. See text for explanation of Systems 1 & 2

baseline. Ten µg of sugar, or one µL of a 1% solution, is sufficient for standards. Once the spots have air-dried, the sheet is rolled into a cylinder and the very bottom edge is placed into a solvent inside a large glass vessel (Fig. 16). Prior experimentation is necessary to ensure that solvent height in the vessel is well below the cylinder's baseline bearing the spotted nectar and standards.

The paper cylinder incubates inside the chamber, sealed on top with a glass plate and plasticine or vaseline, for 24 h at 25 to 30°C. The solvent front gradually rises and carries sugars with it. The distance travelled up the cylinder by the sugars, and the degree of separation of the nectar's sugars, depend on the height of the paper cylinder and the type of solvent used.

Satisfactory separation of the sucrose, glucose and fructose of nectars has been achieved using either of the following solvent systems: a) butanol/ethanol/water in a ratio of 4:2:1, by volume, and b) n-propanol/ethyl acetate/water in a ratio of 7:1:2, by volume.

After the solvent front has reached the top of the chromatogram (paper cylinder), the cylinder can be removed from the sealed chamber, stood upright, and allowed to air dry for 48 hours at room temperature. Thereafter, the chromatogram is mist-sprayed with a reagent consisting of a 1% solution of p-anisidine hydrochloride in absolute

Fig. 19. open flowers of *Echium plantagineum* labelled with identifying tape at bases; left is a flower open for insect visitation; at right is a bagged flower protected from visits by large insects

ethanol (e.g., 0.5 g p-anisidine hydrochloride dissolved in 50 mL ethanol). The chromatogram is then heated in a 75°C oven for one-half hour.

Apart from the distances they have travelled from the baseline, identification of the sugars is assisted by their characteristic color reactions with the reagent. The spot of sucrose, which travels least, and that of fructose, which migrates farther than the glucose standard, both stain yellow. In contrast, the spot of glucose stains dark brown. Example chromatograms are drawn in Fig. 17. Judging from the size of the sugar spots of the nectar, the exudate analyzed on the left chromatogram is found to be sucrose dominant, with minor amounts of glucose and fructose, whereas the nectar assayed on the right chromatogram contains some sucrose but is dominated by the monosaccharides glucose and fructose. Because in both chromatograms the monosaccharides have similar-sized spots, their proportions in the nectars are therefore approximately equal.

If the spots stain weakly, it may be possible to better elucidate the chromatogram by examining it exposed to the bulb of an ultraviolet lamp, under which the spots will fluoresce. ***To prevent eye damage, wear a recommended safety shield for the eyes, or take care not to stare at the bulb.***

2.3.3 Measuring Rates of Nectar Secretion & Resorption

Plant species can vary tremendously in flower size and morphology, and in the location and anatomy of their floral nectaries. Thus it is not surprising that characteristic properties of floral nectars, such as volume and sugar concentration, also can differ dramatically. Even the relative quantities of vascular tissue (phloem, less often accompanied by xylem) that enter the glands of different species, can influence the sugar concentration of freshly secreted nectar. Each species has its own *initial nectar sugar concentration*, commonly in the order of 5 to 20%. Nectars taken by foragers at flowers are usually between 15 and 60% sugar. Besides the properties inherent to individual species, many environmental factors influence nectar sugar production and eventual concentration, including sunlight, temperature, humidity and plant nutrition.

To understand more about nectar production, other aspects of flower "behaviour" are worthy of investigation. Answers to particular questions — 'When does nectar secretion commence?', 'Is secretion continuous or does it stop at some time?', and 'Are there differences in the rate of nectar secretion or reabsorption during the life of a flower?' —

can provide meaningful details about patterns of nectar production in the species under investigation.

The question of changes in rates of floral nectar secretion and reabsorption now has been addressed in several species. A research strategy particularly suited to investigate these rates under field conditions has been employed by Corbet, Delfosse and Búrquez. This methodology involves two systems utilized in tandem. The systems are similar in that they both involve some sort of exclusion of flower visitors, so as to avoid unwanted nectar removal. The systems differ in that one of then requires careful, repeated sampling of the nectary of an individual flower — a practice that may result in injury to the nectar gland, especially to those nectaries situated out of sight, at the flower base.

In *System 1*, many flowers of a young and identical stage or age from several plants are labeled or tagged (with pieces of drinking straw, coloured tape, string, etc.), and thereafter caged or bagged in the early morning, before flower visitors arrive. For plants with flowers that open and attract visitors at night, caging is performed before nightfall. After the initial caging, nectar is collected from 8 to 10 previously undisturbed tagged flowers from the cage at each of several sampling intervals (e.g. each 2 hours) throughout the flowering lifetime.

System 2 involves the sampling of a fresh batch of 8 to 10 flowers in previously undisturbed plants in the field. Here, the stage of flower sampled should closely match the stage attained at that time by flowers of *System 1.* After all nectar is removed, the flowers are tagged and covered (bagged), before they are sampled once again, about 20 to 90 min later. The volume and concentration of nectar initially gathered to empty these flowers yields an estimate of total standing crop of nectar at the sampling time.

Using this technique, three components require measurement:

1) The standing crop of nectar in flowers exposed to visitors. Changes in this component at different times of the day or night reflect the relative rates of gain by secretion and loss by reabsorption and/or removal by visitors.
2) The rate of nectar secretion in flowers of *System 2.* This measurement is likely to be an underestimate if some nectar is reabsorbed during the brief bagging period, or the nectary is injured during initial gathering of nectar.
3) The rate of change in quantity of nectar, and therefore the net rate of secretion and/or resorption, in flowers of *System 1*. Since each flower was sampled for nectar only once, there are no injury effects here.

Plots of these components, like the one (Fig. 18) performed for *Echium plantagineum* (adapted from Corbet and Delfosse) allow periods of high rates of secretion and reabsorption to be recognized by this indirect method.

Various physiological studies involving the application of radio-labeled sugar to nectaries or to nectar *in situ* have given direct confirmation that reabsorption of sugars by glands can take place. However, in some species, reclamation of all uncollected nectar sugar may not occur. This can occur when the corolla of an old flower finally falls off and carries with it the residual nectar that no longer contacts the flower's persistent nectary.

Other constituents besides the sugars of nectar can be reabsorbed from standing nectar. Radio-labeled forms of some amino acids, for example, have been shown to re-enter the flower after being experimentally added to floral nectar. However, low quantities of amino acids in nectar may not always be the result of reabsorption — highly developed nectaries, consisting of aggregations of trichomes, tend to produce nectars low in amino acid content.

2.3.4 Selection & Breeding for High Nectar Sugar Production

Varieties or lines of crop plants can differ markedly in many traits, including several aspects of flowering. Selection of varieties or lines that yield high amounts of floral nectar can be useful for increasing food production. A higher quantity of nectar sugar can increase attractiveness to pollinators, resulting in improvement of pollination, along with potential increases in fruit and seed set. Honey production, bee colony strength, and pollinator populations also benefit from resources of a higher quality.

When making nectar comparisons between plant varieties or selections, Shuel (1989) emphasized the need to grow the plants (preferably simultaneously) under controlled conditions. This helps to eliminate variation due to environmental factors. Also, it is essential to standardize procedures of nectar collection and measurement. Flowers should be sampled for nectar at the same time each day and must be of the same age, to ensure unbiased comparisons.

To avoid errors in measurement of nectar yields, which may result from nectar withdrawal by insects, plants for such comparisons are often grown in greenhouses. The highest yields may be lower here than under field conditions, but relative performance between variants is usually maintained. Alternative methods to exclude nectar collection by flower visitors include the caging of several plants in the field or plot, and even the bagging of many flowers, on inflorescences or as individuals (e.g *Echium*, Fig. 19). Because each exclusion technique introduces unique variables in *macro-* and *microclimate*, comparisons are meaningful only if the same procedure has been utilized for all selections. For comparative tests, nectar yields usually are based on nectar sugar accumulated over a certain period (generally 24 hours). The choice of duration will vary with species and depends on factors like floral longevity and nectar sugar reabsorption.

Although there is still great potential in this field of research, significant discoveries have already been made. More applications for *nectar breeding* will surface when our knowledge of the process of nectar secretion becomes more complete. Most attempts at selection for nectar sugar production have been made empirically, and these include assessments of the number of flowers per plant, the number of flowering periods per year, flower color, aroma, size, receptacle diameter, volume of functional phloem in the flower stalk (pedicel), and number of nectary stomata. Species of the Leguminosae have been studied most extensively.

Regarding flower color, honeybees were found to prefer clones of *Medicago sativa* that had purplish flowers rather than yellowish-green ones (the latter tended to yield lower amounts of nectar). Florets with yellow-green petals were also found to have a musty smell, which may account for their relative unattractiveness to *Apis mellifera*. Also, weakly-scented flowers of *Lotus corniculatus* produce relatively little nectar. They can be rejected early in the artificial selection process for improved nectar sugar production.
There is a method of selection for nectar production, *based on flower size,* that is very practical and may be suitable for a wide variety of species. By inserting the bases of flowers into a template (previously drilled to provide holes of increasing diameter), one can rapidly screen lines with the largest flowers. Flowers of *Lotus corniculatus* that produced the most nectar had the largest diameter. Alfalfa clones with the greatest receptacle diameter also yielded the most nectar.

The final two parameters, nectary stomata and phloem in the flower stalk, require microscopy and hence are not as practical, but still potentially very useful. In selections of birdsfoot trefoil a strong relationship has been demonstrated between nectar yield and the quantity of phloem in cross-sections of the flower stalk. That is, flowers producing the most nectar were those that apparently could receive the greatest quantities of photosynthate.

Investigations of the relationship between number of nectary stomata and floral nectar production now have been conducted in three legume species. Results obtained with *Lotus corniculatus* and *Medicago sativa* were inconclusive. In *Vicia faba*, however, an inverse relationship was detected — those plants with the *largest number* of nectary stomata per gland produced the *lowest* quantities of nectar sugar. Therefore, for breeding purposes, one should not seek plants whose floral nectaries bear relatively large numbers of stomata. High stomatal number might best be avoided, if the results for the faba bean are applicable elsewhere.

As well as selection for total nectar sugar production, similar efforts can be directed at selecting for desirable sugar types in nectar. For instance, sugar solutions (mimicking nectar) with very high proportions of glucose and fructose, but very low in sucrose, have tended to be less attractive to honeybees (*Apis mellifera*) in feeding tests. Plants within a species might have greater appeal to these insects if their nectar were not so rich in hexose. Furthermore, nectar of Brassicaceae is notorious for high glucose content. Honey produced by *A. mellifera* from such nectar crystallizes (granulates) quickly, a physical property that can be a nuisance for honey extraction from the comb, and also causes a reduced shelf life of honey in the liquid state.

2.3.5 Pollen, Anthers & Dehiscence

Anthers are the floral organs that produce and house pollen grains which contain the male gametophytes or nuclei. Although technically incorrect, the pollen grains may be considered roughly the plant equivalent to animal spermatozoa without tails. Early within the floral bud stage, pollen microsporogenesis from mother cells occurs within the young anther locules. The pollen grains are even nourished by the anthers. It is at this time that grains also take on their final structural details and acquire a lipoidal/carotenoid coating known as the surface pollenkitt. Anthers may be relatively sessile or attached to long thin filaments that extend the anthers beyond the floral perianth. They almost always consist of two chambers, the locules, in which the pollen grains are produced and from which they are released and carried away by vectors such as wind and water or animals including bees, wasps, flies, butterflies, beetles, birds or bats.

1) Pollen release.

During the floral life history the anthers split (known as ***dehiscence***), opening along a predetermined line of weakness, the stomium, to release their precious cargo. This longitudinal stomial dehiscence occurs and the locule walls fold back exposing the pollen to insects and other floral visitors. Pollen is then *shed* (remember that anthers dehisce but pollen is shed).

Anther dehiscence is largely mediated by environmental humidity levels such that dehiscence often takes place during the driest times of the day. A little-understood and equally important factor that controls anther dehiscence is the amount of water (turgor pressure) mediated by the supporting anther filaments (see Schmid & Alpert, 1977 for a detailed explanation of this, called Burck's hypothesis for anther dehiscence).

Rarely does anther dehiscence depart from the typical full-length stomial rupture pattern. Some flowering plants have anthers that dehisce by means of valves or flaps of tissue (such as in the Berberidaceae). The only other major form of anther dehiscence considered here is the widespread type known as poricidal dehiscence, which occurs in many crop plants. About 8-9% of the world's flowering plants have highly specialized anthers that do not dehisce along the entire length of the locules. Instead, these floral anthers

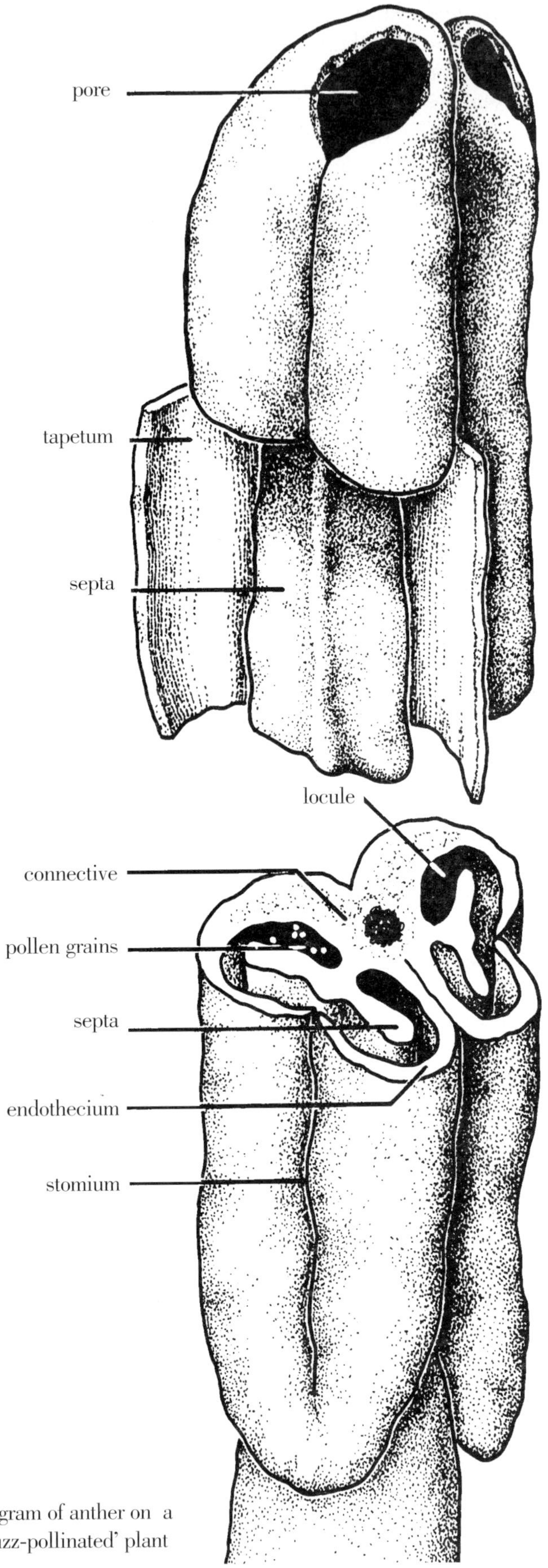

diagram of anther on a 'buzz-pollinated' plant

dehisce by means of two small pores, or tear-drop shaped ruptures — the only openings through which pollen can be shed and made available to the bee pollinators. Tropical and temperate blossoms of this type (exemplified by the genera; *Bixa, Cassia, Cyphomandra, Lycopersicon, Melastoma, Solanum, Vaccinium*) are usually ***"pollen-only"*** flowers which give no floral nectar but have abundant and protein-rich small, "dry" pollen grains.

Pollinating bees, mainly belonging to the Antho-phorinae (as well as tropical *Bombus, Melipona, Euglossa, Eulaema,* and *Eufriesea, Amegilla,* some colletids and many halictids) are the usual visitors of flowers bearing poricidal anthers. For as yet unknown reasons the honeybee genus *Apis* and most stingless bees do not use this form of pollen harvesting and collect it using other methods. The large, robust bees use their indirect flight muscles to literally become living tuning forks. Thus, the small, dry pollen is forcibly ejected or ***"sonicated"*** in an efficient manner by bees that rapidly harvest the pollen. Pollen from *Solanum*, deadly nightshades, has the highest protein level (48 to 56%) of any angiosperm pollen analyzed to date. While sonicating blossoms the bees utilize strong vibrations which, as a by-product, produce amusing loud sounds — hence the name "buzz ***pollination***". Anthers of buzz-pollinated plants are large and may form a cone or brush around the pistil. The anthers are often a bright yellow, contrasting against a blue or white perianth. This is probably a mechanism used to attract pollinators, since the enlarged anthers always give the appearance of abundant yellow pollen, even when they contain none.

Pollen is not always released from dehiscing anthers as single grains, or monads, but is often shed as a unit of adhering grains (such as the polyads in the genus *Acacia*) or in even more elaborate shrink-wrapped packages called ***"pollinia"*** found in orchids and milkweeds. This pollen is not used by bees or other pollinators as food but the pollinia are transferred by insects and birds as they fly between flowers. The study of pollen dispersal and receipt are greatly facilitated in angiosperm taxa bearing pollinia, often identifiable to species even without the aid of a microscope.

2) Pollen Viability and Germination.

Many floral biologists assume that pollen is always viable, each grain having the same chances of germinating on a stigma, sending down a pollen tube, and fertilizing ovules. This is often not the case. Workers studying pollination systems should strive to understand how long pollen grains remain viable within anthers and while residing on various animal pollen vectors. When pollen is shed from anthers following anther dehiscence, the individual grains, polyads or pollinia begin a complex journey that may last many days. The usual sequence of events for pollen grains occurs in the following order: *a)* pollen arrives at a stigma via abiotic or biotic vectors, *b)* attachment and adhesion of grains to stigmatic surface is controlled by chemical composition of both stigma and pollen, *c)* hydration of pollen grains occurs and stigmatic papillae recognize the grain as compatible or foreign, *d)* the pollen germinal pores (apertures) open, and full hydration takes place, *e)* the pollen tube emerges and begins growing down the stylar tissue (if no incompatibility mechanisms prevent such

halictid bee buzz-collecting pollen from anthers of a poricidally-dehiscent plant

growth), and *f)* the growing tube reaches the ovary and nuclei are delivered to the ovules to produce a fertilized zygote which will become a seed within a fruit.

One cannot tell just by looking at pollen grains whether they are viable or capable of germinating upon a stigma and effecting fertilization. During its long journey from protective anthers to a receptive stigma the pollen grain may be exposed to hostile external environments that dry or wet the grain. However, dehydrated, shrunken grains often land upon a moist stigma, rehydrate and usually send down healthy pollen tubes and nuceli. In most cases the key to understanding whether a grain can successfully fertilize and ovule is ***pollen viability/germinability***. Various ***in vivo*** germination assays and ***in vitro*** chemical tests have been developed to gauge these aspects of pollen vitality. The most obvious and probably the most accurate method is simply to perform hand pollinations by placing fresh pollen onto receptive stigmas **for stigmatic receptivity tests see Section 2.4)**. One needs only to know that the stigmas are indeed receptive. Several hours to many days later, depending upon the crop species, pollen tubes can be seen within the stylar tissues and counted. Alternatively, following hand pollinations, the resulting fruit can be weighed or counted, as well as the seeds within.

3) Minimum grains per fruit.

Another assessment that is rarely made but nonetheless valuable is to determine the *minimum* number of pollen grains per stigma that will result in setting a fruit. With a microscope for counting pollen grains, one can determine the number of grains needed to produce a marketable fruit. Using a fine paint brush, or toothpick, various amounts of pollen are added to receptive stigmas on flowers within bags, which are then rebagged to exclude pollinators. Wait until the flower begins to wilt and then harvest these stigmas. They can be acetolyzed or simply placed on a microscope slide, with an appropriate pollen stain, and the stigmatic pollen load quantified. The fruit resulting from the same flower is left in place and labeled, in order to compare the determined pollen grain load and quality of the fruit produced by this amount of pollen.. A "stigmatic loading curve" might be useful to assess the minimum number of grains needed to effect fertilization or to produce a fruit or seeds of the desired size. This information is still very scarce for many crops.

Traditionally, floral biologists have seldom performed hand pollinations (see Section 2.3.4) — which provide a way to assess not only pollen function but stigmatic receptivity as well. All too often, ***in vitro*** chemical assays of pollen germinability (= stainability) are relied upon, with mixed results. Pollen germination also can be studied by placing test pollen on various agar-based culture media or 5-60% sucrose solutions. These are relatively elaborate tests and will not be discussed in this chapter. For many years one chemical assay was used very widely to determine pollen viability. This is the so-called "Cotton Blue" (aniline blue in lactophenol) test. This widely used chemical assay supposedly stains viable pollen a bright blue, based on staining the callose layer, and inviable grains either do not stain, or they take up the dye very slightly. *Under close scrutiny, this test fails to yield accurate results for many flowering plants.* We do not recommend its use.

This and other chemical tests may exhibit a low correlation with real pollen germinability ***in vivo***. Furthermore, they completely miss the important role of the pistillate flower in determining whether or not pollen germination and fertilization occur. In summary, the use of various vital dyes (e.g. 1% methylene blue, neutral red or aniline blue) should be used with the utmost caution since pollen stainability depends not on the true viability but on the grains' content. In some cases even long-dead pollen from herbarium specimens produce reactions with certain dyes!

Several additional chemical tests have been used to measure pollen viability/germinability with better results. One such chemical assay, ***Alexander's stain,*** has been used to distinguish aborted from non-aborted pollen grains. (Only the non-aborted grains have the chance to fertilize ovules). A recipe for the ***modified Alexander's stain*** and discussion of the modified procedure are given here —

- 10 ml 35% ethanol
- 10 ml Malachite Green (1 ml of 1% solution in 95% ethanol)
- 50 ml distilled water
- 25 ml Glycerine (glycerol)
- 5 g Phenol (**use caution as phenol can cause severe skin burns*)
- 5 g Chloral Hydrate (a controlled chemical substance in some countries)
- 50 ml Fuchsin (5 ml of 1% aqueous solution)

- 5 ml Orange G (0.5 ml of 1% aqueous solution)
- 1-4 ml Glacial Acetic Acid

Method

a) Mix all of the above ingredients in the order given and store the resulting mixture in a brown, or aluminum foil-wrapped bottle, at room temperature. The stain solution should last about one month at room temperature, or longer if refrigerated. It has been suggested that more glacial acetic acid should be used depending upon the thickness of the pollen grain walls, up to 4 ml for thick-walled polle.

b) Pollen grains to be tested are placed on a microscope slide to which is added a drop of stain. A coverslip is added and the slide may be gently warmed over a small flame from an alcohol lamp. Non-abortive grains should stain red while aborted grains, lacking functional cytoplasm, will appear greenish.

c) At this time the grains should be examined under a microscope. If no color differentiation is apparent, then the slides should be kept at 50°C for 24 h and re-examined later. This chemical test often gives highly positive results for germinability, which occurs because cytoplasm within grains is no guarantee that the grains are indeed viable and would germinate and produce pollen tubes on a stigma.

Recently, a newer chemical test, the ***Fluorochromatic Reaction (FCR test)*** has been used by many floral biologists as a dependable test for pollen viability. The materials needed and a brief description of the technique follow —

Materials for FCR test:

- 20 mg of fluorescein diacetate in 10 ml acetone
- 15% sucrose solution

FCR Method

a) The reagent is prepared by placing 10 ml of the freshly made sucrose solution in a clear glass vial. The fluorescein diacetate in acetone is added (about 1-3 drops) until the solution turns a light milky or greyish color.

b) Pollen grains to be tested must be stored for 10-30 min in a chamber with high relative humidity. This enables the membranes in dehydrated grains to recover and prevents false-negative results in the FCR test procedure.

c) If pollen bursts in the 15% sucrose solution, use a 20 to 30% solution. The pollen sample is dispersed in a drop of fluorescein diacetate on a microscope slide. Place the slide in a Petri dish atop wet filter paper for 10 min and then cover with a cover slip.

d) Examine the pollen under a compound microscope equipped with the appropriate fluorescence filters and illumination. A violet exciter filter should be used. Pollen grains which fluoresce a bright golden-yellow should be scored as viable. Empty, undeveloped grains will not fluoresce at all. Grains that have lost their viability will show only a weak fluorescence. Observations and counts should be completed on the grains no more than 10-15 min after the coverslip is added. The test solution must be fresh.

The FCR test seems to be the most reliable and accurate of all the chemical assays. It is a test for active esterases and plasmalemma integrity of the vegetative cell, but still not a test of viability. Dehydrated grains may give false positive results and should be rehydrated in a small chamber prior to testing. An obvious disadvantage of this otherwise excellent technique is the requirement for the assay to be conducted in a well-equipped laboratory, using an expensive fluorescence microscope and the appropriate filters.

2.3.6 *Pollen Chemistry*

As a floral reward, pollen is far richer in nutritive components than mostly sugar-containing nectar. Pollen is the essential source of proteins, amino acids and lipids for the majority of the worlds' pollinators and their offspring. Bees are the chief arthropod consumers of floral pollen in all tropical environments. All bees achieve high assimilation efficiencies as they rapidly turn pollen nitrogen into body nitrogenous compounds. This is especially true for tropical bees living in massive colonies at relatively high densities. Most of the nutritive-deriving substances within pollen grains are contained within the pollen walls known as the exine and intine. The intine inner walls are cellulosic, but the outer walls, the exine, comprise a tough biopolymer made up of cross-linked carotenoid pigments. The exine is essentially indigestible for most unicellular or metazoan organisms, except for a few collembolans. Exine passes untouched through the intestines, maintaining a recognizable sculptured exine shell in the feces which can be acetolyzed, and identified to determine plant taxa in pollinator diets. The exine is an incredibly resistant biological natural product, very similar in composition to the outer shell of insects and other arthropods. Pollen grains 100 million years old have been recovered from sediments, with little apparent biological degradation.

Inside the exine pollen walls, the cytoplasm contains rich protein, lipoidal and carbohydrate sugar/starch energy reserves for the developing pollen tube, later supplemented by the stylar sporphytic tissue. Bees and other pollinators need not "crack" the tough pollen grain walls, as often suggested in the literature, to release the nutrients within. Pollen within animal guts is exposed to high osmotic pressures, acids and various degradable enzymes. Thus the pollen grains swell, then 'pseudo-germinate', making thin membranes extrude from germinal apertures. Osmotic shock and the bees' normal digestive biochemistry do the rest. Surface oils, as colorful pollenkitt, are also important nutritionally, especially in taxa with oily pollen such as the Asteraceae, are also efficiently removed within pollinator digestive tracts.

The diversity of nutritious, and sometimes non-nutritive, chemicals within pollen is staggering; we have just begun to understand the complex array of substances present within and on the surface of pollen grains and their relative importance for pollinator diets. A summary of commonly reported values for the chemical composition of pollen harvested by honeybees is given in Table 1. Pollen typically contains from 20 to 50% water and the indigestible exine comprises about 8 to 35% of the pollen dry weight. Protein levels for pollen harvested by bees typically ranges from 7 to 35% (see Table) but levels are much lower (e.g. from 5 to 20%) in pollen from anemophilous plant species. The highest yet recorded nitrogen (7 to 9%) and protein levels (48 to 56%) in any type of pollen were found in a buzz-pollinated *Solanum*.

It is apparent from the above **Table** that while pollen may be a perfect food for certain pollinators, such as bees, it is far from perfect as a diet for humans, as sometimes claimed in the modern health food literature. Pollen contains none of the important lipid-soluble vitamins (A, E, and K). The impact of these and other pollen nutritional findings for pollinators is largely unknown at the present time, but a fertile area for new research. With an average value of almost 25%, protein is a major component of pollen and clearly the most important for pollinator nutrition.

Pollen protein levels are especially high in pollen from bee-pollinated plants (25 to 56% protein) those pollinated by bats (40 to 44% protein). This is likely due to the fact that plants are very dependent upon these pollinators and have 'upped the stakes', providing rich protein for their pollinators. All bees, even those that are parasites and collect no pollen themselves, and many flower-visiting bats, are dependent on nitrogen and protein in pollen to satisfy their adult and juvenile demands for amino acids and proteins. Certain aminoid acids (e.g. proline and tyrosine) are very rich in pollen from bat-pollinated tropical plants and have been hypothesized to be important in forming strong wing membranes. Roles for certain amino acids for invertebrate pollinators are unknown. Most nutritional studies have revealed that bees, and most other insects, require the same so-called 'rat/mouse' 10 essential amino acids, as do humans.

General chemical composition of pollen

Component	No. Spp. Analyzed	Average Levels	Typical Ranges
Protein	277	24%	7.5-35%
Lipids	52	5%	1-15%
Carbohydrates	47	27%	15-45%
Phosphorous	54	0.5%	0.1-0.6%
Ash	60	3.1%	1-5%
Potassium	56	0.6%	0.2-1.1%
Calcium	60	0.2%	0.1-0.5%
Magnesium	60	0.2%	0.1-0.4%
Sodium	30	0.04	0.15-0.8%
Iron	51	140µg/g	wide
Manganese	28	100µg/g	wide
Zinc	21	78µg/g	wide
Copper	27	14µg/g	6-25µg/g
Nickel	23	5µg/g	0-?µg/g
Boron	?	trace	?
Iodine	?	?	4-10µg/g
Thiamin	8	9µg/g	4-22µg/g
Niacin	6	157µg/g	130-210µg/g
Riboflavin	8	19µg/g	?
Pyridoxine	2	9µg/g	?
Pantothenic acid	33	28µg/g	5-50µg/g
Folic acid	8	5µg/g	?
Biotin	4	0.3µg/g	0.2-0.6µg/g
Vitamin C	7	350µg/g	0-740µg/g
Vitamin A	?	0	0
Carotenes	4	95µg/g	50-150µg/g
Vitamin D	4	0	0
Vitamin E	4	14µg/g	?
Vitamin K	4	0	0

1) Determining Protein Levels in Pollen.

Getting reliable and accurate nitrogen, protein and amino acid determinations can only be performed in a well-equipped chemical laboratory and often using expensive specialized modern equipment. Nitrogen determinations are usually performed on pollen, or food or soils, with the use of an traditional technique known as the ***micro-Kjeldahl reaction.*** The method requires high temperature cooking, a mercury catalyst, and a specialized apparatus (e.g. manufactured by the Braun & Luebbe corporation in Elmsford, N.Y). Recently, progress has been made in combusting small amounts of a sample in a furnace and then examining the pyrrolysis products for nitrogen. Also available are very expensive, dedicated amino acid analyzers (for both hydrolysates and native proteins/peptides— the so-called physiological "free" amino acids) for the quantification of amino acids in pollen. These are usually found only in the most modern biochemistry department laboratories, where samples can sometimes be run on a fee basis. Alternatively, amino acids can be qualitatively, and somewhat quantitatively, determined from pollen and nectar samples using a miniaturized ***2-dimensional TLC (thin layer chromatography)*** methodology perfected by the late I. Baker, a pioneer in experimental pollination biology (see account on pp. 156-159 in Dafni, 1992). Readers are directed to the nectar analysis section for a brief account of semi-quantitative amino acid concentration technique, called the histidine scale (also developed by Baker and Baker) using ninhydrin reagent and sucrose solutions.

2) *Assay for Starch in Pollen.*

Many pollens contain large amounts of starch in the form of microscopic starch granules (amyloplasts) similar to those found in some plant stems and root storage tissues. As an energy storage reserve substrate for developing pollen tubes, starch contains far less energy (15,562 to 16,638 joules/g) than pollenkitt lipids (which range from 38,883 to 42,248 joules/g), as determined by bomb calorimetry. Starchy pollen is largely absent in flowers pollinated by animals, with a few tropical plants as exceptions. Pollen with high levels of starch are representative of wind-pollinated plant taxa.

A very old chemical test which is still used is the iodine reaction given by starch granules. This test has been used by floral biolotists to assess whether starch is present or absent from pollen grains. The materials and procedure for the test are as follows —

Iodine Pollen Starch Test (IKI)

Materials:

- 0.2 g Potassium iodide
- 1 g iodine crystals (use caution as these stain everything and the fumes are toxic)
- 100 ml distilled water

The following procedure should be conducted under a chemical fume hood. The potassium iodide is dissolved in a small amount of water, then the iodine crystals are added while stirring. Once dissolved, the remainder of the water should be added. The resulting solution should be kept in a tightly stoppered brown, or aluminum-wrapped, glass bottle. The solution will last for many months. Pollen grains thought to be starchy are placed on a microscope slide to which 1 or 2 drops of the IKI test solution are added. Wait 3-5 minutes for the grains to stain, then add a coverslip and examine under the microscope. Starchy pollen grains will look black because the starch amyloplasts will stain blue-black or occasionally red to purple.

More elaborate spectrophotometric quantitative assays (e.g. using o-toluidine) for starches and other complex carbohydrates are available but beyond the needs of most floral biologists. The IKI test is best used with mature pollen grains freshly removed from dehisced anthers. Normally starchless mature pollen grains may contain some starch when still immature. Even old, dried pollen removed from herbarium sheets or after many years storage yield reliable positive data for the presence of starch. Baker and Baker (1979) have shown that starchless pollen grains (normally containing lots of lipids — another energy storage reserve) are typical of bee-pollinated angiosperms, especially in ***pollen-only*** flowers, and for some examples of fly pollination. Starch grains, which invariably contain some lipids, are typical of species that are autogamous, anemophilous or pollinated by some butterflies and birds.

3) *Lipids in Pollen Grains.*

Most, if not all, pollen grains contain a diversity of lipids, of various classes and in widely varying amounts. A very simple test for the presence of abundant lipids in pollen is to place some pollen, or bee-collected loads upon a piece of brown absorbent paper, such as a lunch bag. After several hours a noticeable greasy dark spot will appear. One simple staining technique for the presence of lipids, especially as abundant surface pollenkitt, uses Sudan IV (see following method). The stain should be made up fresh each time. A red color indicates the presence of lipids. The pollen grains should be examined microscopically within 3 to 4 min after the stain has been applied as the red color disappears in as little as 10 min in some cases. Other lipid assays will be given in a following section on floral lipids (produced by specialized floral glands as rewards for pollinating bees). Various extraction schemes have been devised for extracting, isolating and identifying the various lipids found in pollen and other botanical sources. Most of these involved the use of extraction with a non-polar organic solvent (chloroform: methanol often at reduced pressure and with heating (as with the time-honored Sohxlet apparatus). Lipid classes may be identified using relatively simple ***Silica gel TLC plates and reagent sprays***. Exact identifications of lipids, especially degree of un-saturations are best done in collaboration with a trained lipid chemist. These methods are beyond the scope of the present account.

2.3.7 *Collecting & Identifying Pollen*

1) *Pollen Collection.*

Pollen is best when harvested fresh from flowers then used immediately, if for chemical assays. If it cannot be used right away, then it should be frozen at -20°C in a deep freeze, lyophilized or frozen at -70° C in an ultra-cold freezer as found in biochemistry labs, for archival storage. For feeding/nutritional tests with bees and other insects pollen should again be used fresh or frozen/lyophilized for later use. Pollen stored at room temperatures rapidly loses its full nutritional value and may spoil (lipids become rancid).

Fresh flowers are brought into the laboratory and the anthers separated from the rest of the perianth parts by stripping them off by hand or with forceps. The anthers plus pollen should be placed upon a brass screen (such as the standardized soil sieve series) which can then be vibrated on a shaker table or the anthers can be gently stirred with a camel's hair or nylon bristle paint brush. Several screens, the finest having a mesh of only 100-300 microns, can be used to clean up the pollen which can then be stored in individual glass or plastic vials in the freezer. Often it is easier to dislodge pollen, even large oily grains, from flowers using the trick the bees learned first, that of floral sonication. In this case, a poricidal blossom or a "shaving brush" type of dozens or hundreds of stamens, is held by one hand or in a clamp

SEM of a pollen grain (cucurbit species) trapped among the branched hairs on a bee's body

a few cm above a glass plate resting upon a black or white piece of paper (depending upon the pollen color) and a vibrating tuning fork (512 Hz, or middle 'C', works nicely) is touched to the anthers. Usually a large cloud of pollen results which, after several dozen flowers are vibrated, can be scraped into a pile using a clean razor blade or glass microscope slide. This method also allows hidden pollen feeders (notably thrips) to be ejected along with the pollen and removed by the experimenter. This procedure can even be by automated using battery-powered vibratile devices.

Pollen harvested by hand using the above method is very clean and can be used for a myriad number of chemical and nutritional assays. Obviously, great amounts of clean pollen are also available by this method for any number of palynological microscope slide vouchers, for simple staining, or for acetolysis.

Pollen can also be collected from bees after they have done the work for you. This is often extremely helpful in the case of plants with very small flowers or ones that produce tiny amounts of pollen. Most native ground-nesting bees (e.g. in the Anthophorinae) carry their pollen dry, unwetted with nectar, in specialized hair patches known as *'scopae'*. Bees can be captured alive directly from flowers into vials, or with a net or into a killing tube (using ethyl acetate or sodium or potassium cyanide). If you are making pollen reference collection, pollen can be removed directly from flowers or indirectly after bees have harvested it. If bee scopal or corbicular loads are used as pollen vouchers, every effort should be made to collect bees which have pure loads from only a single plant species.

Large amounts of pollen (many kilograms) can be easily collected with the use of pollen traps installed on honeybee colonies in Langstroth wooden equipment. The details of constructing and using these traps has been reported elsewhere. Our preference is for the use of a bottom type trap known as the ***modified O.A.C. pollen trap.*** This trap removes about 60-65% of the incoming pollen, which the bees offset by sending out more pollen foragers. The drawer can be opened and pollen removed at hourly, daily or weekly intervals to suit the needs of the user. These are excellent devices not only to examine the diet of honeybees living in various ecosystems, but also useful for determining phenological patterns in the local plant community. One honeybee colony over the course of a year harvests pollen and nectar from a territory at least 80-100 km^2 in extent. Pollen traps have also been useful in the study of diet breadth in African honeybees in the tropics and European honeybees in the Sonoran desert and for studies of competition between honeybees and native bees).

2) Pollen Identification (Palynology).

Pollen identification is not easy but can be learned by floral biologists having access to a laboratory and a good compound microscope, and who are willing to take the time to learn methods necessary to prepare, slide-mount and identify the grains.

Pollen can be prepared for light microscopy and easily studied requiring only a good compound microscope (needed magnifications of 400-1000 X) and a few common laboratory chemicals. Airborne pollen is often collected by aerobiologists on sticky tapes or on a glass "gravity slide" to which petroleum jelly or Silicone Oil has been smeared. Medical allergists routinely examine and identify this "raw" pollen without staining or chemical preparation. ***This method should not be used by bee or floral biologists since the surface pollenkitt oils and cytoplasm obscure important taxonomic details necessary to identify the pollen grains to family, genus or species.*** Pollen, whether fresh from flowers, on bees or from larval provisions or feces, should be minimally de-greased and treated with a common pollen stain (e.g. Safranin-O, Basic Fuchsin and similar red stains are good because microscope optics are color-corrected for green light) to enhance resolution of the morphological surface features of the grains. One of the simplest techniques for staining pollen is to use the following recipe for glycerine jelly which can be made ahead of time, stored and even carried into the field (see recipes at the end of this section). A semi-permanent slide mount of these samples can be made by adding a coverslip and ringing the edges with beeswax plus resin or one of the commercially available "metal flake" clear fingernail polishes. Ideally, access to a chemical laboratory or fume hood is required where pollen samples can be treated with solvents and acids. Most entomologists have used potassium hydroxide (KOH) to clear insect genitalia, and this chemical can be used to as a treatment for modern pollen grains to help distinguish morphological features. No matter what pollen preparation techniques are used, collection of pollen directly from plants also requires the taking of a pressed ***plant voucher specimen,*** identified by an expert botanist, which can be deposited in a public herbarium along with a specimen number that can be referenced in subsequent publications.

A simple technique involves the collection of ***undehisced anthers*** from large floral buds and subsequent softening of these tissues in Lactic Acid for several hours. They are then transferred without the acid to a test tube containing 10% KOH at room temperature for 24 h. Alternatively, the pollen can be boiled for 5-10 min in KOH. This results in the hydrolyzation of the cellulose and lysing the protoplast. The resulting pollen grain walls (exines) are now free of obscuring pollenkitt lipids, but are practically colorless and must be stained prior to microscopy. Caution should be used with pollen grains and KOH because pollen is rapidly destroyed by strong bases (such as KOH) although it can withstand immersion in very strong acids. Any of the red stains for pollen mentioned previously will work here.

3) Pollen obtained from bee nest samples or bees.

The collected material can go directly into the KOH solution. Very detailed recommendations and methods (standardized quantitative analytical techniques for bee-collected pollen samples, minimum sample sizes and appropriate statistical procedures) are given in a recent paper by O'Rourke and Buchmann (1991) and will not be discussed in this treatment. We and others also discuss the importance of not neglecting disparate pollen volumes when considering the nutritional value to pollinators from pollen of various plant taxa, and also present formulae for calculating volumes for pollen grains of different shapes. These are among the subjects of ***melissopalynology,*** a discipline that is starting to come of age, under increasing critical scrutiny, in study of bees, pollen and pollination.

The standard pollen treatment technique used by professional palynologists worldwide is the traditional ***Acetolysis technique*** devised many years ago by Erdtman (see Faegri and Iversen, 1989). This technique should be applied with caution since acetolysis involves caustic acids in a highly exothermic chemical reaction. It should always be done in an approved chemical fume hood while wearing a protective laboratory coat, gloves and safety glasses or face shield. The complete recipe for the Acetolysis reagent and procedure is given at the end of this section. The chemical reaction progresses rapidly and the pollen samples must not contain any water. Water in the samples can provide a violent explosion of boiling acid to vent from the open test tubes. Floral tissue or bee materials can be placed directly into concentrated glacial acetic acid as a pretreatment dehydration step prior to acetolysis.

Acetolyzing pollen grains (from pure pollen, anthers, flowers, bee materials etc.) involves gently boiling the materials in a mixture of nine parts acetic anhydride (anhydrous, concentrated acetic acid) and one part concentrated sulfuric acid for several minutes. This caustic reagent dissolves away the interior grain cytoplasm and the surface oils leaving the pollen in a "fossilized" state. The sporopollenin polymer constituting the pollen grain exine walls survive unchanged (for exceptions in tropical pollens, such as Musaceae, Lauraceae, and others) but do take on a darker or brownish-black coloration. Such acetolyzed pollen is thus ideal for light microscopy and usually requires no additional staining. Grains that become too dark can easily be bleached using sodium hypochlorite (laundry bleach). Acetolyzed pollen is washed and then stored in glycerin or silicone oil. For reference slides the best technique is to remove anthers directly from flowers on dried herbarium voucher specimens collected for this purpose and which correspond to field notes and specimen numbers and have been identified by a trained plant taxonomist.

Recipe for Glycerine Jelly Mounting Medium/Stain

Materials:

- 7 g gelatine
- 42 ml cold distilled water, in which gelatine is mixed and warmed gently with stirring
- 50 ml glycerine and 0.5 g Phenol (or 10 drops of 80% phenol solution)

Procedure:

Dissolve 0.1 g Basic Fuchsin stain in 10 ml ethanol. Taking 1/3 of the glycerine jelly, slowly add the stain, drop by drop, until a clear pink color is produced (Approximately 1/10,000 of solution is the powdered stain). The glycerine jelly plus stain can now be stored in dark glass vials or jars at room temperature for many months until needed. It is easy to produce jelly of various color intensity. Some experimentation on pollen may be necessary to achieve the desired results.

Small bits of stained glycerine jelly are sticky and can be picked up on the tip of an insect pin or dissecting probe then applied to different body areas of pollinators to selectively sample the adhering pollen grains. The blob with pollen is simply placed on a microscope slide then heated over a small flame and a coverslip added. This is an easy and inexpensive technique.

Acetolysis Method

Materials:

- Concentrated Sulfuric Acid (>98%)
- Acetic Anhydride (anhydrous)

Procedure:

CAUTION-NO WATER MUST BE PRESENT INSIDE THE REACTION TEST TUBE.
WEAR A LAB COAT, GLOVES AND PROTECTIVE EYEWEAR

a) After the KOH treatment, or without, the pollen residue in a centrifuge tube is dehydrated with glacial acetic acid, centrifuged and the supernatant discarded.

b) Treat sample with a fresh (made daily) mixture of 9 parts acetic anhydride (concentrated anhydrous acetic acid) and 1 part concentrated sulfuric acid. This reaction is highly exothermic so mixing should be slow and with stirring. The tube can be cooled by partial immersion in an external beaker of cold water. Heat the tube gently in the same test tube in a beaker partly full of boiling water on a hot plate under the fume hood. Heat to the boiling point (e.g. by immersion for some 3-5 min. in the boiling water bath). Allow solution, now black, to cool, centrifuge for 3-5 minutes (at least 3-5,000 rpm) then decant supernatant and discard. The pollen residue will be a blackened mass in the bottom of the tube.

c) Wash the pollen residue 1-2 times with glacial acetic acid, centrifuge and decant supernatant.

d) Wash 1-2 times with water, centrifuge and decant. At this time you may want to put the sample, with traces of water, into a desiccant environment or vacuum oven to remove any water before further processing.

e) Store the acetolyzed pollen residue in glycerine (glycerol) or mount in glycerine on microscope slides under a coverslip. Silicone oil can also be used but requires additional steps.

f) Glycerol slide mounts should be water-free and then sealed with fingernail polish to make semipermanent mounts. These slides will last for many years, stored horizontally in a darkened slide cabinet. They also have the advantage (for glycerine or Silicone Oil) that the pollen grains can be rotated while viewed at high magnification simply by taping gently on the coverslip with the point of a lead pencil.

The necessity of a ***fume hood, following strict safety precautions, etc.*** will be amply repaid the floral biologist who uses this technique. Important morphological details necessary for accurate identifications can only be seen (often at 1000 X) in acetolyzed pollen grains. Acetolysis also allows so-called LO analysis (focus changed slightly from the surface to the interior of the grains) whereby one can see structural wall elements within the exine which may be important diagnostic features allowing specific identification.

4) Counting Pollen Grains.

Diverse methods have been developed for counting pollen grains in anthers, from sediments, or on bees and then establishing ***relative pollen frequencies.*** Perusal of modern palynology textbooks is often an excellent place to become acquainted with these methods and their statistical analysis. If one has access to sensitive electronic microbalances (Cahn-type electrobalances), then gravimetric methods can be utilized for estimating pollen grain numbers. One of the oldest, and least expensive, accurate techniques is a combination of volumetric dilution and counting with the aid or a ***haemocytometer*** (blood-cell counting reticle microscope slide). In this scenario, an unknown number of pollen grains, as from a honeybee corbicular load, are dispersed into a known volume of water or ethanol (often with the aid of a surfactant and ultrasonic dispersing probe). Then, a subsample, perhaps 10 µl, of the swirling mixed solution plus pollen grains (not allowing them to settle out of solution) is placed on the haemocytometer counting grid and counts are made according to standardized procedures used by medical technologists for decades in counting red blood cells. This is a handy technique for pollen grains.

If no haemocytometer counting slide is available, then there is another simple counting method which can be used in its place. Using about 400 X magnification, one first calculates the area of the microscopic visual field. This area is then divided into that of the area of the coverslip. This will yield a number on the order of 4,000 which will remain a constant for future use with the same microscope objective and ocular lenses. Count the pollen grains (usually a count of 300-500 even up to 700 is necessary) in 10 randomly- selected visual fields and compute the mean. Multiply the mean by the constant determined above, and since the pollen "stew" does not cover the entire area under the cover glass, also by an estimate of the fraction of the area which is covered.

Palynologists use exotic pollen or spores (usually from *Eucalyptus*, *Zea* or *Lycopodium*) as an internal calibration method. This calibration technique also allows the summation of results from different microscope slides, prepared from the same pollen mixture. Methods such as the ***"Lycopodium Method"*** are preferable for achieving repeatable quantitative results. Floral biologists can use this relatively simple and inexpensive technique to good advantage in their own studies. The technique is based upon an original method developed by Stockmarr (1971). One or more commercial *Lycopodium clavatum* spore tablets [available commercially from Lunds University, Department of Quaternary Geology, Tornavagen 13, S-223 63, Sweden].

Each tablet contains approximately 10,000 acetolyzed spores. The number per tablet, based upon a referenced batch number and accompanying calibration sheet, is stated to an accuracy of 3.3%;. [The price is about U.S. $10.00 per bottle of 500 tablets; they also come in larger tablets with more spores per tablet]. The user selects how many tablets to use and first dissolves these completely in 10% HCL. The sample is centrifuged and the supernatant decanted. The unknown test sample is added to the tube with the internal spore standard. The sample is thoroughly mixed, acetolyzed or otherwise processed carrying the spores throughout. A microscopic count (usually 500-1,000 grains) is made and the ***number of pollen grains and spores*** is tabulated. This functions like a "mark-release/recapture" in animal population studies and gives an accurate estimate of the original number of pollen grains present in a mixture, without the tedium of counting them all. Further, when used to equate pollen type counts on several slides, the spores counted on each slide provides a weighting factor, by which the number of each pollen type can be altered to represent the portion of the entire pollen mixture that it actually represents.

For example, if two kinds of pollen are found on two slides prepared from a single mixture, and the number of spores counted on slide A is twice that counted on slide B, then the two pollen types on slide B comprise two-thirds the sample represented by the two slides. (If the spore counts were identical on both slides, then numbers of pollen grains on each slide would each represent half of the total sample). Statistical reliability demands that the number of spores used should not be less than 10-20% of the expected total pollen grains present. However, this can prove impractical when dealing with giant pollen grains or polyads, because of their large size difference from the spores. Nonetheless, because the numerical relationship between the pollen and spore internal standard is constant throughout the procedure, the total number of pollen grains present in the original sample is given by the following formula;

Total Pollen Grains =

$$\frac{\text{Grains Counted X Total } \textit{Lycopodium} \text{ spores added}}{\text{No. spores counted in subsample}}$$

Within the past 5 to 10 years, automated electronic counters have been developed for dealing with particle counts in the thousands to hundreds of thousands range. These have largely grown out of the biomedical need for rapidly counting large numbers of red blood cells in diverse samples, and for assessing contaminant particles in automotive hydraulic fluids etc. One such counter, the ***Coulter Counter***, has been used recently by some floral biologists. Few floral biologists have the monetary resources necessary (typically 10,000-20,000 USD) to acquire and service these high-technology instruments. One drawback of the Coulter Counter is the requirement that pollen grains be dispersed and counted in particle-free expensive electrolytic aqueous solvents. Other particle counters measure the pollen grains by the shadows they cast upon a sensor/photomultiplier and samples can be run in pure water or ethanol. Since these instruments are expensive and rather exotic they will not be discussed further. They have, however, revolutionized counting pollen grains in the last 10 years for myopic-strained pollen-counting biologists!

2.3.8 Oils Produced as Floral Rewards by Tropical Flowers

In the late 1960s floral biologists were surprised to learn of the discovery by S. Vogel that many tropical plants, especially belonging to the family Malpig-hiaceae, produced specialized calycine glands (he termed "***elaiophores***") that secreted novel free fatty acids as floral rewards for their specialized bee pollinators. =Some tropical plants which bear elaiophores and offer lipids to their pollinating bees include the nance (*Byrsonima crassifolia*), acerola and Barbados cherry (*Malpighia glabra, M. punicifolia*) in the Neotropics (see **Appendix I**). In the tropical Malpighiaceae, the oil glands are large oil-filled "blisters" on the calyx. In other flowers, such as some tropical cucurbits (e.g. *Momordica* and *Thladiantha* spp.) they form dense mats of oil-secreting trichomes. Elaiophore floral oils are energetically rich (about 3,300 j/g) foodstuffs, usually substituted for nectar in the provision masses of bees. Many other details are reviewed elsewhere.

Chemical Testing For Floral Oils

Suspected oil glands (trichome or epithelial elaiophores) on flowers can be tested by simple chemical means to determine, at least, if true floral lipids are present. One such test is with the use of the aforementioned ***Sudan IV stain*** (used for pollenkitt lipids). Vogel proposed that essential oils (e.g. terpenes and other components of floral scents) produced by specialized areas of the perianth known as osmophores could be visualized by soaking fresh flowers, often with a surfactant and under reduced atmospheric pressure, in a 1:10,000 solution of Neutral Red dye to distilled water. The authors have also used the Neutral Red test on regular elaiophores whose lipids also stain red. Others have suggested the use of 4% osmium tetrachloride which stains the oils and elaiophore tissues black almost immediately. Since this chemical is very dangerous and can cause eye damage or blindness if not handled properly under a fume hood wearing "water type" safety goggles, we do not recommend the use of osmium for detecting floral oils. Obviously, if one is chemical inclined, there are TLC and GC/mass spectral techniques for use on derivatized floral oils (usually producing methyl esters for GLC analysis) which can be used to quantify the lipid classes present and to identify relcalcitrant oily molecules.

2.3.9 Plant Breeding Systems & Pollination

1) Introduction.

Pollination is the first step in the process of sexual reproduction in plants. The pollination requirements of plants are varied. Knowing how they are met, and how the process of sexual reproduction continues, requires an understanding of plant breeding systems and reproductive strategies. Tropical crops represent the gamut of possibilities known to science. Thus, if one is to be effective in practical issues such as pollination technology in tropical crop production, plant reproductive systems must be understood. Unfortunately, this subject involves a suite of technical terms (see ***Glossary***) unfamiliar to many people involved in pollination and agriculture.

As flowers develop, the sexual organs or sporophylls undergo genetic changes in some of their cells. In the anthers, special cells called "***pollen mother cells***" are formed (see also Section 2.3.2). These go through the process of reduction-division, or meiosis, during which the number of chromosomes in the resulting cells is halved. From each pollen mother cell, four pollen grains are generated, each with half the number of chromosomes (i.e. haploid) of the parent plant (which is diploid). In the ovary, similar, but more complex, events take place which give rise to a multicelled ovule in which each cell is haploid.

After pollination has taken place, the pollen grains on the receptive stigma germinate. They produce a pollen tube which grows down through the style and into the plant's ovary. While the tube is growing, it contains two nuclei. The one at the tip of the pollen tube is the *tube nucleus* and the one that follows is called the *generative nucleus*. As growth of the tube proceeds, the generative nucleus divides into two *sperm nuclei*. These two nuclei are liberated into the ovary to fertilize a single ovule. Each ovule in the ovary comprises several cells, one of which is the egg nucleus. One of the sperm nuclei unites with the egg nucleus eventually to give rise to the embryo, while the other sperm nucleus unites with one or more of the other nuclei in the ovule to give rise eventually to the endosperm. This is referred to as "***double fertilization***" and is peculiar to the flowering plants (Angiospermae), to which nearly all crop plants belong. Through this process, the embryo becomes diploid and ready to grow into another plant. The endosperm is often rich in stored resources from which the growing embryo can draw during its growth. However, the relative importance of the endosperm and other tissues in providing nutrients to the growing embryo and seedling varies much between plants.

After fertilization has been accomplished, and in some plants even before, the fruit starts to develop. There are numerous kinds of fruits (such as pomes, drupes, berries, siliques, follicles,) depending on how they are formed and which of the parent plant's tissues expand to give the seed-containing fruit. That, however, is not of direct importance to the issues of pollination, but it behoves pollination biologists to understand the nature of the fruits and seeds with which they are concerned. Further aspects of the mechanics of pollination have been reviewed in Section 2.3.1.

2) Sexuality in Plants.

Most crop plants are ***hermaphroditic***, i.e. with *both male and female reproductive organs* functional on every plant, but not necessarily at the same time. When the male and female flowers are separated the plants are termed *monoecious*. Some monoecious, hermaphroditic plants self-pollinate readily, and some of these may be self-fertile. Some monoecious plants have pollen-producing flowers on one part of the plant, but ovary-bearing flowers on other parts. For some, there may be difference in the timing of maturation of the two kinds of flowers. For hermaphroditic plants with hermaphroditic (the so-called "perfect") flowers, there may be spatial or temporal separation of the sexual function as well.
Some plants, however are ***unisexual***. The plants bear flowers that produce only pollen, with others producing flowers which bear the ovaries, and eventually the fruits and seeds. Clearly, *self-pollination* or *self-fertilization* is impossible.

If one considers the pathway taken by pollen during the natural process of pollination, and notes whether or not the pollination results in fertilization and seed-set, the spectrum of possibilities can be summarized as in the above ***Figure*** (after Richards, 1987). This diagram also helps introduce some of the technical terms used in pollination biology.

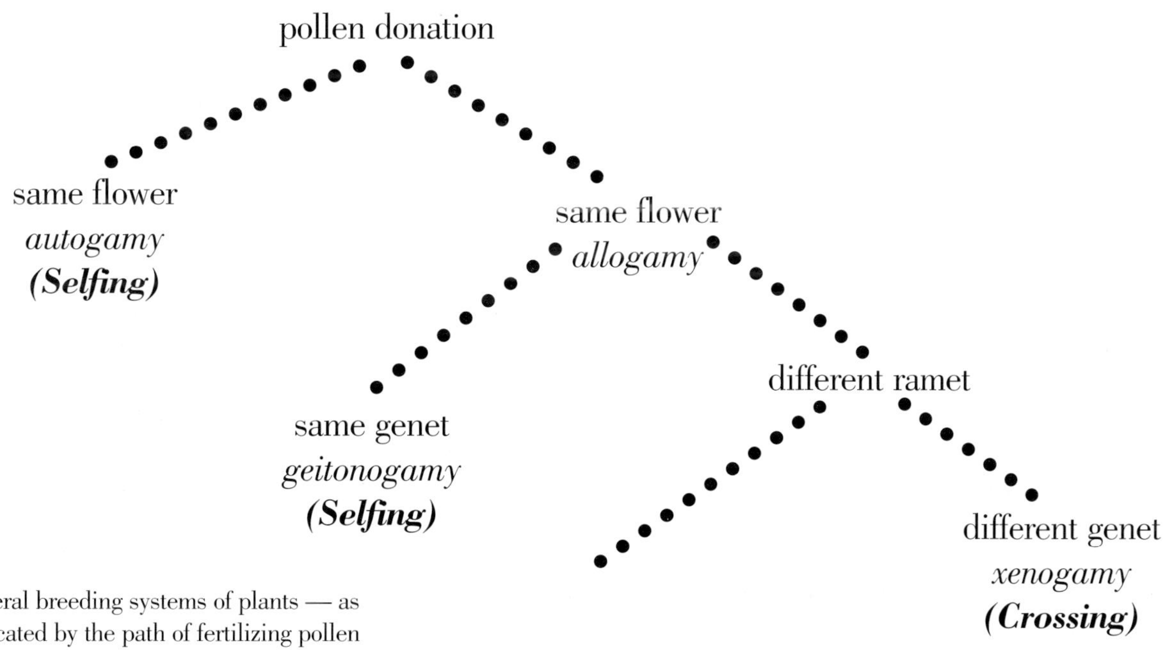

general breeding systems of plants — as indicated by the path of fertilizing pollen

Selfing and Crossing (see text)

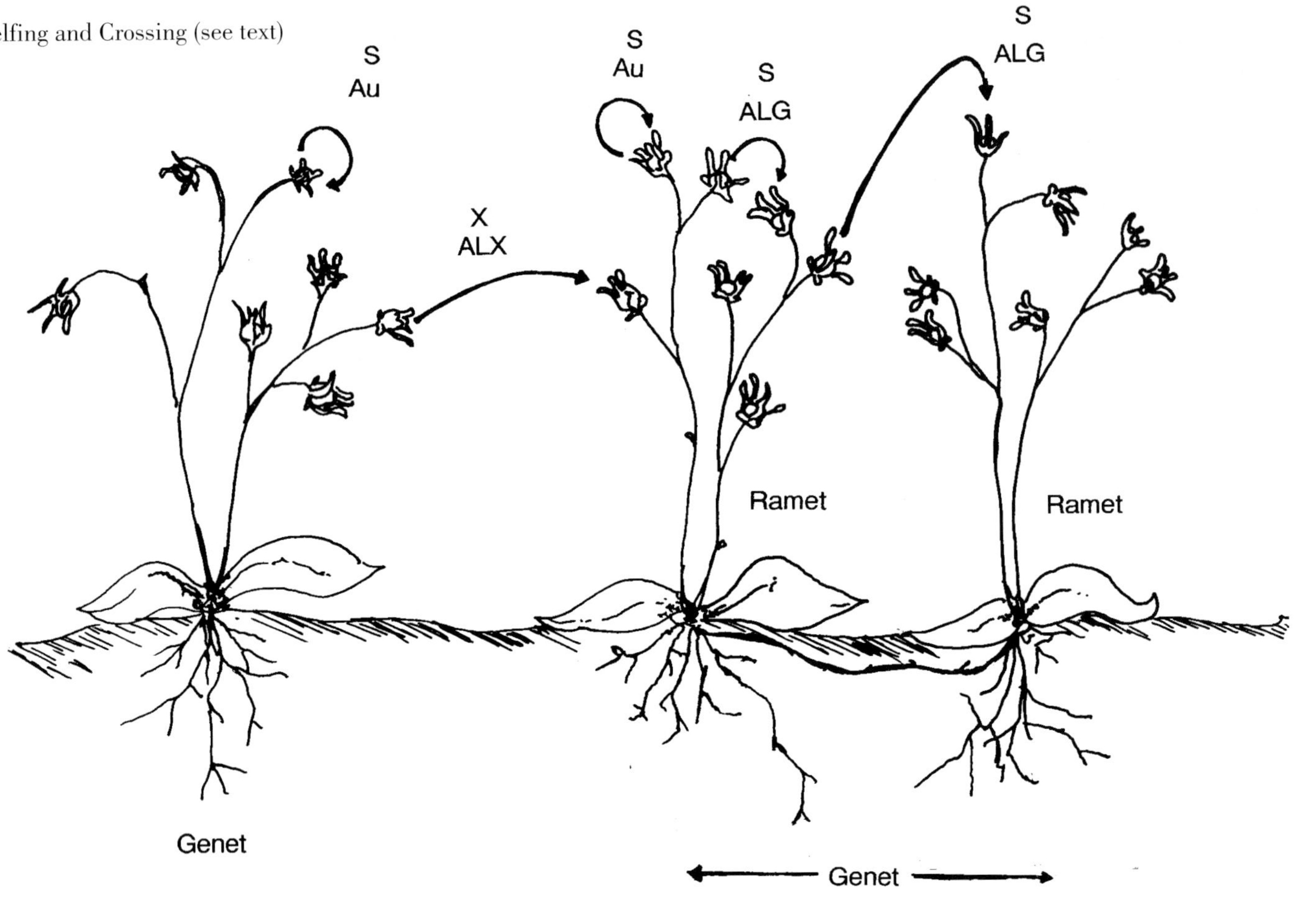

Autogamy, or self-pollination results from the pollen fertilizing the ovules of the *same flower*. The opposite of this process is ***cross-pollination, or outcrossing*** (technically, *xenogamy*). Many crop plants are autogamous (***tomatoes***, some ***citrus***, ***peanuts*** and many other cultivated ***legumes***, see also **Appendix I**). Plant breeders have selected for autogamy in some plants that are naturally outcrossing. Nevertheless, it is well worth remembering that many self-compatible crop plants may still produce a higher quality crop if cross-pollination takes place (***sunflowers*** and ***oil-sed varieties*** of ***Brassica***). Autogamy and xenogamy are shown in the next ***Figure.***

Geitonogamy refers to *self-fertilization*, in that the pollen fertilizing the ovules originates from the *same parent plant*, or from a genetically identical clone (called a *ramet*). The second Figure also presents this situation. Geitonogamy and autogamy are genetically exactly, or almost, equivalent in that both result in self-fertilization and ***inbreeding***. However, it is known that minor genetic differences can occur between flowers of the same plant or clone. Geitonogamy is common in self-fertile plants with flowers that are of different sexes, or that have flowers in which the anthers and stigmata mature at different times.

Xenogamy takes place when pollen from one plant (or *genet*) results in the fertilization of the ovules of a different plant (another genet). Of course, most plants which are self-fruitful, can be xenogamous as well. These may be referred to as facultative inbreeders or facultative out-crossers, depending on one's point of view. In fact, in many self-fertile plants, xenogamy is encouraged by differences in the timing of maturation of the anthers and stigmata, or by the structure of the flowers. Further, as noted above, the quality of the crop may be higher if natural '*selfers*' are '*outcrossed*'.

However, there are many crop plants that require exchange of pollen with other plants of the same species to set fruit. These are referred to as "***obligate outcrossers***". Obligate out-crossers have various mechanisms by which self-fertilization is prevented

- *In some monoecious plants, the staminate and pistillate flowers may be produced at different times (e.g. in **oil palm**),*
- *in others the flowers are produced at about the same time but mature separately (e.g. in **coconut**),*
- *in others, staminate flowers and pistillate flowers are produced at different places (and sometimes times) on the plant (e.g. **squashes, gourds, pumpkins, melons, cucumbers**).*

In those with hermaphroditic flowers, there may be spatial separation of the anthers and stigmata which prevents self-pollination (this is called *herkogamy*), in others there may be a temporal separation of maturation (called *dichogamy*). In *dichogamous* plants, the anthers maturing first is called protandry and the stigmata maturing first is called protogyny.

A special case of herkogamy is known in starfruit and bilimbi. In these species, different plants produce flowers with different forms in a system called ***heterostyly***. This refers to the flowers of one kind having long styles and short stamens, while flowers of the other kind are the reverse. Not only are the sporophylls of different sizes on the different plants of the same species, but the pollen from the long stamens is larger than that from the short stamens, perhaps a reflection of the length of the style through which the tubes of each must grow, respectively.

In many monoecious plants which produce hermaphroditic flowers, dichogamy and herkogamy may be poorly developed. These plants rely on incompatibility mechanisms (which are often also operating in other obligately xenogamous plants) to prevent self-fertilization. Plants with self-incompatibility can not self-fertilize. If self-pollination takes place, as it often does, seed-set will not eventuate. Self-incompatibility is quite complex. Gametophytic, multiallelic self-incompatibility is the most common form. This sort of incompatibility is genetically controlled by two or more alleles (called S alleles) and is effected by the genetic make-up of the pollen, the tubes of which grow into the stylar tissue, but fail to penetrate the ovary. Sporophytic, multiallelic self-incompatibility is also controlled by two or more alleles, but is effected by the genetic make-up of the anther and the failure of the pollen to germinate on the stigma. This form of self-incompatibility is best known in the Brassi-caceae or cabbage family. Other incompatibility mechanisms are known, involving gametophytic or sporophytic interactions, but with single or double allelic genetic control. Breeders have bred lines of self-incompatible plants as a means to produce hybrid seeds in plants which are normally self-compatible.

The concept of plants of different "sexes" would seem to have been introduced through heterostyly and self-incompatibility. But, even in such plants, each breeding type has both male (pollen donor) and female (seed production) function. In some plants, though, the sexes are separated to greater or lesser extents. The sorts of breeding systems included are included in the term dicliny.

Within dicliny, dioecy is the most marked form of sexual separation with plants which are pollen donors only (males) and others which are pollen recipients (seed producers) only (females). There are few dioecious and widely domesticated crop plants, but some grapes (*Vitis*), jojoba, asparagus, and spinach serve as examples. Clearly if one is interested in harvesting seeds or fruits of such plants, then pollinizer, male, plants are needed is some minimum number to allow for pollination but to maximize crop production. For crops such as asparagus, male plants tend to grow more vigorously and are the plants of choice for crop production. Seed is needed only for breeding or re-planting.

There are examples of plants which are incompletely dioecious, but still diclinous. In species of gynodioecious plants, some individuals are male sterile (i.e. female) while others are hermaphroditic. The combination of female-sterility and hermaphroditism (androdioecy) is almost unknown in nature. Some species show various intergradations of sexuality and are variously referred to as polygamous, polygamo-dioecious, subgynoecious, and so on. Gynodioecy has been bred into some crops, e.g. *cotton* and *oilseed rape*, to produce hybrid seed which is harvested from the male-sterile variety planted with hermaphroditic pollen-donors.

3) Apomixis & Parthenocarpy.

Apomixis refers to asexual reproduction. This category of reproduction includes agamospermy and vegetative reproduction.

Agamospermy is the process by which a plant's ovules develop into seed without fertilization. Pollination is not required, even though flowers, sometimes very showy, are produced. Agamospermy seems rare among crop plants, but it is represented in *Citrus* (oranges, grapefruit, lemons, etc.), mango, mangosteen, some bramble berries (*Rubus*), and some cereals and grasses (**see also Appendix I**).

There are a number of crops which are propagated vegetatively and for which pollination is never, or rarely, a problem. Tropical crops which are propagated vegetatively include some of the important tubers, such as *potatoes, yams, manihot, bananas*, and *sugar cane*. Many trees and shrubs can be propagated vegetatively from cuttings. For these, seeds are not used except in the case of plant breeding for new varieties.

Some plants, which produce harvested fruits (as fruits or vegetables) do not require pollination by any means. The fruits develop without fertilization of the ovules taking place. This is called parthenocarpy. Examples of these are bananas (although wild bananas are pollinated by birds and bats), pineapples, some varieties of citrus (seedless varieties especially), and seedless cucumbers (which if pollinated become misshapen and bitter).

Conclusions.

From the viewpoint of crop pollination biology, the details of breeding systems are important, although the actual mechanisms of self-incompatibility need not be of much concern. Artificial crop selection for breeding systems, however, would necessarily involve the most detailed knowledge obtainable. Understanding breeding systems helps investigators, or the pollination technologist to determine pollination requirements of the plants of interest, and aids in making appropriate and informed recommendations. *The published literature that involves pollination contains many errors and partially correct generalizations.* Much apicultural literature, which encompasses a great deal of pollination, fails to recognize the complexities of plant breeding systems, as does a lot of literature in plant sciences. The latter also tends to oversimplify zoological and physical issues in pollination.

Such understandings are especially important given the increasing complexity of pollination requirements in crop plants as breeders manipulate plants' sexualities to maximize growth or productivity, or to preserve desirable plant characteristics. To date, these include artificial selection —

- ***to bring about apomixis or self-compatibility and self-pollination to circumvent insect pollination,***
- ***introduce self-compatibility and dicliny to allow for the production of hybrid seed, and***
- ***develop parthenocarpic varieties to assure fruit-set and seedlessness.***

In general, it is always useful to experimentally investigate the pollination requirements of crop plants, variety by variety.

2.3.10 Methods for Determining Pollination Requirements

The methods for determining pollination requirements of plants are not complex. The following is a step by step guide —

1) Examine the flowers of a number of plants to determine the structure of the flower and the relative positions of all the floral parts. Note especially if various parts are reduced, absent, whether they mature at different times, or change their positions as they mature. Determine how the nectar and pollen are presented to pollinators, and how the floral mechanism works.

2) Make observations on the kinds of pollinators that seem to be effective at pollen transfer (e.g. wind, insects [and kind], birds, bats).

3) If the flowers are perfect, testing for *self-fertilization or agamospermy* can be done by bagging. The bags should be pollinator-proof. Also, they should not create an overly humid and hot microclimate within. Cheesecloth works well, but keep in mind that wind-borne pollen grains can blow through the mesh. The bags should be anchored firmly to the plant's stem and not allowed to touch the flowers being studied. White paper bags are suitable for use on wind pollinated plants. If the bagged flowers fail to set seed, one can be reasonably sure that *pollination is required*. However, the need for cross-pollination may not have been proven if the flowers are *dichogamous*.

4) To sort out if a plant is *outcrossing* or *selfing*, controlled pollinations must be made. Bagged flowers must be cross-pollinated with pollen from **a)** other plants and **b)** within the same plant. If seeds are set by treatment **b)** then one would conclude that the plant was *geitonogamous*. But, if seeds were set only from treatment **a)** obligate outcrossing would be the explanation. Pollen can be transferred by plucking stamens and touching the anthers to stigmata, brushing pollen from anthers onto an artist's paint brush and then brushing stigmata, or by using a freshly dead bee (impaled on a toothpick or similar sliver) as the paint brush. For small flowers, sometimes using the whole flower instead of the stamens is required.

5) If the bagged flowers in treatment 3) set seed, three explanations can be invoked: **a)** the flowers are self-fertile, **b)** they are agamospermous, **c)** pollen entered the bag.

6) To test for *agamospermy*, the immature stigmata of the flowers can be clipped, or if large enough, covered with aluminium foil. If clipping is used, the wound should be dressed with a little soft wax (just molten bees' wax works well) to prevent drying and infection. *If the flowers set seed, agamospermy would be the explanation*. If they don't, self-fertility would be invoked. If both treatments resulted in seed-set, facultative agamospermy would be presumed, but this is highly unlikely.

7) To test for *self-fertility* and be sure, one would also make controlled self- and cross-pollinations in flowers within bags in a manner similar to treatment 4).

8) For all the treatments noted above, remember to have *open pollinated control, or check, flowers* on the experimental plants and on others. In general, one should have *at least 10 plants for each treatment and several flowers on each plant*. Depending on the size of the plant, one can have one to several treatments on a single plant. Large plants, such as shrubs and trees, can accommodate several treatments. Small, herbaceous plants may have to be bagged in their entirety to accommodate a single treatment. The actual experimental design has to be tailored to the plant, the flowers, and the suspected pollination mechanism.

Additional References

Baker, H. G., Baker, I. 1979. Starch in angiosperm pollen grains and its evolutionary significance. Amer. J. Bot. 66:591-600.

Block, R. J., Durrum, E. L., Zweig, G. 1958. A manual of paper chromatography and paper electrophoresis. Academic Press, New York, USA.

Buchmann, S. L. 1983. Buzz pollination in angiosperms. In: Handbook of experimental pollination biology, C. E. Jones, R. J. Little (eds.). New York: Van Nostrand-Rheinhold, Inc., pp. 73-113.

Buchmann, S. L. 1985. Bees use vibration to aid pollen collection from non-poricidal flowers. J. Kansas Entomol. Soc. 58:517-525.

Buchmann, S. L. 1986. Vibratile pollination in *Solanum* and *Lycopersicon*: a look at pollen chemistry. In: Solanaceae systematics and biology. W. G. D'Arcy (ed.), Columbia University Press, Chapter 18, pp. 237-252.

Buchmann, S. L., Cane, J. H. 1989. Bees assess pollen returns while sonicating *Solanum* flowers. Oecologia 81:289-294.

Buchmann, S. L., O'Rourke, M. K. 1991. Importance of pollen grain volumes for calculating bee diets. Grana 30:591-595.

Buchmann, S. L., Buchmann, M. D. 1981. Anthecology of *Mouriri myrtilloides* Melastomataceae: Memecyleae) an oil flower in Panama. Biotropica 13:7-24, special supplement on Reproductive Botany.

Buchmann, S. L., O'Rourke, M. K., Shipman, C. W., Thoenes, S. C., Schmidt, J. O. 1992. Pollen harvest by honeybees in Saguaro National Monument: potential effects on plant reproduction. In: Proceedings of the symposium on research in Saguaro National Monument, Tucson, AZ., C.P. Stone, E.S. Bellantoni (eds.), pp. 149-156.

Burquez, A., Corbet, S. A. 1991. Do flowers reabsorb nectar? Functional Ecology 5:369-379.

Campbell, N. J., Ayers, G. S. 1989. Selecting for high carbohydrate production in birdsfoot trefoil (*Lotus corniculatus* L.) through direct and indirect evaluation. Journal of Apicultural Research 28(3):121-125.

Corbet, S. A., Delfosse, E. S. 1984. Honeybees and the nectar of *Echium planta-gineum* in southeastern Australia. Australian Journal of Ecology 9:125-139.

Cruden, R. W., Hermann, S. M. 1983. Studying nectar? Some observations on the art. In: The biology of nectaries. B. Bentley and T. Elias (eds) Columbia University Press, New York. pp. 223-241.

Buchmann, S. L. 1987. The ecology of oil flowers and their bees. Ann. Rev. Ecol. Syst. 18:343-69.

Davis, M. B. 1966. Determination of absolute pollen frequency. Ecology 47:310-311.

Davis, A. R., Gunning, B. E. S. 1991. The modified stomata of the floral nectary of *Vicia faba* L. 2. Stomatal number and distribution as selection criteria for breeding for increased nectar sugar production. Acta Horticulturae 288:329-334.

Elias, T. S., Gelband, H. 1976. Morphology and anatomy of floral and extrafloral nectaries in Campsis (Bignoniaceae). American Journal of Botany 63:1349-1353.

Faegri, K., Iversen, J. 1989. Textbook of pollen analysis. (4 ed.), Wiley, N.Y.

Frey-Wyssling, A., Häusermann, E. 1960. Deutung der gestaltlosen Nektarien. Ber. schweiz. Bot. Gesell. 70:150-162.

Hesse, M.,Waha, M. 1989. A new look at the acetolysis method. Pl. Syst. Evol. 163:147-152.

Howell, D. J. 1972. Physiological adaptations in the syndrome of chiropterophily with emphasis on the bat *Leptonycteris* Lydekker. Unpubl. Ph.D. dissertation, Tucson: Dept. Biol. Sci., Univ. Arizona, 217 pp.

Inouye, D. W., Favre, N. D., Lanum, J. A., Levine, D. M., Meyers, J. B., Roberts, F. C., Tsao, F. C., Wang, Y. 1980. The effects of nonsugar nectar constituents on nectar energy content. Ecology 61:992-996.

Käpylä, M. 1978. Amount and type of nectar sugar in some wild flowers in Finland. Annales Botanici Fennici 15:85-88.

Kevan, P.G., Lee, H., Shuel, R.W. 1991. Sugar ratios in nectars of varieties of canola (Brassica napus). Journal of Apicultural Research 30 :99-102.

Kronestedt-Robards, E.C., Greger, M., Robards, A.W. 1989. The nectar of the Strelitzia reginae flower. Physiologia Plantarum 77:341-346.

Kuo, J., Pate, J. S. 1985. The extrafloral nectaries of cowpea (*Vigna unguiculata* (L.) Walp.): I. Morphology, anatomy and fine structure. Planta166:15-27.

Lüttge, U., Schnepf, E. 1976. 5.3. Organic substances. In: Encyclopedia of plant physiology. Transport in plants II. B. Tissues and organs. U. Lüttge, M. G. Pitman (eds.). New York: Springer- Verlag, pp. 244-277.

McKenna, M., Thomson, J. D. 1988. A technique for sampling and measuring small amounts of floral nectar. Ecology 69:1306-1307.

Mesquida, J., Marilleau, R., Pham-Delegue, M-H., Renard, M. 1988. A study of rapeseed (*Brassica napus* L. var. *oleifera* Metzger) flower nectar secretions. Apidologie 193:307-318.

Olesen, J. M., Warncke, E. 1989. Flowering and seasonal changes in flower sex ration and frequency of flower visitors in a population of *Saxifraga hirculus*. Holarctic Ecology 12:21-30.

O'Rourke, M. K., Buchmann, S. L. 1991. Standardized pollen analytical techniques used for various bee-collected samples. Environ. Entomol. 20:508-528.

Ordway, E., Buchmann, S. L., Kuehl, R. O., Shipman, C. W. 1987. Pollen dispersal in *Cucurbita foetidissima* HBK. (Cucur-bitaceae) by bees of the genera *Apis, Peponapis* and *Xenoglossa*. J. Kansas. Entomol. Soc. 60:489-503.

Punchihewa, R. W. K. 1984. Anthophilous insects and the pollination ecology of Asclepias syriaca L. and Asclepias incarnata L. in southern Ontario. M.Sc. Thesis, University of Guelph, Guelph, Canada.

Roberts, R. B. 1979. Spectrophotometric analysis of sugars produced by plants and harvested by insects. Journal of Apicultural Research 18(3):191-195.

Schmid, R. 1976. Filament histology and anther dehiscence. Bot. J. Linnean Soc. 75:303-315.

Schmid, R. 1988. Reproductive versus extra-reproductive nectaries - historical perspective and terminological recommendations. Botanical Review 54:180-232.

Severson, D. W., Erickson, E. H. 1984. High performance liquid chromatography of carbohydrates in cucumber nectar. Journal of Apicultural Research 22:158-162.

Schmid, R., Alpert, P. H. 1977. A test of Burck's hypothesis relating anther dehiscence to nectar secretion. New Phytologist 78:487-498.

Schmidt, J. O., Buchmann, S. L. 1992. Other products of the hive. In: The hive and the honeybee, J. Graham (ed.) Hamilton, Illinois: Dadant and Sons, Inc., pp. 927-988.

Schuman, G. E., Stanley M. A., Knudsen, D. 1973. Automated total nitrogen analysis of soil and plant samples. Soil Science Society of America Proc. 37:480-481.

Shuel, R.W. 1989. Improving honey production through plant breeding. Bee World 70:36-45.

Shuel, R.W. 1992. The production of nectar and pollen. In: The hive and the honeybee. J. M. Graham (ed.), Hamilton, Illinois, USA: Dadant & Sons, pp. 401-436.

Stanley, R.G., Linskens, H. F. 1974. Pollen: biology, biochemistry, management. Berlin: Springer-Verlag, 307 pp.

Stockmarr, J. 1971. Tablets with spores used in absolute pollen analysis. Pollen et Spores. 13:615-621.

Teuber, L. R., Barnes, D. K. 1979. Breeding alfalfa for increased nectar production. In: Proceedings of the 4th International Symposium on Pollination, Maryland Agricultural Experimental Station, Special Miscellaneous Publication 1:109-116.

Teuber, L. R.,Albertsen, M.C., Barnes, D.K., Heichel, G. H. 1980. Structure of floral nectaries of alfalfa (*Medicago sativa* L.) in relation to nectar production. American Journal of Botany 67:433-439.

Vogel, S. 1969. Flowers offering fatty oil instead of nectar. XI Proc. Intl. Bot. Congr., Seattle, p. 229.

Vogel, S. 1974. Olblumen und olsammelnde Bienen. Akad. Wissenschaften u. Literature Mainz. Wiesbaden: Franz Steiner, 547 pp.

Waller, G. D. 1980. A modification of the O.A.C. pollen trap. Amer. Bee J. 120:119-121.

Wykes, G. R. 1952. The preferences of honeybees for solutions of various sugars which occur in nectar. Journal of Experimental Biology 51:294-300.

2.4 POLLINATOR BEHAVIOUR & FLOWER PHENOLOGY

The behaviour of potential pollinators on flowers is a crucial step in the transfer of pollen from the anthers to the stigma of flowers of different plants, of different flowers of the same plant, or within the same flower (see Sections 1.2, 2.3). Many writers refer to flower visitors as pollinators without knowing whether or not the visitors actually *effect* pollination. There are numerous examples of this sort of error in the literature on crop pollination.

The behaviour of flower visitors can be broken down into activities at different scales. At the largest scale is the movement of flower visitors between plants on the landscape or between areas under crop production. On a smaller scale is the movement of flower visitors between plants in patches or within areas under crop production. On yet a smaller scale is the movement of flower visitors within a particular plant. The actual dimensions of these scales are variable, depending on the spacing of areas, on the spacing of the plants, and on the size of plants. For the latter two, small herbaceous plants contrast with large orchard trees. On the finest scale is the movement of flower visitors within flowers.

The most effective pollinators are those moving between plants and flowers in such a way as to transfer pollen to satisfy the breeding system of the plant. From the viewpoint of agriculture, and the production of most crops requiring pollination by insects worldwide, honeybees are the pollinators of choice. They are well understood biologically and they can be managed for crop pollination.

Thus, honeybee behaviour in relation to crop pollinaton will be discussed before other pollinators are introduced and the roles of floral attractants, and morphology in pollinator behaviour considered.

2.4.1 Dance language

All species of honeybees have well developed dance language of communication through which they impart to their nest (hive) mates information about the distance and direction to various resources, usually floral nectar and pollen. When that information is exchanged in the colony, the richness of the resource is also conveyed along with chemical information about it (e.g. floral scents carried on the bees' bodies).

In general the dance-language consists of honeybee foragers returning to their colonies and performing a dance on the wax comb of the nest. If resources are beyond a particular distance from the colony, the dance comprises two loops, to left and to right, with a central run during which the bee wags its abdomen (see Figure). The directional component of the dance is indicated by orientation of the central run with respect to the direction to the sun, and the distance indicated by the tempo of the dance. In the darkness of the hive for *Apis mellifera*, *A. cerana* and *A. koschevnikovi*, the orientation of the dance is such that vertically upwards represents the position of the sun. Thus, if the forager dances its central run vertically, recruits to foraging interpret this as an instruction to go to the resource by flying in the horizontal direction of the sun. If the dancer orients its central run 60° to the left of verically upwards, the recruits fly 60° to the left of the horizontal direction to the sun. The distance component of the instructions is given by the tempo of the dance, the closer the faster. If resources are nearby, the wagging dance is not used; a simple round dance conveys the message that resources are closeby. Details of the dance-language can be found in most beekeeping and bee biology books.

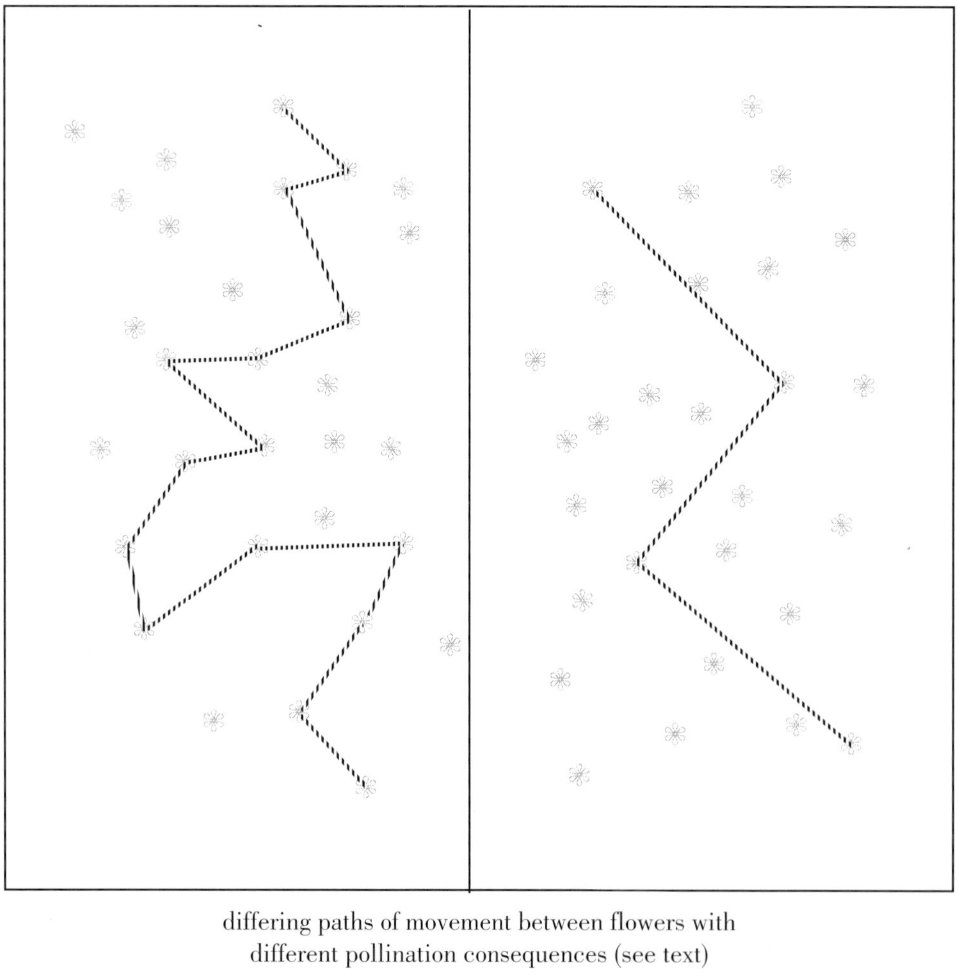

differing paths of movement between flowers with different pollination consequences (see text)

Apis mellifera performing the waggle dance for nestmates

The dance-language of honeybees varies from species to species, and within species it varies between races. It is important to realize that knowledge of the dance-language has great practical importance in understanding the general foraging ranges one can expect from a particular species and race of honeybee. Further, the information is important to apiary planning in terms of the number of hives that can be supported for honey production or to remain strong in pollination activity on crops. Clearly, if the foraging range is small, large numbers of honeybees would deplete the resources available and soon be competing strongly with each other. This situation detracts from honey production, and can cause competing colonies to decline in population as the total population of honeybees exceeds the carrying capacity (i.e. the number of honeybees that the area can support) of the region within their foraging range.

The tempos of the dance language vary among species and races of honeybees. The differences can be seen to relate roughly to the sizes of the bees; the smaller the bee, the faster the change in dance tempo with distance and the shorter the distance at which the wagging dance becomes operational. The general distances over which the different species and races of honeybee normally forage has been studied in many crop situations but reveals wide variability in both natural environments and agricultural areas. Honeybees may forage less than a km or up to several km from their nests, or more. Circles of those diameters (e.g. 1 km, 7 km, 10 km) would enclose their foraging ranges, which can be enormous, exceeding a few hundred square kilometers (see Roubik, in Spivak et al., 1991). Knowledge of the production of nectar or pollen or both in that area would provide an estimate of the density of honeybees that could be supported in the habitat (*not* necessarily pertaining to a specific apiary). For crop pollination, the foraging range dictates the extent to which the honeybees could be expected to be effective.

Given that most cropping systems use areas that are smaller than the foraging areas of *Apis mellifera* and *A. cerana*, the issue is not that serious. However, overcrowding of crops with bees can quickly cause a decline in the pollinator force as they move off to other forage or even debilitate each other's colonies by depleting resources.

Assuming that the density of pollinators on a crop is appropriate to the needs of the crop and to the needs of the pollinators, the adequate distribution of the pollinators is important. For *A. mellifera* it has been found that pollinating foragers spread out widely over a given crop. Although one might expect more activity closer to the hives than at distance, one often finds more or less even dispersion within an orchard or field. The foragers are more dense near the hive, but the areas covered are far greater at increasing distance from the hive. Thus, from the results of field experiments on oilseed rape and in orchards, it seems that it is not necessary to place colonies of pollinating honeybees in small groups (about 4) throughout the crop. Larger groupings of hives seem to have the same effect and ease the job of the beekeepers providing pollination services.

2.4.2 Patterns of Movement Between Plants

The subject of optimal foraging, by which animals expend as little energy as they can to obtain resources (food) which return to them the greatest amount of energy in a given period of time, has been applied to pollinator foraging. Most research has been done using bumblebees, but the same patterns have been noted for honeybees.

In general, bees foraging at a patch of flowers rich in rewards (nectar or pollen) visit many plants in that patch, skipping over few and visiting near neighbours. They also tend to switch direction between one plant and the next so that their path of progress is a tight zigzag, often crossing itself. Nevertheless, the overall progress is generally forward. If the patch of flowers is poor in rewards, bees tend to skip over plants as they sample, and progress in a shallower zigzag path. The result of this behavioural pattern is shown in the first Figure, where one can see that a pollinator on a rich patch will accomplish more floral visits per unit area more quickly, and presumably effect more pollination, than one on a poor patch. If the patch is poorer, the mobile pollinator tends to move relatively directly and quickly through it.

The clues a forager uses to decide to change patches are not fully understood. The relative richness of the rewards in various patches are presumably compared in some way so that the pollinating forager can track the sources of the richest returns. Some experiments on honeybees strongly suggest that individuals do not do that — They remain constant to relatively poorly rewarding sources or to relatively richly rewarding sources and do not switch as long as some reward, at least, is available. This sort of behaviour may have to do with the sociality of honeybees and the overall foraging behaviour of the foraging population of a colony to maximize resource accumulation. It does not seem to apply to other bees or pollinators.

in inflorescences, often with their openings or petal access platforms pitched slightly downwards. Thus approaching from below is easier. Further, upward flight and crawling in combination is more easily controlled than

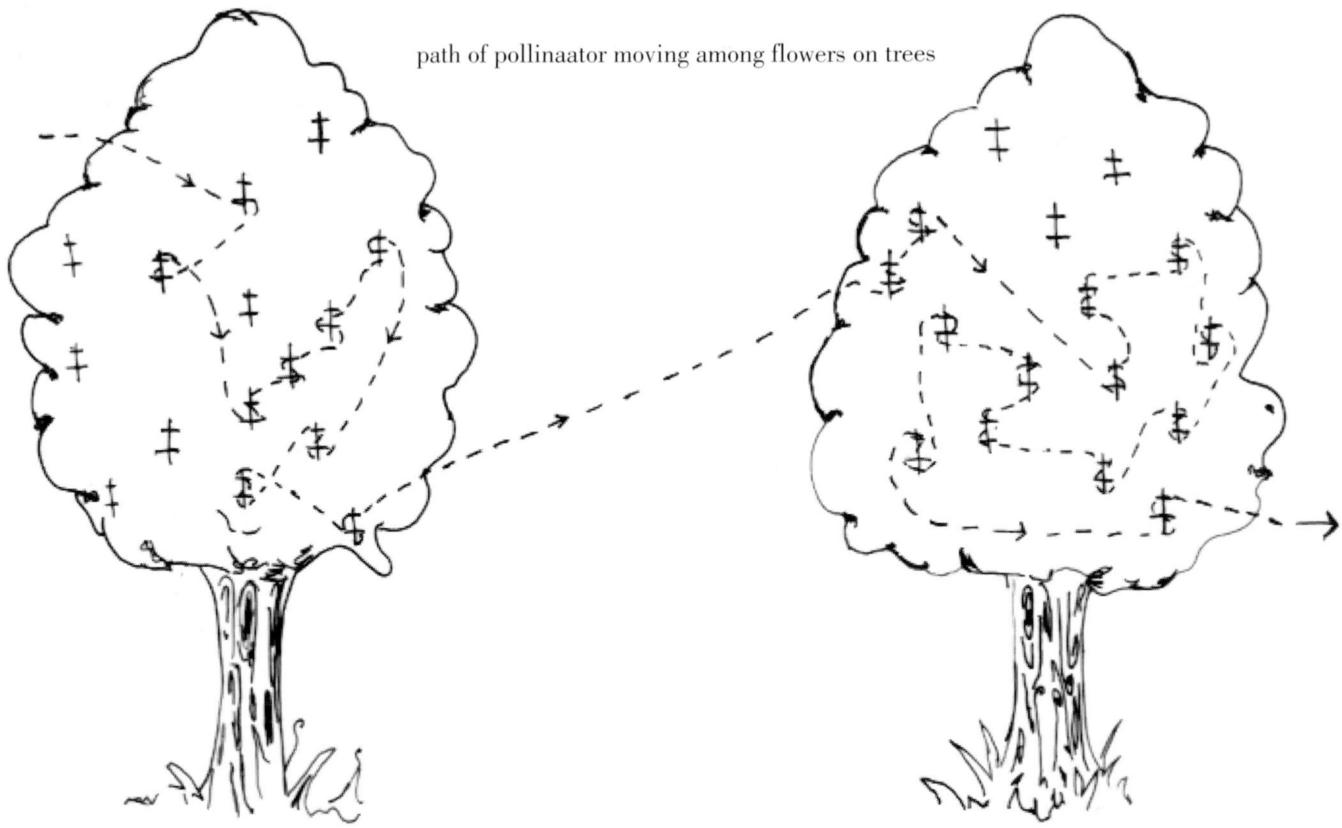

downward motion by either, and crawling is much more energy- conserving that highly expensive flight. Often the volume of nectar in lower flowers is higher than in upper ones, but is a less concentrated solution of sugars. Taste satiation by going from relatively insipid to relatively strong would not happen and allow the forager to recognize the quality of reward as it progressed. From the plants' standpoint, it is important to note that many plants which produce inflorescences do so with flowers that are first in male phase, producing pollen, and then in female phase with receptive stigmas. Thus, as the youngest flowers are at the tops of inflorescences, pollen is removed from male-phase flowers of a given plant after the pollinator has already visited the female-phase flowers. Self-pollination is, thereby, discouraged.

2.4.3 Movements on Individual Plants

Tree crops, which are particularly important in the tropics, when in bloom, present to pollinators vast arrays of flowers and copious quantities of resources. The problems of pollination for blooming trees are poorly understood. In general, on a given tree, the flower-visiting insects tend to work their way down the tree over the period of a day or half-day. The reasons for that presumably have to do with the sequence of the opening of flowers and the way the rewards are distributed in the flowers over the crown of the tree. Individual bees generally enter a flowering tree at a higher level than they leave. This pattern has been attributed to the energetics of flight and carrying the foraging load — it being more conserving of energy to start 'high and light' and to descend while becoming loaded with nectar or pollen. Movements onto the next tree have not been studied. However, recent studies in Malaysia (Roubik, unpublished) show that *Apis koschevnikovi* and *A. dorsata* readily move between large trees, staying at roughly the same level above the ground.

The pattern of movement of bees, moths, and birds on smaller plants with inflorescences is generally similar. The forager ascends the plant, visiting the flowers as it does so. Several reasons have been given for this behaviour and concern the presentation of the flowers, the amount of resource within the flowers, and the energetics of flight and crawling. The flowers are usually fairly close together,

2.4.4 Behaviour on Flowers

It is crucial that pollinators work at flowers in such a way as to cause the transfer of pollen. If they fail to do that, they may not be pollinators at all, or they may be inefficient to greater or lesser extents. The least efficient flower visitors as pollinators are nectar or pollen robbers. These visitors cause damage to the flowers they visit, remove nectar or pollen, and cause no pollination themselves. They have been shown to discourage visits by the legitimate pollinators, either by attacking them, or by making flowers less rewarding. Some carpenter bees and bumblebees are notorious for making punctures in

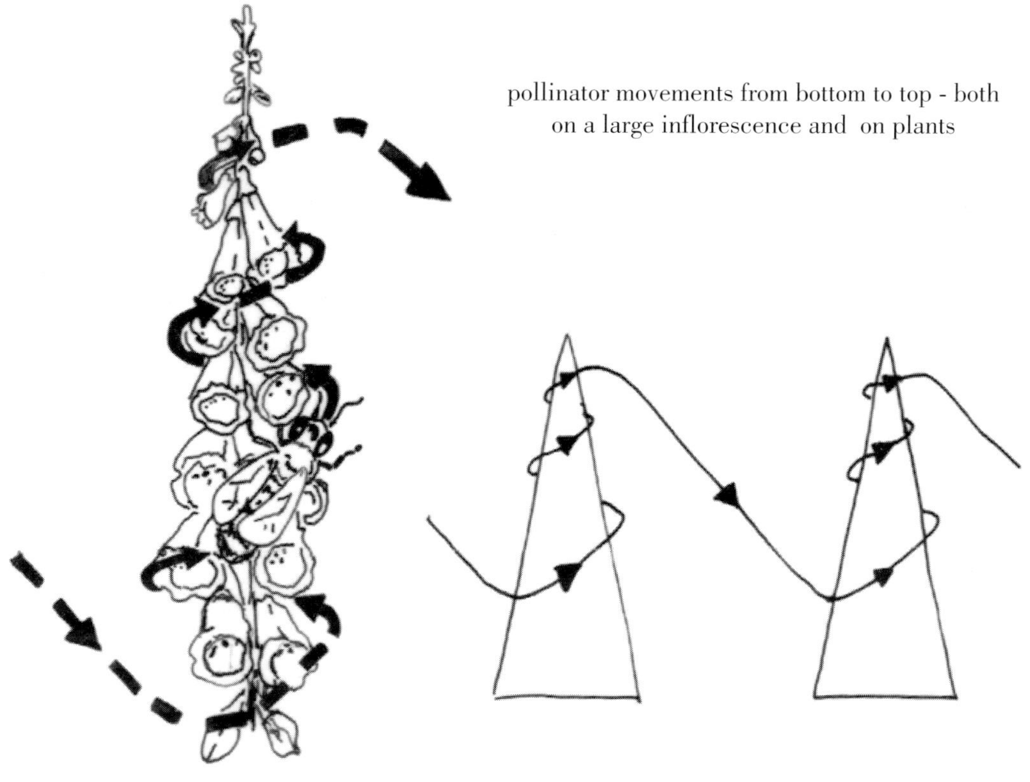

flowers in which the rewards are hidden within, such as in nectar the tubes of flowers of the mint family (Lamiaceae) and legumes (Leguminosae) or hidden pollen in the blueberry family (Ericaceae) or tomato family (Solanaceae), require special skills for the pollinator. Researchers have found that pollinators (e.g. bumblebees, honeybees, and leaf cutter bees) must invest some time in learning how to manipulate such complex flowers to become speedy and efficient in extracting the rewards they seek, and, coincidentally, causing pollination.

the bases of flowers and removing nectar that way. Several Old-World and New-World *Trigona* are also nectar and pollen robbers, taking many minutes to chew holes in anthers or corollas. Such bees never even enter the flower to touch the stigmata or anthers.

Other visitors may just be thieves. They do no physical damage to the flower, but remove pollen or nectar without bringing about pollination. Small bees, particularly *Trigona* and small Halictidae in the tropics, visit a wide diversity of flowers, removing pollen from the anthers or nectar from the nectaries of large flowers without touching the stigmata. Even honeybees can be nectar thieves. They have been recorded often as visitors to flowers of passion fruit, from which they remove nectar, but, because of the large size of the flowers they very rarely touch the anthers nor the stigmata. Honeybees also are known for "side-working" flowers by which they remove nectar by inserting their probo-scides between floral parts from the side. This is a common problem with some varieties of apple, some legumes, and cole crops.

Such 'floral larceny' detracts from the activities of legitimate pollinators by depleting the resources that they are seeking as they visit flowers and touch both anthers and stigmata. Careful observation of flower visiting behaviour can allow discrimination between effective pollination and mere floral visitation. Unfortunately, there are many published accounts about "pollination" that do not attend to this important detail.

Many crop plants have relatively simple flowers from the standpoint of a pollinator's being able to obtain the reward it seeks. Open, bowl shaped flowers, and the complex inflorescences of the sunflower (Asteraceae) and carrot (Apiaceae) families pose little challenge to flower visitors extracting the resources they seek. However, the complex

Watching naive bees fumble with complex flowers during their first few encounters with them can be quite amusing. It is worth keeping in mind that the more complex flowers usually produce more reward of a higher quality, obtainable only by a restricted guild of pollinators. The value of good education is suggested to humans from the study of pollination.

2.4.5 *Memory & Constancy*

As noted above, bees and other pollinators in general are capable of learning to recognize various cues and to perform various tasks. Although the dance language is of prime importance in orientation and the transfer of information among honeybees, foragers are able to memorize the landscape over which they forage. They develop what are called cognitive maps which enable them to orientate themselves independently of, and to supplement information from, the position of the sun. These cognitive maps use visual and olfactory landmarks. Other pollinators are also know to use their familiarity of landscape in their home ranges for orientation.

The advantage of such behaviour to accuracy in homing of bees, or other pollinators, as they return to their nests or roosts is plain, but it also works on outbound foraging flights. This results in individuals having a tendency to return to the same foraging sites, especially if the resources there are plentiful. The phenomenon of site constancy can detract from pollination efficiency, especially if the site happens to be a single, large tree that requires cross pollination.

The importance of floral constancy by pollinators is clear from the plants' viewpoint: the pollinator visits the flowers of only that species and so is more efficient at transferring

female *Xylocopa*, robbing the flower of *Centrosema*

pollen. From the standpoint of the pollinator, familiarity with the flowers of a given species, and how to most efficiently obtain the resources sought, is enhanced by learning (as noted above) and by floral constancy.

2.4.6 Floral Phenology

Floral phenology refers to the blooming sequence of flowers and their development in time. Both concepts are important to pollination and its management.

In any given location, plants flower in more or less the same sequence from year to year. How this sequence is governed by nature is not fully understood. Some plants bloom with the stimulus of lengthening days (as in the early part of the year in the northern hemisphere) or declining day length (as in the latter part of the year in the northern hemisphere). Others are unaffected by day length and bloom once the plant reaches a certain size and are daylength neutral.

Also affecting the rate at which plants come into bloom is the amount of heat that has accumulated during the growing season. This is often measured in 'Growing Degree Days' above a certain threshold temperature. The details of how to accomplish these measurements and how to establish the threshold temperatures are well known, but beyond the scope of pollination per se. For some crops, such as maize, 'Maize Heat Units' (which are GGDs) are well known for many varieties and useful for predicting flowering and ear ripening dates. The flowering sequence of apple varieties can also be predicted by GGDs and the sequence, asynchronicity and synchronicity of bloom between varieties is important in considerations of orchard planning for interplantings of pollinizer varieties for the main crop. One variety of pollinizer potentially may be able to fertilize another variety, but if they bloom at different times separated by a little as a week or so, cross-pollination by insects cannot occur. In many parts of the world, especially the seasonal tropics (wet and dry seasons), presence or absence of rainfall may override any effects of day length or temperatures, or both. Thus, growers and pollination managers must be able to "read" the seasons and keep in close contact to coordinate their activities to ensure pollination is at its maximum.

Beekeepers and pollinator managers must be aware that their wards require food at times other than when the crop to be pollinated is in bloom. Thus, knowledge of flowering phenology in a region is important. Good pollen and nectar availability stimulates active foraging and brood rearing. Thus, to provide a strong pollinator force for a crop, pollinator managers should keep their animals in resource-rich areas for some time before moving them to the crop. Further, if there is a period of dearth (diminished availability of resources) managers should know when that occurs and its duration. They should thereby know how much supplemental feeding, or migratory practices, are required. Pollinator managers may also be able to provide services to a sequence of different crops. Clearly they need to know when each blooms, for how long, and what types (nectar or pollen or both), quality, and quantity of resources the crop offers to the pollinators.

female *Xylocopa* visiting the flower of *Canavalia*

Once pollination services are complete, honeybee colonies may be weakened. Often the aim of providing pollinators is to saturate the blooming crop to assure maximum pollination. This creates intense competition amongst the pollinators so that their populations exceed the carrying capacity and so weaken. A pollinator population, such as of honeybees, so weakened should be moved to a resource-rich environment for recovery. One of the most serious cases requiring this sort of management is in pollination in greenhouses.

The situations described above vary from place to place and crop to crop. Thus, it is difficult to arrive at any but the most general of suggestions. Nevertheless, the value of the ***"floral calender"*** can be appreciated. The floral calender is just a listing of which flowers bloom when and for how long, coupled (for the sake of practicality) with their abundance and value to pollinators. Floral calenders must be prepared specifically for particular regions for what may apply in one place may not apply in another, even sometimes quite nearby.

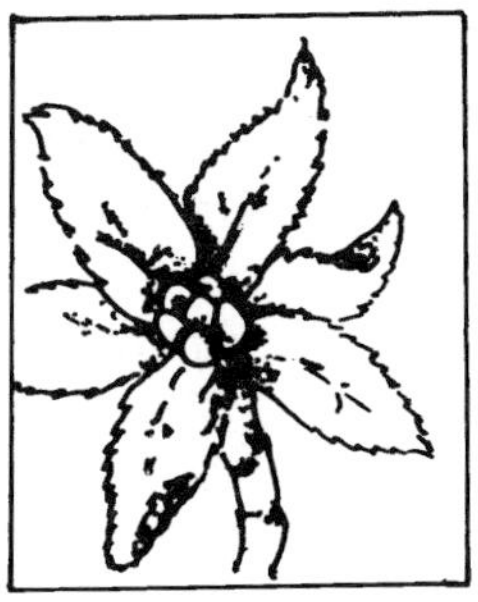
young bud

bud

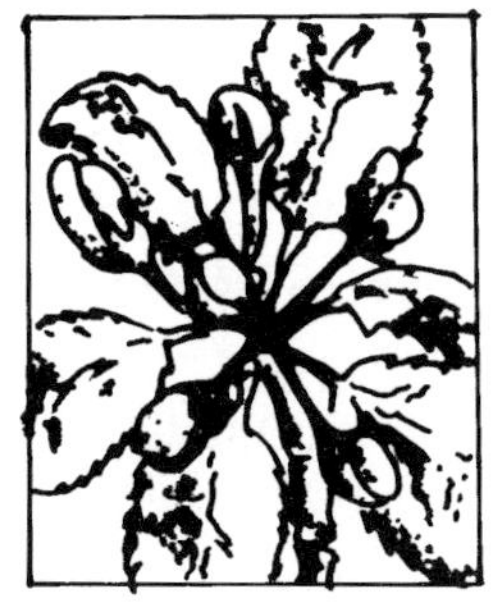
full-pink bud

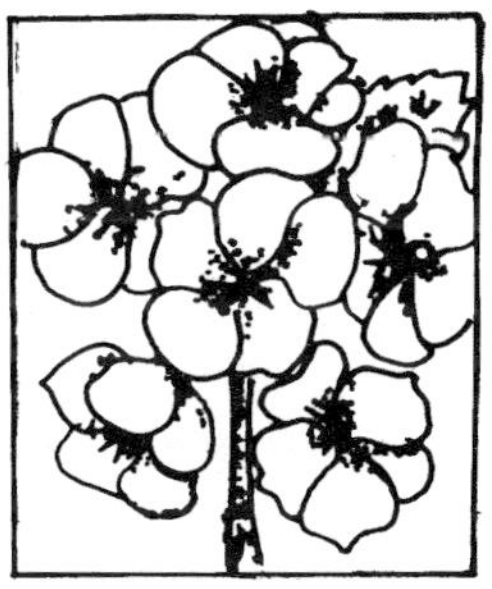
in flower

after petal fall

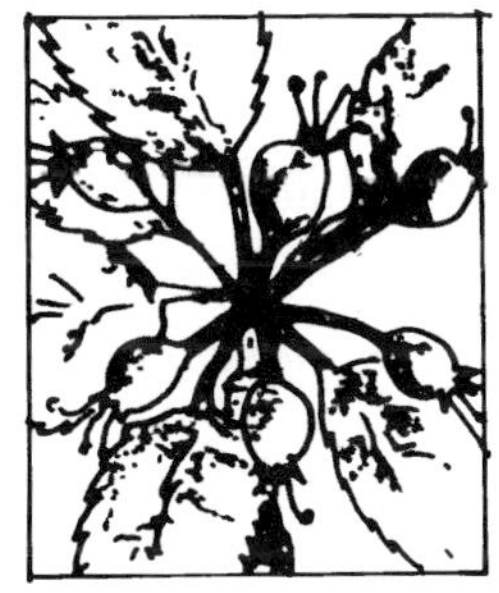
young fruit

At a finer level is the developmental sequence of flowering on individual plants. Some plants are ***"indeterminate bloomers"*** but others are ***"determinate"***. Indeterminate means that these plants continue to produce flowers over rather ill-defined periods. Some continue to produce flowers especially if pollination has been lacking. Most insect-pollinated annual crops (except cereals) fit this category. For such plants the timing for pollination to take place is not so critical because the plants, more or less, compensate for unpollinated flowers by producing more. In the tropics, some perennial crops are indeterminate bloomers. For these, the pollination season is long. Determinate bloomers are best illustrated by perennial crops, especially fruits. Pome and stone fruits are good examples. They have a fixed number of flowers which will open over a relatively short period. Thus, the period for pollination is short and critical to getting good yields. With such crops, grower and pollinator-servicing efforts must be closely coordinated for the best results.

In the wet tropics, determinate bloomers may produce inflorescences that are determinate, but the plant may continue to bloom almost or completely year-round. Oil palm is an excellent example (2.5.2).

At a still finer level is the phenology within individual flowers. It is important to understand the stages in flower maturation, especially for pesticide applications. For example, on pome fruits, insecticides and miticides for control of leaf rollers, leaf miners, fruitworm, red mites are applied up until the first flowers come into "full pink" bud (see Figure). Once the flowers are open, the risk of poisoning pollinators is great. Even after "petal fall" or "calyx" stage, when spraying against other insect and mite pests is recommended, pollinator populations (honeybees or other removable bees) should be taken away. The preceding presents the stages that have been adopted for pome fruits. Some hermaphroditic flowers shed pollen before the stigmata are receptive. These are termed "protandrous". Others have mature stigmata before the anthers dehisce. These are termed "protogynous". Others have the differences in maturation so slightly separated, or mature stigmata and anthers simultaneously that they are termed "homogamous". Most crop plants produce homogamous flowers. Protogyny is quite uncommon. The importance of understanding these fine differences lies in also understanding the activities of the pollinators. Pollen-collecting bees learn quickly to avoid flowers lacking in pollen, i.e. flowers in the female stage. Fortunately, for crop pollination, dichogamy (protandry or protogyny) is generally weak or absent in domesticated plants. However, much remains to be learned about the floral biology of tropical, perennial crops.

2.5 PREPARING POLLINATORS FOR USE IN TROPICAL AGRICULTURE: CASE STUDIES

Insect pollinators have been recognized as critical to pollination of agricultural crops in the tropics. Choosing an appropriate pollinator among myriad flower visitors, and for mass-rearing, are important for success of managed pollination (see Sections 1.3 to 1.5, 2.1,). The subject must be taken beyond the European honeybee — rearing — pollination paradigm in order to select appropriate pollinators for diverse tropical plants. Honeybees are not a solution to every pollination problem, as demonstrated by oil palm, passion fruit, durian, and other widely-cultivated tropical crops discussed here as examples.

In agriculture where assisted pollination is required, an understanding of pollination efficacy is closely aligned with understanding the flowering tempo of crops. The existence of a ***dual flowering tempo*** among tropical crops, and contrasting life cycles among pollinators (bats, bees, flies and others) demands different management strategies. Management of honeybees and carpenter bees, for example, follows a pattern shown by different flowering tempos of their tropical resources. Timing is always crucial for the stocking of abundant pollinators to provide pollination services. Although we do not understand all the mechanisms that may operate, pollinators likely time their life cycles to coincide with the availability of floral resources, which in turn are usually regulated by moisture, temperature and daylength.

The two most distinctive categories in flowering can easily be recognized:

Type I — ***Periodic flowering***, or the 'boom-and-bust tempo' **(1 to 2 cycles/year)** — seen in *mango, litchi, rambutan, durian* and *rubber*

Type II — ***Sustained flowering***, or the 'hum-drum tempo' **(6 to 7 cycles/year)** — seen in *oil palm, cocoa, coconut, starfruit, watermelon, passion fruit* and *soursop*

These two flowering 'types' correspond to different pollinator management schemes:

Type A — ***migratory management***, or the movement of colonies or pollinators into crop fields as they are needed

Type B — ***stationary management***, or the year-round maintenance of pollinators in the field

2.5.1 Varied Pollinators for Southeast Asian Crops

For crops that bloom continuously, about every two months, the use of hived colonies of *Apis cerana* is recommended for establishment in the crop area. In contrast, the management of bee colonies for crops that flower periodically should rely on *A. mellifera*; for which migratory beekeeping is more suitable. Between the two extremes there are other flowering tempos, and pollinator types that are adapted to them. The local flora clearly will affect whether colonies or other pollinators can be sustained (see Sections 1.6 and 2.1).

The three examples chosen of bees managed for pollination, involve one of the Asiatic honeybees, *Apis cerana*, a large carpenter bee *Xylocopa (Platy-nopoda) latipes*, and stingless bees of the genus *Trigona*. There are more than a hundred bees of these genera in southeast Asia (see Introduction), and many may prove adaptable to similar management techniques. However, the other species of Asian honeybees, *Apis koschevnikovi, Apis dorsata, Apis andre-niformis, Apis florea* and giant honeybees of the Philippines, Sulawesi and the Himalayas, though recognized as significant to some fruit crops, are still not easily manipulated for pollination services.

1) *Apis cerana* for pollination of coconut and starfruit.

- ***Appropriate equipment***

The major principle for selecting appropriate materials

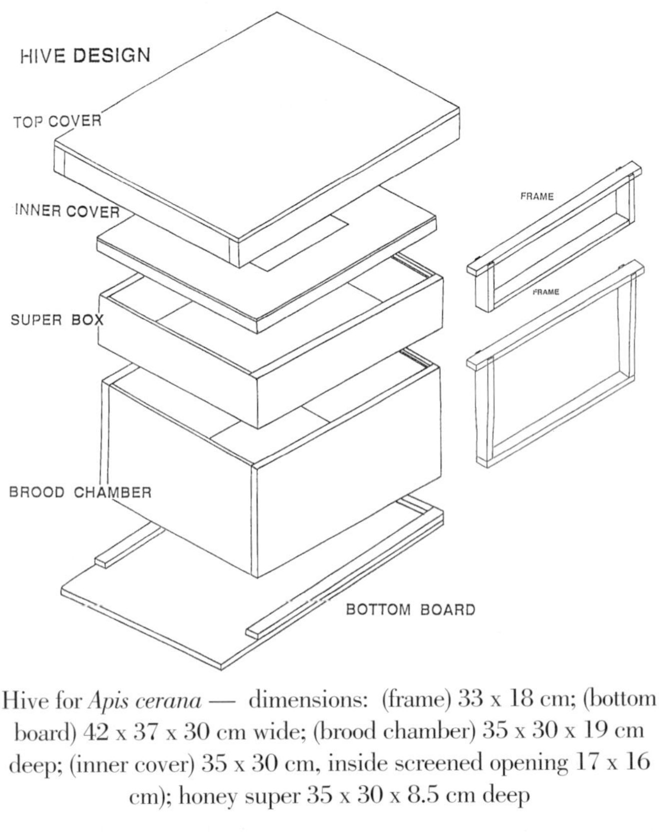

Hive for *Apis cerana* — dimensions: (frame) 33 x 18 cm; (bottom board) 42 x 37 x 30 cm wide; (brood chamber) 35 x 30 x 19 cm deep; (inner cover) 35 x 30 cm, inside screened opening 17 x 16 cm); honey super 35 x 30 x 8.5 cm deep

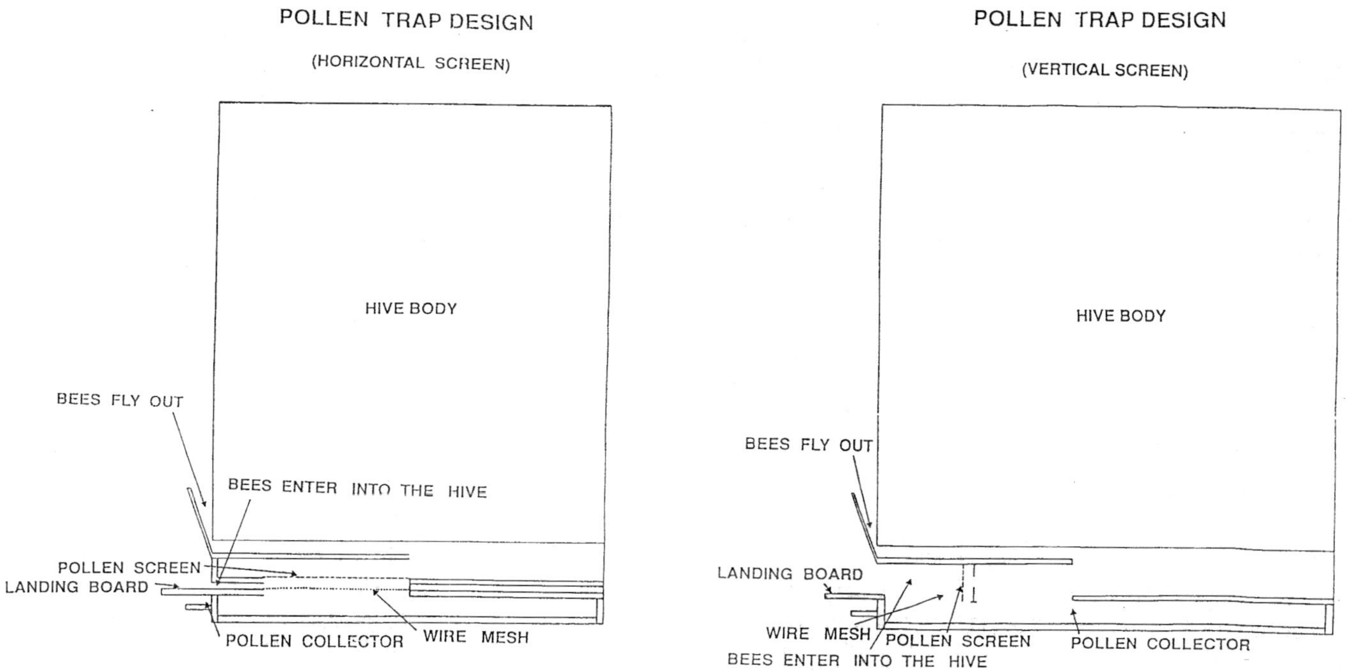

two pollen traps for the hive base: one with horizontal or vertical pollen trap screens

to maintain *A. cerana* is to incorporate the salient biological aspects of the species and environmental conditions in hive design and other equipment. By that measure, adoption and modification of the *A. mellifera* hive design for *A. cerana* must of course consider the smaller bee and colony size of *A. cerana*. In addition, the takeoff and landing behaviour of *A. cerana* at the hive, humid weather conditions or heavy rains, and protection from predators at the hive entrance all must be considered.

Many of the successful traditional southeast Asian hives "*gelodogs*" from hollow tree trunks, and the modified Langstroth hive design (see Figure) incorporate such features. Traditional hives may be suitable for permanent apiaries. The Taiwanese hive design — with wire mesh windows for ventilation, and adjustable entrance gaps — was developed for migratory beekeeping and the confinement of colonies during chemical spraying. Superior ventilation of the hive is given by the removable wire mesh which is critical for the bees during transportation under hot and humid conditions.

In all the above hive designs, control of hive humidity and predator access at the nest entrance can be achieved by constructing or adjusting nest entrances to provide an unexposed landing board. This both reduces exposure to torrential rains and also reduces detection by natural enemies, like wasps, toads and birds. The landing board should not project beyond the hive platform. Thus bees are able to fly rapidly once outside the hive. Small hive entrances will assist the bees in regulating temperature, humidity and the evaporation of water from nectar to make honey in the hive, especially under hot and humid conditions. The choice of a relatively small volume of the hive, and small frames and honey 'super', allow for the intermittent and small volume of floral resources harvested by bees from the crop plants having a sustained flowering tempo. A pollen trap, which sequesters some of the colony's incoming pollen, can also be designed such that the landing board is unexposed. The trap is placed beneath the hive, like the well known 'O. A. C. pollen collector' (see Section 2.3.2), rather than exposed on the landing board. Pollen collection by the colony is stimulated, when some of the normal pollen income is removed.

- ***Colony preparation for migratory beekeeping***

Colonies selected for moving are to be prepared in advance and a schedule should be followed. In general, the management procedure is similar to the management of *A. mellifera*, i.e. boosting colony population, re-queening if necessary, and feeding.

- ***Stocking density***

Numbers of colonies needed to service crops like watermelon, coconut and starfruit varies from 4-8 colonies/hectare. Although no proper studies have been made, this suggested stocking density has been widely used in southeast Asia. The distribution of colonies in the crop area is influenced by the availability of shade, water sources, resources that can be reached by the bees in a foraging radius of about 400 meters, protection from natural enemies, and public safety.

- ***Vehicle***

For a short distance, the practical mode of transporting colonies through narrow and difficult access roads is a three-wheel motorbike rather than a truck or lorry. The small vehicle is easily maneuvered. This method is widely used in the coconut-growing areas of peninsular Malaysia and Vietnam. Small trucks (2-ton) can also be used for longer distances if good roads are available.

- ***Agreement***

Making an agreement for pollination services requires that

Modified Langstroth hive containing colony of *Apis cerana*, and hives placved in coconut palm plantation

the beekeeper and grower or plantation owner sign a document (see sample contracts in **Appendix II**). For large plantation owners, small-scale beekeepers can best work through beekeeping cooperatives to provide the required number of bee colonies for pollination service. Agricultural institutions can provide organizational skills and extension services. Beekeeping cooperatives can negotiate for a collective bargain, for the interest of semi-literate beekeepers and act on their behalf in drawing up a pollination contract). This is seen as an essential step in facilitating migratory beekeeping and pollination work (see Section 2.1).

- ***Management team***

The management team of the beekeeping cooperative will initially conduct an evaluation on the condition of bee colonies (see **Appendix II**), in terms of colony size and health. They may also set standards for management and a work schedule. ***Standard hives***, in this context, are a great aid to management efforts. Beekeepers can either participate by renting out their colonies, selling them, or invest fully by collecting the dividends after the season. A management team, consisting of few good beekeepers from the cooperative, can be responsible for the maintenance and transportation of colonies. This form of management helps to avoid confusion that may arise from involvement of all participating beekeepers. Work schedules, such as feeding, boosting colony strength, colony division, re-queening, etc. must be left to the discretion of the management team.

- ***Selecting the new location***

In selecting a new location for the bee colonies, sources of running water (e.g. streams), dry ground conditions, exposure to natural enemies (birds, toads, wasps, and weaver ants are the usual nuisances to the bees), shade, accessibility and convenience — away from the public and possible vandalism — traffic, distribution of potential foraging areas and many other factors must be taken into consideration. The location must be accessible, at least by road or footpaths, to be reached by vehicles. Placing colonies under camouflage or plant leaves near the entrance of hives will afford protection from birds and wasps. In addition, shade from bushes and tall trees will keep colonies from experiencing too much heating. However, too much shade will make the area unsuitably damp after rains.

- ***Moving colonies***

Transport is usually done at night for several reasons — such as eliminating the problem of overheating during travel by day, and to ensure that colonies are confined within their hives. The hives are closed immediately following sunset, when all the bees have returned. Timing of colony transport is critical and depends on the crop (see Sections 1.2, 2.1). The Taiwanese hive design has a collapsible entrance — a board used either to close the entrance or the window. When moving colonies the entrance is closed and the wire-mesh screen window allows ventilation on both sides of the hive. Upon arrival the entrance can be opened and the window is closed. Confinement of bees in this type of hive is also easy to achieve at times when pesticides are applied — especially useful for hives that are kept in starfruit fields.

When colonies are stacked on top of each other during moving, sufficient ventilation must be provided between adjacent colonies. Shaking during transportation can cause combs in the hive to break and therefore the combs should have complete frame and wire cross-support (Section 2.1) and care must be taken to minimize vibration during moving. Water is sprayed occasionally on the hives to prevent overheating. The floor of the truck, which vibrates and is a source of heat, must be properly ventilated from the front of the vehicle, and preferably insulated. An electric fan may be used.

- ***Managing the colonies***

Upon arrival, the hives are unloaded immediately into the new apiary site and, if necessary, colonies are fed with sugar-syrup to reduce the chance of absconding. While feeding with sugar syrup or granulated sugar, comb conditions may be inspected, to remove or repair damaged combs shaken out of frames during transport. Utmost care and attention must be given to ants near the hive-support or on the ground. The ants might attack the colony during the first day. The colony will usually be able to take care of ants after being settled more than a day.

Weeds are omnipresent in tropical orchards and herbicide application or cutting should be considered. At times, strip-cutting of flowering weeds is necessary to induce visitation by bees at crop flowers, by eliminating competing blooms. Management of weeds can be done by cutting before introduction of the bee colonies.

2) Carpenter bees (Xylocopa) for pollination of passion fruit.

Pollination of passion flowers by carpenter bees is a classical example of a crop plant that has a consistent relationship with its insect pollinators, primarily the carpenter bees. In southern Asia the primary pollinating *Xylocopa* is often *X. latipes* (subgenus *Platynopoda*). Passion fruit has a largely self-incompatible breeding system, although fruit production is also limited by factors other than pollination. Despite the high fidelity between the pollinator and plant, interestingly, carpenter bees do not collect the flower's pollen, but only nectar. The same is true for most other visitors. Pollen sources are sought from other flowers. The pollen grains are deposited on the smooth and shiny dorsum of the carpenter bees while the bees place their heads between the stamens and the floral base, where nectar is presented in a trough (see Section 1.4).

From a practical standpoint, the abundance of carpenter bees must be augmented to meet the pollination demand (unless self-compatible cultivars are developed). Otherwise, commercial planting of passion fruit requires manual or hand-assisted pollination (using three fingers), to pollinate the flowers of the self-incompatible passion flowers — as practiced in passion fruit plantations in Bahia, Brazil.

Xylocopa (carpenter bee) nest boxes; dimensions: 32 x 28 x 18 cm deep

Carpenter Bee Nest Hives

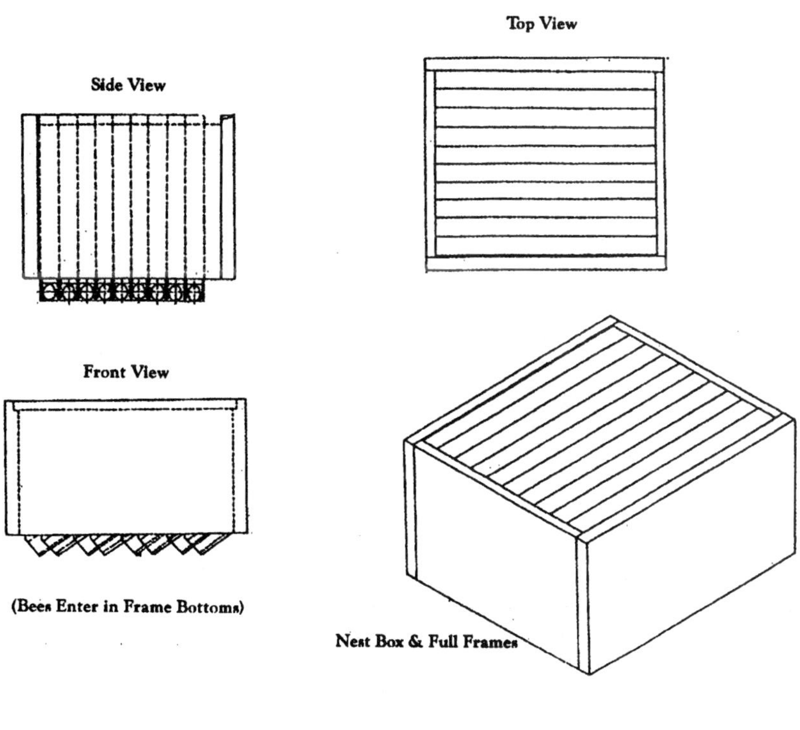

The passion fruit usually requires a large bee for pollination. Because passion fruit blooms throughout the year, it benefits from the continued presence of carpenter bees, which are largely seasonal. Certain management techniques can help to create an adequate environmental niche for *Xylocopa*:

- ***Location of the nesting site***

In choosing a nesting site of carpenter bees one must consider: ***a)*** shelter from rain, ***b)*** trellis orientation, ***c)*** availability of pollen sources, and ***d)*** mating requirements. Since carpenter bees only build their nest in dry woods, protection from rain must be assured by providing a roof, such as attap (palm branches), but not metallic materials that become heated during hot days. The shelter should be walled to reduce heat radiation, but must not impede the bees in their flight orientation. If walls are needed to protect nests from the drifting moisture of heavy downpours, spaces for light and flight pathways for the bees should be considered.

Cross-pollination in passion fruit can be encouraged by placing the nests in the ***middle of the passion flower vines' trellis***, rather than at the beginning. When carpenter bees forage away from the nesting site, moving from one trellis to the next accomplishes outcrossing. Foraging from one flower to the next on the same trellis will produce selfing.

Availability of pollen sources in the morning hours is crucial for the bees. Passion flowers in southeast Asia offer their nectar after midday. Bees need a pollen source that is available at a different time. The flowers are needed in the vicinity of the nesting site. *Thunbergia*, *Jacaranda* (Bignoniaceae) *Pelto-phorum*, *Glyricidia* (Leguminosae) and many others can be planted as pollen sources for the carpenter bees.

Mating in carpenter bees takes place near the nesting site in an open space. To accommodate perching behaviour of the males, and provide a flight path for females, an open area should be available near the bee nests. Males need perch posts to wait for females. Perches can be provided by erecting poles around the nest sites. During copulation (in the air) the flight taken by the couples is towards the sun; an open space near the nesting shelter will help the carpenter bees during mating flight.

- ***Establishing nests***

The recommended ***hive design*** for rearing carpenter bees is similar to the multiple-comb hive of the honeybee because the "bee space" concept of Langstroth is incorporated. The bee space for *Xylocopa latipes* is 2.1 cm

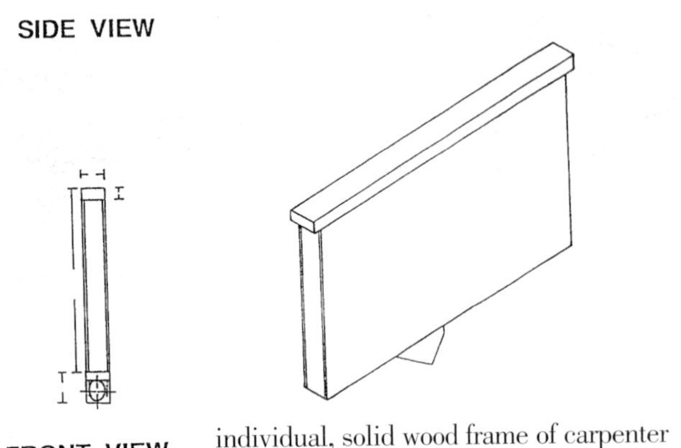

individual, solid wood frame of carpenter bee nest box (see text)

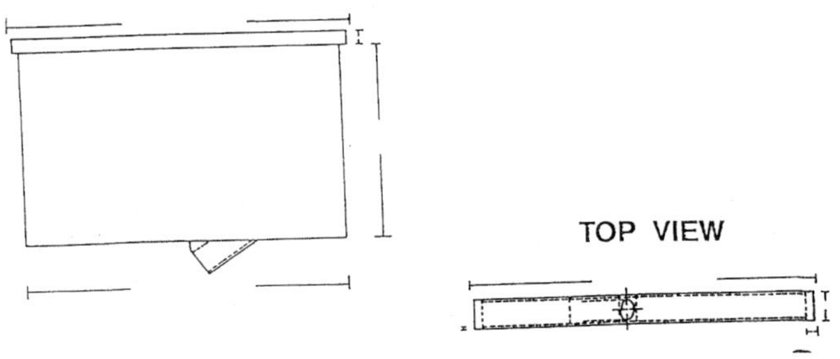

individual frame for carpenter bee nest hive; dimensions: 32 x 18 x 17 cm; note entrance area protrudes 3 cm beyond hive and frame, and HIVES ARE HUNG VERTICALLY; entrance positions are intercalated (see other Figures)

— the diameter of tunnels that the bees excavate in wood. Multiple removable frames are positioned in full contact in a hive with a top cover (see Figure). The entrance of the hive is located on the individual frames, at the bottom. No bottom boards are needed. Multiple frames are for the purpose of colony expansion. Further, the frames are slightly larger than the bee space, three centimeters in thickness, in order to restrict horizontal extension of the carpenter bees' galleries.

The carpenter bee hive design consists of ***"dummy" entrance, solid frame of dry wood, hive body and top cover***. Soft, dry wood containing not more than 21% moisture is a requirement. Moisture exceeding 21% may cause the pollen provisions in the brood cells to ferment. Nesting wood can be cut and left to dry in the sun, then later placed in an oven to reduce moisture to < 21% if the air is humid. Soft and dry wood is preferred to reduce the labor of the bee when constructing its nest galleries. The dimensions of the hive are shown in the Figure.

For nesting substrate, the ***grain of the frame should be horizontal***. The carpenter bees will construct galleries along the grain. If the wood grain is vertical, galleries will be made in vertical position too, making it difficult for the carpenter bees to remove excavated wood materials upwards.

A 'dummy' entrance is placed at the middle bottom of the drywood frame to entice female bees to select a new nesting site. The trapezoidal shape is to allow placement of the dummy entrance at an angle to the frame wood, like that of the natural nest, and the depth of 3 cm for the gallery is sufficient for a carpenter bee to stay in the gallery in a curled position — during the initial days of founding a new nesting site.

• *Additional techniques*

For promoting new nest establishment by carpenter bees, locating wild nests for initial build up of the population is usually quite easy. Apparently, lack of nesting substrates and floral resources for pollen and nectar are problems for the bees. Techniques for colony establishment, and a colony division technique to increase the number of nests, have been successfully developed, as described below.

To induce bees to begin new nests, a box of uncolonized frames with dummy entrances can be placed next to an established carpenter bee nest site, as follows. First, the nesting substrate is jarred until the bees move out of their tunnels. Quickly, entrances of the abandoned galleries are sealed to prevent the bees from returning. Pebbles or any other hard material can be used for this purpose. Sometimes, when the carpenter bees refuse to leave the nest, the wooden substrates will have to be broken open to force them to leave. For shelter during the night, the homeless carpenter bees will take refuge in the dummy entrances. Make sure that more than sufficient nest substrates are given when the old nesting substrates are removed or made inaccessible. Next, check that disturbances from ants do not cause total abandonment of the area. Within days, the carpenter bees will start to construct new galleries, beginning in the dummy entrances.

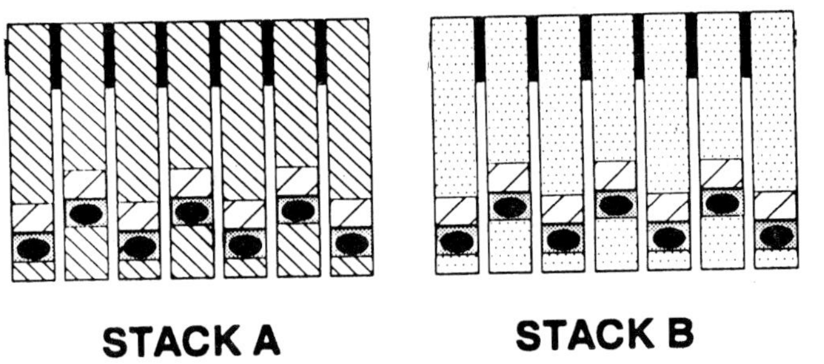

two comb stacks: stack A (above) has been colonized by carpenter bees, and stack B has not

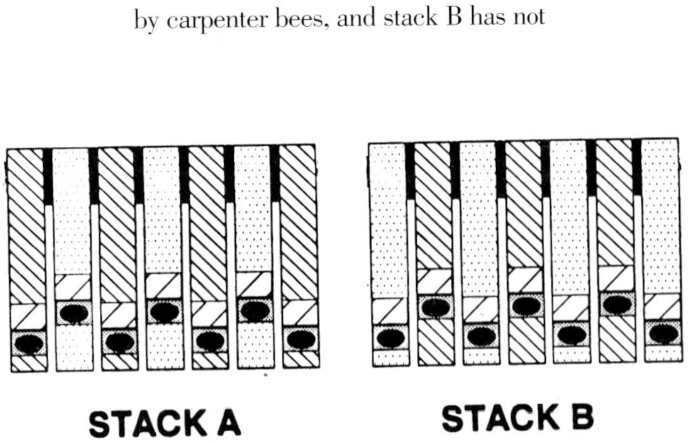

procedure for switching colonized and uncolonized frames between nest hives; this provides new nests in frame stacks, so that bees will progressively colonize the frames

Nest hive boxes for *Xylocopa*, showing individual frames, tunnels made by bees (exposed for photograph) within a frane and placementof nest boxes in a bee shelter, suspended by wires

If the wood is relatively soft, gallery construction can be accomplished within a week to ten days.

Splitting a hive of carpenter bees is recommended to expand and spread the population in new hives so that it can be distributed. Expansion can be done by alternate mixing of colonized and uncolonized frames together in the 'mother' hive. This is achieved by replacing alternate frames with uncolonized frames in the hive, thus two hives can be made. The remainder of the colonized frames from the mother hive are also mixed alternately with uncolonized frames in the new hive (see Figure). Similar inducement of carpenter bees to build nests in introduced new frames can be accomplished as described previously, but it must be done only on frames that are filled with bee nests. To accelerate occupation of new frames in the hive, the sealing of nest entrances can be performed, as described above. The hived bees must be supplied with water if the weather is dry. This is used by the bees to soften and excavate wood while building their tunnels. When moving an artificial nest (the hive) one must ensure that it is done at night, exactly in the way performed for honeybees.

- ***Managing nests for pollination***

The management of carpenter bees for passion fruit pollination considers both the passion fruit plant and the carpenter bees. All sources of bee forage must be available, weeds and pests must be controlled, and there should be supplements provided to encourage the growth of established 'colonies'.

Competition from most weeds is not a problem for pollination using carpenter bees because they are not like honeybees, which often visit a wider variety of flowers at any given time. Plants that bloom in the afternoon compete with the passion flowers.

Pest management of certain mites, ants, flies, parasitic beetles (Meloidae) and others is difficult because carpenter bees build tunnels that are not easily inspected. Little is known on the extent of damage, if any, caused by mites that are abundant in bee cells. Many *Xylocopa* carry mites in special pockets on the body, so they are apparently beneficial to the bees. Ants, small flies and wasps have been observed to co-inhabit carpenter bee nests, but their activities in the cavities are not known. However, it has been observed that some of them are endoparasites that multiply in the young larvae. Dead larvae are found in some cells, especially during the wet season. To control pests such as ants the nest area should be made inaccessible by applying a sticky compound or repellent at the base of the hive stand. Since no chemical pesticides can be used, it is best that the nest boxes are kept isolated from each other, to prevent rapid spread of any pests.

3) Other animals for pollination.

The diversity of crops for pollination follows the diversity of insects and other organisms associated with pollination. Some that have been actively bred or augmented for pollination in the tropics are flies (mango), stingless bees (rubber), carpenter bees (passion fruit) and midges (cocoa). Although the technology and approaches are rather rudimentary and not well established, they are evolving. Their time has come as a recognized facet of diversification in tropical agriculture. I believe there are many more species of insects, especially the beetles, that will be utilized for pollination services of economically important fruit crops. Pollination technology and pollination services will figure prominently in sustainable development of tropical orchards.

• *Pollination of mango by the Dipteran complex*

Flies and bees are considered primary pollinators of mango (*Mangifera indica*), but wasps, butterflies, beetles, bats and ants visit flowers, and the plant is also self-pollinating (Figure in Section 1.5.1). Mango flowers seem to attract honeybees only when there is abundant available nectar. Flies seem to be the main visitors of many cultivars, presumably when they have less nectar. Several mango plantations in Malaysia breed flies to bring about pollination, using fish refuse as substrates, during the short flowering season. Compared to honeybees, the technique of breeding dipterans is relatively easy and needs little maintenance. However, this practice does have the drawback that it cannot be done very close to human settlements — many flies also are vectors of human diseases. Large plantations would be the only situations where such stocking with flies would be useful.

A myriad of dipteran species (families Calliphoridae, Sarcophagidae & others) breed quickly using the protein provided in fish refuse, or dead and decaying organisms. Discarded plastic bottles can be used to contain the breeding substrates for flies. Alternatively, plastic shopping bags are perforated on all sides to allow the fish to decompose. Flies may lay their eggs there. In general, half a kilo of fish is placed in each plastic container ***after first having been placed in piles in the sun for a few days***, to allow flies to lay eggs.

The plastic bags of dead fish with dipteran eggs are hung on lower branches of mango trees in the plantation, at a density of one bag for about 8 to 10 trees. Adult flies generally emerge in less than a week during the warmest time of the year. In less than two weeks the number of flies increases dramatically.

As a general guideline, 10-14 days of substrate introductions in plastic bags can produce sufficient flies to pollinate a surge of mango blooming within a plantation area. Unlike bees, the flies do not return to a fixed domicile, and they disperse rapidly throughout the crop. It is recommended that undergrowth in the orchards should be cut before the flies emerge, so that there will not be other flowers that they visit instead of mango inflorescences. Open understory conditions also favor a decrease in humidity and may lessen the spread of sooty mold or fungi that are pests of mango. This problem is prevalent in areas where drip irrigation is used.

Post-pollination practices are as follows: ***a)*** The fly-breeding containers need to be removed after pollination because they may fill with water and provide breeding sites of mosquitoes; remains of the fish refuse can be collected and turned into compost; ***b)*** between two to three weeks after blooming finishes, excess fruit needs to be thinned to prevent branch breakage or the production of small fruit; excess fruit may be taken to indicate greater than optimum pollination and, as in most agricultural practices, adjustments will then be required between pollination and local requirements for a marketable product. In southeast Asia, mango fruit are individually wrapped in paper bags after thinning has been done, and fruits are ripe in about 100 days; ***c)*** application of fertilizer after fruit growth has begun is critical to final fruit formation — fertilizers should be applied at least 50 days before the end of the full ripening process.

• *pollination of cacao by midges*

Of the 78 known genera of Diptera known as midges (Ceratapogonidae) the genus *Forcipomyia* is the best known of those visiting the flowers of *Theobroma cacao* (see 1.1.1, 1.5.3). These soft-bodied midges breed in moist plant tissues such as banana plant parts and cocoa pod husks. Substrates or shelters for the midges are prepared from banana plants that have been cut up to provide attractive decomposing food material for use by the fly larvae. This technique stresses the use of 'unclean' understory areas in and around cacao plantations. However, too much humidity under the canopy may increase the spread of fungal diseases like 'vascular streak dieback'. Again, the breeding of pollinators must be adjusted to the cultivation practices required for the plant.

• *preparing an environment conducive to pollination by bats*

Fruit production by the durian is accomplished through outcrossing by cave bats (*Eonycteris spelea*) which range widely and visit the large, rewarding flowers for nectar. The branch can support the bats that hang on the flowers, and the stigma touches the chest of the animal, thus acquiring pollen. The following cultural practices to encourage bat pollination are worth considering: ***a)*** planting distance should not be too close, to obstruct the flight path of the bats; an unbroken, solid canopy may also be difficult for the bats to navigate in; pruning of some branches to open gaps is recommended; **b)** the planting of other bat-pollinated plants, such as *Musa*, *Oroxylum*, and *Ceiba* in the vicinity of durian plantations may increase visitation to trees of the plantations or provide supplementary feeding for bats; **c)** roosting places for the cave bats are ordinarily provided in the wild areas, but it may be possible to provide smaller shelters, of attap, to provide roosting places for the cave bats or other small bats.

2.5.2 *Oil Palm Pollination & Weevils*

The Oil Palm is native to perhumid, tropical west Africa. It is one of the most important sources of vegetable oil in the world and is now grown in plantations in West Africa, Southeast Asia, and tropical America. The oil is produced by the fruits and seeds. Despite its importance, the means of pollination have been elucidated only as recently as 1979 when the importance of insects, rather than wind, was proven.

The main natural pollinators of oil palm are three species of weevils (Coleoptera: Curculionidae) of the genus *Elaeidobius*. The one that has received the most attention is *E. kamerunicus*. This weevil was introduced, after careful quarantine and study of its potential to become a pest of non-target plants, from its native habitat in West Africa into Malaysia in 1981. There it immediately became established and spread rapidly throughout the plantations of the country, resulting in millions of dollars in profit to the industry. Pollination of oil palm was previously done by people blowing pollen, mixed with talc or another carrier/diluent, onto the receptive female flowers.

To understand how oil palm is pollinated by these beetles, it is necessary to examine the sexuality of the palms. Oil palms are hermaphroditic plants, but normally only reproduce through male or female function at any one time. The inflorescences may be either male or female. The male inflorescence consists of numerous long, finger-like spikelets, each comprising thousands of pollen-producing florets. The bulky, female inflorescence consists of hundreds of densely packed florets, each of which is an egg-shaped ovary tipped by a trifid stigma 0.5 cm across.

The weevils breed in male inflorescences of the oil palm. While the spikelets are shedding their copiously produced pollen, female weevils visit them. It is here they become dusted with the pollen and lay their eggs. The female weevils, and the males which also visit the spikelets — presumably in search of mates — move from palm to palm. They thus effect outcrossing. During some such movements they visit palms with female inflorescences. The reasons they visit the female inflorescences is not fully understood, but the characteristic scent of the beetles, *shared by the male inflorescences*, is the main attractant.

The spikelets remain on the male inflorescences, attached to the palms and protected in the axils of the fronds, after they have expended their pollen. However, their role in pollination has not yet finished because the next generation of weevil grubs grows and develops within them.

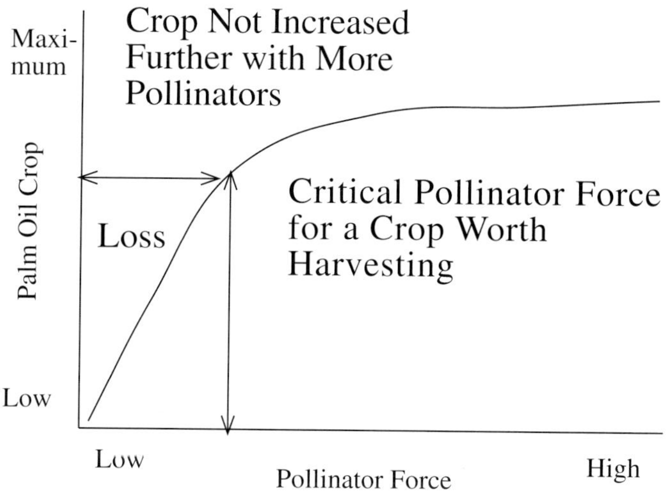

The 'Pollinator Force' concept, applied to African oil palm

From the viewpoint of pollination management, the population of weevils is clearly important. However, the density of male inflorescences relative to the density of female inflorescences is also important, as is the probability that a given weevil will transfer pollen. One must also keep in mind that the male inflorescences do not contribute directly to oil yield in a plantation and do occupy the place of female inflorescences on palms in male phase. Thus they can be considered to remove palms in male phase from oil productivity for about a year.

Thus, the aim of plantation management to maximize fruit and oil yield must optimize the ratio of expression of male and female inflorescences to allow for complete pollination of female inflorescences. That, in turn, requires that the population of weevils not drop below the threshold for complete pollination. Of course, absolutely complete pollination should not be expected (Section 2.2).

Kevan and coworkers proposed "Pollinator Force" (see Figure) as a practical measure for monitoring pollinators for oil palm. To calculate the pollinator force, the following information is required:

1) The density of male inflorescences or spikelets

2) The number of weevils emerging from those spikelets

3) The proportion of weevils not visiting male inflorescences which are shedding pollen

4) The proportion of weevils not visiting a receptive female inflorescence.

The values for 3) and 4) are unknown, and would require great effort in research to estimate. One may assume that these values would be low if 1) were high. However, if 1) is too high, at the expense of palms' production of female inflorescences, then crop yield would suffer. Unfortunately, the optimum ratio of palms in male phase to palms in female phase is not known and the mechanisms within the palms which cause the production of inflorescences of one sex or the other are incompletely understood.

The preceding Figure presents the concept of "Pollinator Force" for oil palms and points up its practical importance. In Malaysia, a decade after the introduction of *E. kamerunicus*, there is no indication that the Pollinator Force has dropped below critical level for economic harvesting. Some minor pests of the weevils have been recorded, e.g. rats feeding at the grub-rich spikelets and some low incidence of diseases. Nevertheless, weevil populations continue to be monitored in various plantations in order to keep a 'weather-eye' on potential pollination.

As oil palms continue to be improved through modern breeding and genetic techniques, the concept of pollinator force may become vital. It may be that with the development of lines of perennially female (oil producing) palms which rarely enter the male phase, plantations will have to include pollinizer (male) palms.

2.5.3 Stingless Bee Colonies for Pollination

The idea of maintaining colonies of stingless bees — (Meliponinae) for pollination outdoors is relatively new. For greenhouse pollination some have been brought to the temperate zone and overwintered successfully in Japan. These were small colonies of *Plebeia* from subtropical southern Brazil, and there are others that do well in cool, subtropical climates (some Mexican *Melipona* and Taiwanese *Lepi-dotrigona*), but these are exceptional. There are between four and five hundred different species of stingless bees. *Meliponines* inhabit virtually all of the tropics and subtropics except some islands (see Section 1.4). Most of their species live in the wetter lowland forests. At least 300 species (over 20 genera) and perhaps another 50 geographic races are confined to the Neotropics. About two dozen species live in Australia, New Guinea, and the Philippines (areas east of Wallace's line), about 40 in Africa and as many as 50 are native to continental Asia and Malesia (Malaysia and Indonesia).

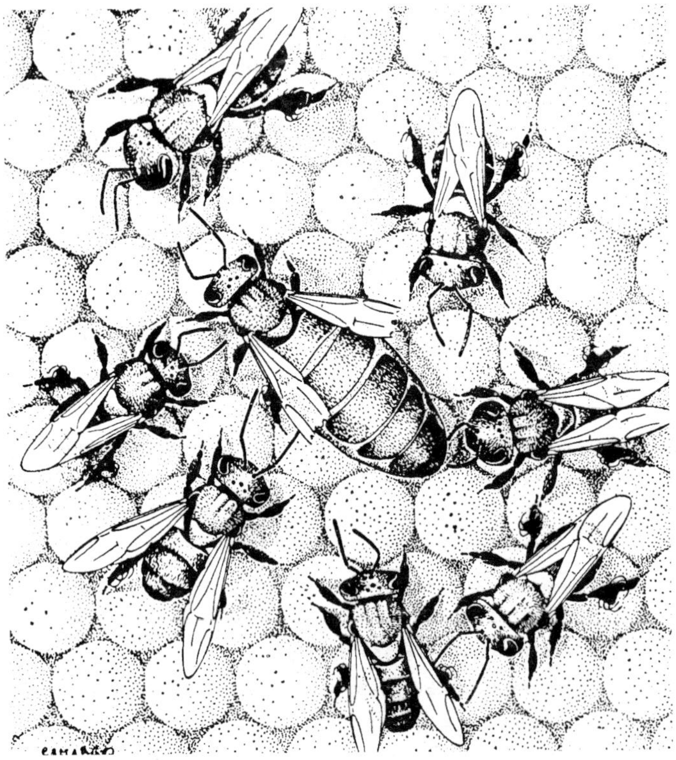

queen stingless bee, *Scaptotrigona*, surrounded by workers on brood comb

1) Pollination potential of stingless bees.

There are many kinds of stingless bees that would be impractical to manage as pollinators. Some reasons for this are the following:

- ***extremely small colony size (less than a few hundred bees), coupled with extremely small forager size (2 to 3 mm in length)***

- ***extremely defensive colony behaviour —a large colony of some Trigona, Partamona, or fire bees, Oxytrigona, can constitute a hazard; these species attack, biting relentlessly, and fire bees cause skin irritation with an acid secretion***

- ***restrictive nesting habit or habitat — many species nest only in the ground, in termite nests, or in ant nests, or they only survive in rain forests; an attempt to transfer colonies to hives, agricultural areas, or cities will only kill them, albeit slowly***

- ***destructive use of flowers, sap and resin sources — a few dozen Neotropical species and at least a few in southeast Asia make holes in flowers and damage fruit, buds and branches, which makes them pests in orchards***

Bearing these facts in mind, there are other stingless bees that have the right characteristics to be managed — reasonably large colonies (a few 100s to several 1,000s of workers), docile temperaments, adequate flight ranges, and adaptability to artificial domiciles (hives). More important, they are able to maintain their original nests in logs cut from trees.

It is worth mentioning that some of the small species with restricted flight ranges (about 500 meters from the nest) could well serve the needs of producing hybrid seed, or be reliable pollinators for crops planted in small patches. Some may also be maintained experimentally indoors, or in greenhouses. All stingless bees except those that repeatedly damage flowers (*Trigona* s. str.) or are obligately parasites of other stingless bees (*Lestrimelitta*, a neotropical genus of seven species, and *Cleptotrigona*, an African genus of one species), contribute on a daily basis to the pollination of flowers.

A few species of stingless bees are highly adapted to disturbed habitats such as cities, or areas with little forest. They are 'weedy' in the sense that they propagate rapidly when able to find nesting sites — often provided by holes in cement block, stone buildings or walls, structural timbers with hollow sections, or in bamboo internodes. These are the Neotropical *Trigona (Tetragonisca) angustula* (formerly called *Trigona 'jaty'*) — this bee also is a parasite of other stingless bee colonies may usurp their nests; and the southeast Asian *Trigona (Heterotrigona) biroi*, *Trigona (Heterotrigona) fuscobalteata*, *T. (Heterotrigona) iridipennis*, *T. (Heterotrigona) geissleri*, *T. (Heterotrigona) itama*, and *T. (H.) minangkabau*. Other genera that inhabit marginal secondary growth habitats or disturbed areas (as well as forests), utilize artificial domiciles, and do not present the drawbacks described above are: *Melipona* (a Neotropical genus of about 45 species), *Nannotrigona* (Neotropical, 15 species), *Trigona* (subgenus *Frieseomelitta*) — Neotropical, 20 species, *Trigona* (subgenus *Tetragona*) — Neotropical, 10 species, *Scaptotrigona* (Neotropical, 40 species) and *Cephalotrigona* (Neotropical, 4 species).

A few Neotropical genera, such as *Cephalotrigona* and *Melipona*, have been "domesticated" (actually, maintained in sections of tree trunks, close to their natural forest habitat) for hundreds of years by avid beekeepers like the Maya of southern Mexico and Central America. They produce abundant honey (a few kilograms), which is normally harvested once a year. A significant advance in management has been made with these bees: their colonies are propagated artificially (see following).

2) Natural enemies of stingless bees.

The most significant enemies of stingless bees are *ants, phorid and other flies, termites* and *parasitic stingless bees* (*Lestrimelitta* and *Cleptotrigona*). Specific management problems caused by each, and their suggested remedies, are given here. The larger animals, which attack colonies of honeybees as well (tamanduas, tayras, honey badgers and the weasel family in general, bears, and civets), do not often attack natural nests of meliponines in the forest. The same can be said of toads, lizards and birds. However, any hive is much more vulnerable and should be protected. In Africa, primates such as chimpanzees attempt to obtain food from stingless bee nests by inserting flexible sticks in nest entrances. There is little that the bees can do to defend from their attacks. Many African species of stingless bees avoid such harassment by nesting deep underground.

• *Ants attack frequently at night* and attempt to overcome the defenses of guard bees at the nest entrance, or find unprotected openings through which they can enter. Water in a container will keep ants from crawling up a hive stand. Suspending the hive by wire or rope from a roof can also prevent ants from gaining access to the bee nest.

• *Phorid, stratiomyid and some syrphid flies* have species that apparently do little more than seek damaged stingless bee nests, which flies lay their eggs in. In the Neotropics, the phorids mount a pheromone-guided attack and deposit eggs throughout the nest, always laying in areas too small to be reached by the bees. Within a few hours or overnight, a nest having *any* openings, including the entrance, will be thoroughly parasitized, and the fly larvae will kill the colony in a few days. The flies find the nests by smell, particularly the acetic acid of fermenting pollen and the smell of honey. *Plugging all the holes with cotton, and keeping all edges sealed with tape*, will temporarily help to keep flies out. The best defenses are intact nests and nest contents (which the bees repair in a few days after a major disturbance, such as moving the nest). The other important defense is a *restricted nest entrance*, one of the proper size for the bees in a natural nest. A gaping hole for a nest entrance is an invitation for disaster.

• *Termites* constantly attack the wood of hives and logs. The bees can defend themselves by applying resin to the inside of the cavity, which will sometimes hold the hive together even if the wood is consumed by termites. Termites also enter the nest entrance of stingless bees, building their galleries through the bees' only exit. This may kill the colony, but it is likely that some termites also attack bees and their food stores. Chemical deterrence of termites is inadvisable. Such insecticides are extremely toxic and, even if applied outside the hive and only to one extreme, such as the base, the bee colony will usually succumb to the poison. The same methods used to prevent ants from reaching the nest will work for protection from termites.

• *The parasitic stingless bees ('robber bees')* are ordinarily the only bees that can overcome stingless bee defenses and enter their nests. Although their attacks do not usually result in death of the host colony, it is debilitated and will be attacked again. Because the robbers, *Lestrimelitta* and *Cleptotri-gona*, are small bees, less than 7 mm long, constriction of the host nest entrance to keep them out would be impractical. To protect stingless bee nests from the robbing parasites, the only permanent solution is the location and destruction of the robber colony.

3) Hiving stingless bees.

The stingless bees suitable for '*meliponiculture*' normally build their nests in hollows found in living trees, or those in dead trees of extremely hard wood. A few species, apparently only *Trigona*, can variously nest in the ground, in tree hollows, or in artificial domiciles like buildings or

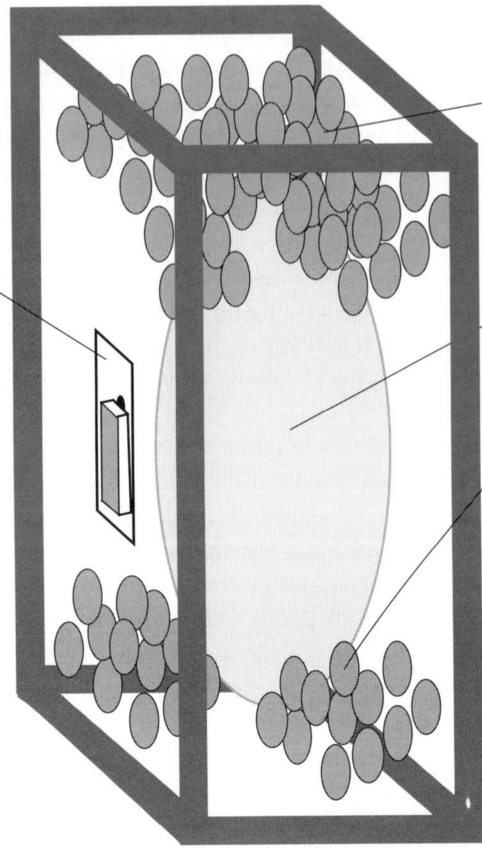

Basic stingless bee hive design, showing position of nest elements

The hive walls are made of smooth hardwood at least 2.5 cm thick. Dimensions vary with the size of the bee nest and bees. Hive dimensions of 20 x 10 x 10 cm are usually adequate. If honey 'supers' are added, they should be no less than 5 cm high and firmly fastened to the brood chamber with hooks or other hardware.

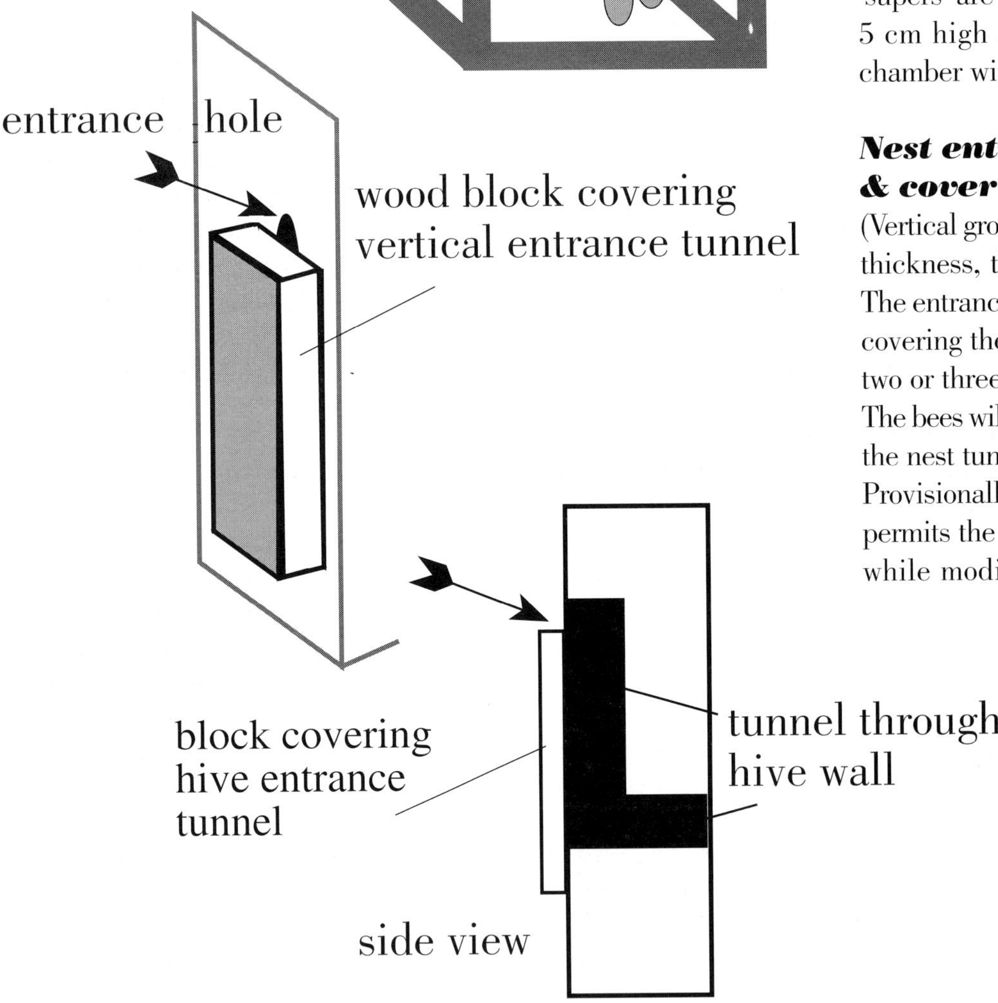

Nest entrance tunnel design & cover plate

(Vertical groove rooted to about half the hive wall thickness, than drilled through at the bottom). The entrance hole at the top of the block or plate covering the tunnel should be large enough for two or three bees.

The bees will reduce the entrance size and modify the nest tunnel with resin.

Provisionally, an entrance tunnel of 2 cm width permits the bees to adequately defend the nest, while modifyng the hive to suit their needs.

containers. Most stingless bees, however, have a surprisingly limited range of suitable nesting sites — although their dimensions can vary substantially. The stingless bees nesting in tree hollows generally use cylindrical cavities. Because of this, the nests may be extensive and range from about 3 to 30 cm in diameter. Many of the species can close off spaces that are too large. Minimum space is the more important consideration and depends upon the bee species. Adequate space for a nest is generally 1 to 10 liters. Trees are felled to obtain nests of stingless bees, either by axe or chain saw. The nest entrance is sealed before or immediately after the tree is cut. A tree too large in girth to permit transport of the section with the bee nest can be reduced in size with the same tools, leaving at least 5 cm of wood around the nesting cavity. Closing a part of the nesting cavity exposed by cracks in the wood, or by cutting in the wrong place, can be accomplished quickly with leaves or vegetation, cloth, newspaper, plastic sheet stapled in place, or with mud. A stronger and more permanent closure must be provided if the bees are to be kept in the log hive.

During the felling of trees and transportation of the nest, all of the younger brood and eggs will be lost. This results from strong shocks and vibration. The bees will open some brood cells, then reuse their contents and the cell wall material. More extensive damage, such as spillage of honey or loosening of honey and pollen storage pot masses which will crush brood, will result in more serious damage to the colony. After a tree is felled, it is desirable to place the log hive upright in its former position, as close to the original nest location as possible. In this way, many of the 'field bees' and returning foragers will locate and enter their nest. Also for this reason, the original nest entrance tube or structure should be maintained as intact as possible. However, when the phorid flies start to arrive, it is best to seal all nest entrances.

The bees are much better off left in their tree trunks. Unless taking honey stores, or artificial feeding within the hive and colony propagation are essential, there is no reason to remove the often delicate and complicated bee nests from the tree cavities and place them in hives. It is not a rational practice and often will result in loss of bee colonies. The most successful stingless bee management in the world has relied on using hollow tree trunks. To complete this choice of hive structure, the two extremes of the nest should be sealed with a rock-hard, thick material that fits as exactly as possible the inner contour of the hive. The Maya accomplish this with pieces of limestone rock, chiseled to form a tight-fitting plug in each end of the hive. The hives can be suspended, or stacked horizontally, or stood on one end.

Wooden hives should be constructed of smooth hardwood at least 2.5 cm (one inch) thick. It may be painted on the outside, preferably with a latex or a water-base paint. The hive should not be painted inside. After seeing many different hive designs for different purposes, I am convinced that most stingless bees adapt to the container they are given, provided it is not far too large or wet. They may, however, abscond (leaving their mated queen behind) if their container is greatly inferior to the natural nesting site. Hives of *metal cans*, *hollow gourds*, or *styrofoam coolers* also work well. These materials have the advantage that, while seldom as strong as the wooden hive, they are not attacked by termites or much damaged by water. Hive dimensions may be vertically elongate (see Figure), in which case they can be of a single or two to three compartments. The upper compartment (a honey 'super', like that used for the three species of hive-*Apis*) is for the stored honey and pollen, or for artificial feeding. The next lower compartment is for the brood, and, if another compartment is attached, it is for colony refuse and, for some species, a drainage or ventilation area. Hives may also be horizontally elongate, in which case there is little need to make more than one section of the hive. It is advisable, however, to make more than one hive cover section — so that only one extreme of the nest is opened at a time, leaving the rest intact. Honey can be removed, bees fed, or colony divisions made either by removing the top cover or, for the long log lives, by removing one end cover. Some species do require additional holes in a lower and/or upper corner of the hive, either so they can dispose of waste water or to let moisture and carbon dioxide escape. These modifications can be provided by the beekeeper, but should be made after the nest and colony are well established. Any new openings that are made should first be sealed with a fine mesh metal screen or substantial cotton, to prevent the entry of parasites.

4) Propagating stingless bee colonies.

In order to reproduce, stingless bee colonies first find a new nesting site and make part of a new nest — complete with stored honey and pollen. After several days to several months, the new nest is occupied by adult bees from the mother nest, including a new queen. This event does not take place on a regular basis. It is not seasonal or yearly, because without the new nesting site, there is no colony reproduction. Artificial division of colonies is essential to increase the numbers of bees available for management.

Colony divisions should be made with great care, following a relatively simple procedure:

- ***Divisions should be made when the weather and bee forage are optimal***

- ***Carefully remove half of the brood combs (both mature pupae and young brood), along with workers on the combs, and some of the stored pollen and honey. Place them in a new hive box. Make sure that the laying queen is not removed. Place the mother and daughter hives next to each other, allowing foraging bees to enter each hive.***

- ***That night, place the new hive with its half nest in the position formerly occupied by the mother colony. Move the hive of the mother colony, with its queen, to a new location, at least 200 meters but not more than a kilometer away. Seal each nest with cotton or some other material that will allow air but not parasites to pass through.***

- ***The next day and thereafter, let the bees forage freely. Do not open the hives unless absolutely necessary. Some of the foragers from the mother colony will return to their former site, thereby strengthening the new***

colony. Many, however, will remain in the nest at its new position. A relative advantage of the mother colony is the intact nest and likely more young bees. The relative advantage to the new colony is the presence of many of the older bees, which will forage and work to defend and complete the new nest.

Within a few days of this operation, the new nest, which lacked a laying queen, should have male (drone) bees flying and landing near the nest entrance. They quickly find sites with virgin queens. If no males are present, the area may not have enough conspecific colonies, or there is no virgin queen present. In one to two weeks, depending on the species, a mated queen should be laying eggs in both the new and the old hive. Both colonies should be actively foraging and collecting pollen, unless they lack queens.

As of this writing, the processes by which a stingless bee queen successfully mates are still not understood. Only one drone mates with her, and the queen does not mate more than once during her lifetime — usually one to two years. What has been clearly established is that queens very seldom mate with brothers or closely-related drones. To do so it detrimental to the welfare of the colony. Successful mating ordinarily requires that at least several colonies of the same species be within several 100 meters of the nest having a virgin queen. Several to many virgin queens may leave the nest and not return, either being caught by some enemy or prevented from reentering by worker bees in the hive. Finally, one queen mates and then returns to the nest to lay eggs.

A few stingless bees will reproduce to utilize suitable nest sites that are close to their hives. This is especially true in orchards or habitats having many flowers and available resources for the bees — where the forest bee species are unable to forage. Thus, colony number can be increased slightly through natural occupancy of domiciles left in the vicinity of meliponaries.

2.5.4 Management & Utilization of Two Honeybees in the Tropics

1) Introduction.

Biology and behaviour of honeybees have been very extensively investigated and are now becoming relatively well known. This has enabled man to effectively exploit their worth in agriculture. The value of honeybee pollination in agriculture is much greater than the value of other hive products. There are large number of crops where cross-pollination is essential. Without pollinators either there is no seed set (e.g. in several cruciferous and some leguminous crops) or seed yield is exceptionally low (e.g. in several cross-pollinated crops). Moreover, rapid changes in pollinators' environments, due to over-ploughing, excessive irrigation, deforestation and indiscriminate use of toxic agrochemicals is leading to the complete destruction of the habitats of wild pollinators has resulted in their large-scale demise. This has made cross-pollination of crops by wild pollinators completely unrealistic and unreliable. The situation is more alarming in the tropics where there are enormous numbers of agricultural pests and there is no legislation of pesticidal usage. Under such conditions, honeybees are among the only reliable pollinators whose population densities at the target crops can be manipulated relatively easily and effectively. In the tropics, where the local and regional economy is not well settled, it is not feasible either to maintain programs for attempting to breed specialized pollinators, or to attempt to manage the crop environment. However, management of honeybee colonies and their employment in crop areas should easily be taken up by the seed growers in the tropics, as elsewhere. This section presents an account of management and utilization practices of two honeybee species, *Apis mellifera ligustica* and *Apis cerana indica*, for pollination of crops. Specific needs for crop pollination in the humid southeast Asian tropics have already been mentioned in Section 2.5.1.

Two species and several geographic races of honeybees (from among at least 7 species and over 30 races) are currently managed on a large scale in the tropics: *Apis mellifera* and *A. cerana*. All but Himalayan honeybees, *Apis laboriosa*, are tropical bees, secondarily adapted to temperate environments and habitats. *A. mellifera ligustica* and *A. mellifera scutellata* have been introduced almost in the whole of the world (Sections 1.3-1.5), whereas *A. cerana indica* is indigenous to tropical (and parts of temperate) Asia. On the basis of the following factors, these two honeybee species can effectively be utilized for pollination of crops. These are: **a)** these species are kept in the hives and do not require excessive management; **b)** their abundance on the crop can be manipulated; **c)** due to their tendency to be generalized foragers, honeybees visit large number of flowering plant species. — they can therefore pollinate a wide variety of crops; **d)** individual foragers display high floral fidelity or constancy — when a forager begins foraging on a plant species, it continues to do so until the resource is exhausted; e) behaviour of honeybees (and other bees) can also be manipulated by modifying the reward system of the plant (see Section 2.3.1).

2) General management practices.

In order to use honeybees as pollinators of crops, management practices should be followed at three different levels. These are:

- *Management of an agro-environment suitable to honeybees* — crops might be bred to be attractive to honeybees. For example, crop breeders do have the ability to change floral fragrance, nectar characteristics, pollen abundance, and also the cultural practices needed for increased fruit or seed production (see also Sections 2.1, 2.3).

• *Prudent experimental use of sprays, bee attractants and repellents* — Some sprays and attractants are used to lure the bees or to increase the attractiveness of target crops during flowering. Foragers are first trained to visit the crop by exposing the recruits to the flowers soaked in certain attractants, like scented sugar syrup; the scent of course, should be from the target crop. Carbohydrate sprays on the crops do sometimes increase the bee visits, but unfortunately do not necessarily increase pollination. Likewise, pollen and certain phytosterols attract foraging honeybees. Nasanov gland secretion components, citral and geraniol in a sucrose solution, sprayed on crops were found to mask the repellent odor (purportedly applied by bees to flowers in order to avoid re-visitation of unrewarding flowers), and may attract bees to crops. It is doubtful that they generally improve yield. Anethole (1-methoxy-4(1-propenyl) benzene) also attracts honeybees. Its utility has yet to be demonstrated.

Certain repellent chemicals are used to reduce the bee poisoning from insecticides and also to reduce the attractiveness of some competing crops. Guidelines for pesticide compatibility with beekeeping and managed pollination are given in **Appendix III.**

• *Removal of floral competitors* — The best way of avoiding negative impact from floral competitors is through their elimination. For example, for the effective pollination of *Brassica*, adjacent chick pea is removed. For effective pollination of alfalfa, adjacent mustard crops are cut. In the second case, pollen foraging on alfalfa by bees, and seed set, are increased.

3) Breeding efficient pollinators and specialists.

Selection and breeding of honeybees through artificial insemination is possible in *A. mellifera* and in *A. cerana*. To date, this technology has failed to produce the desired specialist pollinators (Free, 1993), and now agronomists look more seriously at alternative pollinator species. The introduction of African *Apis mellifera scutellata* has created many significant changes in the Neotropics (Section 1.5.3). For beekeeping, their main advantage is in the prevention of hive theft, and their wide abundance and availability. They are, however, difficult to manage because of often unpredictable behaviour and frequent absconding (Section 2.1). Their use in apiculture, however, is widespread, and beekeepers can deal with them.

When bees of desired pollination and other characteristics are selected, these should include: ***a)*** traits promoting strong colonies (unfortunately, this is incompatible with bees that are unaggressive, specialists on a given plant species and slow to reproduce); ***b)*** colonies having high resistance to diseases or pesticide poisoning; ***c)*** colonies that intensively collect pollen (hence reproduce rapidly); ***d)*** bees with flight range and tongue length matched to target crops, e.g. bees with short flight range and long tongue length are more useful in the pollination of red clover; ***e)*** specialist strains of bees have been bred for alfalfa, and it may still be possible to arrive at strains that are both viable and vigorous for other crops (although the inbreeding needed to produce specialists generally reduces fitness and vigor, see Section 1.2), and ***f)*** bees that fit into an agricultural setting, by not stinging viciously after disturbances.

4) Management of honeybee colonies.

The equatorial tropics witness high humidity and high temperature, while the subtropics and semi-arid tropics present low humidity, high temperatures during summer and lower temperatures during winter. General management practices for honeybees in the tropics therefore vary from place to place and mainly relate to the respective conditions. In the tropics, queens engage in brisk egg-laying activity after mating and become exhausted within a year. Therefore the queen is replaced every year. Other conditions required in the colonies used for pollination (see Appendix II) are:

• *presence of a newly mated and laying queen*
• *presence of an ample quantity of brood and honey (specified below)*
• *feeding with sugar syrup during dearth*
• *prevention of diseases, pests and enemies*

General management practices for beekeeping with European *A. mellifera ligustica* in tropical Australia and Asia, and that of *A. mellifera scutellata*, *A. m. adansonii* and other races in Africa and tropical America are now known. *A. cerana indica* is found in much of tropical Asia and its management practices for this region, though a bit different from *A. mellifera*, are also known.

Specific management techniques for *A. mellifera* in tropical Asia include avoidance of the ectoparasitic mite *Tropilaelaps clareae*, which normally (originally) was a parasite solely of *Apis dorsata*, the common giant honeybee.

Mites that kill or debilitate brood are not easy to control and require a mixture of chemical and cultural management. Another mite, *Varroa jacobsoni*, is also a pest of *Apis mellifera*, but not quite as serious. It was originally associated only with *Apis cerana*. These mites infest brood, feeding on the pupae and killing them or preventing their normal development. *Central to control schemes is the need to break the brood rearing cycle of the honeybees.* In Burma, this is done by caging the queen for about three weeks and fumigating the hive with phenothiazine. In Thailand, brood is removed and allowed to die, and fine sulphur and naphthalene is applied as a fumigant in hives.

In many tropical regions, management against American and European foulbrood diseases is required. In the Indian subcontinent, *A. cerana indica* is severely threatened by Thai sac brood virus due to the recent recurrence of the epidemic. In other parts, this honeybee is safe. Though this honeybee has relatively few pests and enemies, absconding is a problem. To prevent the colonies from absconding, the following practices are important:

- *safety against wax moth and predatory wasps (use smaller hive gates)*
- *management against high temperature (use cooling arrangements)*
- *management against prolonged rains (ensure sugar syrup feeding)*

5) *Methods for crop pollination.*

For use of honeybees in pollination, preparation of standard colonies and making a provision of required number of colonies is essential. Although there are no clear colony recommendations for the tropics, standards from temperate areas might be extended to these regions too (see **Appendix II**). The field of oilseed rape (*Brassica juncea*) has about 20 million flowers/hectare when in full bloom. The European honeybee pollinates on an average 6 flowers per min and (at Hisar) remains active for 5 hrs (300 min) on the crop during February (when winter becomes mild). During this period five standard grade-'A' field colonies/hectare are required to daily pollinate each flower. Flowers of ray crops remain open for three days. Therefore, each flower is expected to receive three visits during its blooming period, which should be enough for effective pollination. However, the number of colonies for the same crop should vary from season to season. Early in the season when there is extreme cold, the activity duration of foragers is limited to about 3 h (200 min). During this period, therefore, the number of colonies required should be greater. Likewise, the number of grade-'B' field colonies for this crop, and grades-A and -B orchard colonies for orchards can be determined. Accordingly, this number can be employed and managed.

Ten-frame colonies of *Apis cerana* at full strength will have 8-10,000 bees and from 800 to 1,000 foragers. Their attributes place them as equals to one-third of the standard grade-A field colony of European *Apis mellifera ligustica*. However, the foraging rate of *A. cerana* is about 1.5 times more than *A. mellifera* and the working hours of this bee on a daily basis are also more. Therefore, about seven standard grade-A field colonies (analogous to *A. mellifera*) should be essential for equivalent pollination of one hectare of a *Brassica juncea* crop. Likewise, the number of *A. cerana* colonies for other crops can be determined and employed for pollination near the target crop. To increase honeybee pollination, time, place and ways of colony placement are most essential. In order to establish fidelity and constancy of honeybees to the target crop, their time of introduction is most important. Colonies should not be moved too soon or too late. When blooming intensity has reached 25%, this is the right time for colonies to be brought for pollination. Otherwise, colonies should be moved in small lots with the progression of flowering to its peak. Where to place the colonies? Normally when available, bees tend to forage for food close to their hives. Placement of colonies close to the target crop, therefore, is important, because such colonies collect more pollen and gain more weight through nectar collection. However, at times, bees reject high rewarding crop in favour of less rewarding e.g. *A. mellifera* prefers chick pea over raya and onion (personal observation). Here, as described above, crop environment management practices are needed to discourage the bees from foraging on non-target crops. Colonies are placed in groups near or inside the target crop area. For large crops, and where low wind velocity is experienced, colonies in groups of 20s should be placed in the middle. For small crop areas, 2-5 colonies per group can be distributed evenly through the crop, beginning at the middle. Dispersion of hives in small groups throughout the crop area is not always essential for a uniform availability of pollinators (see Section 2.4). Under high wind velocity conditions, the colonies in small groups should be placed along the edges of the target crop (upwind and downwind). However, selection of colony group size and their dispersion should depend upon production results of their particular management. For *A. cerana*, larger groups should be easier to maintain than the smaller and more scattered groups

6) *Feasibility of honeybee-pollination in the tropics.*

All the management practices for the utilization of honeybees as pollinators of crops, as described above, are not feasible in the tropics at this moment. The simple reason is unavailability of the required facilities. Due to the availability of *A. mellifera* almost everywhere and *A. cerana* in tropical and temperate Asia, both species can be utilized for pollination of crops. *Apis cerana* can still be a better or more easily managed pollinator due to its superior attributes:

- *Colonies of* A. cerana *are smaller than those of A. mellifera, which facilitates their easy handling.*
- *Foragers of* A. cerana *have shorter foraging range than those of* A. mellifera *and therefore, visit the target crop with more reliability*
- *Colonies of* A. cerana *have defenses against very significant Asian colony predators such as certain wasps of the genus* Vespa, *which attack* Apis mellifera *and against which the European honeybee does not defend itself.*
- *Foraging rates of* A. cerana *are higher than those of* A. mellifera. *Therefore, smaller forager populations of the colonies should be equally useful as larger colonies of European honeybees.*

Utilization of honeybees for pollination of crops has not yet become popular in the tropics due primarily to ignorance and unavailability of colonies and other facilities. In many of the countries, there are no beekeeping or pesticide laws, or these are completely ineffective. Besides, honeybees are kept only for honey production. Benefits of honeybee pollination in the tropics are still largely inaccessible to the seed growers. However, the benefits of honeybee pollination are much greater than the benefits from other hive products. In order to tap this underutilized pollination potential in tropics for increasing food production, managed pollination will have to be recognized as one of the several mainstays of agriculture. This is of utmost importance to meet the food demands of a rapidly increasing world population.

Acknowledgments: I wish to thank Prof. R.B. Mathur for the general facilities, Nirmala Sihag for encouragement, Dr. Manju Singh for her valuable suggestions, and Anju Rathee for extending help at various stages of preparation of this article. Funds for this article were utilized from the project '(Cg) Zoo-6-Oh' financed by Ministry of Environment, Government of India, New Delhi.

2.5.5 *Management of Subtropical Solitary Bees for Pollination*

Fig. 1. above: soda-straw nesting sites in boxes, for use by nesting solitary bees; below: nesting tunnels of waste stems *(Arundo)* — sarkanda —for *Xylocopa fenestrata*, a large carpenter bee

Two megachilid species, *Megachile haryanaensis* and *Chalicodoma rubripes*, have been identified as potential specialist pollinators of alfalfa (*Medicago sativa*). Another two megachilid species, *Chali-codoma lanata* and *Chalicodoma cephalotes*, are pollinators of pigeon pea (*Cajanus cajan*), and one large carpenter bee, the xylocopine *Xylocopa fenestrata*, pollinates several cucurbits in semi-arid and subtropical climates. For pollination of these crops, the bees can be managed in artificial nesting devices, using management practices described here. These particular bees are active during high ambient temperature and low relative humidity.

For many crops, honeybees are unreliable pollinators that act mainly as nectar or pollen thieves. Although they do not perforate or rob the flowers, they do not transfer the pollen between anthers and stigmata. Therefore, in such crops they are of no use and also detract from the pollination value of other organisms by competition — making the resource less attractive or available. This is true in several leguminous and other crops such as alfalfa, pigeon pea and some cucurbits (see also Sections 1.2 to 1.5 and **Appendix I**).

These bees are attracted to nest in tunnels. Where megachilid bees accept a wide variety of tunnels for nesting, *Xylocopa fenestrata* accepts limited tunnel types in hollow stems only. However, hollow stems with soft internal pith are easily accepted by all these bees. Nest material includes hollow stems of castor, *Ricinis communis* and sarkanda, *Arundo* (Fig. 1). The internodes of preferred length and internal diameter are taken by females choosing a nesting site. These tunnels are tightly packed in a wooden rectangular basket or hive (Fig. 1). These baskets, with their front a bit tilted downward (to avoid the spilling out of nesting tubes) are put in the bee shelters near the edges of crop areas. Wooden baskets can easily be constructed. The cells of megachilids and immatures of xylocopines from the nest tunnels can be taken out by tearing them open with a knife (Fig. 2). If

1) Bee nests.

Bee nesting behaviour and temperature preferences are shown below in the Tables.

Nesting tunnels & nest block number	% tunnels accepted 1985	1986	1987
Sarkanda (Arundo sp.)			
1	15	34	58
2	17	24	53
3	16	28	52
Castor (Ricinus communis)			
1	12	32	61
2	15	25	52
3	14	30	55

Table 1. Relative acceptance of sarkanda *(Arundo)* and castor *(Ricinus)* tunnels in different years by wild pollinators of crops

Fig. 2. (left) cells of *Chalicodoma rubripes* (formerly called *Megachile flavipes*) and *C. lanata (M. lanata)* in opened nesting tunnels of castor stems (from Sihag, 1992).

Fig. 3. above: particle board nesting domiciles taken apart to show two halves of the structure and the paper soda straws in place within the drilled tunnels, below: drilled nesting tunnels in particle boards, fitted in an iron frame

needed, these cells containing the immature bees can be transported to be stored in the desired places. New tunnels are provided again to replace the used ones.

2) Soda straws (for megachilids only).
This involves the use of paper drinking straws of different diameters. Straws are cut with sharp blades into desired lengths, with one end kept open and other end closed with cotton plugs. The straws are tightly packed in small wooden boxes carrying a thin cotton sheet at the base and closed from one side (see Fig. 1). These boxes are kept in the wooden baskets and placed in bee-shelters. The cells can be removed simply by tearing off these tunnels by hand if need be. The limitation of such nesting tunnels is that they are very easily taken out by the nest predators like birds and reptiles.

3) Wooden block—soda straw complex (for megachilids only).
This nesting complex consists of rectangular wooden blocks having holes, uniformly drilled, 10-12 cm in length and 4.5 to 7.5 mm in diameter, depending upon the size of the bee species to be domiciled (see Figure). These domiciles make a permanent structure. Straws can be put inside the holes for the easy removal of the cells. One end of the holes is kept blind either by taking a block of larger breadth or by applying an iron or metal sheet behind to check the activity of parasites and nest destroyers. Wooden blocks are tightened together with ropes or metal wires and then are placed or hung in the bee shelters (see Fig. 4).

4) Particle board—soda straw complex (for megachilids only).
The most handy and permanent structures are the domiciles of particle board, though these are also the most costly domiciles. The particle boards with holes of desired length and diameter drilled between pairs are put one above the other and fitted with the help of nuts and bolts on the side iron rods in the iron frames. (Fig. 3). The whole

Bee species	*Crops Pollinated*	*Nesting behaviour*	*Nest tunnel accepted* diameter (mm)	length (cm)	*T° for activity*
M. haryanaensis (M. nana)	Alfalfa	Leafcutter bee	4.5-5.5	10-12	26-43°C
C. rubripes (M. flavipes)	Alfalfa	Mason bee	5.5-6.5	10-12	28-43°C
C. lanata (M. lanata)	Pigeon pea	Mason bee	6.5-7.5	10-12	26-38°C
C. cephalotes (M. cephalotes)	Pigeon pea	Utilizes resin/animal faecal material for partitioning and closing the cells/tunnel	5.5-6.5	10-12	27-41°C
X. fenestrata	Cucurbits	Carpenter bee	10-12	23-30	25-48°C
X. pubescens	Cucurbits	Carpenter bee	10-12	23-30	25-48°C

Table 2. Nesting behaviour, nest acceptance and temperature tolerances of sub-tropical megachilid and xylocopine bees.

structure is then hung in the bee shelter with the help of metal wires. After the filling of the tunnels with bee cells, the particle boards can be opened and the cells can be taken out (if need be) with the help of fine forceps. However, if difficulty arises in the removal of cells, the soda straws can be slipped inside the holes (one side kept blind) from which cells can be taken out as described above (Fig. 3). The use of particle boards has an advantage that they can be cleaned very easily and can be reset in the form of blocks for fresh use. For this purpose, individual particle boards of eight blocks are serially numbered.

Fig. 4. permanent bee shelters; above: made from mud and angle-iron, some nests suspended from ceiling, below: shelter erected at the edge of an alfalfa field, showing nests suspended within the shelter

5) The bee shelter.

Bee domiciles are protected from sun and rain in the bee shelters. A moveable bee shelter is of utmost importance for taking the bee domiciles from one place to the other. However, a permanent bee shelter that is erected beside the field, and fixed in one place can also serve the purpose, where the target crop is grown year after year. The shelter can be constructed on angle iron covered on the top with the help of asbestos or cement sheets firmly gripped with the help of iron hooks, or it can be in the form of a simple mud hut. (Fig. 4). In the former type of shelter, between the horizontal angle-irons, a space is left for placing the wooden baskets or hanging the wooden blocks or particle board complexes (Fig. 3). In the latter type of shelters the domiciles are suspended by ropes from the roof supports, or iron rods, and so forth, or placed on the supports positioned for this purpose along the sides of the walls. However, there is a great risk of ants attacking the immature bees. This can be prevented either by placing containers filled with water at the bases or by applying one of the available insecticides that is not highly toxic to bees (see **Appendix III**), around the bottom of nest boxes.

6) Colonization of domiciles by bees.

Megachilid bees can be procured during the dormancy season (at Hisar, India from December to February) as diapausing larvae, whereas xylocopine larvae can be procured when *X. fenestrata* engages in brisk provisioning activity (at Hisar from March to October). The larvae are allowed to hatch in the desired places. These bees construct nest cells after emergence in the same tunnel or in its vicinity. Males are the first to emerge and when about 40% males have emerged, females start emergence. After mating, these bees again build their populations subject to the availability of forage. The crop grower's task is to provide appropriate nesting material in sufficient number in the bee shelters.

Once sufficient numbers of the bees have built on the nesting site these can be dispersed in the desired areas. The population can also be procured or augmented by the method of 'trap-nesting'. Under this technique, small nesting structures are placed in bee shelters near the reference crop grown at different localities. The nesting tunnels with immatures are procured from the hives and taken to the desired place where after emergence these will colonize and establish their population.

In this way, these bees can be domiciled and managed in desired regions. Since these bees work at high-temperature regimes, they can be introduced in many places having conditions necessary for setting and maturation of alfalfa, pigeon pea and cucurbits. *Chalicodoma lanata* is active at relatively lower temperatures which coincide with the blooming period of pigeon pea. *Xylocopa fenestrata* is active at a wide range of temperatures when a large number of cucurbits come in to bloom. Alternate waste stem tunnels with hollow pith may be tried for keeping these and other related bees, depending upon availability in different tropical zones of the globe.

The management and utilization of these pollinators where honeybees are not always useful has become most essential for the pollination and increased seed and fruit production of crops. The practice takes advantage of unused natural resources. Some solitary megachilid and xylocopine bees play an important role in the increase of crop production. Devising a methodology for their domiciling and management will help to solve this problem to a great extent and the implications seem to be far-reaching, especially in the poor developing and undeveloped countries where the economy is shattered and high technology is far from the reach of a common man. Very simple technology suggested here, with marginal or no inputs but immediate utility, shall make a viable tool for the seed and fruit growers to produce their crops in the arid and tropical parts of the world, where these bees are adapted.

Use of hollow stems of castor (*Ricinus*), stout grasses (*Saccharum*) and sarkanda (*Arundo* spp.) and many others (which will simply go waste), along with wooden baskets to hold them require no inputs. Rather, these waste materials shall be utilized for a very useful purpose for keeping the pollinating bees, due to which benefits in crop seed production should be very high. For other techniques, the growers shall have to apply some inputs (see also Sections 1.6, 2.1). The bee houses can even be constructed with mud (see Fig. 4) where, in nature, megachilid bees very often go for nest construction, in crevices of walls and logs. However, the pollination scientists would be required to conduct surveys of the local host plants of these bees. Kapil and Jain, and Sihag (see Additional References) have already listed some host plants of these bees, which can be introduced along with the bees.

Because these bees select tunnels of different diameters with some overlap, and the new emergents tend to nest in or near the parental nests, the bee groups can easily be segregated in separate shelters to minimize competition between species. In addition, many of the natural enemies of these bees are recognized, which should make their management somewhat easier.

Acknowledgment. The author wishes to thank Prof. R. B. Mathur for general facilities. The article was prepared under the project 'C(a)-Zoo-2-NP-Agri.', sponsored by the state government of Haryana.

2.5.6 *Trap-Nesting Solitary Bees in the Humid & Wet Tropics*

Results given here are from using a variety of wooden and paper straw trap nests in central Panama and Yucatan, Mexico during ten years. This is a substantially different habitat from the subtropical and more arid places where trap nesting has been used for pollinators (see Sections 1.7, 2.5.5). Annual rainfall is between two and four meters in these lowland tropical habitats, making them fall within the range of tropical moist and wet forests that have a distinct dry season — usually of three to four months.

• *Material preparation*

The wooden nesting blocks or boards were prepared with non-treated, oven-dried wood, usually pine or medium hard to soft woods with a relatively fine grain. The wooden trap-nest blocks were prepared by drilling holes, in the direction of the grain, 8-10 cm in length. The inner end of the tunnel was left blind when no soda straws were used (see Section 2.5.6). When paper soda straws were placed within the drilled tunnels, in large nesting blocks of approximately one meter length by 15 cm high by 8 cm deep, holes were drilled completely through the block. The paper soda straws were bent closed at one extreme after being inserted in the holes. Holes were drilled of varying diameters (1/8, 1/4, 3/8 and 1/2 inches) or all of a single diameter in each block. Blocks employed were either section of 2 x 4 inch construction timber, or larger wooden blocks, in which several hundred to a few thousand holes could be drilled. Spacing of the holes varied from one to two cm between the individual tunnels.

• *Placement of artificial domiciles for bee capture*

The best areas for successful capture of nesting bees were in protected situations, and where some bees could already be seen nesting. Frequently there were under the edges of wooden or palm thatch roofs, or next to large dead trees. Placement within the forest usually resulted in high colonization rates by various ants and *Trypoxylon* wasps. Therefore, abandoned buildings or rural houses near relatively forested areas were sought for domicile placement sites.

• *Bees commonly using trap nests*

The principal types of bees induced to nest in the blocks were *Centris analis*, *Megachile* spp., *Heriades* and *Paratetrapedia*. The first species is extremely common from Mexico to Brazil and is one of a strictly Neotropical group of anthophorines. This small centris and a few other species that colonized tunnels preferred trap nest tunnels of 1/4 to 3/8 inch diameter. They would be ideal pollinators for buzz-pollinated crops such as acerola, *Malpighia* , which is not visited by honeybees or many other bees. *Centris analis* , unlike many bees of this genus, is not highly seasonal — it has seven yearly generations and a developmental (egg to adult) time of 45 days (see Roubik, 1989). The next two genera are megachilids, with the latter considerably smaller and usually colonizing only the tunnels of 1/8 inch diameter. At least four megachilid species commonly used the tunnels and appeared to be seasonal, largely active during the dry season. The *Paratetrapedia* was strongly seasonal, only seen at nests during the dry season to wet-season transition period. It used tunnels of 1/4 inch diameter.

•*Nest parasites and pests*

The persistent parasites detected for *Centris* were the chalcidoid wasp, *Leucospis*, and a few species of the cleptoparasitic megachilid, *Coelioxys*. Each of these lays its eggs in the nest of a host bee, usually a megachilid or anthophorine. Stingless bees, *Trigona muzoensis* and *T. corvina*, also came to the nest cells of *Centris analis* to collect the resin-like nest entrance plugs deposited by the female bees when their nests were completed — presumably to decrease nest accessibility to the *Leucospis*, which slowly chews through the plugs and then enters the nest to parasitize the cells. The *Trigona* arrived in small groups of 5 -15 bees to interfere with the activities of nesting *Centris* or other bees by attacking the adults outside the nests, but they were never a serious problem. Other small parasitic wasps or mites arrived in the cells of some of the bees but were infrequently seen.

•*Recommendations for future work*

The trap-nesting bees of the lowland Neotropics should be explored as potential pollinators for a variety of crops. Some of the drawbacks are that the same nesting sites attract insects such as *Trypoxylon* wasps, which use potential nesting tunnels and seal them off with mud. In addition, the common *Centris* use a strategy to deceive their natural enemies, primarily *Leucospis*, that uses a large number of nest tunnels — they seal them off with the resin-like plug, even though they have not constructed or provisioned cells within them. This behaviour makes the use of nest blocks with several hundred holes more likely to produce the desired results.

Additional References (see also Section 1.7)

Beekeeping in Rural Development. 1979. Commonwealth Secretariat: International Bee Research Association. London: Marlborough House, Pall Mall, U.K.,196 pp.

Danka, R. G., Gary, N. E. 1987. Estimating foraging populations of honeybees from individual colonies. J. Econ. Ent. 80:544-547.

Jay, C. 1986. Spatial management of honeybees on crops. Ann. Rev. Entomol. 31:49-65.

Kapil, R. P., Jain, K. L. 1980. Biology and utilization of insect pollinators for crop production. Hisar: Haryana Agricultural University, Final Technical Report, 81 pp.

Nogueira-Neto, P. 1970. A criaçao de abelhas sem ferrão. São Paulo: Chacaras e Quintais.

Nogueira-Neto, P., Imperatriz-Fonseca, V. L., Kleinert-Giovannini, A., Viana, B. F., de Castro, M. S de. 1986. Biologia e manejo das abelhas sem ferrão. São Paulo: Tecnapis.

Roubik, D. W. 1979. Nest and colony characteristics of stingless bees from French Guiana. J. Kansas Entomol. Soc. 52:443-470.

Roubik, D. W. 1983. Nest and colony characteristics of stingless bees from Panamá. J. Kansas Entomol. Soc. 56:327-355.

Roubik, D. W. In press. Wild bees of Brunei. In: Proceedings of the Symposium on Current ~Issues in Tropical Rain forest Research, D. Edwards and Lord Cranbrook, (eds.) Royal Geographic Society-University of Brunei Darussalam: The Brunei Rain forest Project, 1991-92.

Sakagami, S. F., Ohgushi, R., Roubik, D. W. (eds.) 1990. Natural history of social wasps and bees in equatorial Sumatra. Sapporo: Hokkaido Univ. Press, 274 pp.

Sihag, R. C. 1983. Life cycle pattern, seasonal mortality, problem of parasitization and sex ratio pattern in alfalfa pollinating megachilid bees. Z. angew. Ent. 96:368-379.

Sihag, R. C. 1988. Characterization of the pollinators of cultivated cruciferous and leguminous crops of subtropical Hisar (India) Bee Wld. 60:153-158.

Sihag, R. C. 1990. Behaviour and ecology of the subtropical carpenter bee, *Xylocopa fenestrata* F. 2. Host plant association. Indian Bee J. 52:38-40.

Sihag, R. C. 1992. Utilization of waste stems of sarkandas and castor as nesting tunnels for culturing/keeping wild pollinators of some crops. Bioresource Technology 42:159-162.

Sihag, R. C. in press. Behaviour and ecology of the subtropical carpenter bee, *Xylocopa fenestrata* F. 6. Foraging dynamics, crop hosts and pollination potential. J. Apic. Res

Sihag, R. C. in press. Behaviour and ecology of the subtropical carpenter bee, *Xylocopa fenestrata* F. 7. Nest preferences and response to nest translocation. J. Apic. Res.

2.6. Resources for Funding Pollination Research

Much of the agricultural potential in temperate climates, both temperate and tropical, has already been exploited. However, much of the tropics remains for rational exploitation that is most urgent because many tropical countries are thickly populated, yet most are poorly developed, and some face food shortages. All possible efforts will have to be made to increase the *efficiency* of agricultural production in the tropics. Many agricultural inputs are now being used or tested in the tropics. Many beekeepers already realize the potential benefits to be realized from lending their colonies for pollination services. However, managed pollination as an input in agriculture has not yet become popular among the crop growers. The simple reasons are lack of knowledge about managed pollination and also of funds, facilities and technical know-how. It is now well established that managed pollination is essential for many tropical crops and its role is extraordinary, and in some cases even indispensable. Efforts for creating facilities and technical training to endeavor to provide adequate pollination service, at least as one of many agricultural inputs, is essential.

Some of the better-known resources for funding pollination research and for securing technical training are listed below. Since managed pollination relates closely to beekeeping, development of beekeeping should mean development of managed pollination. Research in these two areas should be highly interdependent. Most of the funding agencies given here assist and support development projects. Because many tropical countries undeveloped or have a poor economy, these projects are of direct relevance to them.

• Resources for funding pollination research and development projects include:

a) The Food and Agriculture Organization (FAO) of the United Nations is a nodal agency for international funding on agricultural and rural development. Within its Agricultural Development programs, it has i) the FAO Government Cooperative Programme which allows one nation to help another nation directly; ii) direct help through Research and Technical Development Division (RTDD) of FAO, which helps to promote individual and cooperative research efforts among member nations; iii) direct help through 'Interaction Information System for Agricultural Sciences (IGRIS), which organizes meetings or workshops to provide information on agricultural development funding exists also for, iv) rural development, and v) women in agriculture — a role of women in development of beekeeping makes a sound basis for FAO assistance in the tropics.
FAO headquarters
Food and Agriculture Organization
Villa delle Terme di Carcalla
00100, Rome, Italy

b) United Nations Educational, Scientific and Cultural Organization (UNESCO) promotes collaboration among national through education, science and culture. Funding may come from i) technical and financial assistance to regional and subregional networks of educational innovation for development, for which innovations in agricultural education should form a basis; ii) technical assistance to strengthen infrastructure and launch projects on science and technology education; UNESCO manages funds from World bank and UNDP (United Nations Development Programme; technical assistance in pollination research and its development education should form a basis; iii) funding of research projects on ecological sciences, which is of direct relevance to pollination and ecosystems (especially agro-ecosystems).
UNESCO headquarters
UNESCO House, 7 Place de Fontenoy
75700, Paris, France

c) World Bank (WB) extends financial assistance in three forms: loans for agriculture and rural development, revolving funds for exporters of agricultural products, and for continuing support of executing agencies for projects funded by UNDP. Because managed pollination directly comes under all the three funding modalities of WB, this can easily be benefited from the WB financial assistance. Nations can apply.
The World Bank
1818, H Street N.W.
Washington D.C. 20433, USA.

d) International Development Agency (IDA) provides assistance primarily in poorer developing countries, as does WB, but only directly to the governments. In agriculture, funds are provided for agricultural research and development, agro-industry, irrigation and drainage, fisheries, livestock, training and technical assistance. Funds are also given to increase technical labour power of a country. Pollination research training for the utilization of managed pollination by qualified persons should provide a basis in the tropical countries to secure this assistance.
IDA headquarters
International Development Agency
1818, H Street, N.W.
Washington D.C. 20433, USA

e) International Fund for Agricultural Development (IFAD) provides concessional term resources for agricultural development in developing countries, emphasizing self-sufficiency and self-reliance. Through IFAD loans, small farmers, laborers, artisans, fisherman and herdsmen are benefited from the strengthening of existing local organizations, the promotion of new farm groups and use of village associations. Beekeeping, the central power of managed pollination, can be benefited from IFAD assistance.

International Funds for Agricultural Development , 107
Via del Serafico
00142, Rome, Italy

f) World Health Organization (WHO) supports projects on 'food and nutrition' in collaboration with UNICEF/UNDP/FAD. Projects on quality food production and nutrition are accepted. Projects on honey production through beekeeping should find support from WHO.
WHO headquarters
World Health Organization
Avenue Appia, 1211
Geneva 27, Switzerland
liaison Office with the UN:
WHO, New York, New York, 10017, U.S.A.

g) other UN funding sources: i) United Nations Environment Programme (UNEP) supports projects on environmental conservation. Bees and many other pollinators form an important component of environment; ii) United Nations Financing Systems for Science and Technology for Development (UNFESSTD) provides funds for projects in developing countries to enhance their scientific and technical capabilities, especially through training; managed pollination in crop and seed production should find a place to secure funds from UNFESSTD; iii) United Nations Development Programme (UNDP is a model UN organization, funding development programmes the world over, especially in developing countries; UNDP provides assistance to the projects of great national importance including related to agriculture, and development of managed pollination should find high priority.

h) Other regional banks: i) Asian Development Bank (ADB) supports limited programmes of agricultural research, headquarters: Asian Development Bank, P.O. Box 789, Manila, 2800, Philippines; ii) African Development Bank

- International funding resources. Several governments and agencies provide funds for research and development in developing countries. Some of these are: i) USAID: U.S. Aid for International Development (USA); ii) CIDA: Canadian International Development Agency (Canada); iii) DANIDA: Danish International Development Agency (Denmark); iv) Dutch government Assistance for Agricultural Development in Developing Countries; v) German Government Assistance for Agricultural Development in Developing Countries
- Bilateral cooperation and mutual funding between governments on agreed disciplines or priority areas, such as development of managed pollination technical capability and pollination research, through mutual funding and exchange programmes.
- International collaboration, via institutes in tropical countries, may support pollination research. Some of the following may provide information on programmes or support

a) International Institute of Tropical Agriculture, Oyo Road, PMB 5320, Ibadan, Oyo State, Nigeria.
b) Caribbean Agricultural Research and Development Institute, University Campus, St. Augustine, Trinidad
c) Instituto Nacional de Technologia Agropecuaria, Rivadavia 1439, 1033 Buenos Aires, Argentina.
d) International Center for Tropical Agriculture Apartado Aereo 67-13, Cali, Valle del Cauca, Colombia
e) Smithsonian Tropical Research Institute, Apartado 2072, Balboa, Panama.
f) Instituto Agronomico de Campinas (São Paulo Agricultural Institute) C. P. 28, Avenido Barao de Itapura 1481, 13100, Campinas, São Paulo, Brazil.
g) University of Guelph, Department of Environmental Biology, Guelph, Ontario, N1G 2W1, Canada
h) Tropical Agricultural Research and Training Center, Turrialba, Cartago, Costa Rica
i) National Research Centre, El-Tahrir Street, El-Dokki, Cairo, Egypt
j) Universiti Pertanian Malaysia, Faluti Pertanian, 43400 UPM Serdang, Selangor, Darul Ehsan, Malaysia
k) Haryana Agricultural University, Hisar, India.
l) University of Phillippines, Los Baños, 3720, Laguna, Philippines
m) University of Ibadan, Ibadan, Oyo State, Nigeria
m) Sind Agricultural University, Tandojam, Bind, Pakistan
n) Agricultural Research and Extension Institute, Nakornsrithammaraj Agricultural Campus, Tungsong, Nakornsritharnrnaraj, Thailand.
o) University of Zambia, P. O. BOX 32379, Lusaka, Zambia.
p) National Institute of Research on Biological Resources, Apartado Postal 63, Xalapa, Veracruz, Mexico
q) National Agricultural University, Apartado 456, Avenida, La Universidad S/n, La Molina, Lima, Peru
r) University of Natal, Private Bag 375, Pietermaritzburg 3200, Natal, South Africa
s) Board on Science and Technology for International Development, Office of International Affairs, National Research Council, 2101 Constitution Avenue, Washington, D. C. 20418
t) INRA, Laboratoire de Zoologie, Lusignan 86600, France
(see also references in Section 1.6)

- Informal journals specifically for research on pollinators and beekeeping in developing countries also exist, among them are: i) *Sphecos,: a forum for aculeate wasp researchers*, c/o National Museum of Natural History, Smithsonian Institution, Washington, DC 20560, Attn.: Dr. A. Menke, Systematic Agricultural Laboratory, NHB Stop 168; ii) *Melissa: the melittologist's newsletter*, Department of Entomology, Smithsonian Institution, NHB-105, Washington, DC 20560, Attn.: Dr. R. McGinley; iii) *Cornucopia* (International Agency for Apiculture Development, P. O. Box 240, Litchfield, Ohio 44253-0240, USA; iv) *Beekeeping & Development* (International Bee Research Association, 18 North Road, Cardiff CF1 3DY, United Kingdom; Fax 0222-665522)

Glossary

Abiotic: not involving living things

Achene: a small, one-seeded fruit in which the thin walls are dry

Agamospermy: seed formation without sexual reproduction

Aggregate fruit: one consisting of the many separate carpels of one flower

Allogamy: Fertilization between pollen and ovules of different flowers

Androdioecy: where male and hermaphrodite genets coexist

Androecium: the male unit or stamens as a unit of the flower

Andromonoecy: where a hermaphrodite bears male and hermaphrodite flowers

Anemophily/anemophilous: wind-pollination; plants that normally shed pollen carried by wind

Angiosperm: flowering plant — a major group of seed plants in which seeds develop within a closed ovary

Anther: part of the floral stamen that normally produces pollen

Anthesis: time of opening of the flower

Anthophilous: flowering-loving — applied to an animal that can be a pollinator

Apogamety: autonomous development of a nucleus, apart from the egg nucleus into an embryo in an agamosperm

Apomixis/apomictic: non-sexual reproduction of a plant, including both forms in which no seeds are produced *(vegetative reproduction*) and those in which seeds are produced *(agamospermy)*

Apospory: the elimination of spore formation from the life cycle with the formation of the gametophyte from vegetative tissues, usually the nucellus, and not from a spore

Archesporium: tissue within the nucellus of a young ovule that gives rise to the embryo-sac mother cell, female meiosis and the embryo-sac

Aril: the network or covering of a seed from the point of seed attachment

Autogamy/autogamous: self-fertilization (within a flower) without the need of a pollinator

Automixis: fusion of nuclei within the embryo-sac

Berry: a fleshy fruit with skin-like covering, having one to many seeds (but no stone), developed from a single pistil

Bract: a small leaf or scalelike structure near the base of a flower

Calyx: sepals or outer whorl of the perianth

Cantharophily/cantharophilous: pollination by beetles

Caprification: pollination of figs with certain tiny wasps

Capsule: a dry, dehiscent seed pod from a flower with a compound pistil

Carpel: the unit of an ovary within a compound pistil

Catkin: Spike or pendulous type of inflorescence made up of flowers of only one sex

Chalaza: the basal part of an ovule where it is attached to a stalk (funiculus)

Chalazogamy: the entry of the pollen tube through the chalaza of the ovule

Chasmogamous: having flowers in which pollination occurs while the flower is open

Chiropterophily: pollination by bats

Cleistogamy/Cleistogamous: having flowers that are self-fertilized without opening (*close-fertilization*)

Chiropterophily/chiropterophilous: pollinated by bats

Clone: one or more individuals obtained from a single parent by vegetative reproduction, i.e. clone plants are *ramets* that belong to the same *genet*

Compatible: capable of producing fertile offspring between plants

Corolla: the whole sheath of petals of a flower

Cross: the union of two cultivars of the same species

Cross-compatible: capable of being fertilized with pollen of a different variety.

Cross-pollination: pollination in which the pollen is transferred to another flower on a different plant of the same species

Cultivar: a group of cultivated plants which, when reproduced sexually or asexually, retain their distinguishing characteristics (but are not necessarily a distinctive botanical species)

Dehisce: opening of a seed pod or anther and release of contents

Dichogamy: a floral condition in which male and female parts mature at different times, preventing self-pollination — in *synchronous dichogamy* the stamens and pistils reach maturity at different times in definite periods, as in the avocado, *Persea*

Dicliny: where not all genets in a population are hermaphroditic, such that males, females or both occur

Dioecy/dioecious: having separate sexes and two kinds of genets, such that stamens or male parts and pistils or female parts are on different plants.

Diploid: having a double set of chromosomes, usually one set from each parent

Diplospory: the development of an apomictic embryo sac by mitosis or modified meioses of the archesporial cells

Drupe: a succulent or fleshy fruit having one seed enclosed in stony endocarp

Elaiophore: a floral oil-secreting gland

Embryo: a rudimentary organism

Embryo-sac: the female gametophyte of flowering plants contained within the ovule, developing from the surviving megaspore after female meiosis, and containing eight nuclei

Embryony: the development of an embryo

Endocarp: inner layer of the pericarp

Endosperm: the food reserve tissue in a seed, triploid in angiosperms, formed from a fertilizing sperm cell combining with the fused polar nuclei

Entomophily/entomophilous: pollination by insects

Exocarp: the outermost layer of the fruit wall

Exserted: protruding beyond the margin of envelope or corolla

Exrareproductive nectary: a nectary on part of the plant outside of the flower

Female flower: one with a stigma

Fertile: capable of bearing fruit

Fertilization: the union of male and female gametes to produce a zygote

Filament: a hairlike element of stamen supporting the anther

Floral nectary (reproductive nectary): a nectary within a flower

Floret: one small flower

Flower: a much shortened axis (the receptacle) bearing whorls of appendages concerned with reproduction — sepals, petals, stamens and carpels

Fruit: mature ovary with all its parts and adherents

Gamete: sexual cell

Geitonogamy: fertilization between pollen and ovules of different flowers on the same genet

Gametophyte: the gamete-producing phase of algae, bryophytes and ferns, displaying alternation of genetations; composed of haploid cells only

Genet: a genetically individual plant, resulting

from a single sexual fusion or zygote, and consisting of one to many ramets

Germination: development of plant from seed

Gynodioecy:/gynodioecious: where female and hermaphrodite genets coexist

Gynoecious: producing only or predominantly pistillate flowers

Gynoecium: the female parts, the carpels, of a flower

Gynomonoecy/gynomonoecious: where a hermaphrodite bears both female and hermaphrodite flowers

Haploid: having a single set of chromosomes from a single parent; usually refers to a germ cell or gamete

Herkogamy/herkogamous: separation of anthers and stigma within a flower such that autogamy cannot occur in the absence of a pollinator

Hermaphrodite: (see *perfect flower*)

Heterogamy/heterogamous: a plant having two or more kinds of flowers

Heterostyly/heterostylous: a genetically-determined condition in which stamens and styles come in two or three distinctive lengths, and individual flowers have stigmas and styles of different lengths — thus promoting crossing; e.g. distyly and tristyly

Homogamy/homogamous: coincidence of anther dehiscence and stigma receptivity within a flower, so that autogamy is possible if herkogamy does not exist

Indehiscent: not splitting open by its valves

Indeterminate: continuing to grow after flowering starts

Inflorescence: flower or group of flowers on a stem

Intersterile: failure to set fruit when flowers are crossed with pollen of certain other cultivars of the same species

Legume: a one-celled fruit (pod) usually dehiscing down both sutures, and having the seed attached along a ventral suture

Locule: compartment or cell of ovary, anther, or fruit

Megaspore: in plant species producing two kinds of spores, the spore-type that gives rise to the female gametophyte

Meiosis: the reduction division of chromosomes, giving rise to two gametes, each with half the chromosomes of the parent cell

Melittophily/melittophilous: pollination by bees

Mesocarp: the middle layer of pericarp or fruit wall

Micropyle: the pore or hole in end of egg through which sperm enters to fertilize the egg

Mitosis: the ordinary changes through which a cell nucleus passes during cell multiplication, producing daughter cells of chromosome number equal to the parent cell

Mixed inflorescence: a branched inflorescence with both racemose and cymose components, as in grape and mango, in which the main inflorescence axis is racemose, the small ultimate branches cymose

Monoecy/monoecious: having separate male and female flowers, but on the same plant

Multiple fruit: a fruit consisting of the compressed fleshy fruitlets of the many flowers of a compact inflorescence in which the axis also becomes fleshy, as in pineapple, custard apple and *Monstera*

Myophily/myophilous: pollination by flies

Nectar: sweet liquid produced in the nectary of plant, usually within a flower

Nectar guide: Certain ultraviolet-reflectant markings on a flower that guide or direct nectar feeders to the nectar source

Nectary: a plant gland that secretes nectar

Nucellus: the central body of the ovule containing the embryo-sac, which acts as a nurse to the archesporium

Oosphere: the unfertilized female gamete

Ornithophily/ornithophilous: pollinated by birds

Outcrossing: sexual reproduction between different genets , usually plants that are different individuals

Ovary: a seedcase or part of the pistil, bearing ovules that develop into seed or fruit

Ovule: the structure that contains the egg nucleus and develops into a seed after fertilization of the egg-cell inside it

Panicle: an inflorescence with the main axis branched into an open racemose flower cluster

Papilionoceous: butterflylike, pealike flowers, with a large upper petal, two lateral wing petals, and two small united keel petals

Papillae: specialized epidermal cells of the stigma which receive the pollen grains

Parthenogenesis: production of new individuals from unfertilized egg cells

Parthenocarpy/parthenocarpic: development of a fruit without fertilization and therefore without seeds — as in navel orange, some figs, seedless grapes, pineapple and banana

Pedicel: stalk or stem of individual flower of inflorescence

Peduncle: the primary flower stalk of inflorescence

Perfect flower: bisexual or *hermaphroditic*; a flower having both stamens and pistil

Perianth: the entire floral envelope including both corolla and calyx

Pericarp: the ovary or fruit wall

Perisperm: storage tissue similar to the endosperm but formed from the nucellus

Petal: a leaf or unit of a usually colored corolla or inner floral envelope

Petiole: the leafstalk

Pistil: the normally central, seed-producing part of flower; usually consisting of ovary, style, and stigma

Pistiilate: having pistils but no stamens

Placenta: the surface or tissue part of ovary to which ovules become attached

Pollen: the powdery grains produced by angiosperm anthers or the microsporangia of gymnosperms, which contain the nucleus that fertilizes the oosphere to form a seed

Pollen tube: a thin tubular outgrowth of pollen grain usually upon contact with stigma, and which penetrates style to ovary, permitting sperm nuclei to unite with egg cell

Pollenizer: plant source of pollen for fertilizing receptive stigmas

Pollinarium: in orchids and asclepiads, the structure detached from the male plant bearing sacs of pollen grain, united to an adhesive disc that attaches to the pollinator

Pollinating: Transferring pollen from anthers to stigmas.

Pollination: placement of pollen on a stigma, the first step in plant fertilization

Pollinator: an animal that moves compatible pollen to a receptive stigma of the same plant species, such that fertilization and seed production can occur

Pod: a monocarpellary fruit that dehisces down both sutures

Polyembryony: with more than one embryo in a fertilized ovule, e.g. mango

Polygamy/polygamous: having both perfect flowers and those of one sex (staminate or pistillate)

Polyembryonic: the presence of more than one embryo in a fertilized ovule, formed adventitiously from the nucellus, for example mango (*Mangifera*)

Pome: a fleshy fruit derived from several carpels, the receptacle and outer pericarp being fleshy, and the inner pericarp, papery

Poricidal: anther dehiscence through apical pores

Porogamy: entry of the pollen tube into the ovule via the micropyle

Protogynous: dichogamous condition in which stima receptivity precedes anther dehiscence within a flower

Protandry/protandrous: a flower in which anthers mature and release pollen before the stigma is receptive

Protogyny/protogynous: a flower in which the stigma is receptive before the anthers release pollen

Psychophily/psychophilous: pollination by butterflies and moths

Raceme: an unbranched inflorescence

Ramet: a physiologically independent individual (from one to many may compose a genet)

Receptacle: an enlarged end of a pedicel to which one or more flowers are attached

Robbing: in pollination ecology, the destruction of at least part of a flower to obtain nectar, pollen or another resource

Seed: a fertilized and matured ovule or rudimentary plant, and food necessary for its germination

Self-compatible: a plant capable of being fertilized by pollen within its own group of flowers (*inflorescence*)

Self-fertile: a flower or floret capable of being fertilized by its own pollen

Self-fertilizing: a flower capable of becoming fertile by placing its pollen on its own stigma (*self-pollinating* and self-*compatible*, or *autogamous* and cleistogamous)

Selfing: fertilization of an ovule by a pollen grain of the same genet

Self-pollinating: (autogamous) capable of placing its own pollen upon its own stigma

Self-sterile: incapable of becoming fertilized by its own pollen

Semi-compatibility: where two genets share some but not all gametophytic compatibility traits, thus in some crosses some pollen grains can effect fertilization and others cannot

Sepal: the outermost part of flower, the parts of which form the calyx

Sessile: sitting, lacking a stalk or petiole

Sexual reproduction: reproduction through union of male and female gametes, (as opposed to *vegetative reproduction*)

Spadix: the fleshy axis of certain inflorescences, such as those of arum lilies, bearing the small flowers or florets

Spathe: a large bract enclosing the flower cluster

Spike: an inflorescence with elongated main axis and sessile flowers

Sporangium: a spore-producing organ

Sporophyte: the spore-producing, diploid phase of a species displaying alternation of generations

Stamen: the male part of a flower consisting of a filaments and anthers

Staminate: being entirely male, bearing only stamens

Staminode: an abortive stamen

Standard: a large upper petal of a papilionaceous legume flower

Stem: main axis of a plant

Syconium: the multiple fruit of a fig in which the edible receptacle (flower axis) is hollow and lined on the inside with numerous fruitlets and seeds

Thievery: in pollination ecology, removal of a floral reward by an animal, where pollination does not follow as a result (as opposed to *robbing*)

Triocey: where coexisting genets are male only, female only, and hermaphrodite (complete polygamy)

Tripping: the release of sexual column (in legumes)

Tuber: the swollen end of an underground stem containing food reserves

Unisexual or imperfect flowers: flowers in which either stamens or pistils are functional, the nonfunctional member may be completely lacking, as in pistillate flowers of pawpaw, or rudimentary or aborted, as in fig, banana, some grapes, and in the staminate flowers of pawpaw

Sterile: barren, unfruitful, incapable of being fertilized

Stigma: the receptive portion of the female sexual column

Style: the part of the sexual column between the ovary and stigma

Tuber: a much enlarged fleshy underground stem or branch with eyes (buds) on sides.

Vegetative reproduction: asexual reproduction (see *apomixis*)

Visitor: all animals visiting a flower, but not necessarily a pollinator

Xenogamy: fertilization between pollen and ovules between different genets

Zoophily: pollination by animals

Zygote: a product of the two gametes or germ cells

Introduction to Appendices

Appendix I. The number of species grown in tropical countries and harvested in semi-wild habitats is enormous. A list given here includes those widely accepted as economically important; most are planted, while a few are harvested from wild plants that are to some extent tended or planted. A few are used for breeding purposes, crossing them with more familiar crop species of the same genus. I have not included species that are sold as ornamental plants, used for ordinary timber or firewood, or used for home remedies. While a few of the pollinator associations indicated here are unpublished personal observations, most are given in the literature cited in Section 1.1.3. The generic names of pollinators follow their general kind, i.e. bird, bee, bat. Abbreviations for some genera in the Appendix are as follows: Xylo = *Xylocopa* (bee); Pt = *Pteropus* (bat); Mc = *Macroglossus* (bat); Eon = *Eonycteris* (bat); Na = *Nanonycteris* (bat) Ei = *Eidolon* (bat); Rou = *Rousettus* (bat); Bla = *Blastocephala* (wasp), Cera = Ceratopogonidae (fly); Ap = *Apis* (bee); Mel = *Melipona* (bee); Franklin = *Frankliniella* (thrip); Cul = *Culicoides* (fly); Stil = *Stilobezzia* (fly); Ancylo = *Ancyloscelis* (bee); Bo = *Bombus* (bee); Os = *Osmia* (bee); Mega = *Megachile* (bee); Anth = *Anthophora* (bee); Trig = *Trigona* s. str. (bee); Meg = Megachiropteran (bat); And = *Andrena* (bee); Fo + *Forcipomyia* (fly); La = *Lasiohela* (fly). Plant breeding systems are indicated by the following abbreviations: SC = self-compatible; AU = autogamous (and also self-compatible); AG = agamospermous (adventitious embryony indicated by a tilde before this abbreviation); OC = obligately outcrossing (either dioecious, monoecious, self-incompatible or, when used in combination with SC, signifying that the pollen must be carried by a vector — the plant is not capable of selfing) PC = parthenocarpic.

Appendix II. Two sample pollination contracts are given, one and English and one in French. As added appendices to each, hive standards and definitions of terms are provided.

Appendix III. Three lists are given from an extensive review of pesticides and other agricultural chemicals and their known effects on bees. The danger of the chemicals for bees is indicated in each section of the appendix as serious, moderate or slight (when used with recommended precautions). The literature from which the information was taken can be found cited at the end of Section 1.6.

Appendix I

No.	Scientific Name	Common Name	Family	Use	Origin	Breed. Sys.	Pollinators	No.	Common Name
1	*Abelmoschus esculentus*	Okra	MALV	Vegetable	Af-As	AU, OC	bee, Apis, Bombus	879	Abaca
2	*Abelmoschus manihot*	Sunset Hibiscus, Aibika	MALV	Vegetable	SE Asia	AU, OC?		284	Abarco
3	*Abelmoschus moschatus*	Musk Okra	MALV	Essence	S Asia	AU, OC	bee, Apis, Bombus	399	Abata Kola
4	*Abroma augustum*	Perennial Indian Hemp	STER	Fiber	S Asia			609	Abilla
5	*Abutilon indicum*	Country Mallow	MALV	Fiber	S Asia			408	Abyssinian Myrrh
6	*Abutilon theophrasti*	China Jute	MALV	Fiber	Asia			166	Abyssinian Oat
7	*Acacia albida*	Ana Tree	LEG	Forage	Africa	SC, OC?	bee, bird?	1156	Aceituna
8	*Acacia aneura*	Mulga	LEG	Forage	Aust	SC, OC	bee, bird?	811	Acerola
9	*Acacia auriculaeformis*	Tan Wattle	LEG	Timber	Aust	SC, OC	bee, bird?	1299	Adzuki Bean
10	*Acacia catechu*	Curch	LEG	Dye	Neot	SC, OC	bee	1230	Afara
11	*Acacia cyanophylla*	Orange Wattle	LEG	Energy	Aust	SC, OC	bee, bird?	515	African Bitter Yam
12	*Acacia decurrens*	Green Wattle	LEG	Tannin	Aust	SC, OC	bee, bird?	744	African Gum Vine
13	*Acacia decurrens/normalis*Green	Wattle	LEG	Timber	Aust	SC, OC	bee, bird?	955	African Locust Bean
14	*Acacia farnesiana*	Huisache	LEG	Essence	Neot	SC, OC	bee	734	African Mahogany
15	*Acacia graveolens/dulce*	Celery	UMBE	Vegetable	W Asia	SC, OC	bee	329	African Oak
16	*Acacia graveolens/rapaceum*	Celeriac	UMBE	Vegetable	W Asia	SC, OC	bee	554	African Oil Palm
17	*Acacia mearnsii*	Black Wattle	LEG	Tannin	Aust	SC, OC	bee, bird?	932	African Rice
18	*Acacia melanoxylon*	Australian Blackwood	LEG	Timber	Aust	SC, OC	bee, bird?	673	African Spiderflower
19	*Acacia nilotica*	Babul	LEG	Gum	S Asia	SC, OC	bee?	337	African Star-Apple
20	*Acacia pendula*	Weeping Myall	LEG	Forage	Aust	SC, OC	bee, bird?	1114	African Violet
21	*Acacia pycnantha*	Golden Wattle	LEG	Dye	Aust	SC, OC	bee, bird?	1180	African Yam Bean
22	*Acacia rosqedora/amazonica*	Rosewood	UMBE	Essence	Asia	SC, OC	bee, bird?	49	Afzelia
23	*Acacia salicina*	Cooba	LEG	Timber	Aust	SC, OC	bee, bird?	652	Agba
24	*Acacia senegal*	Gum Arabic	LEG	Gum	Africa	SC, OC	bee, bird?	989	Aguacatillo, Coyo
25	*Acacia seyal*	Shittim-Wood	LEG	Forage	Africa	SC, OC	bee, bird?	936	Ahipa
26	*Acacia tortilis*	Umbrella Thorn	LEG	Multiple	Africa	SC, OC	bee, bird?	511	Airpotato
27	*Acacia victoriae*	Narran	LEG	Forage	Aust	SC, OC	bee, bird?	272	Ají
28	*Acanthosicyos horridus*	Naras	CUCU	Nut	W Afr			1251	Ajmud
29	*Acca sellowiana*	Feijoa	MYRT	Fruit	Neot	OC, SC	bee	194	Akee
30	*Acorus calamus*	Sweet Flag	ARAC	Essence	S Asia	SC?	bee, fly?	65	Aleurtes Trisperma
31	*Acrocarpus fraxinifolius*	Shingle Tree	LEG	Timber	S Asia			1142	Alexandrian Senna
32	*Acrocomia aculeata*	Paraguay Palm, Macaya	PALM	Oil	Neot			843	Alfalfa
33	*Acrocomia mexicana*	Coyol, Babosa	PALM	Oil	Neot			238	Algarobilla
34	*Acrocomia sclerocarpa*	Gi-Gi	PALM	Fruit, Oil	Neot			1048	Algarobo
35	*Acrocomia totai*	Mbocayá	PALM	Fruit	Neot			693	Algarrobo
36	*Acrocomia vinifera*	Vinifera Palm	PALM	Fruit	Neot			1049	Algarrobo Blanco
37	*Actinidia arguta*	Tara vine	ACTIN	Fruit	S Asia	OC	bee	68	Alkanna Root
38	*Actinidia deliciosa*	Kiwifruit	ACTIN	Fruit	S Asia	OC	insect, wind, bee, Apis	1010	Allspice
39	*Adansonia digitata*	Baobab	BOMB	Vegetable	Africa		bee, Apis, bat	1053	Almond
40	*Adhatoda vasica*	Malabar Nut	ACAN	Medicinal	S Asia			1058	Almond
41	*Aegle marmelos*	Indian Bael	RUTA	Fruit	S Asia			82	Alyceclover
42	*Aerva tomentosa*	Kapok Bush	AMAR	Fodder	S Asia	OC?	bee?	153	Aman Palm
43	*Aframomum angustifolium*	Madagascar Cardamom	ZING	Spice	Africa	OC?	bee?	85	Amaranth
44	*Aframomum danielli*	Camerouns Cardamom	ZING	Spice	Africa	OC?	bee?	1077	Amboyna
45	*Aframomum granum-paradisi*	Grains of Paradise	ZING	Spice	Africa	OC?	bee?	132	Amendoim
46	*Aframomum korarima*	Korarima Cardamom	ZING	Spice	Africa	OC?	bee?	647	American Licorice
47	*Aframomum melegueta*	Melegueta-Pepper	ZING	Spice	Africa	OC?	bee?	555	American Oil Palm
48	*Aframomum sceptrum*	Tablus Spice	ZING	Medicinal	Africa	OC?	bee?	91	Ammi
49	*Afzelia spp.*	Afzelia	LEG	Timber	Africa			1250	Ammi, Ajowan
50	*Agathis dammara*	Manila Copal	ARAU	Resin	SE Asia			7	Ana Tree
51	*Agathosma betulina*	Buchu	RUTA	Medicinal	Africa			385	Anchoté
52	*Agave angusifolia*	Dwarf Aloe, Babsi	AGAV	Fiber	Neot	OC?	bat, Macroglossus	108	Andrographis Paniculata
53	*Agave cantala*	Cantala	AGAV	Fiber	Neot			789	Angled Luffa
54	*Agave fourcroydes*	Hennequen	AGAV	Fiber	Neot	SC,OC	bat, bee	504	Angleton Bluestem
55	*Agave lecheguilla*	Istle	AGAV	Fiber	Neot			1013	Anise
56	*Agave sisalana*	Maguey, Sisal	AGAV	Fiber	Neot	OC?	wind,bee,bat,Pteropus	193	Annatto, Achiote
57	*Agave spp.*	Pulque	AGAV	Beverage	Neot	SC, OC	bat, bee	1273	Añu
58	*Agave tecta*	Giant Maguey	AGAV	Beverage	Neot			812	Apple
59	*Albizia falcata*	Batai	LEG	Timber	S Asia			813	Apple
60	*Albizia falcataria*	Jeungjing	LEG	Timber	SE Asia			1086	Apple
61	*Albizia lebbek*	Kokko	LEG	Timber	Paleot			1055	Apricot
62	*Aleurites fordii*	Tung-Oil-Tree	EUPH	Oil	SE Asia		bee, wasp	389	Arabica Coffee
63	*Aleurites moluccana*	Candlenut	EUPH	Oil	SE Asia	SC, OC	bee, wasp, thrip	1279	Aramina Fiber
64	*Aleurites montana*	Mu-Oil-Tree	EUPH	Oil	SE Asia	OC	bee, wasp	466	Archucha
65	*Aleurites trisperma*	Aleurtes Trisperma	EUPH	Oil	SE Asia	OC	bee, wasp	543	Argue Pheasant-Tree
66	*Alhagi pseudalhagi*	Camelthorn	LEG	Forage	W Asia			726	Aricauguá
67	*Aliberta edulis*	Trompillo	RUB	Fruit	Neot			1073	Arrayán
68	*Alkanna tuberculata*	Alkanna Root	BORA	Tannin	W Asia			1197	Arrow Poison
69	*Allium ampeloprasum*	Leek	LILI	Spice	Asia	~AG,S-OC	bee, Apis, fly	1265	Arrowleaf Clover
70	*Allium cepa*	Onion	LILI	Vegetable	Asia	~AG,S-OC	bee, Apis, fly	831	Arrowroot, Aru-Aru
71	*Allium chinense*	Rakkyo	LILI	Vegetable	Asia	~AG,S-OC	bee, Apis, fly	607	Asafetida Gum
72	*Allium fistulosum*	Welsh Onion	LILI	Vegetable	Asia	~AG,S-OC	bee, Apis, fly	518	Asian Bitter Yam
73	*Allium sativum*	Garlic	LILI	Spice	Asia	~AG,S-OC	bee, Apis, fly	516	Asiatic Yam
74	*Allium schoenoprasum*	Chives	LILI	Vegetable	Medit	~AG,S-OC	bee, Apis, fly	150	Asparagus
75	*Allium tuberosum*	Chinese Chives	LILI	Vegetable	Asia	~AG,S-OC	bee, Apis, fly	226	Asparagus Broccoli
76	*Alocasia macrorrhiza*	Giant Taro	ARAC	Starch	S Asia			1233	Asparagus pea
77	*Aloe barbadensis*	Barbados Aloe	AGAV	Medicinal	Africa	OC?	bee, Apis, bat, Rousettus	18	Australian Blackwood
78	*Aloe ferox*	Cape Aloe	AGAV	Medicinal	Africa	OC?	bee, Apis, bat, Rousettus	203	Australian Bluestem
79	*Aloe perryi*	Socotrine Aloe	AGAV	Medicinal	Africa	OC?	bee, Apis, bat, Rousettus	569	Australian Desert Lime
80	*Alpinia galanga*	Greater Galanga	ZING	Spice	S Asia			1161	Australian Nightshade
81	*Alpinia officinarum*	Lesser Galanga	ZING	Spice	S Asia			984	Australian Sheepbush
82	*Alysicarpus vaginalis*	Alyceclover	LEG	Forage	Af-As			764	Australian Tea-Tree
83	*Amaranthus caudatus*	Inca-Wheat	AMAR	Cereal	Neot			906	Australian Tobacco
84	*Amaranthus cruentus*	Spanish Greens	AMAR	Cereal	Neot			861	Australian Wild Limes
85	*Amaranthus gangeticus* *	Amaranth	AMAR	Vegetable	S Asia			301	Avaram
86	*Amaranthus hypochondriacus*	Princess-Feather	AMAR	Vegetable	Neot			987	Avocado
87	*Amaranthus leucocarpus*	Grain Amaranth	AMAR	Grain	Neot			154	Awarra Palm
88	*Amaranthus lividus*	Bondue, Bayam Betul	AMAR	Vegetable	S Asia			537	Ayan
89	*Amaranthus mantegazzianus*	Chaquillón	AMAR	Cereal	Neot			452	Ayote

No.	Scientific Name	Common Name	Family	Use	Origin	Breed. Sys.	Pollinators	No.	Common Name
90	*Amaranthus tricolor*	Chinese Amaranth	AMAR	Vegetable	S Asia			280	Babaco
91	*Ammi copticum*	Ammi	UMBE	Medicinal	S Asia			927	Babassu
92	*Ammodaucus leucotrichus*	Cafoun	UMBE	Spice	N Afr			928	Babassu
93	*Amomum aromaticum*	Bengal Cardamom	ZING	Spice	S Asia	OC?	bee, Apis, Amegilla	19	Babul
94	*Amomum compactum*	Large Cardamom	ZING	Spice	SE Asia	OC?	bee, Apis, Amegilla	919	Bacaba
95	*Amomum krervanh*	Cambodian Cardamom	ZING	Spice	SE Asia	OC?	bee, Apis, Amegilla	186	Bacco
96	*Amomum subulatum*	Nepal Cardamom	ZING	Spice	S Asia	OC?	bee, Apis, Amegilla	1031	Bacuri do Parana
97	*Amomum xanthioides*	Tavoy Cardamom	ZING	Spice	S Asia	OC?	bee, Apis, Amegilla	1015	Bakek
98	*Amorphophallus konjac*	Elephant Yam	ARAC	Tuber	SE Asia			826	Balata
99	*Amorphophallus rivieri*	Elephant Yam	ARAC	Tuber	S Asia			1152	Balau, Meranti
100	*Amyris balsamifera*	Balsam Amyris	RUTA	Essence	Neot			911	Balsa
101	*Anabasis aphylla*	Leafless Anabasis	CHEN	Pesticide	Asia			100	Balsam Amyris
102	*Anacardium giganteum*	Wild Cashew	ANAC	Fruit	Neot	SC, OC	fly, bee, wasp, butterfly	863	Balsam-Apple
103	*Anacardium occidentale*	Cashew, Marañon	ANAC	Nut, Fruit	Neot	SC, OC	fly, bee, wasp, butterfly	864	Balsam-Pear
104	*Anacardium rhinocarpus*	Mijao	ANAC	Fruit	Neot	SC, OC	fly, bee, wasp, butterfly	233	Balsamo
105	*Anacolosa luzoniensis*	Galo	OLAC	Nut	SE Asia			1315	Bambara Groundnut
106	*Ananas comosus*	Pineapple, Piña	BROM	Fruit	Neot	OC, PC	bird	489	Bamboo Betoong
107	*Ananas spp.*	Wild Pineapple	BROM	Breed., Fru	Neot	OC, PC	bird	184	Bambusa
108	*Andrographis paniculata*	Andrographis Paniculata	ACAN	Medicinal	S Asia			400	Bamenda Cola
109	*Anethum graveolens*	Dill	UMBE	Spice	Medit	SC, OC	fly, bee	877	Banana
110	*Angelica archangelica*	Garden Angelica	UMBE	Essence	Asia			878	Banana
111	*Aniba duckei*	Rosewood Oil	LAUR	Essence	Neot			966	Banana Passion Fruit
112	*Anisoptera spp.*	Mersawa	DIPT	Timber	SE Asia			1324	Bangle
113	*Annona cherimola*	Cherimoya	ANNO	Fruit	Neot	OC?	beetle	632	Baniti
114	*Annona diversifolia*	Ilama	ANNO	Fruit	Neot	OC?	beetle	39	Baobab
115	*Annona montana*	Cimarrona	ANNO	Fruit	Neot	OC?	beetle	1330	Barba de Burro
116	*Annona muricata*	Soursop, Guanabana	ANNO	Fruit	Neot	OC?	beetle	529	Barbacoa
117	*Annona purpurea*	Torete, Custard Apple	ANNO	Fruit	Neot	OC?	beetle	77	Barbados Aloe
118	*Annona reticulata*	Sweetsop	ANNO	Fruit	Neot	OC?	beetle	783	Barbasco
119	*Annona squamosa*	Sugar Apple	ANNO	Fruit	Neot	OC?	beetle	1199	Barbatimao
120	*Anogeissus latifolia*	Gum Ghatti	COMB	Gum	S Asia			1293	Bard Vetch
121	*Antelaea azadirachta*	Neem	MELI	Oil	S Asia			687	Barley
122	*Anthemis nobilis*	Roman Chamomile	COMP	Essence	Medit			1079	Barwood
123	*Anthocephalus chinensis*	Kadam	RUB	Timber	SE Asia			912	Basil
124	*Anthriscus cerefolium*	Chervil	UMBE	Spice	Medit			526	Basralocus
125	*Anthyllis vulneraria*	Kidney-Vetch	LEG	Forage	Medit			59	Batai
126	*Antidesma bunius*	Bignai	EUPH	Fruit	S Asia	OC	bee?	1153	Batavian Dammar
127	*Apios spp.*	Potato Beans	LEG	Tuber	N Amer			949	Bauh Keluak
128	*Crotalaria*	Celery	UMBE	Vegetable	Eur-As	SC, OC	bee, fly	1012	Bay
129	*Arachis glabrata*	Florigraze Peanut	LEG	Forage	Neot	AU,~OC	bee	886	Bayberry
130	*Arachis hypogaea*	Peanut, Maní	LEG	Oil	Neot	AU,~OC	bee, Megachile, Lipotriches	1062	Beach Plum
131	*Arachis prostrata*	Ground Nut	LEG	Forage	Neot	AU,~OC	bee	990	Bean
132	*Arachis repens*	Amendoim	LEG	Forage	Neot	AU,~OC	bee	192	Beet
133	*Areca catechu*	Betel Palm	PALM	Multiple	S Asia	OC	bee, fly, thrip	498	Beggarlice
134	*Arenga pinnata*	Sugar Palm	PALM	Mult, Sweet.	S Asia	OC	fly, bee, beetle?	162	Belladonna
135	*Armoracia rusticana*	Horseradish	CRUC	Spice	Medit	OC	insect	390	Bengal Almond
136	*Arnica montana*	European Arnica	COMP	Medicinal	Europe			1227	Bengal Almond
137	*Arracacia xanthoriza*	Peruvian Carrot	UMBE	Tuber	Neot	SC, OC?	fly, bee	874	Bengal Bean
138	*Artemisia absinthium*	Wormwood	COMP	Essence	Asia			93	Bengal Cardamom
139	*Artemisia dracunculus*	Taragon	COMP	Spice	Asia			869	Benoil Tree, Murungai
140	*Artemisia maritima*	Maritime Wormwood	COMP	Medicinal	Asia			305	Berangan
141	*Artemisia vulgaris*	Mugwort	COMP	Spice	Asia			362	Bergamot
142	*Artocarpus altilis*	Breadfruit	MORA	Vegetable	Polyne.	OC, PC	wind, fly, beetle	475	Bermudagrass
143	*Artocarpus heterophyllus*	Jackfruit	MORA	Vegetable	SE Asia	OC, PC	wind, fly, beetle	1257	Berseem
144	*Artocarpus integer*	Chempedale	MORA	Fruit	S Asia	OC, PC	wind, fly?	133	Betel Palm
145	*Artocarpus odoratissima*	Morang	MORA	Fruit	SE Asia	OC, PC	wind, fly?	1014	Betel Pepper
146	*Artocarpus rigida*	Monkey Jack	MORA	Fruit	SE Asia	OC, PC	wind, fly?	810	Bhanbhendi
147	*Asimina grandiflora*	Wooly Pawpaw	ANNO	Fruit	Neot	SC, OC	fly?	202	Bible Frankincense
148	*Asimina triloba*	Pawpaw	ANNO	Fruit	Neot	SC, OC	fly?	788	Big Trefoil
149	*Aspalathus linearis*	Rooibos Tea	LEG	Beverage	A Afr			126	Bignai
150	*Asparagus officinalis*	Asparagus	LILA	Vegetable	Medit	OC	bee, Apis	171	Bilimbi
151	*Asperula odorata*	Sweet Woodruff	RUBI	Spice	Medit			805	Bindi, Kandela
152	*Astragalus gummifer*	Gum Tragacanth	LEG	Elastomer	Africa	OC?	bee, Megachile, Bombus	917	Binuang
153	*Astrocaryum aculeatum*	Aman Palm	PALM	Multiple	Neot	OC?	bee	214	Bird Rape
154	*Astrocaryum jauary*	Awarra Palm	PALM	Oil	Neot	OC?	bee	228	Bird Rape, Pak-Choi
155	*Astrocaryum murmuru*	Murmuru Palm	PALM	Fruit	Neot	OC?	bee	786	Birdsfoot Trefoil
156	*Astrocaryum tucuma*	Tucum-Assu Palm	PALM	Fruit	Neot	OC?	bee	1105	Biriba
157	*Astrocaryum vulgare*	Cumara Palm, Tucumao	PALM	Fruit	Neot	OC?	bee	1054	Bitter Almond
158	*Astronium balansae*	Urunday	ANAC	Tannin	S Amer			739	Bitter Lettuce
159	*Atriplex hortensis*	Garden Orach	CHEN	Vegetable	S Asia			365	Bitter Orange
160	*Atriplex spp.*	Saltbush	CHEN	Ero., For.	Aust	OC?	bee, fly?	747	Bitter Pea
161	*Atropa acuminata*	Indian Belladonna	SOLA	Medicinal	S Asia			1291	Bitter Vetch
162	*Atropa belladonna*	Belladonna	SOLA	Medicinal	Asia			1089	Bitter Wood
163	*Attalea funifera*	Coquilla Nut	PALM	Oil	Neot			1037	Black Beniseed
164	*Attalea spectabilis*	Cohune Nut Palm	PALM	Oil	Neot			908	Black Cumin
165	*Aucoumea klaineana*	Okoumé, Gaboon	BURS	Timber	W Afr	OC?	fly? bee?	507	Black Fonio
166	*Avena abyssinica*	Abyssinian Oat	POA	Cereal	S Asia	AG, OC?	wind, bee	1301	Black Gram
167	*Avena byzantine*	Red Oat	POA	Cereal	S Asia	AG, OC?	wind, bee	695	Black Henbane
168	*Avena fatua*	Wild Oat	POA	Cereal	S Asia	AG, OC?	wind, bee	841	Black Medic
169	*Avena sativa*	Common Oat	POA	Cereal	S Asia	AG, OC?	wind, bee	872	Black Mulberry
170	*Avena sterilis*	Sterile Oat	POA	Cereal	S Asia	AG, OC?	wind, bee	219	Black Mustard
171	*Averrhoa bilimbi*	Bilimbi	OXAL	Fruit	SE Asia	SC, OC	bee, Apis, fly	1029	Black Phyllium
172	*Averrhoa carambola*	Starfruit, Carambola	OXAL	Fruit	SE Asia	SC, OC	bee, Apis, fly	528	Black Sapote
173	*Axonopus afrinis*	Carpetgrass	POA	Forage	Neot			17	Black Wattle
174	*Axonopus compressus*	Tropical Carpetgrass	POA	Forage	Neot			234	Black Zira
175	*Azadirachta indica*	Nim Tree	MELI	Med., Timb.	S Asia			1110	Blackberry
176	*Baccaurea dulcis*	Kapoondong	EUPH	Fruit	SE Asia			1295	Blackpod Vetch
177	*Baccaurea motleyana*	Rambai	EUPH	Fruit	SE Asia			1030	Blonde Phyllium
178	*Baccaurea racemosa*	Menteng	EUPH	Fruit	S Asia			383	Blue Clitoria

No.	Scientific Name	Common Name	Family	Use	Origin	Breed. Sys.	Pollinators
179	*Baccaurea ramiflora*	Mafai	EUPH	Timber	SE Asia		
180	*Bactris gasipaes*	Peach Palm, Pejibaye	PALM	Fruit	Neot	OC	beetle
181	*Baeckea frutescens*	Udjung Atup	LEG?	Essence	SE Asia		
182	*Balanites aegyptiaca*	Desert Date	PALM	Fruit	Af-As	SC?	fly?
183	*Balanocarpus heimii*	Thai Dammar	DIPT	Resin	SE Asia		
184	*Bambusa spp.*	Bambusa	. POA	Timber	Asia		
185	*Baphia nitida*		Camwood	LEG	Dye	Africa	
186	*Barosma betulina*	Bacco	CHEN	Essence	S Afri		
187	*Barringtonia procera*	Nua	BARR	Nut	Solo. Is		
188	*Basella alba*	Malabar Spinach	BASE	Vegetable	Africa		
189	*Bauhinia esculenta*	Camel's-Foot	LEG	Beverage	Africa	OC?	bat, Pteropus
190	*Benincasa hispida*	Waxgourd, Kundor	CUCU	Vegetable	S Asia	OC,SC?	bee?
191	*Bertholletia excelsa*	Brazil-Nut	LECY	'Nut'	Neot	OC	bee, euglossine, Epicharis
192	*Beta vulgaris*	Beet	CHEN	Vegetable	Asia	OC	bee, thrip
193	*Bixa orellana*	Annatto, Achiote	BIXA	Dye, Misc.	Neot	OC, SC	bee, Melipona, Eulaema
194	*Blighia sapida*	Akee	SAPI	Fruit	Africa	SC, OC	bee, wasp, fly?
195	*Boesenbergia rotundata*	Chinese Keys	ZING	Spice	SE Asia		
196	*Bombacopsis quinata*	Pochote	BOMB	Timber	Neot	OC	bat, bird, bee
197	*Bombax buonopozense*	Silk Cotton	BOMB	Fiber	Africa		
198	*Bombax ceiba*	Red Silk-Cotton	BOMB	Fiber	Neot	OC	bat Pteropus,bird,bee moth
199	*Bombax malabaricum*	Indian Silk Cotton	BOMB	Fiber	S Asia	OC	bat, bird, bee, moth
200	*Borassus flabellifer*	Palmyra Palm	PALM	Sweetener	S Asia		
201	*Boroja patinoi*	Borojó	RUBI	Beverage	Neot		
202	*Boswellia sacra*	Bible Frankincense	BURS	Resin	W Asia		
203	*Bothriochloa intermedia*	Australian Bluestem	POA	Forage	Aust	~AG	
204	*Bothriochloa pertusa*	Comagueyana	POA	Forage	Neot	~AG?	
205	*Bouea macrophylla*	Kundang	ANAC	Fruit	SE Asia		
206	*Bowdichia nitida*	Sucupira	LEG	Timber	Neot		
207	*Brachiaria brizantha*	Signal Grass	POA	Forage	Africa		
208	*Brachiaria decumbens*	Surinam Grass	POA	Forage	Africa		
209	*Brachiaria deflexa*	Yague	POA	Cereal	Africa		
210	*Brachiaria mutica*	Pará Grass	POA	Forage	Africa		
211	*Brachiaria ramosa*	Browntop Millet	POA	Forage	S Asia		
212	*Brachylaena hutchinsii*	Muhuhu Oil	COMP	Essence	E Afr		
213	*Brassica algogabra*	Chinese Kale	CRUC	Vegetable	S Asia	OC	insect, bee
214	*Brassica campestris*	Bird Rape	CRUC	Oil	S Asia	OC, AU	bee, Apis
215	*Brassica carinata*	Ethiopian Mustard	CRUC	Oil	Medit	OC, AU	bee, Apis
216	*Brassica juncea*	Brown Mustard	CRUC	Oil	S Asia	OC, AU	bee, Apis
217	*Brassica napus*	Rape	CRUC	Oil	Medit	OC, AU	bee, Apis
218	*Brassica napus/napobrassica*	Rutabaga	CRUC	Vegetable	Medit	OC, AU	bee, Apis
219	*Brassica nigra*	Black Mustard	CRUC	Spice	Medit	OC, AU	bee, Apis
220	*Brassica oleracca/botrytis*	Broccoli	CRUC	Vegetable	Medit	OC, AU	bee, Apis
221	*Brassica oleracca/gemmifera*	Brussels-Sprouts	CRUC	Vegetable	Medit	OC, AU	bee, Apis
222	*Brassica oleracea*	Broccoli, Cauliflower	CRUC	Vegetable	Medit	OC, AU	bee, Apis
223	*Brassica oleracea/acephala*	Collards	CRUC	Vegetable	Medit	OC, AU	bee, Apis
224	*Brassica oleracea/capitate*	Cabbage	CRUC	Vegetable	Medit	OC, AU	bee, Apis
225	*Brassica oleracea/gongylodes*	Kohlrabi	CRUC	Vegetable	Medit	OC, AU	bee, Apis
226	*Brassica oleracea/italica*	Asparagus Broccoli	CRUC	Vegetable	Medit	OC, AU	bee, Apis
227	*Brassica pekinensis*	Chinese Cabbage	CRUC	Vegetable	S Asia	OC, AU	bee, Apis
228	*Brassica rapa*	Bird Rape, Pak-Choi	CRUC	Oil, Veget.	S Asia	OC, AU	bee, Apis
229	*Brosimum alicastrum*	Ramón	MORA	Fruit, Fora.	Neot	OC	wind, bee
230	*Brosimum utile*	Cow-Tree	MORA	Gum	Neot	OC	wind, bee
231	*Broussonetia papyrifera*	Paper Mulberry	MORA	Fiber	SE Asia		
232	*Buchanania latifolia*	Cuddapah Almond	ANAC	Oil	S Asia		
233	*Bulnesia sarmienti*	Balsamo	ZYGO	Essence	S Amer		
234	*Bunium bulbocastanum*	Black Zira	UMBE	Spice	E Asia		
235	*Bursera penicillata*	Linaloe Oil	BURS	Essence	Mexico		
236	*Byrsonima crassifolia*	Nance, Murici	MALP	Fruit	Neot	SC,OC	bee, Centris, Epicharis
237	*Caephaelis ipecacuanha*	Ipecac	RUBI	Medicinal	Neot		
238	*Caesalpinia brevifolia*	Algarobilla	LEG	Tannin	Neot	SC, OC	bee
239	*Caesalpinia coriaria*	Divi-Divi	LEG	Tannin	Neot	SC, OC	bee
240	*Caesalpinia echinata*	Brazilwood	LEG	Dye	Neot	SC, OC	bee
241	*Caesalpinia sappan*	Sappanwood	LEG	Dye	Neot	SC, OC	bee
242	*Caesalpinia spinosa*	Tara	LEG	Tannin	Neot	SC, OC	bee
243	*Cajanus cajan*	Pigeonpea	LEG	Vegetable	Africa	AU, OC	bee, Chalicodoma, Xylo
244	*Calamus spp.*	Rattan	PALM	Multiple	SE Asia		
245	*Calathea allouia*	Cornroot, Larién	MARA	Wax,Tuber	Neot		
246	*Calocarpum mammosum*	Mamey Sapote	SAPO	Multiple	Neot		
247	*Calophyllum ionophyllum*	Domba Oil, Penaga Laut	GUTT	Oil	S Asia		
248	*Calopogonium coeruleum*	Calopo	LEG	Forage	Neot		
249	*Calopogonium mucunoides*	Prisolilia	LEG	Erosion	Neot		
250	*Caltha palustris*	Cowslip	RANU	Vegetable	Europe		
251	*Camellia sasanqua*	Camellia	THEA	Oil	S Asia	SC, OC	insect, fly, bee
252	*Camellia sinensis*	Tea	THEA	Beverage	S Asia	SC, OC	insect, fly, bee
253	*Campanula rapunculus*	Rampion	CAMP	Vegetable	Medit		
254	*Campnosperma brevipetiolata*	Terentang	OCH	Timber	Oceania		
255	*Cananga odorata*	Ylang-Ylang	ANON	Essence	SE Asia	OC?	moth
256	*Canarium album*	Chinese White Olive	BURS	Fruit	SE Asia		
257	*Canarium indicum*	Java Almond	BURS	Nut	SE Asia		
258	*Canarium indicum*	Java-Almond	BURS	Nut	SE Asia		
259	*Canarium luzonicum*	Pili Nut	BURS	Nut	SE Asia		
260	*Canarium luzonicum*	Pili Nut	BURS	Resin	SE Asia		
261	*Canarium ovatum*	Pili Nut	BURS	Nut	Philipp.		
262	*Canarium pimela*	Chinese Black Olive	BURS	Fruit	S Asia		
263	*Canarium vulgare*	Kenai Nut	BURS	Oil	SE Asia		
264	*Canavalia ensiformis*	Jackbean, Frijól Burro	LEG	Vegetable	Neot	AU, OC	bee
265	*Canavalia gladiata*	Swordbean, Poroto	LEG	Forage	Neot	AU, OC	bee
266	*Canavalia maritima*	Canavalia Maritima	LEG	Erosion	Neot	AU, OC	bee
267	*Canavalia plagiosperma*	Giant Bean, Poroto	LEG	Vegetable	Neot	AU, OC	bee

No.	Common Name
793	Blue Lupine
565	Boer Lovegrass
377	BolaBolo
433	Bologi
88	Bondue, Bayam Betul
1008	Bonete
201	Borojó
740	Bottle Gourd
1123	Bowstring Hemp
191	Brazil-Nut
1069	Brazilian Guava
1201	Brazilian Lucerne
240	Brazilwood
142	Breadfruit
1075	Breadroot
753	Broad-Leaved Lavender
1292	Broadbean
583	Broadleaf Peppermint
220	Broccoli
222	Broccoli, Cauliflower
216	Brown Mustard
211	Browntop Millet
221	Brussels-Sprouts
51	Buchu
523	Buck Yam
604	Buckwheat
449	Buffalo Gourd, Chilicote
562	Buffalo Spinach
313	Buffelgrass
1276	Bullrush
319	Bungu
310	Burma Cedar
643	Burmese Lacquer
1271	Burweed
382	Butterfly Pea
296	Butternut, Almendra
842	Button Medic
224	Cabbage
1107	Cabbage Rose
420	Cabbage Tree
599	Cabeludinha
888	Cabreuva Oil
631	Cabuya
627	Cabuya, Pia Floja
1237	Cacao Blanco, Pataste
1235	Cacao Cimarrón
1241	Cacao de Mico
1236	Cacao Rana
92	Cafoun
846	Cajeput, Gelam
435	Calabash Tree
248	Calopo
95	Cambodian Cardamom
585	Camden Wollybut
189	Camel's-Foot
251	Camellia
66	Camelthorn
44	Camerouns Cardamom
351	Camphor-Tree
883	Camu-Camu
185	Camwood
266	Canavalia Maritima
601	Candelilla
63	Candlenut
957	Candletree
974	Canelilla
53	Cantala
446	Cantaloupe
78	Cape Aloe
1004	Cape Gooseberry
269	Caper
1065	Capulín
414	Caranday
292	Caraway
558	Cardamom
571	Caribgrass
287	Carissa Lunatus
289	Carissa Oxycantha
415	Carnauba Wax Palm
318	Carob
173	Carpetgrass
488	Carrot
1094	Cáscara
103	Cashew, Marañon
824	Cassava
822	Cassava, Yuca
303	Cassia
352	Cassia
306	Castilloa Rubber

No.	Scientific Name	Common Name	Family	Use	Origin	Breed. Sys.	Pollinators	No.	Common Name
268	*Canna edulis*	Edible Canna	CANN	Starch	Neot			1104	Castorbean
269	*Capparis spinosa*	Caper	CAPP	Spice	Europe			1306	Catjang
270	*Capsicum annuum*	Red Pepper	SOLA	Spice	Neot	AU, OC	bee	16	Celeriac
271	*Capsicum baccatum/pendulum*	Pimentchien	SOLA	Spice	Neot	AU, OC	bee	15	Celery
272	*Capsicum chinense*	Ají	SOLA	Spice	Neot	AU, OC	bee	128	Celery
273	*Capsicum frutescens*	Chili Pepper	SOLA	Spice	Neot	AU, OC	bee	315	Centro
274	*Capsicum pubescens*	Goat Chili	SOLA	Spice	Neot	AU, OC	bee	1215	Ceylon Spinach
275	*Carapa guianensis*	Crabwood, Tangare	MELI	Timber	Neot			539	Ceylon-Goosebeny
276	*Carica cauliflora*	Wild Papaya	CARI	Fruit	Neot	OC	moth, bird, bee	993	Chaguita
277	*Carica microcarpa*	Wild Papaya	CARI	Fruit	Neot	OC	moth, bird, bee	89	Chaquillón
278	*Carica monoica*	Wild Papaya	CARI	Fruit	Neot	OC	moth, bird, bee	691	Chaulmoogra
279	*Carica papaya*	Papaya	CARI	Fruit	Neot	SC, OC	moth, bird, bee	384	Chaya
280	*Carica pentagona*	Babaco	CARI	Fruit	Neot	OC, PC?	moth, bird, bee	1137	Chayote
281	*Carica pubescens*	Mountain Papaya	CARI	Fruit	Neot	SC, OC	moth, bird, bee	144	Chempedale
282	*Carica quercifolia*	Naracatia	CARI	Fruit	Neot	SC, OC	moth, bird, bee	113	Cherimoya
283	*Carica stipulata*	Ecuadorian Papaya	CARI	Fruit	Neot	SC, OC	moth, bird, bee	1061	Cherry Laurel
284	*Cariniana pyriformis*	Abarco	LECY	Timber	Neot			124	Chervil
285	*Carissa carandas*	KarAnda	APOC	Fruit	S Asia	OC?	moth, bee?	935	Chestnut of America
286	*Carissa edulis*	Egyptian Carissa	APOC	Fruit	Africa	OC	moth, bee?	326	Chia
287	*Carissa lunatus*	Carissa Lunatus	APOC	Oil	Af-Asia	OC?	moth, bee?	830	Chicle, Níspero
288	*Carissa macrocarpa*	Natal-Palm	APOC	Fruit	Africa	OC?	moth, bee?	341	Chicory
289	*Carissa oxyacantha*	Carissa Oxycantha	APOC	Oil	Af-Asia	OC?	moth, bee?	827	Chicozapote
290	*Carludovica palmata*	Panama Hat Palm	CYCL	Fiber	Neot	SC, OC	beetle, bee, Trigona	761	Chijura
291	*Carthamus tinctorius*	Safflower	COMP	Oil	W Asia	SC, OC	insect, bee, Apis	621	Chilean Strawberry
292	*Carum carvi*	Caraway	UMBE	Spice	Europe		bee, fly	273	Chili Pepper
293	*Carya illinoensis*	Pecan	JUGL	Nut	Neot	OC	wind, insect?	6	China Jute
294	*Caryocar amygdaliferum*	Seaving	CARY	Oil	Neot			90	Chinese Amaranth
295	*Caryocar brasiliensis*	Piqui Brava	CARY	Oil	Neot			1113	Chinese Arrowhead
296	*Caryocar nuciferum*	Butternut, Almendra	CARY	Nut, Mult.	Neot			1188	Chinese Artichoke
297	*Caryocar villosum*	Pequiá	CARY	Oil	Neot			262	Chinese Black Olive
298	*Caryodendron orinocense*	Orinoco Nut, Cacay	EUPH	Nut	Neot			227	Chinese Cabbage
299	*Caryota urens*	Fishtail Palm	PALM	Sweetener	SE Asia			304	Chinese Chestnut
300	*Casimiroa edulis*	Mexican Apple	RUTA	Fruit	Neot			75	Chinese Chives
301	*Cassia auriculata*	Avaram	LEG	Tannin	Africa	SC, OC?	Xylocopa, Amegilla	1328	Chinese Date
302	*Cassia fistula*	Indian Laburnum	LEG	Medicinal	S Asia	SC, OC?	Xylocopa, Amegilla	564	Chinese Ephedra
303	*Cassia siamea*	Cassia	LEG	Timber	SE Asia	SC, OC?	Xylocopa, Amegilla	424	Chinese Hazelnut
304	*Castanea mollissima*	Chinese Chestnut	FAGA	Nut	Asia	OC	wind, insect, bee	213	Chinese Kale
305	*Castanopsis sumatrana*	Berangan	FAGA	Nut	SE Asia			195	Chinese Keys
306	*Castillia elastica*	Castilloa Rubber	MORA	Resin	Neot			1247	Chinese Lacquer Tree
307	*Castillia ulei*	Gaucho Rubber	MORA	Resin	Neot			871	Chinese Mulberry
308	*Catha edulis*	Khat	CELA	Medicinal	Africa			530	Chinese Persimmon
309	*Catharanthus roseus*	Periwinkle	APOC	Medicinal	S Asia			1098	Chinese Rhubarb
310	*Cedrela toona*	Burma Cedar	MELI	Timber	SE Asia			1126	Chinese Tallow Tree
311	*Ceiba pentandra*	Kapok	BOMB	Fiber	Neot?	SC, OC	bat,Pt,Mc,Eon,Pt,Na,Ei,bird	557	Chinese Waterchestnut
312	*Celosia argentea*	Mirabel, Soko	AMAR	Vegetable	W Afr			256	Chinese White Olive
313	*Cenchrus ciliaris*	Buffelgrass	POA	Forage	Neot	AU		522	Chinese Yam
314	*Centrosema acutifolium*	Vichada	LEG	Forage	Neot	SC, OC	bee	1026	Chios Mastic
315	*Centrosema pubescens*	Centro	LEG	Forage	Neot	SC, OC	bee	74	Chives
316	*Cephaelis ipecacuanha*	Ipecac	RUBI	Medicinal	Neot	OC?	butterfly, bee	719	Chlabato
317	*Cephalonema polyandrum*	Punga	TILI	Fiber	Africa			626	Chuchao
318	*Ceratonia siliqua*	Carob	LEG	Gum, Mult.	W Asia	OC		477	Chufa, Earth Almond
319	*Ceratotheca sesamoides*	Bungu	PEDA	Oil	Africa			834	Chupa-Chupa
320	*Ceroxylon alpinum*	Wax Palm	PALM	Wax	Neot			115	Cimarrona
321	*Ceroxylon utile*	Wax Palm	PALM	Wax	Neot			356	Cinnamon
322	*Chamaedora spp.*	Palmheart, Palmito	PALM	Vegetable	Neot			372	Citron
323	*Chamaerops humilis*	Dwarf Fan Palm	PALM	Fiber	N Afr			358	Citropus
324	*Chenopodium album*	Pigweed	CHEN	Vegetable	S Asia			378	Clapper Polyandra
325	*Chenopodium ambrosioides*	Wormseed, Epazote	CHEN	Medicinal	Neot			379	Clausena Dentata
326	*Chenopodium nuttaliae*	Chia	CHEN	Vegetable	Neot			1209	Clove
327	*Chenopodium quinoa*	Quinoa	CHEN	Cereal	Neot			575	Coca
328	*Chloris gayana*	Rhodesgrass	POA	Forage	Africa	AU	wind, insect?	909	Cochineal Cactus
329	*Chlorophora excelsa*	African Oak	MORA	Timber	Africa			1238	Cocoa
330	*Chlorophora tinctoria*	Fustic-Mulberry	MORA	Tannin	Neot			1174	Cocona, Peach Tomato
331	*Chrysanthemum balsamita*	Costmary	COMP	Spice	W Asia	OC, AU?	insect, bee	388	Coconut
332	*Chrysanthemum cinerariifolium*	Pyrethrum	COMP	Pesticide	Africa	OC, AU?	insect, bee	838	Cocorita Palm, Jaguá
333	*Chrysanthemum coccineum*	Persian Insect-Flower	COMP	Pesticide	W Asia	OC, AU?	insect, bee	629	Cocuiza
334	*Chrysanthemum coronarium*	Garland Chrysanthemum	COMP	Vegetable	Asia	OC, AU?	insect, bee	393	Coffea Excelsa
335	*Chrysanthemum spatiosum*	Crown Daisy	COMP	Medicinal	Europe	OC, AU?	insect, bee	394	Coffea Liberica
336	*Chrysophyllum cainito*	Star Apple, Cainito	SAPO	Fruit	Neot			395	Coffea Pugenoides
337	*Chrysophyllum delevoyi*	African Star-Apple	SAPO	Multiple	Africa			164	Cohune Nut Palm
338	*Cicca distichus*	Otaheite Gooseberry	EUPH	Fruit	W Asia			926	Cohune Nut Palm
339	*Cicer arietinum*	Garbanzo, Chickpea	LEG	Vegetable	Asia	AU, OC	bee	403	Colchicum
340	*Cichorium endivia*	Endive	COMP	Vegetable	Medit		bee, Apis	223	Collards
341	*Cichorium intybus*	Chicory	COMP	Beverage	Medit		bee, Apis	359	Colocynth
342	*Cinchona calisaya*	Quinine	RUBI	Medicinal	Neot	OC	bee, butterfly, fly	576	Colombian Coca
343	*Cinchona lansifolia*	Quinine	RUBI	Medicinal	Neot	OC	bee, butterfly, fly	578	Color Azafrán
344	*Cinchona ledgeriana*	Quinine	RUBI	Medicinal	Neot	OC	bee, butterfly, fly	1274	Coltsfoot
345	*Cinchona officinalis*	Quinine	RUBI	Medicinal	Neot	OC	bee, butterfly, fly	204	Comagueyana
346	*Cinchona pitayensis*	Quinine	RUBI	Medicinal	Neot	OC	bee, butterfly, fly	996	Common Bean
347	*Cinchona pubescens*	Quinine	RUBI	Medicinal	Neot	OC	bee, butterfly, fly	611	Common Fig
348	*Cinchona succiruba*	Quinine	RUBI	Medicinal	Neot	OC	bee, butterfly, fly	1068	Common Guava
349	*Cinnamomum aromaticum*	Saigon Cinnamon	LAUR	Spice	SE Asia	OC	fly	710	Common Indigo
350	*Cinnamomum burmannii*	Padang-Cassia	LAUR	Spice	SE Asia	OC	fly	751	Common Lavender
351	*Cinnamomum camphora*	Camphor-Tree	LAUR	Medicinal	S Asia	OC	fly	646	Common Licorice
352	*Cinnamomum canis*	Cassia	LAUR	Essence	SE Asia	OC	fly	1093	Common Mignonette
353	*Cinnamomum massoia*	Massoia	LAUR	Essence	Papua	OC	fly	169	Common Oat
354	*Cinnamomum oliveri*	Oliver's Bark	LAUR	Essence	Aust	OC	fly	1309	Common Periwinkle
355	*Cinnamomum tamala*	Indian-Bark	LAUR	Spice	S Asia	OC	fly	532	Common Persimmon
356	*Cinnamomum verum*	Cinnamon	LAUR	Spice	S Asia	OC	fly	1111	Common Rue

No.	Scientific Name	Common Name	Family	Use	Origin	Breed. Sys.	Pollinators	No.	Common Name
357	*Cistus ladaniferus*	Labdanum	CIST	Essence	Medit			1244	Common Thyme
358	*Citropus gilletriana*	Citropus	RUTA	Rootstock	Africa			1296	Common Vetch
359	*Citrullus colocynthis*	Colocynth	CUCU	Medicinal	Africa	SC, OC	bee	513	Composite Yam
360	*Citrullus fistulosus*	Indian Squash Melon	CUCU	Vegetable	S Asia	SC, OC	bee	392	Congo Coffee
361	*Citrullus lanatus*	Watermelon	CUCU	Fruit	Africa	SC, OC, PC	bee, Apis	514	Convolvulacca Yam
362	*Citrus aumntium/bergamia*	Bergamot	RUTA	Essence	S Asia	AU, ~OC	bee, Apis	23	Cooba
363	*Citrus aurantifolia*	Lime	RUTA	Fruit	S Asia	AU, ~OC	bee, Apis	412	Copaiba
364	*Citrus aurantium*	Sour Orange	RUTA	Essence	S Asia	AU, ~OC	bee, Apis	670	Copal Noir
365	*Citrus aurantium/amara*	Bitter Orange	RUTA	Essence	S Asia	AU, ~OC	bee, Apis	485	Copal, Ogea, Insenso
366	*Citrus grandis*	Pummelo	RUTA	Fruit	S Asia	AU, ~OC	bee, Apis	163	Coquilla Nut
367	*Citrus hystrix*	Pummelo	RUTA	Fruit	SE Asia	AU, ~OC	bee, Apis	551	Coquirana
368	*Citrus latipes*	Pummelo	RUTA	Fruit	S Asia	AU, ~OC	bee, Apis	422	Coriander
369	*Citrus limetta*	Sweet Lime	RUTA	Fruit	S Asia	OC, PC	bee, Apis	546	Corkwood
370	*Citrus limon*	Lemon	RUTA	Fruit	SE Asia	AU, ~OC	bee, Apis	853	Corn Mint
371	*Citrus macroptera*	Lemon	RUTA	Fruit	SE Asia	AU, ~OC	bee, Apis	1323	Corn, Maiz
372	*Citrus medica*	Citron	RUTA	Fruit	SE Asia	AU, ~OC	bee, Apis	245	Cornroot, Larién
373	*Citrus paradisi*	Grapefruit	RUTA	Fruit	S Asia	AU, AG	bee, Apis	1067	Costa Rican Guava
374	*Citrus reticulata*	Mandarin	RUTA	Fruit	S Asia	SC, OC	bee, Apis	331	Costmary
375	*Citrus sinensis*	Sweet Orange	RUTA	Fruit	S Asia	AU,AG,PC	bee, Apis	1129	Costus, Kuth
376	*Citrus tachibana*	Tachibana Orange	RUTA	Fruit	S Asia	AU, OC	bee, Apis	658	Cotton
377	*Clappertonia ficifolia*	BolaBolo	TILI	Fiber	Africa			659	Cotton
378	*Clappertonia polyandra*	Clapper Polyandra	TILI	Fiber	Africa			660	Cotton
379	*Clausena dentata*	Clausena Dentata	RUTA	Fruit	S Asia			661	Cotton
380	*Clausena lansium*	Wampi	RUTA	Fruit	S Asia			662	Cotton
381	*Claytonia perfoliata*	Winter Purslane	PORT	Vegetable	N Amer			663	Cotton
382	*Clitoria laurifolia*	Butterfly Pea	LEG	Erosion	Neot	AU, OC	bee	664	Cotton
383	*Clitoria ternatea*	Blue Clitoria	LEG	Forage	Neot	AU, OC	bee	665	Cotton
384	*Cnidoscolus chayamansa*	Chaya	EUPH	Vegetable	Neot			5	Country Mallow
385	*Coccinia abyssinica*	Anchoté	CUCU	Vegetable	E Afr			230	Cow-Tree
386	*Coccinia grandis*	Ivy Gourd, Kundiri	CUCU	Vegetable	SE Asia			1305	Cowpea
387	*Coccoloba uvifera*	Seagrape	POLYG	Erosion	Neot			250	Cowslip
388	*Cocos nucifera*	Coconut	PALM	Oil, Bev.	SE Asia	SC, OC	wind, fly,bee,bat,Eony,Mac,Pt	431	Cowtree
389	*Coffea arabica*	Arabica Coffee	RUBI	Beverage	Africa	AU, OC	fly, bee, Apis, Melipona	33	Coyol, Babosa
390	*Coffea bengalensis*	Bengal Almond	RUBI	Beverage	Africa	AU, OC	fly, bee	275	Crabwood, Tangare
391	*Coffea canephora*	Robusta Coffee	RUBI	Beverage	Africa	OC	wind, bee, Creightonella	432	Crambe
392	*Coffea congensis*	Congo Coffee	RUBI	Beverage	Africa	AU, OC	fly, bee	1260	Crimson Clover
393	*Coffea excelsa*	Coffea Excelsa	RUBI	Beverage	Africa	AU?	fly, bee	453	Crookneck Pumpkin
394	*Coffea liberica*	Coffea Liberica	RUBI	Beverage	Africa	OC	wind, fly, bee	438	Crotalaria
395	*Coffea pugenoides*	Coffea Pugenoides	RUBI	Beverage	Africa	AU?	fly, bee	335	Crown Daisy
396	*Coffea racemosa*	Wild Coffee	RUBI	Beverage	Africa	AU?	fly, bee	423	Crownvetch
397	*Coffea stenophyla*	Wild Coffee	RUBI	Beverage	Africa	AU?	fly, bee	956	Cuajilote
398	*Coix lacryma-jobi*	Job's Tears	POA	Multiple	SE Asia	SC, OC	bee?	519	Cuculmeca
399	*Cola acuminata*	Abata Kola	STER	Nut	Africa	OC	fly, insect	447	Cucumber
400	*Cola anomala*	Bamenda Cola	STER	Nut	Africa	OC	fly, insect	232	Cuddapah Almond
401	*Cola nitida*	Gbanja Kola	STER	Nut	Africa	OC	fly, insect	157	Cumara Palm, Tucumao
402	*Cola verticillata*	Owe Kola	STER	Nut	Africa	OC	fly, insect	535	Cumaru
403	*Colchicum autumnale*	Colchicum	LILI	Medicinal	Europe			457	Cumin
404	*Coleus parviflorus*	Ratala	LABI	Tuber	SE Asia			1240	Cupuacu
405	*Colocasia esculenta*	Dasheen, Taro	ARAC	Tuber	Af-As		fly	10	Curch
406	*Colocaasia es./antiquorum*	Eddoe	ARAC	Tuber	Af-As		fly	799	Currant Tomato
407	*Colubrina reclinata*	Snake Bark, Maubi	RHAM	Medicinal	Neot			875	Curry-Leaf-Tree
408	*Commiphora abyssinica*	Abyssinian Myrrh	BURS	Resin	Africa			1154	Curuba, Melocotón
409	*Commiphora erythraea*	Opopanax Gum	BURS	Resin	Africa			1160	Cut-Egg Plant
410	*Commiphora opobalsamum*	Mecca Myrrh	BURS	Resin	Africa			1226	Dafo
411	*Conocarpus lancifolius*	Damas	COMB	Multiple	Africa			1025	Dahoma
412	*Copaifera officinalis*	Copaiba	LEG	Resin	Africa			960	Dallisgrass
413	*Copaifera palustris*	Sepetir	LEG	Timber	SE Asia			1300	Dalrymple Vigna
414	*Copernicia alba*	Caranday	PALM	Wax	Neot			411	Damas
415	*Copernicia prunifera*	Carnauba Wax Palm	PALM	Wax	Neot			1108	Damask Rose
416	*Corchorus capsularis*	White Jute	TILI	Fiber	S Asia	AU	bee	1060	Damson Plum
417	*Corchorus olitorius*	Tussa Jute	TILI	Fiber	Af-Asia	AU	bee	1220	Dandelion
418	*Cordeauxia edulis*	Jicib Bean, Ye-eb	LEG	Multiple	Africa	AU, OC?	bee?	405	Dasheen, Taro
419	*Cordia spp*	Laurel, Frejo	BORA	Timber	Neot	SC, OC	bee, fly	997	Date Palm
420	*Cordyline australis*	Cabbage Tree	ASPH	Tuber	Aust.	SC, OC?	bee?	531	Date Plum, Kaki
421	*Cordyline fruticosa*	Ti Palm	ASPHO	Multiple	S Asia	SC, OC?	bee	490	Dendrocalamus
422	*Coriandrum sativum*	Coriander	UMBE	Spice	S Asia	OC	bee, fly	182	Desert Date
423	*Coronilla varia*	Crownvetch	LEG	Forage	Europe	OC	bee, Bombus	493	Desmanthus Virgatus
424	*Corylus chinensis*	Chinese Hazelnut	BETU	Nut	Asia	OC	bee	496	Desmodium Distortum
425	*Corylus colurna*	Turkish Hazel	BETU	Nut	W Asia	OC	bee	497	Desmodium Gyroides
426	*Corylus ferox*	Himalayan Hazel	BETU	Nut	Asia	OC	bee	500	Desmodium Ovalifolium
427	*Corylus heterophylla*	Siberian Hazel	BETU	Nut	Asia	OC	bee	1147	Dhaincha, Jantar
428	*Corylus maxima*	Giant Filbert	BETU	Nut	W Asia	OC	bee	503	Diaz Bluestem
429	*Corylus tibetica*	Tibetan Hazelnut	BETU	Nut	Asia	OC	bee	109	Dill
430	*Couma macrocarpa*	Sorva Gum	APOC	Elastomer	Neot			239	Divi-Divi
431	*Couma utilis*	Cowtree	APOC	Beverage	Neot			247	Domba Oil, Penaga Laut
432	*Crambe abyssinica*	Crambe	CRUC	Oil	Medit			780	Double Coconut
433	*Crassocephalum biafrae*	Bologi	COMP	Vegetable	W Afr			484	Dragon's Blood
434	*Crategus pubescens*	Tejocote	ROSA	Fruit	Neot	SC, OC?	bee?	541	Dragon's Blood
435	*Crescentia cujete*	Calabash Tree	BIGN	Multiple	Neot	OC	bat, Macroglossus	542	Dragon's Blood
436	*Crocus sativus*	Saffron	IRID	Dye	Medit	OC?	bee?	741	Dragon's Head
437	*Crotalaria brevidens*	Slenderleaf Crotalaria	LEG	Forage	Africa	SC, OC	bee	698	Dum Palm
438	*Crotalaria intermedia*	Crotalaria	LEG	Forage	S Asia	SC, OC	bee, Megachile	548	Durian
439	*Crotalaria juncea*	Sunn Hemp	LEG	Fiber	S Asia	SC, OC	bee, Chalicodoma, Xylocopa	547	Durians
440	*Crotalaria lanceolata*	Lanceleaf Crotalaria	LEG	Erosion	Africa	SC, OC	bee	52	Dwarf Aloe, Babsi
441	*Crotalaria pallida*	Smooth Crotalaria	LEG	Forage	Africa	SC, OC	bee	876	Dwarf Banana
442	*Crotalaria spectabilis*	Showy Crotalaria	LEG	Erosion	S Asia	SC, OC	bee	323	Dwarf Fan Palm
443	*Croton tiglium*	Purging Croton	EUPH	Medicinal	SE Asia			459	East Indian Arrowroot
444	*Crytostegia sp.*	Rubber	ASCL	Elastomer	Africa			470	East Indian Lemongrass
445	*Cucumis auguria*	Gherkin	CUCU	Vegetable	Neot	OC	bee, Apis	1099	East Indian Rhubarb

No.	Scientific Name	Common Name	Family	Use	Origin	Breed. Sys.	Pollinators	No.	Common Name
446	*Cucumis melo*	Cantaloupe	CUCU	Fruit	Africa	SC, OC	bee, Apis	1119	Eastern Elderberry
447	*Cucumis sativus*	Cucumber	CUCU	Vegetable	S Asia	OC, PC	bee, Apis	524	Eboe Yam, ñame blanco
448	*Cucurbita ficifolia*	Malabar Gourd	CUCU	Oil	Neot	SC, OC	bee, insect	728	Ecuador Walnut
449	*Cucurbita foetidissima*	Buffalo Gourd, Chilicote	CUCU	Oil	Neot	SC, OC	bee, Peponapis, Xenoglossa	283	Ecuadorian Papaya
450	*Cucurbita llundelliana*	Peten Gourd	CUCU	Vegetable	Neot	SC, OC	bee, Peponapis, Xenoglossa	406	Eddoe
451	*Cucurbita maxima*	Zapallo, Winter Squash	CUCU	Vegetable	Neot	SC, OC	bee, Peponapis, Xenoglossa	268	Edible Canna
452	*Cucurbita mixta*	Ayote	CUCU	Vegetable	Neot	SC, OC	bee, Peponapis, Xenoglossa	1045	Eggfruit Tree, Canistel
453	*Cucurbita moschata*	Crookneck Pumpkin	CUCU	Vegetable	Neot	SC, OC	bee, Peponapis, Xenoglossa	1169	Eggplant
454	*Cucurbita pepo*	Vegetable Marrow	CUCU	Vegetable	Neot	SC, OC	bee, Peponapis, Xenoglossa	286	Egyptian Carissa
455	*Cucurbita pepo/medullosa*	Marrow	CUCU	Vegetable	Neot	SC, OC	bee, Peponapis, Xenoglossa	694	Egyptian Henbane
456	*Cucurbita pepo/melopepo*	Marrow	CUCU	Vegetable	Neot	SC, OC	bee, Peponapis, Xenoglossa	784	Ekki, Eba
457	*Cuminum cyminum*	Cumin	UMBE	Spice	S Asia	AU, OC	bee, Apis	508	Elephant apple
458	*Curcuma amada*	Mango Ginger	ZING	Spice	S Asia		bee	1042	Elephant Bush
459	*Curcuma angustifolia*	East Indian Arrowroot	ZING	Vegetable	S Asia		bee	981	Elephant Grass
460	*Curcuma aromatica*	Yellow Zedoary	ZING	Dye	S Asia		bee	98	Elephant Yam
461	*Curcuma longa*	Turmeric	ZING	Spice	S Asia	OC?	bee?	99	Elephant Yam
462	*Curcuma mangga*	Temu Mangga	ZING	Spice	SE Asia		bee	1001	Emblic
463	*Curcuma xanthorrhiza*	Temu Lawak	ZING	Med., Spice	SE Asia		bee	650	Emping Great Fruit
464	*Curcuma zedoaria*	Zedoary	ZING	Essence	S Asia	OC?	bee?	340	Endive
465	*Cyamopsis tetragonoloba*	Guar, Clusterbean	LEG	Gum	Af-As	AU, OC	bee, Apis	499	Engorda Caballo
466	*Cyclanthera pedata*	Archucha	CUCU	Fruit	Neot			560	Ensete
467	*Cydonia oblonga*	Quince	ROSA	Starch	Medit			577	Eschweilera Odorata
468	*Cylicodiscus gabunensis*	Okan	LEG	Timber	Africa			800	Esparto
469	*Cymbopogon citratus*	West Indian Lemongrass	POA	Essence	Africa		wind, insect?	1193	Esparto
470	*Cymbopogon flexuosus*	East Indian Lemongrass	POA	Essence	Africa		wind, insect?	215	Ethiopian Mustard
471	*Cymbopogon martinii*	Rosha Grass	POA	Essence	SE Asia		wind, insect?	589	Eucalyptus
472	*Cymbopogon nardus*	Nardus Grass	POA	Essence	SE Asia		wind, insect?	592	Eucalyptus
473	*Cymbopogon winteranus*	Winter's Grass	POA	Essence	Africa		wind, insect?	136	European Arnica
474	*Cynara scolymus*	Globe Artichoke	COMP	Vegetable	Africa	OC	bee, Bombus	855	European Pennyroyal
475	*Cynodon dactylon*	Bermudagrass	COMP	Erosion	Africa			622	European Strawberry
476	*Cynodon plectostachyus*	Stargrass	COMP	Forage	Africa			29	Feijoa
477	*Cyperus esculentus*	Chufa, Earth Almond	CYPE	Forage	Africa	SC, OC	bee?	619	Fennel
478	*Cyperus papyrus*	Papyrus	CYPE	Fiber	Africa	SC, OC	bee?	1266	Fenugreek
479	*Cyperus rotundus*	Nutgrass	CYPE	Tuber	Africa	SC, OC	bee?	613	Fig
480	*Cyphomandra betacea*	Tree Tomato	SOLA	Fruit	Neot			559	Finger Millet
481	*Cyrtosperma chamissonis*	Giant Swamp Taro	ARAC	Tuber	Polynes		fly	630	Fique
482	*Dacrydium spp.*	Huon Pine	PODO	Resin,Timb.	Asia	OC?	wind, bee?	299	Fishtail Palm
483	*Dacryodes edulis*	Safou, Eben	BURS	Fruit	W Afr.			573	Fitweed
484	*Daemonorops draco*	Dragon's Blood	PALM	Resin	SE Asia			776	Flax
485	*Daniellia spp.*	Copal, Ogea, Insenso	LEG	Resin,Timb.	Africa			129	Florigraze Peanut
486	*Datura metel*	Hindu Datura	SOLA	Medicinal	S Asia	SC, OC?	bee	1222	Fluted Gourd
487	*Datura stramonium*	Jimson Weed	SOLA	Medicinal	Mexico	SC, OC?	bee	1149	Foxtail Millet
488	*Daucus carota*	Carrot	UMBE	Vegetable	N Amer	SC, OC	fly, bee, Apis	754	French Lavender
489	*Dendrocalamus asper*	Bamboo Betoong	POA	Vegetable	SE Asia		wind, insect?	330	Fustic-Mulberry
490	*Dendrocalamus spp.*	Dendrocalamus	POA	Timber	S Asia		wind, insect?	1033	Gala Dinich
491	*Derris elliptica*	Tuba Root	LEG	Pesticide	Papua	OC?	bee?	732	Galanga
492	*Derris spp.*	Luba Root	LEG	Pesticide	SE Asia			608	Galbanum, Kasnih
493	*Desmanthus virgatus*	Desmanthus Virgatus	LEG	Forage	Neot			105	Galo
494	*Desmodium barbatum*	Tick Clover	LEG	Forage	Neot	AU?,SC,OC	bee	1278	Gambier
495	*Desmodium discolor*	Horse Marmalade	LEG	Forage	Neot	AU?,SC,OC	bee	636	Gamboge
496	*Desmodium distortum*	Desmodium Distortum	LEG	Forage	Neot	AU?,SC,OC	bee	638	Gamboge Tree
497	*Desmodium gyroides*	Desmodium Gyroides	LEG	Forage	SE Asia	AU?,SC,OC	bee	339	Garbanzo, Chickpea
498	*Desmodium intortum*	Beggarlice	LEG	Forage	Neot?	AU?,SC,OC	bee	639	Garcinîa Xanthocarpus
499	*Desmodium nicaraguense*	Engorda Caballo	LEG	Forage	Neot	AU?,SC,OC	bee	110	Garden Angelica
500	*Desmodium ovalifolium*	Desmodium Ovalifolium	LEG	Forage	Neot	AU?,SC,OC	bee	762	Garden Cress
501	*Desmodium tortuosum*	Mozoton	LEG	Forage	Neot	AU?,SC,OC	bee	891	Garden Myrrh
502	*Desmodium uncinatum*	Silverleaf Desmodium	LEG	Forage	Neot	AU?,SC,OC	bee	159	Garden Orach
503	*Dichanthium annulatum*	Diaz Bluestem	LEG	Forage	SE Asia			1100	Garden Rhubarb
504	*Dichanthium aristatum*	Angleton Bluestem	LEG	Forage	SE Asia			624	Garden Strawberry
505	*Digitalis purpurea*	Purple Foxglove	SCRO	Medicinal	Medit		bee	334	Garland Chrysanthemum
506	*Digitaria exilis*	Hungry-Rice	POA	Cereal	Africa	SC, OC	bee?	73	Garlic
507	*Digitaria iburua*	Black Fonio	POA	Cereal	Africa	SC, OC?	bee?	307	Gaucho Rubber
508	*Dillenia indica*	Elephant apple	DILL	Fruit	S Asia			401	Gbanja Kola
509	*Dioscorea abyssinica*	Rikua	DIOS	Tuber	E. Afr	OC	beetle, fly, bee?	640	Gendarussa vulgans
510	*Dioscorea alata*	Winged Yam	DIOS	Tuber	S Asia	OC	beetle, fly, bee?	733	Geocaipa
511	*Dioscorea bulbifera*	Airpotato	DIOS	Tuber	SE Asia	OC	beetle, fly, bee?	975	Geranium
512	*Dioscorea cayensis*	Yellow Guinea Yam	DIOS	Tuber	Africa	OC	beetle, fly, bee?	445	Gherkin
513	*Dioscorea composita*	Composite Yam	DIOS	Medicinal	Neot	OC	beetle, fly, bee?	1117	Ghia, Chia
514	*Dioscorea convolvulacea*	Convolvulacea Yam	DIOS	Tuber	Neot	OC	beetle, fly, bee?	34	Gi-Gi
515	*Dioscorea dumetorum*	African Bitter Yam	DIOS	Tuber	Africa	OC	beetle, fly, bee?	267	Giant Bean, Poroto
516	*Dioscorea esculenta*	Asiatic Yam	DIOS	Tuber	S Asia	OC	beetle, fly, bee?	428	Giant Filbert
517	*Dioscorea floribunda*	Wild Yam	DIOS	Medicinal	Neot	OC	beetle, fly, bee?	967	Giant Granadilla
518	*Dioscorea hispida*	Asian Bitter Yam	DIOA	Tuber	S Asia	OC	beetle, fly, bee?	58	Giant Maguey
519	*Dioscorea macrostachya*	Cuculmeca	DIOS	Vegetable	Neot	OC	beetle, fly, bee?	481	Giant Swamp Taro
520	*Dioscorea mexicana*	Wild Yame	DIOS	Medicinal	Neot	OC	beetle, fly, bee?	76	Giant Taro
521	*Dioscorea nummularia*	Kerung	DIOS	Tuber	SE Asia	OC	beetle, fly, bee?	1163	Gilo
522	*Dioscorea opposita*	Chinese Yam	DIOS	Tuber	S Asia	OC	beetle, fly, bee?	1326	Ginger
523	*Dioscorea pentaphylla*	Buck Yam	DIOS	Tuber	SE Asia	OC	beetle, fly, bee?	641	Ginko
524	*Dioscorea rotundata*	Eboe Yam, ñame blanco	DIOS	Tuber	W Afr.	OC	beetle, fly, bee?	941	Ginseng
525	*Dioscorea trifida*	Yampi	DIOS	Tuber	Neot	OC	beetle, fly, bee?	1302	Giri Yam Pea
526	*Dicorynia guianensis*	Basralocus	LEG	Timber	Neot			603	Gissard, Nanac
527	*Dioscoreophyllum cumminsii*	Serendipity-Berry	MENI	Sweetener	Africa			474	Globe Artichoke
528	*Diospyros digyna*	Black Sapote	EBEN	Fruit,Timb.	Neot	OC	bee, wasp, fly	274	Goat Chili
529	*Diospyros ebenaster*	Barbacoa	EBEN	Fruit	Neot	OC	bee, wasp, fly	1150	Golden Timothy Grass
530	*Diospyros kaki*	Chinese Persimmon	EBEN	Fruit	S Asia	OC, PC	bee, wasp, fly	21	Golden Wattle
531	*Diospyros lotus*	Date Plum, Kaki	EBEN	Fruit	S Asia	OC	bee, wasp, fly	1308	Gombi Bean
532	*Diospyros virginiana*	Common Persimmon	EBEN	Fruit	Neot	OC	bee, wasp, fly	797	Gow-Kee, Box Thorn
533	*Dipterocarpus costatus*	Gurjun	DIPT	Resin	SE Asia			87	Grain Amaranth
534	*Dipterocarpus kerrii*	Keruing Gondol	DIPT	Timber	SE Asia			45	Grains of Paradise

No.	Scientific Name	Common Name	Family	Use	Origin	Breed. Sys.	Pollinators
535	*Dipteryx odorata*	Cumaru	LEG	Spice	Neot	AU, OC?	insect
536	*Dipteryx oppositifolia*	Tonka Bean	LEG	Spice	Neot	AU, OC?	insect
537	*Distemonanthus benthamianus*	Ayan	LEG	Timber	Africa		
538	*Dolichos uniflorus*	Horsegram	LEG	Cereal	S Asia	AU, SC	bee, Apis
539	*Doryalis hebecarpa*	Ceylon-Goosebeny	FLAC	Fruit	S Asia	OC	bee?
540	*Doryalis spp.*	Kitembilla, Kei Apple	FLAC	Fruit	SE Asia	OC	bee?
541	*Dracaena cinnabari*	Dragon's Blood	AGAV	Resin	Africa		
542	*Dracaena draco*	Dragon's Blood	AGAV	Resin	Africa		
543	*Dracontomelon mangiferum*	Argue Pheasant-Tree	ANAC	Fruit	SE Asia		
544	*Drimys winteri*	Winter's-Bark	MAGN	Medicinal	Neot		
545	*Dryobalanops spp.*	Kapur	DIPT	Timber	SE Asia		
546	*Duboisia leichhardtii*	Corkwood	SOLA	Medicinal	Aust		
547	*Durio spp.*	Durians	BOMB	Fruit	SE Asia	OC	bat, Eon, Pt, Mac
548	*Durio zibethinus*	Durian	BOMB	Fruit	SE Asia	OC	bat, Eon, Pt, Mac
549	*Dyera costulata*	Jelutong	APOC	Elast,Timb.	SE Asia		
550	*Ecballium elaterium*	Squirting Cucumber	CUCU	Medicinal	Medit		
551	*Ecclinusa balata*	Coquirana	SAPO	Elastomer	Neot		
552	*Elaeagnus angustifolia*	Russian Olive	ELAE	Timber	Asia	OC?	bee?
553	*Elaeagnus philippensis*	Lingaro	ELAE	Fruit	S Asia	OC?	bee?
554	*Elaeis guineensis*	African Oil Palm	PALM	Oil	Africa	OC	beetle, Elaeidobius, insect
555	*Elaeis oleifera*	American Oil Palm	PALM	Oil	Neot	OC	beetle, insect
556	*Elaeocarpus dentatus*	Hinau	ELAE	Tannin	Aust	OC?	bat, Pteropus
557	*Eleocharis dulcis*	Chinese Waterchestnut	CYPE	Tuber	S Asia		
558	*Elettaria cardamomum*	Cardamom	ZING	Spice	S Asia	OC	bee, Apis, Amegilla
559	*Eleusine coracana*	Finger Millet	POA	Cereal	S Asia	AU	wind, bee
560	*Ensete ventricosum*	Ensete	PALM	Fiber, Veg.	Africa		
561	*Entandrophragma utile*	Utile	MELI	Timber	Africa		
562	*Enydra fluctuans*	Buffalo Spinach	COMP	Vegetable	SE Asia		
563	*Ephedra gerardiana*	Pakistan Ephedra	GNET	Medicinal	Asia		
564	*Ephedra sinica*	Chinese Ephedra	GNET	Medicinal	Asia		
565	*Eragrostis chloromelas*	Boer Lovegrass	POA	Erosion	Africa		
566	*Eragrostis curvula*	Weeping Lovegrass	POA	Forage	Africa		
567	*Eragrostis lehmanniana*	Lehmann Lovegrass	POA	Erosion	Africa		
568	*Eragrostis tef*	Teff	POA	Cereal	Africa	AU	wind, bee?
569	*Eremocitrus glauca*	Australian Desert Lime	RUTA	Fruit	Aust		
570	*Eriobotrya japonica*	Loquat	ROSA	Fruit	S Asia	AU, OC	bat, Rousettus, bee, Apis
571	*Eriochloa polystachya*	Caribgrass	POA	Forage	Neot		
572	*Eruca vesicaria*	Roquette	CRUC	Oil	S Asia	OC	bee, Apis
573	*Eryngium foetidum*	Fitweed	UMBE	Spice	S Asia		
574	*Erythea salvadorensis*	Palma	PALM	Oil	Neot		
575	*Erythroxylum coca*	Coca	ERYT	Medicinal	Neot		
576	*Erythroxylum novogratense*	Colombian Coca	ERYT	Medicinal	Neot		
577	*Eschweilera odorata*	Eschweilera Odorata	LECY	Nut	Neot		
578	*Escobedia scabrifolia*	Color Azafrán	SCRO	Dye	S Amer		
579	*Eucalyptus astringens*	Mallee	MTRT	Tannin	Aust	SC, OC	bee
580	*Eucalyptus camaldulensis*	Red River Gum	MYRT	Multiple	Aust	SC, OC	bee
581	*Eucalyptus citriodora*	Lomon-Scented Gum	MTRT	Essence	Aust	SC, OC	bird, bee
582	*Eucalyptus cladocalyx*	Sugar Gum	MYRT	Multiple	Aust	SC, OC	bee
583	*Eucalyptus dives*	Broadleaf Peppermint	MYRT	Essence	O.Wld.	SC, OC	bat, bee?
584	*Eucalyptus globulus*	Tasmanian Blue Gum	MYRT	Essence	Aust	SC, OC	bird, bee
585	*Eucalyptus macarthurii*	Camden Wollybut	MYRT	Essence	OldWorld	SC, OC	bird, bat, bee?
586	*Eucalyptus radiata*	Narrowleaf Peppermint	MYRT	Essence	Aust	SC, OC	bird, bat, bee?
587	*Eucalyptus saligna*	Sydney Blue Gum	MYRT	Timber	Aust	SC, OC	bee
588	*Eucalyptus smithii*	Gully Gum	MYRT	Essence	Aust	SC, OC	bat, bee?
589	*Eucalyptus spp.*	Eucalyptus	MYRT	Timber	O.Wld.	SC, OC	bird, bat, Pteropus, bee
590	*Eucalyptus sideroxylon*	Red Iron Bark	MYRT	Timber	O.Wld.	SC, OC	bird, bee
591	*Eucalyptus staigeriana*	Lemon Iron Bark	MYRT	Essence	Aust	SC, OC	bat, bee?
592	*Eucalyptus wandoo*	Eucalyptus	MYRT	Tannin	O.Wld.	SC, OC	bat, bee?
593	*Euchlaena mexicana*	Teosinte	POA	Forage	Neot		
594	*Eugeissona utilis*	Kadjatoa	PALM	Multiple	SE Asia		
595	*Eugenia brasiliensis*	Pitomba	MYRT	Fruit	Neot	SC?, OC	bee
596	*Eugenia dombeyi*	Grumichana	MYRT	Fruit	Neot	SC?, OC	bee
597	*Eugenia myrcianthes*	Iba-jayi	MYRT	Fruit	Neot	SC?, OC	bee
598	*Eugenia pungens*	Guayabo	MYRT	Fruit	Neot	SC?, OC	bee
599	*Eugenia tomentosa*	Cabeludinha	MYRT	Fruit	Neot	SC?, OC	bee
600	*Eugenia uniflora*	Surinam-Cherry	MYRT	Fruit	Neot	SC?, OC	bee
601	*Euphorbia antisyphilitica*	Candelilla	EUPH	Wax	Neot		
602	*Euterpe edulis*	Palmheart, Palmito	PALM	Vegetable	Neot	SC, OC?	fly, beetle?
603	*Euterpe oleracea*	Gissard, Nanac	PALM	Vegetable	Neot	SC, OC?	fly, beetle?
604	*Fagopyrum esculentum*	Buckwheat	POLG	Cereal	S Asia	OC	bee, Apis
605	*Fagopyrum tataricum*	Tartary Buckwheat	POLG	Cereal	S Asia	OC	bee, Apis
606	*Feronia limonia*	Indian Wood-Apple	RUTA	Fruit	S Asia		
607	*Ferula assa-foetida*	Asafetida Gum	UMBE	Resin	W Asia		
608	*Ferula gummosa*	Galbanum, Kasnih	UMBE	Resin	W Asia		
609	*Feviella cordofolia*	Abilla	CUCU	Oil	Neot		
610	*Fibraurea tinctoria*	Huang T'eng	MENI	Medicinal	SE Asia		
611	*Ficus carica*	Common Fig	MORA	Fruit	SE Asia	PC, OC	wasp, Bla, Cera
612	*Ficus elastica*	India Rubber Fig	MORA	Resin	SE Asia	OC	wasp, Bla, Cera
613	*Ficus roxburghii*	Fig	MORA	Fruit	Neot	PC, OC	wasp, Bla, Cera
614	*Ficus vogelii*	Vogel Fig	MORA	Resin	Africa	OC	wasp, Bla, Cera
615	*Fimbristylis globulosa*	Tikus	CYPE	Fiber	S Asia		
616	*Flacourtia cataphracta*	Runeala Plum	FLAC	Fruit	SE Asia	OC?	bee
617	*Flacourtia rukam*	Rukam	FLAC	Fruit	S Asia	OC?	bee
618	*Flemingia vestita*	Soh-phlong	LEG	Tuber	S Asia		
619	*Foeniculum vulgare*	Fennel	UMBE	Spice	S Asia	AU, OC	bee, Apis
620	*Fortunella spp.*	Kumquat	RUTA	Fruit	S Asia	~AG, OC	bee
621	*Fragaria chiloensis*	Chilean Strawberry	ROSA	Fruit	Neot	AU, OC	bee, fly
622	*Fragaria vesca*	European Strawberry	ROSA	Fruit	Europe	AU, OC	bee, fly
623	*Fragaria virginiana*	Virginia Strawberry	ROSA	Fruit	Nearc	AU, OC	bee, fly

No.	Common Name
373	Grapefruit
80	Greater Galanga
1047	Green Sapote
12	Green Wattle
13	Green Wattle
916	Greenheart
131	Ground Nut
596	Grumichana
714	Guabo Real
668	Guaiac
713	Guamo
465	Guar, Clusterbean
969	Guaraná
669	Guarea
598	Guayabo
959	Guayule
1320	Guinea Pepper
588	Gully Gum
24	Gum Arabic
120	Gum Ghatti
152	Gum Tragacanth
533	Gurjun
972	Gutta Sundek
939	Gutta-Percha
708	Hairy Indigo
1297	Hairy Vetch
973	Harmala Shrub
1178	Hausa Potato
681	Hemigraphia Alternata
755	Henna
54	Hennequen
426	Himalayan Hazel
556	Hinau
486	Hindu Datura
637	Hire Tai
696	Hirta Grass
913	Hoary Basil
1184	Hog Plum, Jobo
914	Holy Basil
1155	Holy Thistle
642	Honeylocust
806	Horse Gram
495	Horse Marmalade
538	Horsegram
135	Horseradish
610	Huang T'eng
860	Huemega
14	Huisache
1294	Hungarian Vetch
506	Hungry-Rice
482	Huon Pine
1259	Hybrid Clover
700	Hyssop
597	Iba-jayi
920	Icaua
711	Icecream Bean, Sipo
1229	Idigbo
114	Ilama
701	Ilex Argentina
634	Imbe
1186	Imbu
83	Inca-Wheat
612	India Rubber Fig
41	Indian Bael
161	Indian Belladonna
1032	Indian Borage
1283	Indian Copal
1329	Indian Jujube
302	Indian Laburnum
737	Indian Lettuce
1020	Indian Long Pepper
1109	Indian Madder
868	Indian Mulberry
199	Indian Silk Cotton
360	Indian Squash Melon
850	Indian Sweetclover
606	Indian Wood-Apple
355	Indian-Bark
716	Inula
237	Ipecac
316	Ipecac
55	Istle
1321	Itabo, Izote
1006	Ivory-Nut Palm
386	Ivy Gourd, Kundiri
1009	Jaborandi
882	Jaboticaba
264	Jackbean, Frijól Burro
143	Jackfruit

No.	Scientific Name	Common Name	Family	Use	Origin	Breed. Sys.	Pollinators
624	*Fragaria X ananassa*	Garden Strawberry	ROSA	Fruit	Nearc	AU, OC	bee, fly
625	*Funtumia elastica*	Lagos Silk-Rubber	APOC	Elastomer	Africa		
626	*Furcraea andina*	Chuchao	AGAV	Fiber	Neot		
627	*Furcraea cabuya*	Cabuya, Pia Floja	AGAV	Fiber	Neot		
628	*Furcraea gigantea*	Mauritius Hemp	AGAV	Fiber	Neot		
629	*Furcraea humboldtiana*	Cocuiza	AGAV	Fiber	Neot		
630	*Furcraea macrophylla*	Fique	AGAV	Fiber	Neot		
631	*Furcraea quicheensis*	Cabuya	AGAV	Fiber	Neot		
632	*Garcinia dulcis*	Baniti	GUTT	Fruit	SE Asia	AG, OC?	
633	*Garcinia lateriflora*	Kariis	GUTT	Fruit	SE Asia	AG, OC?	
634	*Garcinia livingstonei*	Imbe	GUTT	Beverage	Africa	AG, OC?	
635	*Garcinia mangostana*	Mangosteen	GUTT	Fruit	SE Asia	AG, OC?	
636	*Garcinia morella*	Gamboge	GUTT	Resin	SE Asia	AG, OC?	
637	*Garcinia multiflora*	Hire Tai	GUTT	Fruit	SE Asia	AG, OC?	
638	*Garcinia tinctoria*	Gamboge Tree	GUTT	Fruit	SE Asia	AG, OC?	
639	*Garcinia xanthocarpus*	Garcinia Xanthocarpus	GUTT	Fruit	SE Asia	AG, OC?	
640	*Gendarussa vulgaris*	Gendarussa vulgans	ACAN	Medicinal	S Asia		
641	*Ginko biloba*	Ginko	GINK	Nut	E Asia		
642	*Gleditsia tricanthos*	Honeylocust	LEG	Multiple	Nearc	OC	bee
643	*Gluta usitata*	Burmese Lacquer	ANAC	Resin	SE Asia		
644	*Glycine max*	Soybean	LEG	Oil	S Asia	AU, ~OC	bee
645	*Glycine wightii*	Perennial Soybean	LEG	Forage	S Asia	AU, ~OC?	bee
646	*Glycyrrhiza glabra*	Common Licorice	LEG	Multiple	Medit		
647	*Glycyrrhiza lepidota*	American Licorice	LEG	Spice	N Amer		
648	*Gmelina arborea*	Yemane	VERB	Timber	S Asia	SC, OC?	bee?
649	*Gmelina leichhardtii*	White Beech	VERB	Timber	Aust	SC, OC?	bee?
650	*Gnetum gnemon*	Emping Great Fruit	GNET	Nut	S Asia		
651	*Gonystylus bancanus*	Ramin	THYM	Timber	SE Asia		
652	*Gossweilerodendron balsamiferum*	Agba	LEG	Timber	Africa		
653	*Gossypium anomalum*	Wild Cotton	MALV	Fiber	Africa	AU, OC	bee
654	*Gossypium arboreum*	Tree Cotton	MALV	Fiber	S Asia	AU, OC	bee
655	*Gossypium barbadense*	Sea Island Cotton	MALV	Fiber	Neot	AU, OC	bee
656	*Gossypium herbaceum*	Levant Cotton	MALV	Fiber	Africa	AU, OC	bee
657	*Gossypium hirsutum*	Upland Cotton	MALV	Fiber	Neot	AU, OC	bee
658	*Gossypium klotzchianum*	Cotton	MALV	Fiber	Neot	AU, OC	bee
659	*Gossypium raimondii*	Cotton	MALV	Fiber	Neot	AU, OC	bee
660	*Gossypium robinsonii*	Cotton	MALV	Fiber	Aust	AU, OC	bee, bird
661	*Gossypium somalense*	Cotton	MALV	Fiber	Africa	AU, OC	bee
662	*Gossypium stocksii*	Cotton	MALV	Fiber	Neot	AU, OC	bee
663	*Gossypium thurberi*	Cotton	MALV	Fiber	Neot	AU, OC	bee
664	*Gossypium trilobum*	Cotton	MALV	Fiber	Neot	AU, OC	bee
665	*Gossypium vitifolium*	Cotton	MALV	Fiber	Neot	AU, OC	bee
666	*Grevillea robusta*	Silky Oak	PROT	Timber	SE Asia	OC?	bat, Pteropus
667	*Grewia asiatica*	Phalsa	TILI	Beverage	S Asia	AU, OC	bee, bird, butterfly
668	*Guajacum officinale*	Guaiac	ZYGO	Resin,Mult.	Neot		
669	*Guarea cedrata*	Guarea	MELI	Timber	Africa		
670	*Guibourtia demeusii*	Copal Noir	LEG	Resin	W Afr		
671	*Guizotia abyssinica*	Niger-Seed	COMP	Oil	Africa	SC, OC	bee, Apis
672	*Gustavia superba*	Membrillo	LECY	Fruit	Neot	SC, OC	bee, Xylocopa, Melipona
673	*Gynandropsis gynandra*	African Spiderflower	CAPP	Vegetable	Africa		
674	*Haematoxylon campechianum*	Logwood	LEG	Dye	Neot	SC, OC?	bee
675	*Hedysarum coronarium*	Spanish Sainfoin	LEG	Forage	Medit	OC, SC	bee, Apis
676	*Helianthus annuus*	Sunflower	COMP	Oil	Neot	SC, OC	bee, Ap, Mel, fly
677	*Helianthus annuus X tuberosus*	Sunchoke (311)	COMP	Multiple	Neot	SC, OC	bee, Ap, Mel, fly
678	*Helianthus tuberosus*	Jerusalem-Artichoke	COMP	Tuber	Neot	SC, OC	bee, Ap, Mel, fly
679	*Helminthostachys zeylanica*	Kamraj	POLYP	Vegetable	SE Asia		
680	*Hemarthria altissima*	Limpo Grass	POA	Forage	Af-As		
681	*Hemigraphia alternata*	Hemigraphia Alternata	ACAN	Medicinal	SE Asia		
682	*Heritiera spp.*	Mengkulang	STER	Timber	SE Asia		
683	*Hevea benthamiana*	Rubber	EUPH	Elastomer	Neot	SC, OC	fly,Cul,Stil,thrip,Franklin
684	*Hevea spp.*	Para Rubber	EUPH	Elastomer	Neot	SC, OC	thrip, Frankliniella, etc.
685	*Hibiscus cannabinus*	Kenaf	MALV	Fiber	Africa	SC, OC	bee
686	*Hibiscus sabdariffa*	Rosella	MALV	Bev., Fiber	Africa	AU, OC?	insect?
687	*Hordeum bulbosum*	Barley	POA	Cereal	W Asia	AU	wind, bee?
688	*Houttuynia cordata*	Tsi	SAUR	Vegetable	SE Asia		
689	*Hovenia dulcis*	Japanese Raisin Tree	RHAM	Fruit	S Asia		
690	*Hydnocarpus anthelmintica*	Lukrabao	FLAC	Medicinal	SE Asia		
691	*Hydnocarpus kurzii*	Chaulmoogra	FLAC	Medicinal	SE Asia		
692	*Hylocereus ocamponis*	Pitaya roja	CACT	Fruit	Neot		
693	*Hymenaea courbaril*	Algarrobo	LEG	Resin	Neot		
694	*Hyoscyamus muticus*	Egyptian Henbane	SOLA	Medicinal	N Afr		
695	*Hyoscyamus niger*	Black Henbane	SOLA	Medicinal	N Afr		
696	*Hyparrhenia hirta*	Hirta Grass	POA	Erosion	Af-As		
697	*Hyparrhenia rufa*	Jaragua Grass	POA	Forage	Af-As		
698	*Hyphaene thebaica*	Dum Palm	PALM	Vegetable	Africa		
699	*Hyptis spicigera*	Kindi, Benefing	LABI	Oil	Africa	SC, OC	bee, wasp, bird?
700	*Hyssopus officinalis*	Hyssop	LABI	Spice	Eur-As		
701	*Ilex argentina*	Ilex Argentina	AQUI	Timber	Neot	OC	
702	*Ilex brevicuspis*	Yerba Mate Mimic	AQUI	Beverage	Neot		
703	*Ilex dumosa*	Kaa-Chiri	AQUI	Beverage	Neot	OC	
704	*Ilex guayusa*	Yerba Mate Mimic	AQUI	Beverage	Neot	OC	
705	*Ilex paraguariensis*	Yerba Mate	AQUI	Beverage	S Amer	OC	
706	*Illicium verum*	Star-Anise Tree	MAGN	Essence	S Asia		
707	*Indigofera arrecta*	Natal Indigo	LEG	Dye	Africa	OC	bee
708	*Indigofera hirsuta*	Hairy Indigo	LEG	Dye	S Asia	OC	insect
709	*Indigofera spicata*	Spicata Indigo	LEG	Dye	Africa	OC	bee?
710	*Indigofera tinctoria*	Common Indigo	LEG	Dye	Africa	OC	bee, Apis
711	*Inga edulis*	Icecream Bean, Sipo	LEG	Fruit	Neot	SC, OC	insect, bird
712	*Inga feuillei*	Pakay	LEG	Fruit	Neot	SC, OC	insect, bird

No.	Common Name
1097	Jacomico
725	Jagua
720	Jalap
865	Jamaica Nutmeg
1210	Jambolan
1208	Jambu Ayer
1246	Japan Tallow
1103	Japan Wax
768	Japanese Lespedeza
1064	Japanese Plum
689	Japanese Raisin Tree
697	Jaragua Grass
722	Jasmine
257	Java Almond
1213	Java Apple
1092	Java Devil Pepper
258	Java-Almond
1023	Javanese Long Pepper
549	Jelutong
678	Jerusalem-Artichoke
1254	Jesuit Nut, Four Horns
60	Jeungjing
938	Jicama
418	Jicib Bean, Ye-eb
487	Jimson Weed
398	Job's Tears
1157	Jojoba
1192	Kaa He'e
703	Kaa-Chiri
123	Kadam
594	Kadjatoa
1034	Kafir Potato, Dazo
679	Kamraj
1167	Kangaroo Apple
311	Kapok
42	Kapok Bush
176	Kapoondong
545	Kapur
285	KarAnda
1191	Karaya Gum
633	Kariis
1189	Kataya Gum
1190	Kataya Gum
1128	Katuk
1021	Kava
803	Kawakawa
1183	Kedondong
735	Kempas
685	Kenaf
263	Kenai Nut
534	Keruing Gondol
521	Kerung
308	Khat
125	Kidney-Vetch
979	Kikuyu Grass
699	Kindi, Benefing
1285	Kinka Oil Ironweed
971	Kiri
743	Kirk's Gum Vine
540	Kitembilla, Kei Apple
38	Kiwifruit
225	Kohlrabi
61	Kokko
46	Korarima Cardamom
767	Korean Lespedeza
1080	Kudzu
1081	Kudzu Vine
931	Kumis Kuching
620	Kumquat
205	Kundang
818	Kwini
357	Labdanum
736	Lablab
1135	Lac-tree
625	Lagos Silk-Rubber
904	Lajalu
440	Lanceleaf Crotalaria
742	Landolphia Rubber
745	Langsat, Duku
1272	Lapulla
94	Large Cardamom
748	Lathyrus Pubescens
419	Laurel, Frejo
752	Lavandin
101	Leafless Anabasis
69	Leek
567	Lehmann Lovegrass
370	Lemon
371	Lemon

No.	Scientific Name	Common Name	Family	Use	Origin	Breed. Sys.	Pollinators	No.	Common Name
713	*Inga nobilis*	Guamo	LEG	Fruit	Neot	SC, OC	insect, bird	591	Lemon Iron Bark
714	*Inga spectabilis*	Guabo Real	LEG	Fruit	Neot	SC, OC	insect, bird	1327	Lempoyang
715	*Intsia palembanica*	Merbau	LEG	Timber	SE Asia			759	Lentil
716	*Inula helenium*	Inula	COMP	Essence	W Asia			81	Lesser Galanga
717	*Ipomoea aquatica*	Water Spinach	CONV	Vegetable	Pan-tro.	OC	bee, Melitoma, Ancylo.	738	Lettuce
718	*Ipomoea batatas*	Sweet Potato	CONV	Tuber	Neot	OC	bee, Melitoma, Ancylo.	769	Leucaena Diversifolia
719	*Ipomoea hispida*	Chlabato	CONV	Vegetable	Neot	OC	bee, Melitoma, Ancylo.	770	Leucaena Esculenta
720	*Ipomoea purga*	Jalap	CONV	Medicinal	Neot	OC	bee, Melitoma, Ancylo.	772	Leucaena Glauca
721	*Iris pallida*	Oris Root	IRID	Essence	Medit			773	Leucaena Leucocephala
722	*Jasminum officinale*	Jasmine	OLEA	Essence	Asia			774	Leucaena Macrophylla
723	*Jatropha curcas*	Physic Nut	EUPH	Oil	Neot			771	Leucaena-Salvadorian
724	*Jessenia bataua*	Patauá	PALM	Oil	Neot			656	Levant Cotton
725	*Jessenia polycarpa*	Jagua	PALM	Oil	Neot			778	Levant Storax
726	*Jessenia repanda*	Aricauguá	PALM	Oil	Neot			994	Lima Bean
727	*Jubaea chilensis*	Wine Palm	PALM	Beverage	S Amer			363	Lime
728	*Juglans neotropica*	Ecuador Walnut	JUGL	Nut	S Amer			1267	Limeberry
729	*Juglans regia*	Persian Walnut	JUGL	Nut	W Asia			680	Limpo Grass
730	*Juncus maritimus*	Sea Rush	JUNC	Erosion	Neot			235	Linaloe Oil
731	*Justicia insularis*	Tettu	ACAN	Vegetable	W Afr			1253	Ling, Two Horns
732	*Kaempferia galanga*	Galanga	ZING	Spice	S Asia			553	Lingaro
733	*Kerstingiella geocarpa*	Geocaipa	LEG	Vegetable	Africa			899	Litchi
734	*Khaya ivorensis*	African Mahogany	MELI	Timber	Africa			951	Little Millet
735	*Koompassia malaccensis*	Kempas	LEG	Timber	SE Asia			674	Logwood
736	*Lablab purpureus*	Lablab	LEG	Vegetable	S Asia			581	Lomon-Scented Gum
737	*Lactuca indica*	Indian Lettuce	COMP	Vegetable	SE Asia	AU, OC	insect	900	Longan
738	*Lactuca sativa*	Lettuce	COMP	Vegetable	Medit	AU, OC	insect	791	Loofah
739	*Lactuca virosa*	Bitter Lettuce	COMP	Medicinal	Medit	AU, OC	insect	570	Loquat
740	*Lagenaria siceraria*	Bottle Gourd	CUCU	Vegetable	Af, As	OC	bat, moth, bee	785	Lotonaris
741	*Lallemantia iberica*	Dragon's Head	APOC	Oil, Veg.	Asia			897	Lotong
742	*Landolphia heudelotii*	Landolphia Rubber	APOC	Elastomer	Africa			492	Luba Root
743	*Landolphia kirkii*	Kirk's Gum Vine	APOC	Elastomer	Africa			1044	Lucuma
744	*Landolphia owariensis*	African Gum Vine	APOC	Elastomer	Africa			690	Lukrabao
745	*Lansium domesticum*	Langsat, Duku	MELI	Fruit	SE Asia			1284	Ma-ha-wa-soo
746	*Lathyrus hirsutus*	Rough Pea	LEG	Forage	W Asia	AU		801	Macadamia
747	*Lathyrus latifolius*	Bitter Pea	LEG	Cereal	Neot			43	Madagascar Cardamom
748	*Lathyrus pubescens*	Lathyrus Pubescens	LEG	Forage	Neot			179	Mafai
749	*Launaea taaraxacifolia*	Wild Lettuce	COMP	Vegetable	W Afr			56	Maguey, Sisal
750	*Laurus nobilis*	Sweet Bay	LAUR	Spice	Medit			1205	Mahogany, Caoba
751	*Lavandula angustifolia*	Common Lavender	LABI	Essence	Medit	OC, SC	bee, Bombus, Apis	807	Mahua
752	*Lavandula hybrida*	Lavandin	LABI	Essence	Medit	OC, SC	bee, Bombus, Apis	896	Makimbeira
753	*Lavandula latifolia*	Broad-Leaved Lavender	LABI	Essence	Medit	OC, SC	bee, Bombus, Apis	448	Malabar Gourd
754	*Lavandula stoechas*	French Lavender	LABI	Essence	Medit	OC, SC	bee, Bombus, Apis	40	Malabar Nut
755	*Lawsonia inermis*	Henna	LYTH	Dye, Orna.	W Asia			188	Malabar Spinach
756	*Lecythis elliptica*	Monkey Pot, Sapucaja	LECY	'Nut'	Neot	OC	bee, eulgossine, Bo, Xy	579	Mallee
757	*Lecythis minor*	Sapucaja	LECY	'Nut'	Neot	OC	bee, eulgossine, Bo, Xy	814	Mamey
758	*Lecythis ollaria*	Monkey-Pot	LECY	Timber	Neot	OC	bee, eulgossine, Bo, Xy	246	Mamey Sapote
759	*Lens culinaris*	Lentil	LEG	Cereal	Medit	AU, OC	insect, bee	374	Mandarin
760	*Leopoldinia piassaba*	Piassaba Palm	PALM	Fiber	S Amer			815	Mango
761	*Lepidium mayennii*	Chijura	CRUC	Tuber	Neot			816	Mango
762	*Lepidium sativum*	Garden Cress	CRUC	Vegetable	Europe			817	Mango
763	*Leptospermum flavescens*	Yellow Tea Tree	MYRT	Essence	Aust			819	Mango
764	*Leptospermum laevigatum*	Australian Tea-Tree	MYRT	Erosion	Aust	SC, OC	bee	458	Mango Ginger
765	*Leptospermum scoparium*	Manuka	MYRT	Essence	Aust	SC, OC	bee	635	Mangosteen
766	*Lespedeza cuneata*	Sericea Lespedeza	LEG	Forage	Asia	AU, OC	bee	1102	Mangrove
767	*Lespedeza stipulacea*	Korean Lespedeza	LEG	Erision	Asia	AU, OC	bee	823	Manicoba
768	*Lespedeza striata*	Japanese Lespedeza	LEG	Forage	Asia	AU, OC	bee	821	Manicoba Rubber, Jequie
769	*Leucaena diversifolia*	Leucaena Diversifolia	LEG	Timber	Neot	SC, OC?	bee	50	Manila Copal
770	*Leucaena esculenta*	Leucaena Esculenta	LEG	Fruit,Timb.	Neot	SC, OC?	bee	825	Manilkara achras
771	*Leucaena glauca*	Leucaena-Salvadorian	LEG	Timber	Neot	SC, OC?	bee	765	Manuka
772	*Leucaena glauca*	Leucaena Glauca	LEG	Forage	Neot	SC, OC?	bee	961	Maracujá
773	*Leucaena leucocephala*	Leucaena Leucocephala	LEG	Forage	Neot	SC, OC?	bee	1275	Marama Bean, Tsi Bean
774	*Leucaena macrophylla*	Leucaena Macrophylla	LEG	Timber	Neot	SC, OC?	bee	837	Maripa
775	*Licania rigida*	Oiticica Oil	CHRY	Oil	Neot			140	Maritime Wormwood
776	*Linum usitatissimum*	Flax	LINA	Fiber	S Asia	AU, OC	thrip, bee, Apis	1139	Marking-Nut Tree
777	*Lippia micromera*	Spanish Thyme	VERB	Spice	Neot			455	Marrow
778	*Liquidambar orientalis*	Levant Storax	HAMA	Resin	SW Asia			456	Marrow
779	*Litsea cubeba*	May Chang Oil	LAUR	Essence	S Asia			828	Massaranduba
780	*Lodoicea maldivica*	Double Coconut	PALM	Medicinal	Seyche			353	Massoia
781	*Lonchocarpus nicou*	Timbó	LEG	Pesticide	Neot			901	Mata Kuching
782	*Lonchocarpus urucu*	Urucu Timbo	LEG	Pesticide	Neot			862	Mauka
783	*Lonchocarpus utilis*	Barbasco	LEG	Pesticide	Neot			628	Mauritius Hemp
784	*Lophira alata*	Ekki, Eba	OCHN	Timber	Africa			779	May Chang Oil
785	*Lotonaris bainesii*	Lotonaris	LEG	Forage	W Asia	SC, OC?	bee, Megachile	35	Mbocayá
786	*Lotus corniculatus*	Birdsfoot Trefoil	LEG	Erosion	W Asia	SC, OC?	bee, Os, Meg, Xy, Anth	410	Mecca Myrrh
787	*Lotus tenuis*	Narrowleaf Trefoil	LEG	Forage	W Asia	SC, OC		47	Melegueta-Pepper
788	*Lotus uliginosus*	Big Trefoil	LEG	Forage	Europe	SC, OC	bee, Bombus	672	Membrillo
789	*Luffa acutangula*	Angled Luffa	CUCU	Fruit	S Asia	SC, OC	moth, bee, Trigona, Apis	682	Mengkulang
790	*Luffa cylindrica*	Smooth Loofah	CUCU	Multiple	Africa	OC	bee, Xy, Ap, Trig	178	Menteng
791	*Luffa operculata*	Loofah	CUCU	Multiple	Neot	OC	bee, Xy, Ap, Trig	715	Merbau
792	*Lupinus albus*	White Lupine	LEG	Forage	Africa	AU, OC	bee	112	Mersawa
793	*Lupinus angustifolius*	Blue Lupine	LEG	Forage	Europe	AU, OC	bee	1050	Mesquite
794	*Lupinus luteus*	Yellow Lupine	LEG	Forage	Europe	AU, OC	bee	300	Mexican Apple
795	*Lupinus mutabilis*	Tarwi	LEG	Oil	Neot	SC, OC	bee	1003	Mexican Husk Tomato
796	*Lupinus termis*	White Lupine	LEG	Forage	Europe	AU, OC	bee	104	Mijao
797	*Lycium chinense*	Gow-Kee, Box Thorn	SOLA	Vegetable	S Asia			950	Millet
798	*Lycopersicon esculentum*	Tomato, Tomatle	SOLA	Fruit	Neot	AU,PC,OC	bee, Bombus, Exomalopsis	312	Mirabel, Soko
799	*L. pimpinellifolium*	Currant Tomato	SOLA	Fruit	Neot	AU, OC	bee, Bombus, Exomalopsis	1234	Miracle-Fruit
800	*Lygeum spartum*	Esparto	POA	Fiber	Africa			851	Molasses Grass
801	*Macadamia integrifolia*	Macadamia	PROT	Nut	Aust	SC, OC	bee, Trig, Ap, beetle, wasp	146	Monkey Jack

No.	Scientific Name	Common Name	Family	Use	Origin	Breed. Sys.	Pollinators	No.	Common Name
802	*Macadamia tetraphylla*	Rough-Shell Macadamia	PROT	Nut	Aust	SC, OC	bee, Trig, Ap, beetle, wasp	756	Monkey Pot, Sapucaja
803	*Macropiper excelsum*	Kawakawa	PIPE	Essence	Aust			758	Monkey-Pot
804	*Macrotyloma atropurpurea*	Siratro	LEG	Forage	Africa	AU?		867	Mora
805	*Macrotyloma geocarpum*	Bindi, Kandela	LEG	Vegetable	Africa	AU?		145	Morang
806	*Macrotyloma uniflorum*	Horse Gram	LEG	Vegetable	S Asia	AU?		836	Moriche
807	*Madhuca longifolia*	Mahua	SAPO	Oil	S Asia			1136	Morula
808	*Madia sativa*	Oil-madi	COMP	Oil	S Amer			1298	Moth Bean
809	*Maesopsis eminii*	Musizi	RHAM	Timber	Africa			281	Mountain Papaya
810	*Malachra capitata*	Bhanbhendi	MALV	Fiber	S Asia			1319	Mountain Plum
811	*Malpighia punicifolia*	Acerola	MALP	Fruit	Neot	SC, OC	bee, Centris	1255	Mozinda
812	*Malus pumila*	Apple	ROSA	Fruit	Medit	~AG, SC, OC	bee, Apis	501	Mozoton
813	*Malus sylvestris*	Apple	ROSA	Fruit	Medit	~AG, SC, OC	bee, Apis	64	Mu-Oil-Tree
814	*Mammea americana*	Mamey	GUTT	Fruit	Neot	OC?	bee, bat?	141	Mugwort
815	*Mangifera caesia*	Mango	ANAC	Fruit	S Asia	OC, SC	fly	212	Muhuhu Oil
816	*Mangifera foetida*	Mango	ANAC	Fruit	S Asia	OC, SC	fly	8	Mulga
817	*Mangifera indica*	Mango	ANAC	Fruit	S Asia	~AG, OC	fly, bee, bat, Pteropus	1303	Mung Bean
818	*Mangifera odorata*	Kwini	ANAC	Fruit	SE Asia	AU, OC	fly	155	Murmuru Palm
819	*Mangifera zeylanica*	Mango	ANAC	Fruit	S Asia	AU, OC	fly	1314	Muscadine Grape
820	*Manicaria saccifera*	Sleeve Palm	PALM	Oil	Neot			809	Musizi
821	*Manihot dichotoma*	Manicoba Rubber, Jequie	EUPH	Resin	Neot	OC	bee	3	Musk Okra
822	*Manihot esculenta*	Cassava, Yuca	EUPH	Energy	Neot	OC	bee	1325	Myoga, Japanese Ginger
823	*Manihot glaziovii*	Manicoba	EUPH	Elastomer	Neot	OC	bee	881	Myoporum
824	*Manihot spp.*	Cassava	EUPH	Multiple	Neot	OC	bee	1228	Myrobalans
825	*Manilkara achras*	Manilkara achras	SAPO	Fruit	Neot	SC, OC	bee	892	Myrtle
826	*Manilkara bidentata*	Balata	SAPO	Elastomer	Neot	SC, OC	bee	236	Nance, Murici
827	*Manilkara chicle*	Chicozapote	SAPO	Elastomer	Neot	SC, OC	bee	282	Naracatia
828	*Manilkara elata*	Massaranduba	SAPO	Elastomer	Neot	SC, OC	bee	1173	Naranjilla, Lulo
829	*Manilkara inundata*	True Balata	SAPO	Elastomer	Neot	SC, OC	bee	1095	Naranjito
830	*Manilkara zapota*	Chicle, Níspero	SAPO	Elastomer	Neot	SC, OC	bee	28	Naras
831	*Maranta arundinacea*	Arrowroot, Aru-Aru	MARA	Starch	Neot	OC	bee	472	Nardus Grass
832	*Marjorana hortensis*	Sweet Marjoram	LABI	Spice	Medit			27	Narran
833	*Marjorana onites*	Pot Majoram	LABI	Spice	Medit			586	Narrowleaf Peppermint
834	*Matisia cordata*	Chupa-Chupa	BOMB	Fruit	Neot			787	Narrowleaf Trefoil
835	*Matricaria chamomilla*	Wild Chamomile	COMP	Essence	Asia			707	Natal Indigo
836	*Mauritia vinifera*	Moriche	PALM	Misc.	Neot			288	Natal-Palm
837	*Maximiliana maripa*	Maripa	PALM	Oil	Neot	SC, OC?	bee	1168	Native Eggplant
838	*Maximiliana regia*	Cocorita Palm, Jaguá	PALM	Oil	Neot	SC, OC?	bee	121	Neem
839	*Medicago arabica*	Spotted Burr Medic	LEG	Erosion	W Asia	AU, OC	bee, Apis	96	Nepal Cardamom
840	*Medicago falcata*	Sickle Medic	LEG	Forage	W Asia	AU, OC	bee, Apis	999	New Zealand Flax
841	*Medicago lupulina*	Black Medic	LEG	Forage	W Asia	AU, OC	bee, Apis	1232	New Zealand Spinach
842	*Medicago orbicularis*	Button Medic	LEG	Forage	W Asia	AU, OC	bee, Apis	847	Niaouli Oil
843	*Medicago sativa*	Alfalfa	LEG	Forage	W Asia	AU, OC	bee, Nomia, Megachile	671	Niger-Seed
844	*Medicago scutellata*	Snail Medic	LEG	Forage	W Asia	AU, OC	bee, Apis	1200	Nigerian Lucerne
845	*Melaleuca alternifolia*	Tea Tree	MYRT	Essence	Aust	OC?	bat, Pteropus	175	Nim Tree
846	*Melaleuca cajuputi*	Cajeput, Gelam	MYRT	Medicinal	SE Asia	OC?	bat, Pteropus	910	Nipa Palm
847	*Melaleuca viridiflora*	Niaouli Oil	MYRT	Essence	Oceania	OC?	bat, Pteropus	187	Nua
848	*Melicocca bijuga*	Spanish Lime, Mamon	SAPI	Fruit	Neot			479	Nutgrass
849	*Melilotus alba*	White Sweetclover	LEG	Forage	Europe	SC, OC	bee, Bombus,Anthidium,Meg	887	Nutmeg
850	*Melilotus indica*	Indian Sweetclover	LEG	Forage	Eur-As	SC, OC	bee, Apis, Bombus	1268	Obeche
851	*Melinis minutiflora*	Molasses Grass	POA	Pesticide	Neot			934	Oca
852	*Melissa officinalis*	Sweet Balm	LABI	Essence	Medit			915	Ocotea, Sassafras Oil
853	*Mentha arvensis*	Corn Mint	LABI	Essence	Medit			921	Oenocarpus Mapora
854	*Mentha cardiaca*	Scotch Spearmint	LABI	Essence	Medit	SC?	fly, bee?	982	Oil-Bean Tree
855	*Mentha pulegium*	European Pennyroyal	LABI	Essence	Medit	SC?	fly, bee?	808	Oil-madi
856	*Mentha spicata*	Spearmint	LABI	Essence	Medit	SC?, OC	fly, bee, Apis	775	Oiticica Oil
857	*Mentha X gentilis*	Scotch Spearmint	LABI	Essence	Medit	SC?	fly, bee?	468	Okan
858	*Mentha X piperita*	Peppermint	LABI	Essence	Medit	SC?	fly, bee	165	Okoumé, Gaboon
859	*Metroxylon spp.*	Sago Palm	PALM	Starch, Fib.	SE Asia	OC	beetle, bee, Apis, Trigona	1	Okra
860	*Micrandra minor*	Huemega	EUPH	Resin	Neot			923	Olive
861	*Microcitrus spp.*	Australian Wild Limes	RUTA	Fruit	Aust			354	Oliver's Bark
862	*Mirabilis expansa*	Mauka	NYCT	Tuber	Neot			70	Onion
863	*Momordica balsamina*	Balsam-Apple	CUCU	Fruit	S Asia	OC, PC	bee, beetle	894	Opepe
864	*Momordica charantia*	Balsam-Pear	CUCU	Vegetable	S Asia	OC, PC	bee, beetle	953	Opium Poppy
865	*Monodora myristica*	Jamaica Nutmeg	ANNO	Spice	W Afr			409	Opopanax Gum
866	*Monopteriix uaca*	Uaku	LEG	Oil	Neot			11	Orange Wattle
867	*Mora excelsa*	Mora	LEG	Timber	Neot			929	Oregano
868	*Morinda citrifolia*	Indian Mulberry	RUBI	Dye	SE Asia			298	Orinoco Nut, Cacay
869	*Moringa oleifera*	Benoil Tree, Murungai	LEG	Oil, Mult.	S Asia			721	Oris Root
870	*Morus alba*	White Mulberry	MORA	Multiple	S Asia	PC?		338	Otaheite Gooseberry
871	*Morus australis*	Chinese Mulberry	MORA	Fruit	S Asia	PC		1000	Otaheite Gooseberry
872	*Morus nigra*	Black Mulberry	MORA	Fruit	S Asia	PC?		1194	Ouabain
873	*Mucuna deeringiana*	Velvetbean	LEG	Forage	SE Asia	OC?	bat, Pteropus, Nan, Mega	1130	Ouricoury
874	*Mucuna pruriens*	Bengal Bean	LEG	Forage	SE Asia	OC?	bat, Pteropus, Nan, Mega	402	Owe Kola
875	*Murraya koenigii*	Curry-Leaf-Tree	RUTA	Multiple	S Asia			1223	Oyster Nut
876	*Musa acuminata*	Dwarf Banana	MUSA	Fruit	SE Asia	PC, OC	bat, Cynopterus, Mac.	1096	Pacuri
877	*Musa balbisiana*	Banana	MUSA	Fruit	SE Asia	PC, OC	bat, bird	350	Padang-Cassia
878	*Musa fehi*	Banana	MUSA	Fruit	Ocea	PC, OC	bat, bird	712	Pakay
879	*Musa textilis*	Abaca	MUSA	Fiber	SE Asia	PC, OC	bat, bird	563	Pakistan Ephedra
880	*Musanga cecropioides*	Umbrella Tree	CERC	Timber	Africa			574	Palma
881	*Myoporum acuminatum*	Myoporum	MYPO	Erosion	Aust			1122	Palma Ixtle
882	*Myrciaria cauliflora*	Jaboticaba	MYRT	Fruit	Neot		bee, Meliponinae?	322	Palmheart, Palmito
883	*Myrciaria dubia*	Camu-Camu	MYRT	Fruit	Neot		bee, Melipona, Scaptotrigona	602	Palmheart, Palmito
884	*Myrica cerifera*	Wax Myrtle, Bayberry	MYRT	Wax	Neot			922	Palmito
885	*Myristica argentea*	Papuan Nutmeg	MYRT	Spice	Papua	OC	moth, fly?, beetle?	200	Palmyra Palm
886	*Myristica caroliniensis*	Bayberry	MYRI	Wax	Neot	OC	moth, fly?, beetle?	290	Panama Hat Palm
887	*Myristica fragrans*	Nutmeg	MYRI	Spice	SE Asia	OC	moth, fly?, beetle?	279	Papaya
888	*Myrocarpus frondosus*	Cabreuva Oil	LEG	Essence	Neot			231	Paper Mulberry
889	*Myroxylon balsamum*	Tolu Balsam	LEG	Resin	Neot			885	Papuan Nutmeg
890	*Myroxylon b./pereirae*	Peru Balsam	LEG	Resin	Neot			478	Papyrus

No.	Scientific Name	Common Name	Family	Use	Origin	Breed. Sys.	Pollinators	No.	Common Name
891	*Myrrhis odorata*	Garden Myrrh	UMBE	Multiple	Asia			1181	Para Cress
892	*Myrtus communis*	Myrtle	MYRT	Spice	Medit			210	Pará Grass
893	*Nasturtium officinale*	Watercress	CRUC	Vegetable	Europe			684	Para Rubber
894	*Nauclea diderichii*	Opepe	RUBI	Timber	Africa			32	Paraguay Palm, Macaya
895	*Nelumbo nucifera*	Sacred Lotus	NYMP	Tuber	S Asia			958	Parmentiera Edulis
896	*Neoglaziovia variegata*	Makimbeira	BROM	Fiber	Neot			1270	Paroquet Bur
897	*Nephelium cuspidatum/petalum*	Lotong	SAPI	Fruit	SE Asia	SC, OC?	bee, Trigona, fly, wasp?	968	Parsnip
898	*Nephelium lappaceum*	Rambutan	SAPI	Fruit	SE Asia	SC, OC	bee, Trigona, fly, wasp	963	Passion Fruit
899	*Nephelium litchi*	Litchi	SAPI	Fruit	SE Asia	SC, OC	bee, Apis, Trigona, fly, wasp	724	Patauá
900	*Nephelium longana*	Longan	SAPI	Fruit	SE Asia	SC, OC	bee, Apis, Trigona, fly, wasp	1035	Patchouli
901	*Nephelium longana/melasianus*	Mata Kuching	SAPI	Fruit	SE Asia	SC, OC	bee, Apis, Trigona, fly, wasp	148	Pawpaw
902	*Nephelium mutable*	Pulasan	SAPI	Fruit	SE Asia	SC, OC	bee, Apis, Trigona, fly, wasp	1028	Pea
903	*Nephelium rambutan-ake*	Pulasan	SAPI	Fruit	SE Asia		bee, Apis, Trigona, fly, wasp	1063	Peach
904	*Neptunia oleracea*	Lajalu	LEG	Vegetable	SE Asia			180	Peach Palm, Pejibaye
905	*Nicotania rustica*	Peruvian Tobacco	SOLA	Multiple	Neot	AU, OC	bee, bird?	130	Peanut, Maní
906	*Nicotania suavolens*	Australian Tobacco	SOLA	Multiple	Aust	AU, OC?	bee, bird?	1084	Pear
907	*Nicotiana tabacum*	Tobacco	SOLA	Multiple	Neot	AU, OC	bee, bird	1085	Pear
908	*Nigella sativa*	Black Cumin	RANU	Spice	W Asia			1087	Pear
909	*Nopalea cochenillifera*	Cochineal Cactus	CACT	Dye, Fruit	Neot			980	Pearl Millet
910	*Nypa fruticans*	Nipa Palm	PALM	Multiple	SE Asia	OC	fly	293	Pecan
911	*Ochroma pyramidale*	Balsa	BOMB	Timber	Neot	OC	bat, Eonycteris	983	Pentadesma Butyraceae
912	*Ocimum basilicum*	Basil	LABI	Essence	S Asia	SC, OC	bee, Apis	1170	Pepino Dulce
913	*Ocimum kilimandscharicum*	Hoary Basil	LABI	Medicinal	Africa	SC?, OC	bee, Apis	1016	Pepper
914	*Ocimum sanctum*	Holy Basil	LABI	Med., Spice	S Asia	SC?, OC	bee, Apis	1017	Pepper
915	*Ocotea pretiosa*	Ocotea, Sassafras Oil	LAUR	Essence	Neot			1018	Pepper
916	*Ocotea rodiaei*	Greenheart	LEG	Timber	Neot			1019	Pepper
917	*Octomeles sumatrana*	Binuang	DATI	Timber	SE Asia			1022	Pepper
918	*Oenanthe javanica*	Water Dropwort	UMBE	Vegetable	SE Asia			1133	Pepper Tree
919	*Oenocarpus bacaba*	Bacaba	PALM	Oil	Neot			858	Peppermint
920	*Oenocarpus distichus*	Icaua	PALM	Oil	Neot			297	Pequiá
921	*Oenocarpus mapora*	Oenocarpus Mapora	PALM	Oil	Neot			985	Perejil
922	*Oenocarpus minor*	Palmito	PALM	Vegetable	Neot			4	Perennial Indian Hemp
923	*Olea europaea*	Olive	OLEA	Oil	Medit	AG, OC	wind, bee, Apis	645	Perennial Soybean
924	*Onobrychis viciaefolia*	Sainfoin	LEG	Forage	Medit	OC	bee, Bombus, Apis	986	Perilla
925	*Opuntia ficus-indica*	Prickly Pear	CACT	Fruit	Neot	AG?, OC	bee	309	Periwinkle
926	*Orbignya cohune*	Cohune Nut Palm	PALM	Oil	Neot			1263	Persian Clover
927	*Orbignya martiana*	Babassu	PALM	Oil	Neot			333	Persian Insect-Flower
928	*Orbignya oleifera*	Babassu	PALM	Oil	Neot			729	Persian Walnut
929	*Origanum vulgare*	Oregano	LABI	Spice	Medit			890	Peru Balsam
930	*Ornithopus sativus*	Serradella	LEG	Forage	Medit			137	Peruvian Carrot
931	*Orthosiphon aristatus*	Kumis Kuching	LABI	SE Asia	Af-As			1134	Peruvian Mastic
932	*Oryza glaberrima*	African Rice	POA	Cereal	Africa	AU, OC	wind, bee	905	Peruvian Tobacco
933	*Oryza sativa*	Rice	POA	Cereal	S Asia	AU, OC	wind, bee ?	450	Peten Gourd
934	*Oxalis tuberosa*	Oca	OXAL	Tuber	Neot			667	Phalsa
935	*Pachira aquatica*	Chestnut of America	BOMB	Seed	Neot	OC?	bat, moth	1002	Phyllanthus Pulcher
936	*Pachyrhizus ahipa*	Ahipa	LEG	Tuber	Neot	SC, OC?	bee?	723	Physic Nut
937	*Pachyrhizus erosus*	Yam Bean	LEG	Tuber	Neot	SC, OC?	bee?	760	Piassaba Palm
938	*Pachyrhizus tuberosus*	Jicama	LEG	Tuber, Ins.	Neot	SC, OC?	bee?	243	Pigeonpea
939	*Palaquium gutta*	Gutta-Percha	SAPO	Elastomer	SE Asia			324	Pigweed
940	*Palaquium oxleyanum*	Pulau Dammar Darat	SAPO	Elastomer	SE Asia			259	Pili Nut
941	*Panax ginseng*	Ginseng	ARAL	Medicinal	E Asia	AU, SC	bee	260	Pili Nut
942	*Pandanus andamanensium*	Screw-Pine	PAND	Fruit	SE Asia	OC	wind, insect?	261	Pili Nut
943	*Pandanus candelabrum*	Screw-Pine	PAND	Fiber	Africa	OC?	wind, insect?	271	Pimentchien
944	*Pandanus houlletii*	Screw-Pine	PAND	Fruit	SE Asia	OC?	wind, insect?	1011	Pimento, Allspice
945	*Pandanus odoratissimus*	Screw-Pine	PAND	Essence	SE Asia	OC?	wind, insect?	106	Pineapple, Piña
946	*Pandanus odorus*	Screw-Pine	PAND	Essence	SE Asia	OC?	wind, insect?	295	Piqui Brava
947	*Pandanus spurius*	Screw-Pine	PAND	Essence	SE Asia	OC?	wind, insect?	1027	Pistachio Nut
948	*Pandanus utilus*	Screw-Pine	PAND	Fiber	Africa	OC?	wind, insect?	692	Pitaya roja
949	*Pangium edule*	Bauh Keluak	FLAC	Vegetable	SE Asia	OC		595	Pitomba
950	*Panicum miliaceum*	Millet	POA	Cereal	As, Af	AU, OC	wind, bee	1059	Plum
951	*Panicum sumatrense*	Little Millet	POA	Cereal	S Asia	AU, OC?	wind, bee ?	196	Pochote
952	*Papaver bracteatum*	Scarlet Poppy	PAPA	Medicinal	S Asia	AU, OC?	bee	1007	Poke
953	*Papaver somniferum*	Opium Poppy	PAPA	Medicinal	SE Asia	AU, OC?	bee	1040	Polyscias Rumphiana
954	*Parashorea spp.*	White Seraya	DIPT	Timber	SE Asia			1083	Pomegranate
955	*Parkia bigibbosa*	African Locust Bean	LEG	Oil, Spice	W Afr	OC?	bat, Eidolon, bee, beetle	1212	Pomerac
956	*Parmentiera aculeata*	Cuajilote	BIGN	Fruit	Neot			833	Pot Majoram
957	*Parmentiera cerifera*	Candletree	BIGN	Forage	Neot			1172	Potato
958	*Parmentiera edulis*	Parmentiera Edulis	BIGN	Fruit	Neot			1176	Potato
959	*Parthenium argentatum*	Guayule	COMP	Elastomer	Neot	OC	insect, bee, Apis	127	Potato Beans
960	*Paspalum dilatatum*	Dallisgrass	POA	Forage	Neot	AG		925	Prickly Pear
961	*Passiflora caerulea*	Maracujá	PASS	Fruit	Neot	OC	bee, Xylocopa, wasp, bird	86	Princess-Feather
962	*Passiflora cincinnata*	Wild Passion Fruit	PASS	Fruit	Neot	OC	bee, Xylocopa, wasp, bird	249	Prisolilia
963	*Passiflora edulis*	Passion Fruit	PASS	Beverage	Neot	OC	bee, Xylocopa, wasp, bird	1051	Prosopis Pallida
964	*Passiflora laurifolia*	Water Lemon	PASS	Fruit	Neot	OC	bee, Xylocopa, wasp, bird	902	Pulasan
965	*Passiflora ligularis*	Sweet Granadilla	PASS	Beverage	Neot	OC	bee, Xylocopa, wasp, bird	903	Pulasan
966	*Passiflora mollissima*	Banana Passion Fruit	PASS	Fruit	Neot	OC	bee, Xylocopa, wasp, bird	940	Pulau Dammar Darat
967	*Passiflora quadrungularis*	Giant Granadilla	PASS	Fruit	Neot	OC	bee, Xylocopa, wasp, bird	57	Pulque
968	*Pastinaca sativa*	Parsnip	UMBE	Tuber	Europe	OC	insect, bee	366	Pummelo
969	*Paullinia cupana*	Guaraná	SAPI	Beverage	Neot	SC, OC	bee, Melipona	367	Pummelo
970	*Paullinia yoco*	Yoco	SAPI	Beverage	Neot	SC, OC	bee, Melipona	368	Pummelo
971	*Paulownia tomentosa*	Kiri	SCRO	Timber	Asia			317	Punga
972	*Payena leerii*	Gutta Sundek	SAPO	Elastomer	SE Asia			443	Purging Croton
973	*Peganum harmala*	Harmala Shrub	ZYGO	Medicinal	Medit			505	Purple Foxglove
974	*Peilanthus pavonis*	Canelilla	EUPH	Wax	Mexico			1289	Purple Vetch
975	*Pelargonium capitatum X radens*	Geranium	GERA	Oil	Africa			977	Purpleheart
976	*Pelargonium graveolens*	Rose Geranium	GERA	Essence	Africa			332	Pyrethrum
977	*Peltogyne purpurea*	Purpleheart	LEG	Timber	Neot			1131	Quebracho
978	*Peltophorum inerme*	Yellow Flame	LEG	Multiple	SE Asia			467	Quince
979	*Pennisetum clandestinum*	Kikuyu Grass	POA	Forage	Africa	SC, OC	insect, bee?	342	Quinine

No.	Scientific Name	Common Name	Family	Use	Origin	Breed. Sys.	Pollinators	No.	Common Name
980	*Pennisetum glaucum*	Pearl Millet	POA	Cereal	Africa	SC, OC	insect, bee?	343	Quinine
981	*Pennisetum purpureum*	Elephant Grass	POA	Forage	Africa	SC, OC	insect, bee?	344	Quinine
982	*Pentaclethra macrophylla*	Oil-Bean Tree	LEG	Oil	Africa	SC, OC?	bee?	345	Quinine
983	*Pentadesma butyraceae*	Pentadesma Butyraceae	GUTT	Oil	SE Asia	OC?	bat, Pteropus	346	Quinine
984	*Pentzia virgata*	Australian Sheepbush	COMP	Forage	Africa			347	Quinine
985	*Peperomia viridispica*	Perejil	PIPE	Vegetable	Neot			348	Quinine
986	*Perilla frutescens*	Perilla	LABI	Oil	S Asia	AU, OC?		327	Quinoa
987	*Persea americana*	Avocado	LAUR	Fruit	Neot	OC	bee, bat, Pteropus, fly	1280	Rabbit-Eye Blueberry
988	*Persea nubigena*	Wild Avocado	LAUR	Fruit	Neot	OC	bee, wasp, fly	1090	Radish
989	*Persea schiedeana*	Aguacatillo, Coyo	LAUR	Fruit	Neot	OC	bee, wasp, fly	1091	Raffia Palm
990	*Phaseolus aborigineus*	Bean	LEG	Vegetable	Neot	AU?	bee	1118	Raintree
991	*Phaseolus acutifolius*	Tepary Bean	LEG	Vegetable	Neot	AU, OC?	bee	71	Rakkyo
992	*Phaseolus coccineus*	Scarlet Runner Bean	LEG	Vegetable	Neot	AU, OC?	bee	177	Rambai
993	*Phaseolus flavescens*	Chaguita	LEG	Vegetable	Neot	AU?	bee	898	Rambutan
994	*Phaseolus inamoenus*	Lima Bean	LEG	Vegetable	Neot	AU?	bee	1318	Rambutan Pachat
995	*Phaseolus lunatus*	Tropical Lima Bean	LEG	Vegetable	Neot	AU, OC	bee	651	Ramin
996	*Phaseolus vulgaris*	Common Bean	LEG	Vegetable	Neot	AU, OC	bee, thrip	229	Ramón
997	*Phoenix dactylifera*	Date Palm	PALM	Fruit	Africa	OC	wind, bee	253	Rampion
998	*Phoenix reclinata*	Wild Date Palm	PALM	Fiber	Africa	OC	bee?	217	Rape
999	*Phormium tenax*	New Zealand Flax	LILI	Fiber	Aust	OC	bird, bee	404	Ratala
1000	*Phyllanthus acidus*	Otaheite Gooseberry	EUPH	Fruit	S Asia	SC, OC?	bee?	244	Rattan
1001	*Phyllanthus emblica*	Emblic	EUPH	Multiple	S Asia	SC, OC?	bee?	1038	Rau Ram
1002	*Phyllanthus pulcher*	Phyllanthus Pulcher	EUPH	Medicinal	S Asia	SC, OC?	bee?	1261	Red Cover
1003	*Physalis ixocarpa*	Mexican Husk Tomato	SOLA	Vegetable	Neot			590	Red Iron Bark
1004	*Physalis peruviana*	Cape Gooseberry	SOLA	Fruit	Neot			1101	Red Mangrove
1005	*Physalis philadelphica*	Tomatillo	SOLA	Fruit	Neot			1185	Red Mombin
1006	*Phytelephas macrocarpa*	Ivory-Nut Palm	PALM	Veg. Ivory	Neot			167	Red Oat
1007	*Phytolacca americana*	Poke	PHYT	Dye	N Amer			270	Red Pepper
1008	*Pileus mexicanus*	Bonete	BOMB	Fiber	Neot			1132	Red Quebracho
1009	*Pilocarpus microphyllus*	Jaborandi	RUTA	Medicinal	Neot			580	Red River Gum
1010	*Pimenta dioica*	Allspice	MYRT	Spice	Neot	OC	insect, bee	1078	Red Sanderswood
1011	*Pimenta officinalis*	Pimento, Allspice	MYRT	Spice	Neot	OC	insect, bee?	198	Red Silk-Cotton
1012	*Pimenta racemosa*	Bay	MYRT	Essence	Neot	OC	insect, bee?	328	Rhodesgrass
1013	*Pimpinella anisum*	Anise	UMBE	Essence	Medit		bee	933	Rice
1014	*Piper betel*	Betel Pepper	PIPE	Multiple	S Asia	OC	bee, fly	1304	Rice Bean
1015	*Piper chaba*	Bakek	PIPE	Multiple	SE Asia	OC	bee, fly	509	Rikua
1016	*Piper clusii*	Pepper	PIPE	Spice	Africa	OC	bee, fly	1140	Ringworm Bush
1017	*Piper cubeba*	Pepper	PIPE	Spice	Af-As	OC	bee, fly	391	Robusta Coffee
1018	*Piper guineense*	Pepper	PIPE	Spice	Africa	OC	bee, fly	122	Roman Chamomile
1019	*Piper longifolium*	Pepper	PIPE	Spice	Neot	OC	bee, fly	149	Rooibos Tea
1020	*Piper longum*	Indian Long Pepper	PIPE	Spice	S Asia	OC	bee, fly	572	Roquette
1021	*Piper methysticum*	Kava	PIPE	Beverage	S Asia	OC	bee, fly	1211	Rose Apple
1022	*Piper nigrum*	Pepper	PIPE	Spice	S Asia	AU, OC	bee, fly	976	Rose Geranium
1023	*Piper officinarum*	Javanese Long Pepper	PIPE	Spice	S Asia	OC	bee, fly	686	Rosella
1024	*Piper saigonense*	Saigon Pepper	PIPE	Spice	SE Asia	OC	bee, fly	22	Rosewood
1025	*Piptadeniastrum africanum*	Dahoma	LEG	Timber	Africa			111	Rosewood Oil
1026	*Pistacia lentiscus*	Chios Mastic	ANAC	Resin	Medit			471	Rosha Grass
1027	*Pistacia vera*	Pistachio Nut	ANAC	Nut	Medit			746	Rough Pea
1028	*Pisum sativum*	Pea	LEG	Vegetable	S Asia	AU, OC	bee, thrip	802	Rough-Shell Macadamia
1029	*Plantago arenaria*	Black Phyllium	PLAN	Medicinal	Medit			444	Rubber
1030	*Plantago ovata*	Blonde Phyllium	PLAN	Medicinal	S Asia			683	Rubber
1031	*Platonia esculenta*	Bacuri do Parana	GUTT	Fruit	Neot			617	Rukam
1032	*Plectranthus amboinicus*	Indian Borage	LABI	Spice	SE Asia			616	Runeala Plum
1033	*Plectranthus edulis*	Gala Dinich	LABI	Tuber	E Afr.			552	Russian Olive
1034	*Plectranthus esculentus*	Kafir Potato, Dazo	LABI	Tuber	S Asia			218	Rutabaga
1035	*Pogostemon cablin*	Patchouli	LABI	Essence	SE Asia	SC, OC	bee, Apis	1112	Ryania
1036	*Polianthes tuberosa*	Tuberose	AGAV	Essence	Neot			895	Sacred Lotus
1037	*Polygala butyracea*	Black Beniseed	POLY	Oil	W Af.r			291	Safflower
1038	*Polygonum odoratum*	Rau Ram	POLYG	Vegetable	SE Asia			436	Saffron
1039	*Polymnia sonchifolia*	Yacon Strawberry	COMP	Veg, Misc.	S Amer.			483	Safou, Eben
1040	*Polyscias rumphiana*	Polyscias Rumphiana	ARAL	Vegetable	SE Asia			859	Sago Palm
1041	*Poncirus trifoliata*	Trifoliate Orange	RUTA	Breed.	N Asia			349	Saigon Cinnamon
1042	*Portulacaria afra*	Elephant Bush	PORT	Misc	Africa		bee	1024	Saigon Pepper
1043	*Pourouma spp.*	Uvilla, Guarumo	MORA	Fruit	Neot			924	Sainfoin
1044	*Pouteria caimito*	Lucuma	SAPO	Fruit	Neot			1151	Sal
1045	*Pouteria campechiana*	Eggfruit Tree, Canistel	SAPO	Fruit	Neot	OC?	bee?	1115	Salak
1046	*Pouteria sapote*	Sapote Mamey	SAPO	Fruit	Neot	OC?	bee?	1252	Salsify
1047	*Pouteria viridis*	Green Sapote	SAPO	Fruit	Neot	OC?	bee?	160	Saltbush
1048	*Prosopis alba*	Algarobo	LEG	Misc	Neot	AU,AG?OC	bee	1120	Sambucus Javanica
1049	*Prosopis chilensis*	Algarrobo Blanco	LEG	Fruit	Neot	AU,AG?OC	bee, Centris	1121	Sambucus Mexicana
1050	*Prosopis juliflora*	Mesquite	LEG	Timber	Neot	AU,AG?OC	bee	1088	Sand Pear
1051	*Prosopis pallida*	Prosopis Pallida	LEG	Misc	Neot	AU,AG?OC	bee	1231	Sandarac Gum
1052	*Prosopis tamarugo*	Tamarugo	LEG	Timber	Neot	AU,AG?OC	bee, Centris	1046	Sapote Mamey
1053	*Prunus amygdalus/dulcis*	Almond	ROSA	Nut	S Asia	SC, OC	bee	241	Sappanwood
1054	*Prunus amygdalus/amara*	Bitter Almond	ROSA	Essence	S Asia	SC, OC	bee	757	Sapucaja
1055	*Prunus armeniaca*	Apricot	ROSA	Fruit	S Asia	SC, OC	bee	1159	Sarsaparilla
1056	*Prunus avium*	Sweet Cherry	ROSA	Fruit	S Asia	SC, OC	bee	952	Scarlet Poppy
1057	*Prunus cerasus*	Sour Cherry	ROSA	Fruit	S Asia	SC, OC	bee	992	Scarlet Runner Bean
1058	*Prunus communis*	Almond	ROSA	Nut	S Asia	SC, OC	bee	854	Scotch Spearmint
1059	*Prunus domestica*	Plum	ROSA	Fruit	S Asia	SC, OC	bee	857	Scotch Spearmint
1060	*Prunus domestica/insititia*	Damson Plum	ROSA	Fruit	S Asia	SC, OC	bee	942	Screw-Pine
1061	*Prunus laurocerasus*	Cherry Laurel	ROSA	Erosion	S Asia	SC, OC	bee	943	Screw-Pine
1062	*Prunus maritima*	Beach Plum	ROSA	Erosion	S Asia	SC, OC	bee	944	Screw-Pine
1063	*Prunus persica*	Peach	ROSA	Fruit	S Asia	SC, OC	bee	945	Screw-Pine
1064	*Prunus salicina*	Japanese Plum	ROSA	Fruit	S Asia	SC, OC	bee	946	Screw-Pine
1065	*Prunus serotina*	Capulín	ROSA	Fruit	S Asia	SC, OC	bee	947	Screw-Pine
1066	*Prunus subcordata*	Sierra Plum	ROSA	Fruit	S Asia	SC, OC	bee	948	Screw-Pine
1067	*Psidium friedrichsthalianum*	Costa Rican Guava	MYRT	Fruit	Neot	SC, OC	bee, Melipona	655	Sea Island Cotton
1068	*Psidium guajava*	Common Guava	MYRT	Fruit	Neot	SC, OC	bee, Mel, Xy, Las, Bo, bat, Rou	730	Sea Rush

No.	Scientific Name	Common Name	Family	Use	Origin	Breed. Sys.	Pollinators	No.	Common Name
1069	*Psidium guineense*	Brazilian Guava	MYRT	Fruit	Neot	SC, OC	bee, Melipona	387	Seagrape
1070	*Psidium littorale*	Strawberry Guava	MYRT	Fruit	Neot	SC, OC	bee, Melipona, Bo., Xy.	294	Seaving
1071	*Psidium microphyllum*	Wild Guava	MYRT	Fruit	Neot	SC, OC	bee, Melipona	1076	Senegal Rosewood
1072	*Psidium montanum*	Wild Guava	MYRT	Fruit	Neot	SC, OC	bee, Melipona	1141	Senna Coffee
1073	*Psidium sartorianum*	Arrayán	MYRT	Fruit	Neot	SC, OC	bee, Melipona	413	Sepetir
1074	*Psophocarpus tetragonolobus*	Winged Bean	LEG	Vegetable	S Asia	AU, OC	bee, Xylocopa	527	Serendipity-Berry
1075	*Psoralea spp.*	Breadroot	LEG	Tuber	Aust			766	Sericea Lespedeza
1076	*Pterocarpus erinaceus*	Senegal Rosewood	LEG	Dye	Africa			930	Serradella
1077	*Pterocarpus indicus*	Amboyna	LEG	Timber	SE Asia			1144	Sesame
1078	*Pterocarpus santalinus*	Red Sanderswood	LEG	Dye	Africa			1312	Sheabutter
1079	*Pterocarpus soyauxii*	Barwood	LEG	Dye	Africa			31	Shingle Tree
1080	*Pueraria lobata*	Kudzu	LEG	Forage	E Asia	SC, OC	bee	25	Shittim-Wood
1081	*Pueraria montana*	Kudzu Vine	LEG	Tuber, For.	S Asia			442	Showy Crotalaria
1082	*Pueraria phaseoloides*	Tropical Kudzu	LEG	Forage	S Asia			1204	Siam Benzoin
1083	*Punica granatum*	Pomegranate	PUNI	Multiple	W Asia	SC, OC	insect?	427	Siberian Hazel
1084	*Pyrus calleryana*	Pear	ROSA	Fruit	S Asia	OC	bee, fly	840	Sickle Medic
1085	*Pyrus communis*	Pear	ROSA	Fruit	S Asia	SC, OC	bee, fly	1066	Sierra Plum
1086	*Pyrus malus*	Apple	ROSA	Fruit	S Asia	OC	bee, Osmia, Anth, And, Apis	207	Signal Grass
1087	*Pyrus pashia*	Pear	ROSA	Fruit	S Asia	SC, OC	bee, fly	197	Silk Cotton
1088	*Pyrus pyrifolia*	Sand Pear	ROSA	Fruit	S Asia	SC, OC	bee, fly	666	Silky Oak
1089	*Quassia amara*	Bitter Wood	SIMA	Misc.	Neot			502	Silverleaf Desmodium
1090	*Raphanus sativus*	Radish	CRUC	Vegetable	W Asia	OC, AU	bee, Apis	1288	Single-Flowered Vetch
1091	*Raphia spp.*	Raffia Palm	PALM	Fiber	Africa			804	Siratro
1092	*Rauwolfia serpentina*	Java Devil Pepper	APOC	Medicinal	SE Asia			820	Sleeve Palm
1093	*Reseda odorata*	Common Mignonette	RESE	Essence	Medit			437	Slenderleaf Crotalaria
1094	*Rhamnus purshianus*	Cáscara	RHAM	Medicinal	Neot			441	Smooth Crotalaria
1095	*Rheedia acuminata*	Naranjito	GUTT	Fruit	Neot			790	Smooth Loofah
1096	*Rheedia brasiliensis*	Pacuri	GUTT	Fruit	Neot			844	Snail Medic
1097	*Rheedia edulis*	Jacomico	GUTT	Fruit	Neot			407	Snake Bark, Maubi
1098	*Rheum officinale*	Chinese Rhubarb	POLYG	Medicinal	S Asia	SC, OC?	insect, fly	1256	Snakegourd, Culebra
1099	*Rheum palmatum*	East Indian Rhubarb	POLYG	Medicinal	S Asia	SC, OC?	insect, fly	1125	Soapberry, Parapara
1100	*Rheum rhaponticum*	Garden Rhubarb	POLYG	Vegetable	S Asia	SC, OC?	insect, fly	79	Socotrine Aloe
1101	*Rhizophora mangle*	Red Mangrove	RHIZ	Timber	Neot		wind, insect	1164	Sodom Apple
1102	*Rhizophoraceae spp.*	Mangrove	RHIZ	Tannin, Dye	S Asia	SC, OC	bat, Eon, Mac, wind, insect	618	Soh-phlong
1103	*Rhus succedanea*	Japan Wax	LEG	Wax	E Asia			1166	Solanum Khasianum
1104	*Ricinus communis*	Castorbean	EUPH	Oil	Africa	OC	wind, insect	1177	Solenostemon
1105	*Rollinia deliciosa*	Biriba	ANON	Fruit	Neot	SC, OC?	beetle	430	Sorva Gum
1106	*Rosa alba*	White Cottage Rose	ROSA	Essence	Medit			1057	Sour Cherry
1107	*Rosa centifolia*	Cabbage Rose	ROSA	Essence	Medit			364	Sour Orange
1108	*Rosa damascena/trigintipelala*	Damask Rose	ROSA	Essence	Medit			1162	Sour-Relish Brinjal
1109	*Rubia cordifolia*	Indian Madder	RUBI	Dye	S Asia			116	Soursop, Guanabana
1110	*Rubus spp.*	Blackberry	ROSA	Fruit	Temp.	~AG, OC	bee, Bombus, Apis	644	Soybean
1111	*Ruta graveolens*	Common Rue	RUTA	Medicinal	Europe		fly	1179	Spanish Broom
1112	*Ryania pyrifera*	Ryania	FLAC	Pesticide	Neot			84	Spanish Greens
1113	*Sagittaria sagittifolia*	Chinese Arrowhead	ALIS	Root Crop	SE Asia			848	Spanish Lime, Mamon
1114	*Saintpaulia ionantha*	African Violet	GESN	Orna.	Africa		bee, Amegilla?	675	Spanish Sainfoin
1115	*Salacca edulis*	Salak	PALM	Fruit	SE Asia			1281	Spanish Tamarind
1116	*Salvia divinorum*	Yerba de María	LABI	Medicinal	S Mex			777	Spanish Thyme
1117	*Salvia hispanica*	Ghia, Chia	LABI	Essence	Neot			856	Spearmint
1118	*Samanea saman*	Raintree	LEG	Timber	Neot			709	Spicata Indigo
1119	*Sambucus canadensis*	Eastern Elderberry	CAPR	Fruit	SE Asia			1182	Spinach
1120	*Sambucus javanica*	Sambucus Javanica	CAPR	Fruit	SE Asia			1187	Spondias venulosa
1121	*Sambucus mexicana*	Sambucus Mexicana	CAPR	Fruit	SE Asia			839	Spotted Burr Medic
1122	*Samuela carnerosana*	Palma Ixtle	AGAV	Fiber	Mexico			550	Squirting Cucumber
1123	*Sansevieria spp.*	Bowstring Hemp	AGAV	Fiber	Africa			336	Star Apple, Cainito
1124	*Santalum album*	White Sandalwood	SANT	Essence	S Asia	SC, OC	bee	706	Star-Anise Tree
1125	*Sapindus saponaria*	Soapberry, Parapara	SAPI	Miac.	Neot			172	Starfruit, Carambola
1126	*Sapium sebiferum*	Chinese Tallow Tree	EUPH	Multiple	S Asia			476	Stargrass
1127	*Satureja montana*	Winter Savory	LABI	Spice	Medit			170	Sterile Oat
1128	*Sauropus androgynus*	Katuk	EUPH	Vegetable	SE Asia			1258	Strawberry Clover
1129	*Saussurea costus*	Costus, Kuth	COMP	Medicinal	S Asia			1070	Strawberry Guava
1130	*Scheelea martiana*	Ouricoury	PALM	Oil	Neot			1196	Strophanthus Kombe
1131	*Schinopsis balansae*	Quebracho	ANAC	Tannin	Neot			1198	Strychnine Tree
1132	*Schinopsis lorentzii*	Red Quebracho	ANAC	Tannin	Neot			1203	Styrax
1133	*Schinus areira*	Pepper Tree	ANAC	Gum, Tan.	Neot			1264	Subclover
1134	*Schinus molle*	Peruvian Mastic	ANAC	Spice	Neot			206	Sucupira
1135	*Schleichera oleosa*	Lac-tree	SAPI	Wax	S Asia			119	Sugar Apple
1136	*Sclerocarya caffra*	Morula	ANAC	Fruit, Mult.	Africa			582	Sugar Gum
1137	*Sechium edule*	Chayote	CUCU	Vegetable	Neot	OC	bee, Trigona, wasp	134	Sugar Palm
1138	*Sechium tacaco*	Tacaco	CUCU	Vegetable	Neot	OC	bee, Trigona, wasp	677	Sunchoke (311)
1139	*Semecarpus anacardium*	Marking-Nut Tree	ANAC	Dye, Mult.	S Asia			676	Sunflower
1140	*Senna alata*	Ringworm Bush	LEG	Medicinal	Neot	SC, OC	bee, Xylocopa, Centris	439	Sunn Hemp
1141	*Senna occidentalis*	Senna Coffee	LEG	Beverage	Africa	SC, OC	bee, Xylocopa	2	Sunset Hibiscus, Aibika
1142	*Senna senna*	Alexandrian Senna	LEG	Medicinal	Africa	SC, OC	bee, Xylocopa	208	Surinam Grass
1143	*Sesamum alatum*	Tacoutta	PEDA	Oil	S Asia	AU, OC?	bee, fly, wasp	600	Surinam-Cherry
1144	*Sesamum indicum*	Sesame	PEDA	Oil	Africa	AU, OC?	bee, fly, wasp	852	Sweet Balm
1145	*Sesamum prostratum*	Wild Sesame	PEDA	Oil	S Asia	AU, OC?	bee, fly, wasp	750	Sweet Bay
1146	*Sesamum radiatum*	Wild Sesame	PEDA	Oil	S Asia	AU, OC?		1207	Sweet Berry
1147	*Sesbania bispinosa*	Dhaincha, Jantar	LEG	Gum, Fiber	S Asia	AU		1056	Sweet Cherry
1148	*Sesbania grandiflora*	Turi	LEG	Misc	S Asia			30	Sweet Flag
1149	*Setaria italica*	Foxtail Millet	POA	Cereal	S Asia	AU, OC	wind, bee	965	Sweet Granadilla
1150	*Setaria sphacelata*	Golden Timothy Grass	POA	Forage	S Asia	AU, OC	wind, bee	369	Sweet Lime
1151	*Shorea robusta*	Sal	DIPT	Resin	S Asia	AG, OC?	thrip, bee?	832	Sweet Marjoram
1152	*Shorea spp.*	Balau, Meranti	DIPT	Oil, Timb.	SE Asia	AG, OC	thrip, bee?	375	Sweet Orange
1153	*Shorea wiesneri*	Batavian Dammar	DIPT	Resin	SE Asia	AG, OC?	thrip, bee?	718	Sweet Potato
1154	*Sicana odorifera*	Curuba, Melocotón	CUCU	Multiple	Neot			151	Sweet Woodruff
1155	*Silybum marianum*	Holy Thistle	COMP	Medicinal	Asia			118	Sweetsop
1156	*Simarouba glauca*	Aceituna	SIMA	Oil, Fruit	Neot			265	Swordbean, Poroto
1157	*Simmondsia chinensis*	Jojoba	BUXA	Wax, Mult.	Neot	OC	bee, wind	587	Sydney Blue Gum

No.	Scientific Name	Common Name	Family	Use	Origin	Breed. Sys.	Pollinators
1158	*Sinapis alba*	White Mustard	BRAS	Forage	S Asia	SC, OC	bee, Apis
1159	*Smilax aristolochiifolia*	Sarsaparilla	SMIL	Medicinal	Neot		
1160	*Solanum aethiopicum*	Cut-Egg Plant	SOLA	Vegetable	Neot	AU, OC	bee
1161	*Solanum aviculare*	Australian Nightshade	SOLA	Medicinal	Aust	AU, OC	bee
1162	*Solanum ferox*	Sour-Relish Brinjal	SOLA	Medicinal	Neot	AU, OC	bee
1163	*Solanum gilo*	Gilo	SOLA	Fruit	Neot	AU, OC	bee
1164	*Solanum incanum*	Sodom Apple	SOLA	Vegetable	Neot	AU, OC	bee
1165	*Solanum indicum*	Tiberato	SOLA	Vegetable	Neot	AU, OC	bee
1166	*Solanum khasianum*	Solanum Khasianum	SOLA	Medicinal	Neot	AU, OC	bee
1167	*Solanum laciniatum*	Kangaroo Apple	SOLA	Fruit	Aust	AU, OC	bee
1168	*Solanum macrocarpon*	Native Eggplant	SOLA	Vegetable	Neot	AU, OC	bee
1169	*Solanum melongena*	Eggplant	SOLA	Vegetable	Neot	AU, OC	bee
1170	*Solanum muricatum*	Pepino Dulce	SOLA	Fruit	Neot	AU, OC	bee
1171	*Solanum nigrum*	Wonderberry	SOLA	Multiple	Neot	AU, OC	bee
1172	*Solanum phureja*	Potato	SOLA	Tuber	Neot	AU, OC	bee
1173	*Solanum quitoense*	Naranjilla, Lulo	SOLA	Fruit	Neot	AU, OC	bee, Bombus, Eulaema
1174	*Solanum toprio*	Cocona, Peach Tomato	SOLA	Fruit	Neot	AU, OC	bee
1175	*Solanum torvum*	Terongan	SOLA	Vegetable	Neot	AU, OC	bee
1176	*Solanum tuberosum*	Potato	SOLA	Tuber	Neot	AU, OC	bee, Bombus
1177	*Solenostemon monostachys*	Solenostemon	LABI	Medicinal	W Afr		bee, Apis
1178	*Solenostemon rotundifolius*	Hausa Potato	LABI	Tuber	Af-As		bee, Apis
1179	*Spartium junceum*	Spanish Broom	LEG	Fiber	Medit	AU	
1180	*Sphenostylis stenocarpa*	African Yam Bean	LEG	Tuber	Africa	SC, OC?	bee?
1181	*Spilanthes acmella*	Para Cress	COMP	Medicinal	N Amer		
1182	*Spinacia oleracea*	Spinach	CHEN	Vegetable	W. Asia	OC	wind, insect
1183	*Spondias cytherea*	Kedondong	ANAC	Fruit	SE Asia		bee
1184	*Spondias mombin*	Hog Plum, Jobo	ANAC	Fruit	Neot		bee
1185	*Spondias purpurea*	Red Mombin	ANAC	Fruit	Neot		bee
1186	*Spondias tuberosa*	Imbu	ANAC	Fruit	Neot		bee
1187	*Spondias venulosa*	Spondias venulosa	ANAC	Fruit	Neot		bee
1188	*Stachys sieboldii*	Chinese Artichoke	LABI	Tuber	Asia		
1189	*Sterculia setigera*	Kataya Gum	STER	Elastomer	Africa		
1190	*Sterculia tragacantha*	Kataya Gum	STER	Elastomer	Africa		
1191	*Sterculia urens*	Karaya Gum	STER	Gum	S Asia		
1192	*Stevia rebaudiana*	Kaa He'e	COMP	Sweetener	Neot		
1193	*Stipa tenacissima*	Esparto	POA	Fiber	N Afr		
1194	*Strophanthus gratus*	Ouabain	APOC	Medicinal	Africa		beetle
1195	*Strophanthus hispidus*	Untsuti	APOC	Medicinal	Africa		beetle
1196	*Strophanthus kombe*	Strophanthus Kombe	APOC	Medicinal	Africa		beetle
1197	*Strophanthus sarmentosus*	Arrow Poison	APOC	Medicinal	Africa		beetle
1198	*Strychnos nux-vomica*	Strychnine Tree	LOGA	Medicinal	S Asia	SC, OC	bee
1199	*Stryphnodendron adstringens*	Barbatimao	LEG	Tan.,Medic.	Neot		
1200	*Stylosanthes erecta*	Nigerian Lucerne	LEG	Forage	Africa		
1201	*Stylosanthes guianensis*	Brazilian Lucerne	LEG	Forage	Neot	SC, OC	bee, Cephalotrigona
1202	*Stylosanthes humilis*	Townsville Stylo	LEG	Forage	Neot	AU	
1203	*Styrax benzoin*	Styrax	STYR	Resin	SE Asia		
1204	*Styrax tonkinensis*	Siam Benzoin	STYR	Medicinal	SE Asia		
1205	*Swietenia macrophylla*	Mahogany, Caoba	MELI	Timber	Neot	OC	bee, fly?
1206	*Syagrus coronata*	Urucuri Palm	PALM	Wax, Oil	Neot		
1207	*Synsepalum dulcificum*	Sweet Berry	SAPO	Sweetener	Africa		
1208	*Syzygium aqueum*	Jambu Ayer	MYRT	Fruit	S Asia	AU, OC	bee, fly?
1209	*Syzygium aromaticum*	Clove	MYRT	Spice	SE Asia	AU, OC	bee, fly?
1210	*Syzygium cumini*	Jambolan	MYRT	Fruit	S Asia	AU, OC	bat?, bee, fly?
1211	*Syzygium jambos*	Rose Apple	MYRT	Fruit	S Asia	AU, OC	bat, Pteropus, bee, fly?
1212	*Syzygium malaccense*	Pomerac	MYRT	Fruit	S Asia	AU, OC	bat,Eon,Mac,Pter,bee,fly?
1213	*Syzygium samarangense*	Java Apple	MYRT	Fruit	S Asia	AU, OC	bat, bee, fly?
1214	*Tacca leontopetaloides*	Tahiti Arrowroot	TACC	Starch	Oceania		
1215	*Talinum triangularae*	Ceylon Spinach	PORT	Vegetable	W Afr		
1216	*Talisia esculenta*	Talisia Esculenta	SAPI	Fruit	Neot		
1217	*Talisia olivaeformis*	Yellow Genip	SAPI	Fruit	Neot		
1218	*Tamarindus indica*	Tamarind	LEG	Multiple	Africa	SC, OC	bee, Apis
1219	*Tamarix articulata*	Tamarisk	TAMA	Timber	Asia		fly, insect
1220	*Taraxacum officinale*	Dandelion	COMP	Beverage	W Asia	AG	bee
1221	*Tectona grandis*	Teak	VERB	Timber	SE Asia	SC, OC	bee, Apis
1222	*Telfairia occidentalis*	Fluted Gourd	CUCU	Vegetable	Africa		
1223	*Telfairia pedata*	Oyster Nut	CUCU	Oil	Africa		
1224	*Tephrosia candida*	White Tephrosia	LEG	Erosion	Neot	AU, OC	bee, Xylocopa, Apis
1225	*Tephrosia vogelii*	Vogel Tephrosia	LEG	Pesticide	Neot	AU, OC	bee, Xylocopa, Apis
1226	*Terminalia brassii*	Dafo	COMB	Timber	Oceania	SC, OC?	fly?
1227	*Terminalia catappa*	Bengal Almond	COMB	Oil	S Asia	SC, OC?	fly, bee?
1228	*Terminalia chebula*	Myrobalans	COMB	Tannin	S Asia	SC, OC?	fly, bee?
1229	*Terminalia ivorensis*	Idigbo	COMB	Timber	Africa	SC, OC?	fly, Lucilia, Chrysomaya
1230	*Terminalia superba*	Afara	COMB	Multiple	Africa	SC, OC?	fly?
1231	*Tetraclinis articulata*	Sandarac Gum	CUPR	Resin	N Afr		
1232	*Tetragonia tetragonioides*	New Zealand Spinach	AIZO	Vegetable	Oceania		
1233	*Tetragonolobus purpureus*	Asparagus pea	LEG	Vegetable	Medit		
1234	*Thaumatococcus daniellii*	Miracle-Fruit	MARA	Sweetener	Africa		
1235	*Theobroma angustifolia*	Cacao Cimarrón	STER	Beverage	Neot	AU, OC	fly, bee
1236	*Theobroma bernouilli*	Cacao Rana	STER	Essence	Neot	AU, OC	fly
1237	*Theobroma bicolor*	Cacao Blanco, Pataste	STER	Spice	Neot	AU, OC	fly
1238	*Theobroma cacao*	Cocoa	STER	Beverage	Neot	AU, OC	fly, Fo, La,
1239	*Theobroma cirmolinae*	Wild Cocoa	STER	Fruit	Neot	AU, OC	fly
1240	*Theobroma grandiflorum*	Cupuacu	STER	Fruit	Neot	AU, OC	fly
1241	*Theobroma simiarum*	Cacao de Mico	STER	Beverage	Neot	AU, OC	fly
1242	*Theobroma speciosum*	Wild Cocoa	STER	Fruit	Neot	AU, OC	fly
1243	*Theobroma subincanum*	Wild Cocoa	STER	Fruit	Neot	AU, OC	fly
1244	*Thymus vulgaris*	Common Thyme	LABI	Spice	Medit	OC?	bee, Apis
1245	*Thymus zygis/gracilis*	Thyme	LABI	Essence	Medit	OC?	bee, Apis
1246	*Toxicodendron succedaneum*	Japan Tallow	ANAC	Wax	Asia		

No.	Common Name
48	Tablus Spice
1138	Tacaco
376	Tachibana Orange
1143	Tacoutta
1214	Tahiti Arrowroot
1216	Talisia Esculenta
1218	Tamarind
1219	Tamarisk
1052	Tamarugo
9	Tan Wattle
242	Tara
37	Tara vine
139	Taragon
605	Tartary Buckwheat
795	Tarwi
584	Tasmanian Blue Gum
97	Tavoy Cardamom
252	Tea
845	Tea Tree
1221	Teak
568	Teff
434	Tejocote
463	Temu Lawak
462	Temu Mangga
593	Teosinte
991	Tepary Bean
254	Terentang
1175	Terongan
731	Tettu
183	Thai Dammar
1269	Thundergod Vine
1245	Thyme
421	Ti Palm
1165	Tiberato
429	Tibetan Hazelnut
494	Tick Clover
615	Tikus
781	Timbó
907	Tobacco
889	Tolu Balsam
1005	Tomatillo
798	Tomato, Tomatle
536	Tonka Bean
117	Torete, Custard Apple
1202	Townsville Stylo
654	Tree Cotton
480	Tree Tomato
1041	Trifoliate Orange
67	Trompillo
174	Tropical Carpetgrass
1082	Tropical Kudzu
995	Tropical Lima Bean
829	True Balata
688	Tsi
491	Tuba Root
1036	Tuberose
156	Tucum-Assu Palm
62	Tung-Oil-Tree
1148	Turi
425	Turkish Hazel
461	Turmeric
417	Tussa Jute
866	Uaku
1310	Ucuba Oiltree
181	Udjung Atup
1277	Ulluque, Papa Lisa
26	Umbrella Thorn
880	Umbrella Tree
1195	Untsuti
657	Upland Cotton
782	Urucu Timbo
1206	Urucuri Palm
158	Urunday
561	Utile
1043	Uvilla, Guarumo
1282	Vanilla
454	Vegetable Marrow
873	Velvetbean
1286	Vernonia Galamensis
1287	Vetiver, Khas-Khas
314	Vichada
36	Vinifera Palm
623	Virginia Strawberry
1311	Virola
1313	Vitex Negundo
614	Vogel Fig
1225	Vogel Tephrosia
380	Wampi
918	Water Dropwort

No.	Scientific Name	Common Name	Family	Use	Origin	Breed. Sys.	Pollinators	No.	Common Name
1247	*Toxicodendron verniciflluum*	Chinese Lacquer Tree	ANAC	Resin	Asia			964	Water Lemon
1248	*Trachucarpus fortunei*	Windmill Palm	PALM	Fiber	E Asia			717	Water Spinach
1249	*Trachylobium verrucosum*	Zanzibar Copal	LEG	Resin	E Afr			893	Watercress
1250	*Trachyspermum ammi*	Ammi, Ajowan	UMBE	Medicinal	S Asia			361	Watermelon
1251	*Trachyspermum roxburghianum*	Ajmud	UMBE	Medicinal	S Asia			884	Wax Myrtle, Bayberry
1252	*Tragopogon porrifolium*	Salsify	COMP	Vegetable	W Asia			320	Wax Palm
1253	*Trapa bicornis*	Ling, Two Horns	ONAG	Medicinal	S Asia			321	Wax Palm
1254	*Trapa natans*	Jesuit Nut, Four Horns	ONAG	Nut	S Asia			190	Waxgourd, Kundor
1255	*Treculia africana*	Mozinda	MORA	Nut	Africa			566	Weeping Lovegrass
1256	*Trichosanthes cucumerina*	Snakegourd, Culebra	CUCU	Med., Veg.	SE Asia	OC	moth	20	Weeping Myall
1257	*Trifolium alexandrium*	Berseem	LEG	Forage	W Asia	OC	bee, Apis, insect	72	Welsh Onion
1258	*Trifolium fragiferum*	Strawberry Clover	LEG	Forage	W Asia	SC, OC	bee, Apis, insect?	469	West Indian Lemongrass
1259	*Trifolium hybridum*	Hybrid Clover	LEG	Forage	W Asia	OC	bee, Apis?	649	White Beech
1260	*Trifolium incarnatum*	Crimson Clover	LEG	Forage	Medit	AU, OC	bee, Apis	1262	White Clover
1261	*Trifolium pratense*	Red Cover	LEG	Forage	W Asia	AU, OC	bee, Bombus, Apis	1106	White Cottage Rose
1262	*Trifolium repens*	White Clover	LEG	Multiple	Medit	SC, OC	bee	416	White Jute
1263	*Trifolium resupinatum*	Persian Clover	LEG	Forage	Europe	OC	bee	792	White Lupine
1264	*Trifolium subterraneum*	Subclover	LEG	Forage	Medit	AU	bee	796	White Lupine
1265	*Trifolium vesiculosum*	Arrowleaf Clover	LEG	Forage	Medit	OC	bee	870	White Mulberry
1266	*Trigonella foenum-graecum*	Fenugreek	LEG	Spice	S Asia	SC	bee, Apis	1158	White Mustard
1267	*Triphasia trifolia*	Limeberry	RUTA	Fruit	SE Asia			1124	White Sandalwood
1268	*Triplochiton scleroxylon*	Obeche	STER	Timber	Africa			954	White Seraya
1269	*Tripterygium wilfordii*	Thundergod Vine	CELA	Pesticide	E Asia			849	White Sweetclover
1270	*Triumfetta bartramia*	Paroquet Bur	TILI	Fiber	Africa			1224	White Tephrosia
1271	*Triumfetta cordifolia*	Burweed	TILI	Fiber	Africa			988	Wild Avocado
1272	*Triumfetta tomentosa*	Lapulla	TILI	Fiber	Africa			102	Wild Cashew
1273	*Tropaeolum tuberosum*	Añu	TROP	Vegetable	Neot			835	Wild Chamomile
1274	*Tussilago farfara*	Coltsfoot	COMP	Medicinal	Asia		bee	1239	Wild Cocoa
1275	*Tylosema esculentum*	Marama Bean, Tsi Bean	LEG	Tuber	Africa			1242	Wild Cocoa
1276	*Typha spp.*	Bullrush	TYPH	Forage	Europe	SC, OC	wind, bee?	1243	Wild Cocoa
1277	*Ullucus tuberosus*	Ulluque, Papa Lisa	BASE	Tuber	Neot	OC?	bee?	396	Wild Coffee
1278	*Uncaria gambir*	Gambier	RUBI	Dye, Tann.	SE Asia			397	Wild Coffee
1279	*Urena lobata*	Aramina Fiber	URTI	Fiber	Africa			653	Wild Cotton
1280	*Vaccinium ashei*	Rabbit-Eye Blueberry	ERIC	Fruit	N Amer			998	Wild Date Palm
1281	*Vangueria madagascariensis*	Spanish Tamarind	RUB	Fruit	Africa			1071	Wild Guava
1282	*Vanilla planifolia*	Vanilla	ORC	Essence	Neot	OC	bee, Eulaema	1072	Wild Guava
1283	*Vateria indica*	Indian Copal	DIPT	Resin	S Asia			749	Wild Lettuce
1284	*Vaupesia cataractarum*	Ma-ha-wa-soo	EUPH	Oil	Neot			168	Wild Oat
1285	*Vernonia anthelmintica*	Kinka Oil Ironweed	COMP	Medicinal	Africa			276	Wild Papaya
1286	*Vernonia galamensis*	Vernonia Galamensis	COMP	Oil	S Asia			277	Wild Papaya
1287	*Vetiveria zizanioides*	Vetiver, Khas-Khas	POA	Essen, Misc	S Asia	SC, OC?	bee?	278	Wild Papaya
1288	*Vicia articulata*	Single-Flowered Vetch	LEG	Erosion	Asia	AU, OC	bee	962	Wild Passion Fruit
1289	*Vicia benghalensis*	Purple Vetch	LEG	Erosion	Medit	AU, OC	bee	107	Wild Pineapple
1290	*Vicia dasycarpa*	Woolypod Vetch	LEG	Forage	Medit	AU, OC	bee	1145	Wild Sesame
1291	*Vicia ervilia*	Bitter Vetch	LEG	Forage	Medit	AU, OC	bee	1146	Wild Sesame
1292	*Vicia faba*	Broadbean	LEG	Vegetable	Medit	AU, OC	bee	517	Wild Yam
1293	*Vicia monantha*	Bard Vetch	LEG	Forage	Medit	AU, OC	bee	520	Wild Yame
1294	*Vicia pannonica*	Hungarian Vetch	LEG	Forage	Medit	AU, OC	insect, bee	1248	Windmill Palm
1295	*Vicia sativa/nigra*	Blackpod Vetch	LEG	Forage	Medit	AU, OC	insect, bee	727	Wine Palm
1296	*Vicia sativa/sativa*	Common Vetch	LEG	Forage	Medit	AU, OC	insect, bee	1074	Winged Bean
1297	*Vicia villosa*	Hairy Vetch	LEG	Forage	Medit	AU, OC	bee	510	Winged Yam
1298	*Vigna aconitifolia*	Moth Bean	LEG	Vegetable	S Asia	AU, OC?	bee	381	Winter Purslane
1299	*Vigna calcaratus*	Adzuki Bean	LEG	Vegetable	S Asia	AU, OC?	bee	1127	Winter Savory
1300	*Vigna luteola*	Dalrymple Vigna	LEG	Forage	Aust	AU, OC?	bee	473	Winter's Grass
1301	*Vigna mungo*	Black Gram	LEG	Vegetable	S Asia	AU, OC	bee	544	Winter's-Bark
1302	*Vigna ornata*	Giri Yam Pea	LEG	Tuber	S Asia	AU, OC	bee	1171	Wonderberry
1303	*Vigna radiata*	Mung Bean	LEG	Vegetable	S Asia	AU, OC	bee	147	Wooly Pawpaw
1304	*Vigna umbellata*	Rice Bean	LEG	Vegetable	S Asia	AU, OC	bee	1290	Woolypod Vetch
1305	*Vigna unguiculata*	Cowpea	LEG	Forage	S Asia	AU, OC	bee	325	Wormseed, Epazote
1306	*Vigna unguiculata/cylindrica*	Catjang	LEG	Vegetable	S Asia	AU, OC	bee	138	Wormwood
1307	*Vigna unguiculata/sesquipedalis*	Yard-Long Bean	LEG	Vegetable	S Asia	AU, OC	bee	1316	Yacca Gum
1308	*Vigna vexillata*	Gombi Bean	LEG	Tuber	Africa	AU, OC	bee	1039	Yacon Strawberry
1309	*Vinca minor*	Common Periwinkle	APOC	Medicinal	W Asia			209	Yague
1310	*Virola sebifera*	Ucuba Oiltree	MYRI	Wax	Neot			937	Yam Bean
1311	*Virola spp.*	Virola	MYRI	Timber	Neot	OC	fly, bee	525	Yampi
1312	*Vitellaria paradoxa*	Sheabutter	SAPO	Wax	W Afr			1307	Yard-Long Bean
1313	*Vitex negundo*	Vitex Negundo	VERB	Medicinal	S Asia			1317	Yautia, Tannia
1314	*Vitis rotundifolia*	Muscadine Grape	VITA	Fruit	S Asia	AU, OC	fly, bee, Apis	978	Yellow Flame
1315	*Voandzeia subterranea*	Bambara Groundnut	LEG	Vegetable	Africa	AU, OC	bee	1217	Yellow Genip
1316	*Xanthorrhoea hastilis*	Yacca Gum	XANT	Resin	Aust			512	Yellow Guinea Yam
1317	*Xanthosoma spp.*	Yautia, Tannia	ARAC	Tuber	Neot			794	Yellow Lupine
1318	*Xerospermum noronhianum*	Rambutan Pachat	SAPI	Fruit	SE Asia			763	Yellow Tea Tree
1319	*Ximenia americana*	Mountain Plum	OLAC	Fruit, Oil	Neot			460	Yellow Zedoary
1320	*Xylopia aethiopica*	Guinea Pepper	ANNO	Spice	W Afr			648	Yemane
1321	*Yucca elephantipes*	Itabo, Izote	AGAV	Vegetable	Neot	SC, OC	moth	1116	Yerba de María
1322	*Yucca filamentosa*	Yucca	AGAV	Fiber	Neot	SC, OC	moth, Pronuba	705	Yerba Mate
1323	*Zea mays*	Corn, Maiz	POA	Cereal	Neot	SC, OC	wind, bee	702	Yerba Mate Mimic
1324	*Zingiber cassumunar*	Bangle	ZING	Medicinal	SE Asia	OC?	bee?	704	Yerba Mate Mimic
1325	*Zingiber mioga*	Myoga, Japanese Ginger	ZING	Spice	Japan	OC?	bee?	255	Ylang-Ylang
1326	*Zingiber officinale*	Ginger	ZING	Spice	S Asia	OC?	bee?	970	Yoco
1327	*Zingiber zerumbet*	Lempoyang	ZING	Medicinal	SE Asia	OC?	bee?	1322	Yucca
1328	*Ziziphus jujuba*	Chinese Date	RHAM	Fruit	S Asia	OC	wasp, bee, fly	1249	Zanzibar Copal
1329	*Ziziphus mauritiana*	Indian Jujube	RHAM	Fruit	W Asia	OC	wasp, bee, fly	451	Zapallo, Winter Squash
1330	*Zornia spp.*	Barba de Burro	LEG	Forage	Neot			464	Zedoary

AGREEMENT FOR POLLINATION SERVICES

This AGREEMENT, made this ____________________ day of ____________________ 19 _____, between

____________________ of____
(Street & No., R.R., P. O. Box No., or Lot, Con. & Township)

____
(City, Town or Village) (Province) (Postal Code)
the producer of the crop to be pollinated, hereinafter known as the PRODUCER, and

____________________ of ____
(Street & No., R.R., P. O. Box No., or Lot, Con. & Township)

____
(City, Town or Village) (Province) (Postal Code)
the beekeeper who will furnish bees for pollination services, hereinafter known as the BEEKEEPER, WITNESS THAT:

A) The PRODUCER hereby agrees to the following stipulations:

1). He/She shall pay the sum of $____________________ per colony amounting to a sum

total of $____________ for pollination services on____________ hectare(s) of ____________________(Crop) at

____ according to the following:
(Location - Street & No., (OR Lot, Con. & Township)

SCHEDULE A: To begin on/before ____________________________ 19 ____ and end on/after
____________________ 19 ________. OR
SCHEDULE B: To begin at the time of ______________(e.g. bud break, 10% or 20% bloom) and end at the time of ____________________ (e.g. 90% petal fall or ccmplete petal fall) as determined by the PRODUCER.

The Producer shall notify the Beekeeper at least______*hours in advance of when he/she estimtes the bees are to be moved.

2). Payments are to be made in __________(e.g. cash or check) to the BEEKEEPER at the rate of _____ ** of the agreed amount when the agreement is signed;
_____** when the bees are moved onto location; and
_____** within _____*** day(s) after the crop is harvested.

3). If any pesticide is to be applied by ground or aerial equipment to any crop, plant, weed or other agricultural product within ______**** km of the location of the bees, the PRODUCER or his/her agent will notify the BEEKEEPER at least______ * hours in advance of the pesticide appication. The method of notification will be by calling collect person-to-person via telephone, or in person.

4). Should the PRODUCER notify the BEEKEEPER to move his/her bees because of pesticides being applied, and the BEEKEEPER elects to move said bees because of the hazard presented, the PRODUCER shall pay the Beekeeper $ ______ per colony for moving the bees out, and $ _____ per colony for moving them back on, with said amount included in the _______ payment to be made ____________________.

5). The PRODUCER agrees to compensate the BEEKEEPER in the amount of $ ______ for each hive damaged by spraying operations over which the PRODUCER has control, including aerial spraying of adjacent fields. Further compensation will be given to the BEEKEEPER for any losses incurred through hives being knocked over by machinery, vehicles, and/or livestock. Compensation to the BEEKEEPER shall be at a rate of $ _____for a lost colony and $ ______ for the total destruction of a hive, with the said amount included in the ________ payment to be made ________________.

6). The PRODUCER agrees to make a count, or have a count made by his/her agent, of the colonies of bees placed on location within _____ day(s) after they are placed in order to ascertain that the stipulated number of colonies have been provided. The BEEKEEPER may be present at such an inspection if he/she so requests.

---

* A usual notification time being forty-eight (48) hours.
** A usual breakdown in payment being one-third (1/3); (1/3); and (1/3).
*** A usual number of days being ten (10).
**** A usual number of kilometers being two or three (2 or 3).

7). The PRODUCER shall provide an adequate, uncontaminated water supply to be kept available for the bees at all times when bees are on location, and kept available in the same location at all times during the life of the AGREEMENT.

8). The PRODUCER shall provide adequate access roads into the field to be pollinated, they they must be suitable for loads weighing up to ____________ # tons.

9). The PRODUCER shall provide all necessary cultural and remedial measures to assure optimum plant growth in said crop.

10). The PRODUCER agrees to assume liability for bee stings while the bees are on said location.

B). The BEEKEEPER hereby agrees to the following stipulations:

1). The BEEKEEPER shall provide ______ colonies per hectare on ________ hectares of the crop herein described, and for the fee established.

2). The BEEKEEPER shall place the colonies in the locations in the field(s) as requested by the PRODUCER. The distribution pattern of the hives shall be_________________.##

3). Having due respect for the property of the PRODUCER, the BEEKEEPER shall properly maintain and care for the said bees providing all necessary beekeeping equipment, transportation, and labor to maintain said bees providing all necessary beekeeping equipment, transportation, and labor to maintain said bees in proper pollinating condition.

4). The BEEKEEPER shall provide a current, valid, official Inspection Certificate showing freedom from disease on all colonies moved on location for pollination.

5). He/she hereby certifies that each colony is queen-right.

6). Each colony to be furnished by the BEEKEEPER shall be composed of at least _______### supers.

7). All colonies placed by the BEEKEEPER shall meet a minimum strength requirement. The PRODUCER reserves the right to engage the services of a Certified Apiary Inspector to assess colony strength according to the colony standards outlines in Appendix 1. Such an inspection shall take place within _____ #### day (s) after the colonies have been placed for pollination.

8. Should the BEEKEEPER fail to deliver the number of colonies of bees as specified in this Agreement, or should the colonies fail to meet the minimum quality requirements as set forth herein, or should the BEEKEEPER fail to deliver the bees on the date specified, he/she agrees to pay the PRODUCER as liquidated damages the rental fee per colony that he/she has failed to deliver; not delivered on the date specified; or which has failed to meet the quality requirements. However, the liquidated damages shall not be imposed if the total quantity promised are delivered within ##### day(s).

9). Upon being notified as set forth in Paragraph 3, Section A above of the imminent application of a pesticide within _____ § km(s) of the location of said bees, the BEEKEEPER shall immediately take action to move, or protect, his/her bees so the pesticide may be applied. Should the BEEKEEPER elect to leave said bees on location after proper notice, he/she shall hold the PRODUCER blameless for any injury which may occur to the bees and/or the hives due to the pesticide application.

10). The BEEKEEPER shall move said bees off location within _______ §§ day(s) from the termination of this Agreement; otherwise, he/she agrees to pay $ _____ per colony per day for each day extending beyond the _______ §§§ day after the termination date of this Agreement, unless otherwise agreed upon and stipulated in writing by the PRODUCER and BEEKEEPER.

A usual load weight being 4 tonnes for a large orchard.

Depending on the crop, distribution may be in groups of four (a pallet load) spread evenly, more or less, throughout, or in larger groups

A usual number of supers being two (2)

####A usual number of days after placement being three to five (3-5).

A usual number of days being three (3).

C). Should any controversy arise hereunder, between the parties hereto, such controversy shall be settled by arbitration. Each party within _____§§§§ day(s) shall appoint one arbitrator and the two so named shall select a third, and the decision of any two such arbitrators shall be binding upon the parties hereto. The cost of such arbitration shall be divided equally between the parties.

D). Title to the said _____________ hectare(s) shall at all times remain in and vested in the PRODUCER Title to the said _____________ colonies of bees and beekeeping equipment shall at all times remain in and vested in the BEEKEEPER.

E). This Agreement, and any appendices or exhibits attached hereto, shall constitute the full and complete Agreement between the parties, and any modification of this Agreement shall be mutually agreed upon and in writing, and signed by all parties to be bound thereby.

SIGNED, SEALED and DELIVERED at ______________________________________, in the Province

of ______________________________________, this ____________ day of ________________, 19 ____.

______________________________	______________________________
(Notary Public or Witness)	(Beekeeper)
______________________________	______________________________
(Address)	(Address)
______________________________	______________________________
(Notary Public or Witness)	(Producer)
______________________________	______________________________
(Address)	(Address)

§ A usual number of kilometers being two or three (2 or 3).
§§ A usual number of days from the termination date being three (3).
§§§ The standard number of days being ten (10).
§§§§ Please refer to the following page for an alternative signatory page.

_______________________________	_______________________________
(Beekeeper)	(Producer)
_______________________________	_______________________________
(Title)	(Title)
Telephone No. _______________	Telephone No. _______________
Province of _______________	County of _______________

On this __________ day of _______________, 19____, _______________________________,

personally appeared before me and is known to me (or satisfactorily proven) to be the person described in and who executed the foregoing instrument and he acknowledged that he executed the same for the purpose therein contained.

____

(Notary Public)

Province of _______________ County of _______________

On this __________ day of _______________, 19____, _______________________________,

personally appeared before me and is known to me (or satisfactorily proven) to be the person described in and who executed the foregoing instrument and he acknowledged that he executed the same for the purpose therein contained.

____

(Notary Public)

* Please note that the above has been taken from USA protocol.

Appendix 1

COLONY STRENGTH STANDARDS

A. FIELD COLONY STANDARDS

1). A STANDARD or GRADE 'A' FIELD colony of bees for field pollination purposes shall be one that contains the following:

(a) 3000 square inches (or 20,000 square centimeters) of comb, 1000 square inches (6500 square centimeters) of comb, which shall be occupied by live brood.

(b) Enough bees to cover all brood well. There shall be enough bees to cover ten (10) standard frames of comb well.

(c) 4.5 kilogram (10 pounds) of honey or its equivalent of suitable bee food other than pollen or pollen substitute.

2). A GRADE 'B' FIELD colony of bees for field crop pollination shall be one that fails to meet the requirements for a standard field colony by not more than 25% on the amount of bees and brood, but does meet the other requirements of a standard field colony.

B. ORCHARD COLONY STANDARDS

1). A STANDARD or GRADE 'A' ORCHARD colony of bees for orchard pollination purposes shall be one that contains the following:

(a) 3000 square inches (or 20,000 square centimeters) of comb space of which 600 square inches (4000 square centimeters) shall be occupied by live brood.

(b) Enough bees to cover all brood well. There shall be enough bees to cover at least six (6) standard frame of comb well.

(c) 4.5 kilograms (10 pounds) of honey or its equivalent in suitable bee food other than pollen or pollen substitute.

2). A GRADE 'B' ORCHARD colony of bees for orchard pollination shall be one that fails to meet the requirements for a standard orchard colony by not more than 25% on the amount of bees and brood, but does meet the other requirements of a standard orchard colony.

Footnote: The different strengths recognize that orchard pollination takes place earlier in the spring, when colonies are not as strong as later in the season, when field crop pollination is required.

Appendix 2

DEFINITION OF TERMS

BROOD: The immature stages of bees — the eggs, larvae and pupae deposited in the cells or comb.

CERTIFIED GOVERNMENT APIARY INSPECTOR: Any qualified person who has a good working knowledge about bees, their habits and diseases, and who is officially appointed by a government agency. In addition, he/she shall have a good knowledge of pollination practices where bees are used.

COLONY: One hive or nest and its contents including frames, comb, honey, bees, brood provisions, pollen and a laying queen bee.

COMB or CELLS: The wax or other framework in the hive or nest in which bees store honey and pollen, or in which the brood is reared.

HIVE: Any recaptacle or container, or part thereof, made or prepared for use by bees, or inhabited by bees.

OFFICIAL INSPECTION CERTIFICATE: A certificate issued by a duly authorized Apiary Inspector stating the results of an inspection for the presence of bee diseases.

PESTICIDE: Any substance or mixture of substances intended to be used for defoliating plants or for preventing, destroying, repelling, or mitigating insects, fungi, bacteria, weeds, rodents, predatory animals or any for of plant or animal life that is a pest, and which may infest or be deterimental to vegetation, man, animal or households, or be present in any environment.

QUEER-RIGHT: A colony with a fertile queen laying worker eggs.

SUPER: A receptacle in which bees store surplus honey; so called because it is placed "over" or above the brood chamber.

ENTENTE DE SERVICES DE POLLINISATION

La présente entente entre

____

(nom) (n° civique, rue. R.R., B. P.)

____

(ville, province, code postal)
producteur de la culture à polliniser, ci-après appelé LE PRODUCTEUR,

et

____

(nom) (n° civique, rue, R.R., B. P.)

____

(ville, province, code postal)
apiculteur fournissant les abeilles pour les services de pollinisation, ci-après appelé L'APICULTEUR,

atteste que las deux parties conviennent de ce qui suit :

A). LE PRODUCTEUR S'ENGAGE A:

1). Payer la somme de _______$ par colonie soit un montant total de ______$ pour les services de pollinisation sur une superficie de ______ hectares où il cultive du/des ________et qui est situeé a l'adresse suivante :

____

(emplacement - numéro civique OU lot), selon le calendrier de service suivant:

CALENDRIER A) :	Service débutant le ou avant le ______ 19 _________ et prenant fin le ou après le _________ 19 _________	
CALENDRIER B) :	Service debutant au moment de ___________	(p. ex. l'éclatement des bourgeons, 10% ou 20% de l'efflorescence)
	et prenant fin au moment de _____________	(p. ex. la chute de 90% des pétales, la chute de tous les pétales)

Le producteur doit aviser l'apiculteur au moins ______[1] heures d'avance lorsqu'il juge le moment venu de retirer les colonies.

2). Payer a l'apiculteur (en argent ou par cheque) le/la _____[2] du montant total à la signature de l'entente; la somme de _______$ au moment ou les abeilles sont amenées chez le producteur; et la somme de __________$[2] dans les _______[3] jours suivant la récolte de la culture pollinisée.

3). Aviser l'apiculteur au moins ________[4] heures d'avance lorsqu'il prévoit l'application terrestre ou aérienne de pesticide sur une culture, une plante, une mauvaise herbe ou sur tout autre produit agricole qui se trouve à moins de ______[5] km de l'emplacement des colonies. L'avis est donné à l'apiculteur par téléphone (frais virés) ou en personne.

4)Payer a l'apiculteur la somme de _______$ par colonie pour le déplacement des abeilles lorsqu'une application de pesticides présente un risque et que l'apiculteur décide de déplacer ses colonies. Le producteur s'engage également à payer la somme de _______$ par colonie pour le retour des abeilles, et d'inclure ce montant dans le _______ paiement qui doit être fait le ___________________________________

---

[1] Un délai raisonnable est de 48 heures.
[2] Un versement raisonnable est un tiers (1/3) pour chaque paiement.
[3] Un délai raisonnable est de 10 jours.
[4] Un délai raisonnable est de 48 heures.
[5] Une distance raisonnable est de 2 à 3 kilomètres.

5). Payer à l'apiculteur un montant compensatoire de ______$ pour chaque ruche endommagée au cours d'une pulvérisation dirigée par le producteur, y compris la pulvérisation aérienne de champs adjacents. Il est possible que des ruches soient détruites par des machines, des véhicules ou des animaux de ferme, et le producteur s'engage à payer à l'apiculteur un montant compensatoire de _______$ par colonie perdue et de ______$ pour l'anéantissement complet d'une ruche, et d'inclure cette somme dans le _________ paiement qui doit être fait le ____

6). Compter ou demander à son mandataire de compter le nombre de colonies placées dans ses champs dans les ______ jours suivant leur installation, afin de s'assurer que le nombre de colonies indiqué a bien été fourni. L'apiculteur peut assister à cette inspection sur demande.

7). Fournir une quantité suffisante d'eau potable à laquelle les abeilles ont accès durant tout la durée de l'entente.

8). Fournir de bonnes voies d'accès dans les champs à polliniser; elles doivent supporter des charges allant jusqu'a ________[6] tonnes.

9). Prendre toutes las mesures culturales et correctives nécessaires pour assurer la croissance optimale de la culture visée.

10). Assumer la responsabilité des piqûres faites par les abeilles pendant leur sejour sur ses terrains.

B. L'APICULTEUR S'ENGAGE A :

1). Fournir ________ colonies par hectare sur une superficie totale de _______ hectares afin de polliniser la culture susmentionnée, au tariff établi.

2). Disposer les colonies aux endroits du champ indiqués par le producteur. Les ruches seront disposées _________[7].

3). Respecter la propriété du producteur; bien s'occuper de ses abeilles et fournir tout le matériel la main-d'oeuvre et le transport requis pour maintenir les abeilles dans de bonnes conditions de pollinisation.

4). Avoir en sa possession un certificat d'inspection officiel récent et valide qui atteste que toutes las colonies d'abeilles louées pour la pollinisation sont exemptes de maladies.

5). Attester que chaque colonie possède une reins féconde.

6). Fournir des ruches dont chacune est composée d'au moins ________[8] hausses.

7). Placer des colonies d'abeilles qui ont toutes la force minimale requise. Le producteur se réserve le droit de faire appel à un inspecteur de ruchers autorisé pour évaluer la force de la colonie, selon les normes établies à l'Annexe 1. L'inspection doit se faire dans les ________[9] jours suivant l'arrivée des abeilles dans les champs a polliniser.

8). En cas d'incapacité de livrer le nombre de colonies d'abeilles indiqué dans l'entente ou de livrer les colonies à la date prévue, ou lorsque les colonies ne répondent pas aux exigences minimales de qualité, l'apiculteur s'engage à payer au producteur. à titre de dommages-intérêts, le montant de la location pour chaque colonie qui n'a pas été livrée, qui n'a pas été livrée à la date prévue dans l'entente ou qui ne répond pas aux exigences de qualité minimales. Il est entendu que les dommages-intétêts ne sont pas exigés lorsque la quantité totale d'abeilles est livrée dans les _______[10] jours.

---

[6] Une charge raisonnable est de 4 tonnes pour un grand verger

[7] Selon la culture, les ruches peuvent étre disposées en groupes de quatre (pleine palette) à égale distance les unes des autres plus ou moins, ou réunies en plus grands groupes

[8] Un nombre raisonnable est de 2 hausses.

[9] Un dalai raisonnable après l'installation est de 3 a 5 jours

[10] Un délai raisonnable est de 3 jours.

9). À la réception de l'avis indiqué au paragraphe 3 de la section A pour avetir qu'une application de pesticide sera faite à moins de _____[11] km de l'emplacement de ruches, l'apiculteur prend immédiatement les mesures nécessaires pour déplacer ou portéger ses abeilles avant que le traitement ait lieu. Lorsqu'il décide de laisser ses abeilles sur place apr`es la réception de l'avis en bonne et due forme, l'apiculteur ne peut tenir le producteur responsable de tout accident dont pourraient être victimes les abeilles ou les ruches par suite de l'application du pesticide.

10). Enlever les abeilles des champs dan les _______[12] jours suivant l'expiration de l'entente; sinon, l'apiculteur paye au producteur la somme de _______$ par colonie par jour pour chaque journée supplémentaire à compter du _______ ième jour après l'expiration de l'entente, à moins que les deux parties aient stipulé d'autres dispositions par écrit.

C. Lorsqu'un point de l'entente est l'objet d'une controverse entre les deux parties, le litige doit être régale par arbitrage. Chaque partie dispose de ________[13] jours pour nommer un arbire, et les deux arbitres nommés doivent en choisir un troisème; la décision rendue par deux des trois arbitres est obligatoire et exécutoire. Les deux parties au litige paient chacune la moitié du coût de l'arbitrage.

D. Le titre de propriété du champ de ________ hectares reste dévolu au producteur pendant toute la durée de l'entente.

Le titre de proprété des ________ colonies d'abeilles louées en vertu de la présente entente et du matériel d'apiculture reste dévolu à l'apiculteur pendant toute la durée de l'entente.

Le présente entente ainsi que toute annexe et pièce jointe constituent l'entente de services complète entre les parties, et toute modification de l'entente doit faire l'objet d'un accord mutuel écrit et signé par toutes les parties à la présente entente.

SIGNÉ, SCELLÉ ET DÉLIVRÉ à:

______________________ (cité OU ville)	______________ (Province)	______________ (jour,mois,année)
______________________ (notaire OU témoin)	______________________ (apiculteur)	
______________________ (numéro civique, R.R., B. P.)	______________________ (numéro civique, R. R., B. P.)	
______________________ (cité OU ville)	______________________ (cité OU ville)	
______________________ (notaire OU témoin)	______________________ (producteur)	
______________________ (numéro civique, R.R., B. P.)	______________________ (numéro civique, R. R., B. P.)	
______________________ (cité OU ville)	______________________ (cité OU ville)	

[11] Une distance raisonnable est de 2 à 3 kilomètres

[12] Un délai raisonnable est de 3 jours après l'expiration de l'entente.

[13] Un délai raisonnable est de 10 jours.

Annexe 1

NORMES RELATIVES À LA FORCE DES COLONIES

A. NORMES POUR LA FORCE DES COLONIES DE RUCHERS DE CHAMPS

1). La colonie d'un rucher de champs STANDARD (CATÉGORIE 'A'), utilisée pour la pollinisation des champs, présente lee caractéristiques suivantes :

a). La superficie des rayons est de 19 300 cm^2 (3 000 po^2), et le couvain occupe une superficie de 6500 cm^2 (1000 po^2).

b). Les abeilles sont présentes en nombre suffisant pour couvrir tout le couvain. Elles sont également assez nombreuses pour bien couvrir de rayons 10 cadres standard.

c). La colonie dispose de 4,5 kg (10 lb) de miel nu d'une quantite equivalents de bonne nourriture pour abeilles autre que la pollen et le succédané de pollen.

2). La colonie d'un rucher de champs de CATEGORIE 'B', utilisée pour la pollinisation des grandee cultures, est une colonie dont la population d'abeilles et ln couvain sont de 25 % inférieurs a ceux d'une colonie de rucher de champs standard, mais elle possède lee autres caractéristiques de la colonie d'un rucher de champs standard.

B. NORMES POUR LA FORCE DES COLONIES DE RUCHERS DE VERGERS

1). La colonie d'un rucher de vergers STANDARD (CATIGORIE 'A'), utilisé pour la pollinisation des vergers, présente les caractéristiques suivantes :

a). La superficie des rayons est de 19 300 cm^2 (3 000 po^2), couvain occupe une superficie de 4 000 cm^2 (600 po^2).

b). Les abeilles sont présentes en nombre suffisant pour couvrir tout le couvain. Elles sont également assez nombreuses pour bien couvrir de rayons 6 cadres standard.

c). La colonie dispose de 4,5 kg (10 lb) de miel nu d'une quantite equivalents de bonne nourriture pour abeilles autre que la pollen et le succédané de pollen.

2). La colonie d'un rucher de vergers de CATEGORIE 'B' utilisée pour la pollinisation des grandes cultures, est une colonie dont la population d'abeilles et le couvain sont de 25 % inférieurs à ceux d'un rucher de verger standard, mais elle possède las autres caractéristiques de la colonie du rucher de vergers standard.

NOTE : L'écart dans la force des colonies est attribuable au fait que la pollinisation des vergers a lieu tôt au printeps lorsque les colonies ne sont pas aussi fortes que plus tard dans la saison, a la période de pollinisation des grandes cultures.

Annexe 2

DÉFINITION DE TERMES UTILISÉS DANS L'ENTENTE

CERTIFICAT D'INSPECTION OFFICIEL: Certificat accordé par un inspecteur de ruchers dûment autorisé et indiquant le résultat d'une inspection pour détecter des maladies chez les abeilles.

COLONIE : Une ruche et son contenu, soit les cadres, lee rayons, la miel, les abeilles et la reine pondeuse.

COUVAIN: Ensemble des oeufs, des larves et des nymphes déposés dans les rayons d'une ruche.

COUVRIR LE COUVAIN: Avoir suffisamment d'ouvrières adultes pour couvrir au moins 90% du couvain qui se trouve dans les rayons.

HAUSSE: Réceptacle dans lequel les abeilles emmagasinent le surplus de miel; partie de la ruche placée au-dessus du nid à couvain.

INSPECTEUR DE RUCHERS AUTORISÉ: Touts personne qualifiée possédant une bonne expérience pratique des abeilles, de leur comportement et de leurs maladies; il est nommé officiellement par le ministre en vertu de la Loi sur l'inspection des ruchers.

PESTICIDE: Substance au mélange de substances servant `a défolier las végétaux ou à éloigner, détruire, repousser au combattre des insectes, des champignons, des bactéries, des mauvais herbes, des rongeurs, des animaux prédateurs ou toute autre forme d'êtres vivants animaux au végétaux nuisibles qui parasitent au gênent la végétation, les humains, les animaux et les biens matériels ou qui sont présents dans un milieu.

RAYONS: Alvéoles de cire construites dans la ruche où les abeilles emmagasinent le miel et le pollen, ou lieu d'élevage du couvain.

REINE FÉCONDE: Reins féconde qui dans une colonie pond un couvain d'ouvrière.

RUCHE: Contenant au récipient construit au aménagé pour accueillir les abeilles ou dans lequel les abeilles habitent.

Appendix III

LIST OF HIGHLY TOXIC CHEMICALS (highly dangerous to bees)

Pesticide name	Pesticide name
ACEFATE	FENTOATO
ALDICARB	FENVALERATE
ALDRIN	FLUCITRINATE
ALFACIPERMETRINA	FORMOTION
AMINOCARB	FOSFAMIDONE
ARSENICO e ARSENIATI	FOSMET
AZINFOS-ETILE	FOXIM
AZINFOS-METILE	GUAZATINA
BENDIOCARB	ISOBENZAN
BHC (gamma-)	ISOCLORTION
BROMOFOS	ISOFENFOS
BROMOFOS-ETILE	LINDANO
BUTOCARBOSSIMA	MALATION
CARBARIL	METALDEIDE
CARBOFENOTION	METAMIDOFOS
CARBOFURAN	METIDATION
CARBOSULFAN	METIL-ETOATO
CIFLUTRIN	METIOCARB
CIPERMETRINA	METOMIL
CLORPIRIFOS	MEVINFOS
CLORPIRIFOS METILE	MONOCROTOFOS
DELTAMETRINA	NALED
DEMETON-S-METILE	OMETOATO
DEMETON-S-METILSOLFONE	PARATION
DIAZINONE	PARATION-METILE
DICLORVOS	PERMETRINA
DIELDRIN	PIRIDAFENTION
DIMETOATO	PIRIMIFOS-METILE
DINOSEB	PRETILACLOR
DIOXACARB	PROPOXUR
DNOC	PROTOATO
EPTACLORO	QUINALFOS
EPTENOFOS	SULFOTEP
ETIOFENCARB	TEPP
FENAMIFOS	TETRACLORVINFOS
FENITROTION	TIONAZIN
FENOXYCARB	TRIAZOFOS
FENPROPATRIN	VAMIDOTION
FENTION	

Appendix III (pt 3)

LIST OF MODERATELY TOXIC CHEMICALS (dangerous to bees, apply at night)

Pesticide name

2,4,5-T
2,4-D
2,4-DB
AMETRINA
AMITRAZ
AMITROL
ARAMITE
BINAPACRIL
BROMOXINIL
CANFECLORO
CARTAP
CLORAMBEN
CLORDANO
CLORFENVINFOS
COUMAFOS
DALAPON SODIUM
DDD
DDT
DEMETON METILE
DIALIFOS
DICAMBA
DICLORAN
DIMETILAN
DINOBUTON
DIOXATION
DISULFOTON
DITAMILFOS
DITIANON
ENDOSULFAN
ENDOTAL
ENDOTION
ENDRIN
EPTC
ETION
ETOFUMESATE
ETOPROFOS
EXITIAZOX

Pesticide name

FENCLORFOS
FLAMPROP-METILE
FLAMPROP-M-ISOPROPILE
FONOFOS
FORATE
FORMETANATO
FOSALONE
ISOBORNIL TIOCIANATO
ISOLAN
MCPA
MECOPROP
MENAZONE
METOSSICLORO
MNFA
OLIO MINERALE
OSSIDEMETON-METILE
OXAMIL
PARAQUAT
PERTANE
PIRAZOFOS
PIRIMICARB
PROFENOFOS
PROPANIL
ROTENONE
SCHRADAN
SULFALLATE
TEMEFOS
TERBUFOS
TIODICARB
TIOMETON
TIOQUINOX
TRANID
TRICLORFON
TRICLORONATO
ZIRAM
ZIREB

Appendix III (cont'd)

LIST OF MODERATELY TOXIC CHEMICALS FOR BEES (use late in the day or when bees not in flight)

Pesticide name	*Pesticide name*	*Pesticide name*
2,3,6-TBA	DICLONE	METALAXIL
ACIDO GIBBERELLICO	DICLORPROP	METAMITRON
ACIDO GIBBERELLICO GA4 + GA7	DICOFOL	METAM-SODIUM
ACIFLUORFEN SODIUM	DIENOCLOR	METIRAM
ALACLOR	DIFENAMIDE	METOBROMURON
ALOSSIFOP-ETOSSIETILE	DIFENZOQUAT	METOLACLOR
ANILAZINA	DIFLUBENZURON	METOXURON
ATRAZINA	DIMETIRIMOL	METRIBUZIN
AZOCICLOTIN	DINOCAP	MOLINATE
BACILLUS THURINGENSIS	DIQUAT	NABAM
BARBAN	DIURON	NAPROPAMIDE
BENFLURALIN	DODEMORF	NAPTALAM
BENOMIL	DODINA	NICOTINA (SOLFATO)
BENTAZONE	EDIFENFOS	OSSICARBOSSINA
BENZOSSIMATO	ETACELASIL	OXADIXIL
BROMACILE	ETEFON	OXIFLUOROFEN
BROMOFENOSSIMA	ETIDIMURON	PENCONAZOLO
BROMOPROPILATO	ETIRIMOL	PENDIMETALIN
BUPIRIMATE	FENAMINOSULF	PICLORAN
BUTILATE	FENARIMOL	PIRETRINE
CAPTAFOL	FENAZAFLOR	PIRIDATE
CAPTANO	FENBUTATIN OSSIDO	POLISOLFURO DI BARIO
CARBENDAZIM	FENCLORIM	POLISOLFURO DI CALCIO
CARBOSSINA	FENMEDIFAM	PROFAM
CHINOMETIONATO	FENPROPIMORF	PROMETRINA
CIANAZINA	FENSON	PROPACLOR
CIEXATIN	FENTIN ACETATO	PROPAMOCARB
CLOFENTEZINE	FENTIN IDROSSIDO	PROPARGITE
CLOPIRALID	FERBAM	PROPICONAZOLO
CLORBENSIDE	FLUAZIFOP-P-BUTILE	PROPINEB
CLORDIMEFORM	FLUBENZIMIN	PROPIZAMIDE
CLORFENETOL	FLUOROCLORIDONE	RAME (COMPOSTI INORGANICI)
CLORFENSON	FLUORODIFEN	SECBUMETON
CLORFENSULFIDE	FLUOSILICATO DI BARIO	SETOSSIDIM
CLORFLURENOL	FLURENOL	SIMAZINA
CLORMEQUAT	FLUVALINATE	TCA-SODIUM
CLOROBENZILATO	FOLPET	TERBUMETON
CLOROPROPILATO	GLIFOSATE	TERBUTRINA
CLOROTALONIL	IDRAZIDE MALEICA	TETRADIFON
CLOROXURON	IMAZAMETABENZ	TIAZAFLURON
CLORPROFAM	IOXINIL	TIRAM
CLORTAL-DIMETILE	IPRODIONE	TRIADIMEFON
CLORTOLURON	ISOPROPALIN	TRICLOPIR
DAZOMET	ISOXABEN	TRIFLURALIN
DESMEDIFAM	LINURON	TRIFORINE
DESMETRINA	MANCOZEB	VINCLOZOLIN
DICLOBENIL	MANEB	ZINEB
DICLOFOP-METILE	MCPB	ZOLFO

FAO TECHNICAL PAPERS

FAO AGRICULTURAL SERVICES BULLETINS

1 Farm planning in the early stages of development, 1968 (E Rep. 1982 – F S)
2 Planning for action in agricultural development, 1969 (E F S)
3 Karakul processing, 1969 (E)
4 Bread from composite flour, 1969 (E* F S)
5 Sun-drying of fruits and vegetables, 1969 (E F S)
6 Cashew nut processing, 1969 (E F S)
7 Technology for the production of protein foods from cottonseed flour, 1971 (E F S)
8 Cassava processing, 1971 (New edition, 1977, available in E, F and S in the FAO Plant Production and Protection Series, No. 3)
9 Worldwide list of food technology institutions, 1971 (E/F/S*)
10 Technology of production of edible flours and protein products from groundnuts, 1971 (E F S)
11 Technology of production of edible flours and protein products from soybean, 1971 (E F S)
12 A guide for instructors in organizing and conducting agricultural engineering training courses, 1971 (E F S)
12 Sup. 1 Elements of agricultural machinery, Vol. 1, 1977 (E S)
12 Sup. 2 Elements of agricultural machinery, Vol. 2, 1977 (E S)
13 Fruit juice processing, 1973 (E S)
14 Environmental aspects of natural resource management – agriculture and soils, 1972 (E F S)
15 Manual on sericulture:
Vol. I – Mulberry cultivation, 1976 (E F)
Vol. 2 – Silkworm rearing, 1973 (E F)
Vol. 3 – Silk reeling, 1972 (E F)
16 The use of aircraft in agriculture, 1972 (New edition, 1974, available in E, F and S in the FAO Agriculture Series, No. 2)
17 Airtight grain storage, 1973 (E F S)
18 Rice testing methods and equipment, 1973 (C E)
19 Cold storage – design and operation, 1973 (E F S)
19/2 Design and operation of cold stores in developing countries, 1984 (Ar E F S)
20 Processing of natural rubber, 1973 (E)
21 Rev. 1 Agricultural residues: world directory of institutions, 1978 (E/F/S)
21 Rev. 2 Agricultural residues: world directory of institutions, 1982 (E/F/S)
22 Rice milling equipment operation and maintenance, 1974 (C E)
23 Rice drying (E**)
24 Worldwide list of textile research institutes, 1974 (E/F/S)
25 Molasses utilization, 1975 (E F S)
26 Tea processing, 1974 (E)
27 Some aspects of earth-moving machines as used in agriculture, 1975 (E)
28 Mechanization of irrigated crop production, 1977 (E)
29 Non-mulberry silks, 1979 (E)
30 Machinery servicing organizations, 1977 (E)
31 Rice-husk conversion to energy, 1978 (E)
32 Animal blood processing and utilization, 1982 (C E S)
33 Agricultural residues: compendium of technologies, 1978 (E/F/S)
33 Rev. 1 Agricultural residues: compendium of technologies, 1982 (E/F/S)
34 Farm management data collection and analysis, 1977 (E F S)
35 Bibliography of agricultural residues, 1978 (E/F/S)
36 China: rural processing technology, 1979 (E)
37 Illustrated glossary of rice-processing machines, 1979 (Multil)
38 Pesticide application equipment and techniques, 1979 (E)
39 Small-scale cane sugar processing and residue utilization, 1980 (E F S)
40 On-farm maize drying and storage in the humid tropics, 1980 (C E)
41 Farm management research for small farmer development, 1980 (C E F S)
42 China: sericulture, 1980 (E)
43 Food loss prevention in perishable crops, 1981 (E F S)
44 Replacement parts for agricultural machinery, 1981 (E F)
45 Agricultural mechanization in development: guidelines for strategy formulation, 1981 (E F)
46 Energy cropping versus food production, 1981 (E F S)
47 Agricultural residues: bibliography 1975-81 and quantitative survey, 1982 (E/F/S)
48 Plastic greenhouses for warm climates, 1982 (E)
49 China: grain storage structures, 1982 (E)
50 China: post-harvest grain technology, 1982 (E)
51 The private marketing entrepreneur and rural development, 1982 (E F S)
52 Aeration of grain in subtropical climates, 1982 (E)
53 Processing and storage of foodgrains by rural families, 1983 (E F S)
54 Biomass energy profiles, 1983 (E F)
55 Handling, grading and disposal of wool, 1983 (Ar E F S)
56 Rice parboiling, 1984 (E F)
57 Market information services, 1983 (E F S)
58 Marketing improvement in the developing world, 1984 (E)
59 Traditional post-harvest technology of perishable tropical staples, 1984 (E F S)
60 The retting of jute, 1985 (E F)
61 Producer-gas technology for rural applications, 1985 (E F)
62 Standardized designs for grain stores in hot dry climates, 1985 (E F)
63 Farm management glossary, 1985 (E/F/S)
64 Manual on the establishment, operation and management of cereal banks, 1985 (E F)
65 Farm management input to rural financial systems development, 1985 (E F S)
66 Construction of cribs for drying and storage of maize, 1985 (E F)
67 Hides and skins improvement in developing countries, 1985 (C E F)
68 Tropical and sub-tropical apiculture, 1986 (E)
68/2 Honeybee mites and their control – a selected annotated bibliography, 1986 (E)
68/3 Honey and beeswax control, 1990 (E** S)
68/4 Beekeeping in Asia, 1986 (E)
68/5 Honeybee diseases and enemies in Asia: a practical guide, 1987 (E)
68/6 Beekeeping in Africa, 1990 (E)
69 Construction and operation of small solid-wall bins, 1987 (E)

70 Paddy drying manual, 1987 (E)

71 Agricultural engineering in development: guidelines for establishment of village workshops, 1988 (C E F)

72/1 Agricultural engineering in development – The organization and management of replacement parts for agricultural machinery, Vol. 1, 1988 (E)

72/2 Agricultural engineering in development – The organization and management of replacement parts for agricultural machinery, Vol. 2, 1988 (E)

73/1 Mulberry cultivation, 1988 (E)

73/2 Silkworm rearing, 1988 (E)

73/3 Silkworm egg production, 1989 (E)

73/4 Silkworm diseases, 1991 (E)

74 Agricultural engineering in development: warehouse technique, 1989 (E F S)

75 Rural use of lignocellulosic residues, 1989 (E)

76 Horticultural marketing – a resource and training manual for extension officers, 1989 (E F S)

77 Economics of animal by-products utilization, 1989 (E)

78 Crop insurance, 1989 (E S)

79 Handbook of rural technology for the processing of animal by-products, 1989 (E)

80 Sericulture training manual, 1990 (E)

81 Elaboración de aceitunas de mesa, 1991 (S)

82 Agricultural engineering in development: design and construction guidelines for village stores, 1990 (E F S)

83 Agricultural engineering in development: tillage for crop production in areas of low rainfall, 1990 (E)

84 Agricultural engineering in development: selection of mechanization inputs, 1990 (E F S)

85 Agricultural engineering in development: guidelines for mechanization systems and machinery rehabilitation programmes, 1990 (E)

86 Strategies for crop insurance planning, 1991 (E S)

87 Guide pour l'établissement, les opérations et la gestion des banques de céréales, 1991 (F)

88/1 Agricultural engineering in development – Basic blacksmithing: a training manual, 1991 (E S)

88/2 Agricultural engineering in development – Intermediate blacksmithing: a training manual, 1991 (E F S)

88/3 Agricultural engineering in development – Advanced blacksmithing: a training manual, 1991 (E F)

89 Post-harvest and processing technologies of African staple foods: a technical compendium, 1991 (E)

90 Wholesale markets – Planning and design manual, 1991 (E)

91 Agricultural engineering in development: guidelines for rebuilding replacement parts and assemblies, 1992 (E S)

92 Agricultural engineering in development: human resource development – training and education programmes, 1992 (E F S)

93 Agricultural engineering in development: post-harvest operations and management of foodgrains, 1994 (E F S)

94 Minor oil crops:
Part I – Edible oils
Part II – Non-edible oils
Part III – Essential oils, 1992 (E)

95 Biogas processes for sustainable development, 1992 (E F)

96 Small-scale processing of microbial pesticides, 1992 (E)

97 Technology of production of edible flours and protein products from soybeans, 1992 (E F)

98 Small-, medium- and large-scale starch processing, 1992 (E F)

99/1 Agricultural engineering in development: mechanization strategy formulation – Vol. 1, Concepts and principles, 1992 (E F S)

100 Glossary of terms for agricultural insurance and rural finance, 1992 (E)

101 Date palm products, 1993 (E)

102 Experiencias de mercadeo de pequeños agricultores en el marco de proyectos de desarrollo rural integrado, 1992 (S)

103 Banking for the environment, 1993 (E S)

104 Agricultural engineering in development: agricultural tyres, 1993 (E)

105 Apicultura práctica en América Latina, 1993 (S)

106 Promoting private sector involvement in agricultural marketing in Africa, 1993 (E F)

107 La comercialización de alimentos en los grandes centros urbanos de América Latina, 1993 (S)

108 Plant tissue culture: an alternative for production of useful metabolites, 1993 (E)

109 Grain storage techniques – Evolution and trends in developing countries, 1994 (E)

110 Testing and evaluation of agricultural machinery and equipment – Principles and practices, 1994 (E S)

111 Low-cost, urban food distribution systems in Latin America, 1994 (E S)

112/1 Pesticide application equipment for use in agriculture – Vol. 1, Manually carried equipment, 1994 (E)

112/2 Pesticide application equipment for use in agriculture – Vol. 2, Mechanically-powered equipment, 1995 (E)

113 Maintenance and operation of bulk grain stores, 1994 (E)

114 Seed marketing, 1994 (E)

115 Selection, testing and evaluation of agricultural machines and equipment – Theory, 1995 (E)

116 Safeguarding deposits – Learning from experience, 1995 (E)

117 Quality assurance for small-scale rural food industries, 1995 (E)

118 Pollination of cultivated plants in the tropics, 1995 (E)

119 Fruit and vegetable processing, 1995 (E)

Availability: April 1995

Ar	– Arabic	Multil	– Multilingual
C	– Chinese	*	Out of print
E	– English	**	In preparation
F	– French		
P	– Portuguese		
S	– Spanish		

The FAO Technical Papers are available through the authorized FAO Sales Agents or directly from Distribution and Sales Section, FAO, Viale delle Terme di Caracalla, 00100 Rome, Italy.